Springer Series on Atomic, Optical, and Plasma Physics

Volume 78

For further volumes:
http://www.springer.com/series/411

The Springer Series on Atomic, Optical, and Plasma Physics covers in a comprehensive manner theory and experiment in the entire field of atoms and molecules and their interaction with electromagnetic radiation. Books in the series provide a rich source of new ideas and techniques with wide applications in fields such as chemistry, materials science, astrophysics, surface science, plasma technology, advanced optics, aeronomy, and engineering. Laser physics is a particular connecting theme that has provided much of the continuing impetus for new developments in the field, such as quantum computation and Bose-Einstein condensation. The purpose of the series is to cover the gap between standard undergraduate textbooks and the research literature with emphasis on the fundamental ideas, methods, techniques, and results in the field.

Sadrilla Abdullaev

Magnetic Stochasticity in Magnetically Confined Fusion Plasmas

Chaos of Field Lines and Charged Particle Dynamics

Springer

Sadrilla Abdullaev
Institute of Energy and Climate Research
Forschungszentrum Jülich
Jülich
Germany

ISSN 1615-5653
ISBN 978-3-319-01889-8 ISBN 978-3-319-01890-4 (eBook)
DOI 10.1007/978-3-319-01890-4
Springer Cham Heidelberg New York Dordrecht London

Library of Congress Control Number: 2013948352

Printed on acid-free paper

Springer is part of Springer Science+Business Media (www.springer.com)

To my parents Sayfulla Abdullaev and Nuri Adilova

Preface

The magnetic confinement concept of thermonuclear plasma is one of the areas of controlled fusion research developing in the past 50 years in a quest for clean and unlimited source of energy. The concept is based on the fact that charged particles predominantly follow the magnetic field and therefore the magnetic field localized in a finite area would be able to contain the hot temperature plasma is a spatially bounded area. In magnetic confinement fusion devices, tokamaks and stellarators, the confinement of charged particles has been achieved by the magnetic fields whose field lines lie on the nested (magnetic) toroidal surfaces. However, from the beginning of these studies it was realized that the magnetic field created in real devices may deviate from the intended ideal toroidal configuration due to technical imperfections or plasma instabilities, which eventually may break the symmetry of magnetic field leading to the destruction of magnetic surfaces and plasma confinement. Therefore the problem of stability of the magnetic surfaces with respect to the small deviations (or perturbations) of the magnetic field from the ideal toroidal configuration had become one of the most important issues in the fusion research.

The important fact used in these studies was that a divergence-free magnetic field is equivalent to a Hamiltonian system with 1 + 1/2 degrees of freedom. Thus the problem of stability of magnetic surfaces has been reduced to the classical problem of the stability of Hamiltonian dynamical systems which has been a subject of many studies in classical and celestial mechanics in the nineteenth and twentieth centuries. Already at that time H. Poincaré noted that the prediction in dynamical systems may become impossible and this problem is not related to the universal laws of motion, but with the specification of the initial conditions, *it may happen that small differences in the initial conditions produce very great ones in the final phenomena. A small error [change] in the former will produce an enormous error [change] in the latter. Prediction becomes impossible . . .* [Poincaré (2009)]. In the modern terminology he discovered the chaotic motion in dynamical systems, now known as a *dynamical chaos*.

Until the 1960s the theory of dynamical systems has been primarily a research area of mathematicians. In the middle of the twentieth century researchers in fusion and accelerator physics revived the interest to this area of classical physics among the physical community whose interest until the 1960s has been mostly occupied by quantum physics. These studies had a crucial impact on the development of the theory of nonlinear dynamics and chaos in physics and beyond it. Particularly, the qualitative criteria of the onset of chaotic motion in physical systems proposed by Chirikov (1959) laid the foundation for the physical theory of chaos.

The phenomenon of chaos of magnetic field lines known as *magnetic stochasticity* (or *magnetic chaos*) plays an important role in the magnetic confinement of fusion plasmas. It concerns with the problems of stability and destruction of the magnetic field surfaces, chaotic behavior of magnetic field lines, and related transport of energy and particles.

In early studies the magnetic stochasticity in fusion devices is considered an undesirable effect which deteriorates the plasma confinement due to the enhanced radial transport of particles and energy along the chaotic field lines. However, at the end of the 1970s it was realized that the phenomenon of magnetic stochasticity can be used to control the transport of energy and particles. Particularly, the *ergodic divertor* concept has been proposed to divert particles and a heat releasing from the plasma to special plates in a controlled way by externally imposed magnetic perturbations which have been implemented in several tokamaks.

At present a magnetic stochasticity has been used that mitigates some undesirable phenomena developing in modern fusion devices. Among them the so-called edge localized modes, known as repetitive heat and particle loading to the divertor targets produced by plasma instabilities at the plasma edge in H-mode regimes. The suppression of these modes is of paramount importance for the International Thermonuclear Experimental Reactor (ITER) to protect wall materials from the damaging effect of huge heat and particles releasing during its operation. The suppression of runaway electrons generated in tokamaks during disruptions of plasma discharges by magnetic perturbations is another example of application of magnetic stochasticity. The runaway electrons with energies up to several tens of MeV may cause severe damage to walls.

These applications show the importance of stochastic magnetic fields to control plasmas in magnetically fusion devices. Therefore, a deeper knowledge of a magnetic stochasticity, its onset, and generic properties would be useful for researchers involved in magnetically fusion studies. There are many textbooks and monographs on chaotic dynamics in which the problems of magnetic stochasticity are presented. However, these presentations are too general and do not take the important features of a magnetic field in fusion devices. At the same time in many books on plasma physics the problem of magnetic stochasticity has been mentioned very briefly and treated only at the elementary level. This description of magnetic field stochasticity became insufficient to describe the present-day experimental observations of very fine patterns of plasma structures related to chaotic magnetic fields. These structures are manifestations of such phenomena as a splitting of separatrices, stable and unstable manifolds, introduced in the mathematical theory of dynamical systems.

The Aim of the Book

This book is intended to fill this gap and to give a systematic theoretical description of magnetic field stochasticity and related charged particle dynamics in magnetically confined fusion devices. The presentation is based on the Hamiltonian formulation of magnetic field lines and charged particles. It employs the classical mathematical tools as well as newly developed methods of Hamiltonian dynamics to study magnetic field lines and particle dynamics in toroidally confined plasmas.

The main intention of this book has been to present generic features of magnetic stochasticity in toroidal plasmas rather than its specific manifestations in a particular fusion device. At present they can be studied by the numerical codes for field line tracing. The knowledge of generic properties of a magnetic stochasticity is much useful to predict and, possible, to effectively control plasma behavior by applied magnetic perturbations. Particularly, we have made the main emphasis on revealing generic and universal features of magnetic perturbations generated by external coils in toroidal plasmas, the properties of chaotic magnetic field lines, and related transport of particles. The asymptotic and mapping methods are intensively employed to describe these generic features of magnetic fields and and their properties.

The choice of materials and the manner of presentation are somehow subjective. The Author tried to cover many issues of a magnetic stochasticity which are well understood. These problems were also studied in tokamaks which are considered the most advanced magnetically confined fusion device. However, some important issues of magnetic stochasticity related to the plasma response to applied magnetic perturbations, the screening of magnetic perturbations, nonlinear transport were outside the scope of the book. At present these problems are not well understood, and they are still under active research.

Structure of the Book

The book consists of 11 chapters and 7 appendices. Chapter 1 presents the Hamiltonian formulation of the equations of magnetic field lines based on the action principle of the classical mechanics. The different forms of the Hamiltonian equations for the magnetic field lines are derived. The magnetic flux coordinates are introduced using the action-angle formalism of classical mechanics. We have considered a model of magnetic field which is used to illustrate the action–angle formalism to study magnetic field lines.

The different analytical models of equilibrium plasmas are presented in Chap. 2. Besides the analytical models for plasmas with circular cross-sections and elongated shapes with magnetic separatrices, we have also described wire models of plasmas. The generic behavior of a magnetic field near the magnetic separatrix and the X-point are also analyzed. Finally, a wire model of a magnetic field with the snowflake divertor configuration is analyzed.

Chapter 3 presents the Hamiltonian equations of field lines in the presence of magnetic perturbations. Particularly, the generic properties of the spectra of magnetic perturbations created by sets of helical coils and saddle coils are analyzed. The corresponding asymptotical formulas for the spectra of the magnetic perturbations are obtained.

The Hamiltonian equations describing a guiding–center motion of charged particles in a toroidal system are presented in Chap. 4. We give an alternative formulation of these equations which uses the cylindrical coordinate system instead of magnetic flux coordinates used in the traditional approach. The Hamiltonian approach is used in Chap. 5 to study the guiding center orbits in equilibrium plasmas. We analyze drift surfaces of passing and trapped particles, and estimate their widths with a number of illustrations. The formulation of Hamiltonian equations in action–angle variables will be given. The behavior of guiding-center orbits near hyperbolic fixed points will be analyzed.

The modern methods to study the dynamics of Hamiltonian systems are presented in Chap. 6. First, we describe the mapping method to integrate Hamiltonian equations based on the canonical transformation of variables. Particularly, the explicit forms of mappings for magnetic field lines are written down. Then we construct the full-turn transfer mapping, the Poincaré mapping. The Poincaré and the Melnikov integrals associated with the full-turn transfer mappings will be introduced and their properties analyzed.

The mathematical aspects of onset of dynamical chaos in Hamiltonian systems are given in Chap. 7. Particularly, we formulate the Kolmogorov's theorem on the conservation of conditionally periodic motion and discuss its implications to a magnetic field system and charged particle motion in a magnetic field. The main obstacles to the integrability of Hamiltonian systems, the creations of isolated periodic orbits, the splitting of separatrices are discussed in detail. They are demonstrated on simple examples. Finally, the onset of chaotic motion due to the creation of infinite number of unstable periodic points and the homoclinic structure of phase space is discussed.

The physical theory of chaos in Hamiltonian systems is presented in Chap. 8. We discuss a nonlinear resonance phenomenon, its features in a magnetic field system, the formation of chaotic layer. The Chirikov's criterion of the onset of global chaos is discussed. We analyze a chaotic layer, its rescaling invariance properties near the hyperbolic fixed points, and their implications on the related transport in plasmas. The features of the onset of chaos in a magnetic system with a reversed shear are also briefly discussed.

The main features of magnetic fields in tokamaks induced by resonant magnetic perturbations are discussed in Chap. 9. First, we study the ergodic divertor concept to control the transport of heat and particle at the plasma edge. Particularly, the implementation of this concept in the Tore Supra and the TEXTOR tokamaks are presented. Then we discuss the features of a magnetic field in the DIII-D tokamak in the presence of magnetic perturbations created by a set of saddle coils. The models of magnetic perturbations in tokamaks with ergodic divertors and set of saddle coils are proposed. We also discuss the generic fractal properties of chaotic field lines at the plasma edge. Finally, some experimental observations of plasma structures induced by resonant magnetic perturbations are briefly discussed.

Chapter 10 describes the transport of field lines and charged particles in a stochastic magnetic field induced by a resonant magnetic perturbations. First, we briefly recall the basics of the kinetic description of the chaotic transport in Hamiltonian systems, particularly, in magnetic system with chaotic field lines. The diffusion of particles in a stochastic magnetic field in collisionless and collisional regimes is studied using analytical and numerical tools. A simple empirical formula for the collisional radial diffusion coefficient of particles in a stochastic magnetic field is presented.

Particle transport in Hamiltonian systems perturbed by a weak turbulent wave field and the transport of runaway electrons caused by a small-scale turbulent magnetic field in tokamaks are studied in Chap. 11. Particularly, we study the fractal nature of radial diffusion of particles and the formation of transport barriers near the low-order rational drift surfaces.

The analytical calculations of magnetic fields generated by sets of helical coils in the dynamic ergodic divertor of the TEXTOR tokamak and in the ergodic divertor of the Tore Supra tokamak are presented in Appendix A. The corresponding calculations of a magnetic field created by a set of saddle coils are given in Appendix B. The method of calculations of the Poincaré integrals for the magnetic field model is presented in Appendix C. In Appendix D the advanced version of the mapping for Hamiltonian systems is derived. The calculations of the Jacobi matrix and its eigenvalues are given in Appendix E. The features of the perturbation Hamiltonian of guiding-center motion in the presence of magnetic perturbations created by a set of saddle coils are studied in Appendix F. The derivation of the full-turn transfer mapping in a many-dimensional Hamiltonian system is described in Appendix G.

Most of the chapters are supplied with bibliographic notes or comments, in which the important references on subjects discussed there are given.

Jülich, Germany Sadrilla Abdullaev

Acknowledgments

The Author gratefully acknowledges Profs. G. H. Wolf, G. Eilenberger, J. Treusch, and U. Samm who made it possible for him to participate in the project "Dynamic Ergodic Diveror" at the Jülich tokamak TEXTOR. It gave him a great opportunity to work in the theory of Hamiltonian chaos and its applications to the problems of magnetic stochasticity. A long–time fruitful collaboration with Dr. K. H. Finken and Prof. K. H. Spatschek in this project has been very crucial and essential.

The Author is also grateful to his teachers and colleagues, (in alphabetic order) R. Balescu, B. V. Chirikov, D. F. Escande, T. Evans, D. del Castillo–Negreto, O. Dumbrajs, M. Forster, G. Fuchs, M. Jakubowski, R. Janev, A. Kaleck, R. Koch, M. Kobayashi, M. Lehnen, P. J. Morrison, A. Nicolai, D. Reiser, D. Reiter, A. Rogister, A. Runov, O. Schmitz, K. H. Spatschek, M. Z. Tokar', B. Unterberg, A. Wingen, R. Wolf, and G. M. Zaslavsky. Their advice, collaboration, many discussions, and support have been essential and highly beneficial for him.

This book as well as my all previous works would not be completed without the support and patience of my family, wife Mavlyuda, and daughters, Temura and Oliya.

Jülich, Germany, May 2013 Sadrilla Abdullaev

Contents

Main Notations

(R, Z, φ)	Cylindrical coordinates
(r, θ, φ)	Quasitoroidal coordinates (Fig. 1.2)
$\mathbf{A}$	Vector potential of a magnetic field
(A_R, A_Z, A_φ)	Components of a vector potential
$\mathbf{B}$	Magnetic field vector
(B_R, B_Z, B_φ)	Components of a magnetic field
R_0	Major radius of torus center
B_0	Magnitude of the toroidal magnetic field at the torus center
I_p	Plasma current
ρ	Minor radius of magnetic surfaces
$\psi \equiv \psi_\varphi$	Toroidal magnetic flux
$\psi_t \equiv \psi_\vartheta$	Poloidal magnetic flux
ψ_N	Normalized poloidal flux (Eq. 2.36)
ϑ	Poloidal angle in which field lines are straight (Eq. 1.32)
φ	Toroidal angle
$q(\psi)$	Safety factor
(m, n)	Poloidal and toroidal mode numbers in (Eq. 3.13)
γ	Coefficient in the asymptotics of $q(\psi)$ in (Eq. 2.51)
$\in$	Perturbation parameter (Eq. 3.11 and Eq. 3.46)
$R_n(\psi) = K_n(\psi) + iL_n(\psi)$	Fourier the coefficients (Eq. 6.77) of the Poincaré integral (Eq. 6.76)
c	Speed of light in vacuum
m_0	Particle mass (used only in Chaps. 4 and 5)
e	Elementary charge
Z_q	Particle charge (in unit e)
$\omega_c = eB_0/m_0c$	Reference gyrofrequency
$E_{ref} = m_0\omega_c^2R_0^2$	Reference energy
ε_0	Normalized energy of a particle at rest (Eq. 4.33)

γ_t	Relativistic factor (Eq. 4.6)
λ_I	Ratio of the action of radial gyro–oscillations I_x to the kinetic energy T_k (Eq.5.3)
$\chi_{\parallel}, \chi_{\perp}$	Parallel and perpendicular diffusion coefficients of particles due to collisions
D_{FL}	Diffusion coefficient of field lines
λ_{mfp}	Mean free path

Chapter 1
Hamiltonian Representation of Magnetic Field

In this chapter we present the Hamiltonian formulation of the equations for magnetic field lines. We specifically consider the magnetic field corresponding the toroidal plasma configuration. Using the action principle of the classical mechanics the Hamiltonian equation for the magnetic field lines will be derived and its different forms will be presented. A particular emphasize will be given to the equations in flux coordinates, in which the field lines are the straight lines. We conclude the chapter introducing the simple model of the magnetic field and study the properties of its magnetic field lines using the methods of classical mechanics.

Let $\mathbf{B} = (B_X, B_Y, B_Z)$ is a static magnetic field in the three-dimensional (3D) space (X, Y, Z). It is divergent-free field satisfying the condition,

$$\nabla \cdot \mathbf{B} = 0. \tag{1.1}$$

The magnetic field lines , $\mathbf{r}(\tau) = (X(\tau), Y(\tau), Z(\tau))$, are three- dimensional curves tangent to the magnetic field $\mathbf{B}$ defined by the equation

$$\frac{d\mathbf{r}}{d\tau} = \mathbf{B}, \tag{1.2}$$

where τ is independent parameter related to the length element of curve $ds = (dX^2 + dY^2 + dZ^2)^{1/2}$: $d\tau = |\mathbf{B}|^{-1} ds$.

The magnetic field lines in the 3D-space may behave in a various way. Depending on the initial conditions they may be confined in a finite domain or may extend to infinity; they may lie on the surfaces or fully cover certain finite 3D domains. Notably that the magnetic confinement of high temperature plasmas for the thermonuclear fusion is based on the creation of the magnetic field whose field lines cover nested toroidal surfaces also known as *magnetic surfaces* which illustrated in Fig. 1.1.

S. Abdullaev, *Magnetic Stochasticity in Magnetically Confined Fusion Plasmas*,
Springer Series on Atomic, Optical, and Plasma Physics 78,
DOI: 10.1007/978-3-319-01890-4_1,

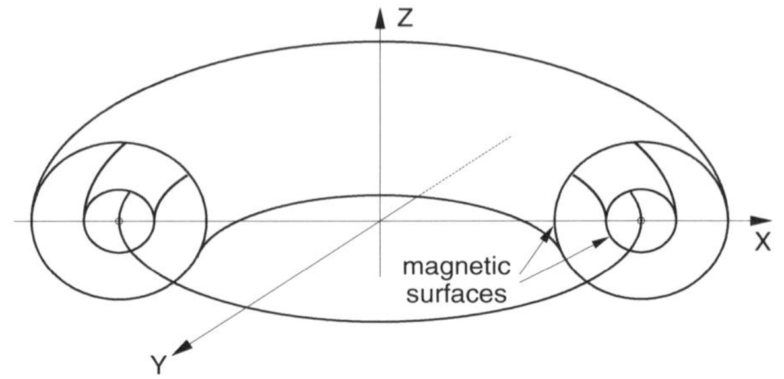

Fig. 1.1 Field lines lie on the nested toroidal surfaces $\psi(X, Y, Z) = \text{const}$

1.1 Hamiltonian Equations for Magnetic Field Lines

The most powerful mathematical tool to study the magnetic field lines is the methods of Hamiltonian dynamics which is the most convenient to describe the regular and chaotic field lines. Below, we formulate the Eq. (1.2) for magnetic field lines in a Hamiltonian form.

The Hamiltonian formulation is based on the presentation of the magnetic field by the vector potential $\mathbf{A}(\mathbf{r})$ related with $\mathbf{B}$ as $\mathbf{B} = \nabla \times \mathbf{A}$. The magnetic field is invariant with respect to the gauge transformation of the vector potential $\mathbf{A} \to \mathbf{A} + \nabla g$, where $g(\mathbf{r})$ is the arbitrary scalar (gauge) function.

The field lines can be derived from a variational principle , that is the action

$$S[\mathbf{r}(\tau)] = \int_{\mathbf{r}_1}^{\mathbf{r}_2} \mathbf{A}(\mathbf{r}(\tau)) \cdot \frac{d\mathbf{r}}{d\tau} d\tau, \tag{1.3}$$

defined as the integral over the vector potential $\mathbf{A}(\mathbf{r})$ along the field line $\mathbf{r}(\tau)$ connecting the endpoints $\mathbf{r}_1$ and $\mathbf{r}_2$ should be minimal [Cary and Littlejohn (1983)]. The action in the classical mechanics is given by

$$S[\mathbf{r}(\tau)] = \int_{\mathbf{r}_1}^{\mathbf{r}_2} (\mathbf{p} \cdot d\mathbf{r} - H dt), \tag{1.4}$$

where $(\mathbf{r}, \mathbf{p})$ are canonical variables, H is the Hamiltonian function, and t is a time. The Hamiltonian equations derived from the variation of the action (1.4) are

$$\frac{d\mathbf{r}}{dt} = \frac{\partial H}{\partial \mathbf{p}}, \qquad \frac{d\mathbf{p}}{dt} = -\frac{\partial H}{\partial \mathbf{r}}. \tag{1.5}$$

Let (u, v, w) be the orthogonal coordinate system, and (A_u, A_v, A_w) be the corresponding components of the vector potential $\mathbf{A}$. By choosing the appropriate gauge function g that $A_w = -\partial g/\partial w$, one can remove the component A_w. Then introducing the momenta $P_u = A_u(U, V, W)$, $P_v = A_v(U, V, W)$, and the Hamiltonian function $H = H(U, V, P_u, P_v) = 0$ the action (1.3) can be presented as

$$S = \int_{\mathbf{r}_1}^{\mathbf{r}_2} (P_u dU + P_v dV - H d\tau) . \tag{1.6}$$

The corresponding field lines are described by the *autonomous* Hamiltonian equation,

$$\begin{aligned} \frac{dU}{d\tau} &= \frac{\partial H}{\partial P_u}, \qquad \frac{dV}{d\tau} = \frac{\partial H}{\partial P_v}, \\ \frac{dP_u}{d\tau} &= -\frac{\partial H}{\partial U}, \qquad \frac{dP_v}{d\tau} = -\frac{\partial H}{\partial V}. \end{aligned} \tag{1.7}$$

The Hamiltonian $H(U, V, P_u, P_v) = 0$ can be chosen as

$$H(U, V, P_u, P_v) = P_u - A_u(U, V, W(P_v, U, V)) = 0, \tag{1.8}$$

by expressing the variable W via (P_v, U, V) $[P_v = A_v(U, V, W)]$ or

$$H(U, V, P_u, P_v) = P_v - A_v(U, V, W(P_u, U, V)) = 0, \tag{1.9}$$

by W via (P_u, U, V) $[P_u = A_u(U, V, W)]$. The choice of the Hamiltonian depends on whether the variable U (or V) can be considered as a time-like independent variable which does not change its direction. Then field lines can be described by the *non-autonomous* Hamiltonian system. In the first case $U = \tau$ the Hamiltonian equations is given by

$$\frac{dV}{dU} = \frac{\partial H_u}{\partial P_v}, \quad \frac{dP_v}{dU} = -\frac{\partial H_u}{\partial V}, \tag{1.10}$$

where $H_u = -A_u(U, V, W(P_v, U, V))$. The similar equation we have in the case $V = \tau$. The two-dimensional system (1.10) is known as *one and half degrees of freedom Hamiltonian system*. This term has a following meaning: the canonically conjugated variables (V, P_v) correspond to "one degree" and the time-like independent variable U corresponds to a half degree since the time runs only in one direction.

Specifically we consider the Hamiltonian equations for field lines in a toroidal system. Let (R, Z, φ) be cylindrical coordinate system , where R is directed along the major radius, Z is the vertical coordinate, and φ is the toroidal angle. They are also related to the quasitoroidal coordinates (r, θ, φ):

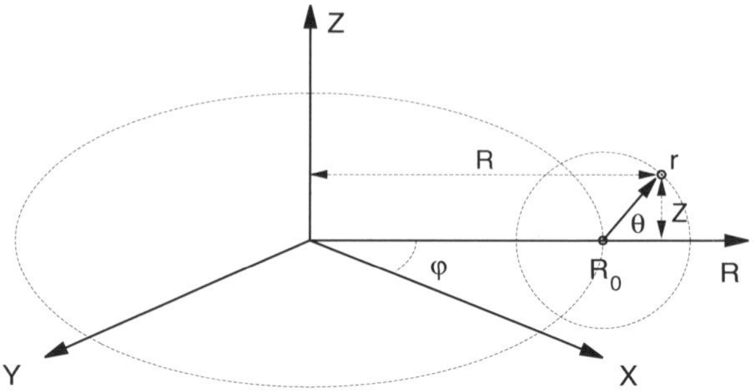

Fig. 1.2 Geometrical coordinates of a toroidal system: (r, θ, φ) is the quasitoroidal coordinates; (R, Z, φ) is the cylindrical coordinate system

$$R = R_0 + r\cos\theta, \qquad Z = r\sin\theta,$$
$$r = \sqrt{(R-R_0)^2 + Z^2}, \quad \theta = \arctan\frac{Z}{R-R_0},$$

shown in Fig. 1.2. Here R_0 is the major radius of the torus center. The ratio $\varepsilon = r/R_0$ is known an inverse aspect ratio (the ratio R/r is called an *aspect ratio*).

Let $\left(A_R, A_Z, A_\varphi\right)$ be the components of the vector potential $\mathbf{A}$ in the cylindrical coordinate system. By choosing $W \equiv R$ we have

$$P_z = A_Z(R, Z, \varphi), \qquad P_\varphi = RA_\varphi, \qquad u = \varphi, \qquad v = Z,$$
$$H = H(Z, \varphi, P_z, P_\varphi) = P_\varphi - R(Z, P_z)A_\varphi\left(R(Z, P_z), Z, \varphi\right) = 0. \quad (1.11)$$

The corresponding Hamiltonian equations are

$$\frac{dZ}{d\tau} = \frac{\partial H}{\partial P_z}, \qquad \frac{dP_z}{d\tau} = -\frac{\partial H}{\partial Z},$$
$$\frac{d\varphi}{d\tau} = \frac{\partial H}{\partial P_\varphi}, \qquad \frac{dP_\varphi}{d\tau} = -\frac{\partial H}{\partial \varphi}. \quad (1.12)$$

The tokamak plasmas the toroidal angle φ can be chosen as an independent time-like variable τ. Then field line equations have the following form

$$\frac{dZ}{d\varphi} = \frac{\partial H_\varphi}{\partial P_Z}, \qquad \frac{dP_Z}{d\varphi} = -\frac{\partial H_\varphi}{\partial Z}, \quad (1.13)$$

where $H_\varphi = -RA_\varphi$.

1.1.1 Equations in Normalized Variables

Introducing the normalized variables,

$$x = \frac{R}{R_0}, \quad z = \frac{Z}{R_0}, \qquad p_z = \frac{P_Z}{B_0 R_0},$$
$$\psi_\varphi = \frac{H_\varphi}{B_0 R_0^2} = -\frac{R A_\varphi}{B_0 R_0^2}, \quad \psi_z = \frac{A_z}{B_0 R_0}, \tag{1.14}$$

Equations (1.12) and (1.13) can be presented in the form

$$\frac{dz}{d\tau} = \frac{\partial h}{\partial p_z}, \qquad \frac{dp_z}{d\tau} = -\frac{\partial h}{\partial z},$$
$$\frac{d\varphi}{d\tau} = \frac{\partial h}{\partial P_\varphi}, \qquad \frac{dp_\varphi}{d\tau} = -\frac{\partial h}{\partial \varphi},$$
$$h = H/B_0 R_0^2, \tag{1.15}$$

and

$$\frac{dz}{d\varphi} = \frac{\partial \psi_\varphi}{\partial p_z}, \quad \frac{dp_z}{d\varphi} = -\frac{\partial \psi_\varphi}{\partial z}. \tag{1.16}$$

The relation between the canonical variables (z, p_z) and to the geometrical coordinates (R, Z, φ) is given by $p_z = A_Z(R, Z, \varphi)/B_0 R_0$. Particularly, when the effect of diamagnetic currents[1] to the toroidal component of magnetic field, B_φ, is negligible, and thus $B_\varphi = B_0 R_0/R$, $A_Z(R, Z, \varphi) = B_0 R_0 \ln(R/R_0)$ one has

$$p_z = \ln \frac{R}{R_0}, \qquad R = R_0 e^{p_z}, \tag{1.17}$$

where B_0 is the magnitude of B_φ at $R = R_0$.

The poloidal magnetic field components, B_R, B_z, and the toroidal field, B_φ, are determined by

$$B_R = -\frac{\partial A_\varphi}{\partial Z} = \frac{B_0 R_0}{R}\frac{\partial \psi_\varphi}{\partial z},$$
$$B_z = \frac{1}{R}\frac{\partial R A_\varphi}{\partial R} = -B_0 \frac{\partial \psi_\varphi}{\partial p_z},$$
$$B_\varphi = -\frac{\partial A_Z}{\partial R} = -B_0 \frac{\partial \psi_z}{\partial x}. \tag{1.18}$$

In the cylindrical coordinate system (R, Z, φ) the equations of field lines (1.2) can be presented as

---

[1] A diamagnetic current is created due to a circular motion of charged particles in an induced magnetic field. It produces a magnetic field which opposes the external magnetic field and thus the total magnetic field is reduced. Therefore a plasma possesses diamagnetic properties. Typically, the magnetic field due to a diamagnetic current is much smaller than the strong toroidal field B_0.

$$\frac{dR}{d\varphi} = \frac{RB_R}{B_\varphi}, \quad \frac{dZ}{d\varphi} = \frac{RB_z}{B_\varphi}. \tag{1.19}$$

1.2 Magnetic Flux Coordinates

The magnetic field can be conveniently presented in the so-called magnetic flux coordinates consisting of one flux coordinate, ψ, and two angle-like coordinates, ϑ, φ. In these coordinates a magnetic field in a equilibrium plasma lie on the closed surface $\Psi(\mathbf{r}) =$ constant, and field lines are given by a straight line, i.e., $\vartheta = \iota(\psi)\varphi + \vartheta_0$, where $\iota(\psi)$ is a function of the flux ψ. These coordinates can be introduced as action-angle variables of the Hamiltonian system (1.12) for magnetic field lines. Below we follow this approach.

1.2.1 Action-Angle Variables

The powerful method to study Hamiltonian system is based on the action-angle variables. The action-angle variables are set of canonical variables useful to study integrable systems. They are also useful in classical perturbation theory of Hamiltonian systems, particularly in the formulation of the Kolmogorov's theorem on conservation of the conditionally-periodic motion under small perturbations, known as the Kolmogorov–Arnold–Moser (KAM) theory (see Sect. 7.1). Below we give a brief description of these variables for a magnetic system.

Consider the equilibrium plasma with nested toroidal magnetic surfaces $\psi(\mathbf{r}) =$ constant as shown in Fig. 1.1. On the torus there are two basic irreducible closed contours C_θ and C_φ that wound torus along the poloidal angle ϑ (the short way around the torus) and the toroidal angle φ (the long way around the torus) as illustrated in Fig. 1.3. We consider these angles, (ϑ, φ), as canonical coordinates of Hamiltonian system. Let ψ_θ and ψ_φ be canonical momenta conjugated to the canonical coordinates (ϑ, φ), respectively, so that the Hamiltonian function H depends only on these variables, $H = H(\psi_\theta, \psi_\varphi)$. The Hamiltonian system (1.12) takes the form

$$\begin{aligned} \frac{d\vartheta}{d\tau} &= \frac{\partial H}{\partial \psi_\theta} = \omega_\theta, & \frac{d\psi_\theta}{d\tau} &= -\frac{\partial H}{\partial \psi_\varphi} = 0, \\ \frac{d\varphi}{d\tau} &= \frac{\partial H}{\partial \psi_\varphi} = \omega_\varphi, & \frac{d\psi_\varphi}{d\tau} &= -\frac{\partial H}{\partial \varphi} = 0, \end{aligned} \tag{1.20}$$

with the Hamiltonian function

$$H = H(\vartheta, \varphi, \psi_\theta, \psi_\varphi) = \psi_\varphi - RA_\varphi(R, Z, \varphi) = 0, \tag{1.21}$$

where the cylindrical coordinates (R, Z) are the functions of (ψ_θ, ϑ): $R = R(\psi_\theta, \vartheta)$ and $Z = Z(\psi_\theta, \vartheta)$. The quantities $\omega_\theta, \omega_\varphi$ are called frequencies of the system.

Formally, the action-angle variables are introduced by the canonical change of variables $(z, \varphi, p_z, p_\varphi) \to (\vartheta, \varphi, \psi_\vartheta, \psi_\varphi)$,

$$\psi_\theta = \frac{1}{2\pi}\oint_{C_\vartheta} p_z(z;\psi_\theta)dz, \quad \psi_\varphi = \frac{1}{2\pi}\oint_{C_\varphi} p_\varphi(z;\psi_\varphi)d\varphi,$$
$$\vartheta_\vartheta = \frac{\partial F(\psi_\vartheta,\psi_\varphi;z,\varphi)}{\partial\psi_\vartheta} \qquad \vartheta_\varphi = \frac{\partial F(\psi_\vartheta,\psi_\varphi;z,\varphi)}{\partial\psi_\varphi}, \tag{1.22}$$

given by the generating function

$$F(\psi_\vartheta,\psi_\varphi;z,\varphi) = \int^{z} p_z(z';\psi_\theta)dz' + \int^{\varphi} p_\varphi(\varphi';\psi_\varphi)d\varphi'. \tag{1.23}$$

In (1.22) the integration is taken along the closed contour C_θ formed by crossing the magnetic surface with the poloidal plane $\varphi =$ const (see Fig. 1.3). Using Eq. (1.14) and the calibration $A_R = 0$, one can easily see that ψ_θ equals to the normalized magnetic flux of the toroidal field B_φ through the surface area S_θ enclosed by the contour C_θ,

$$\psi_\theta = \frac{1}{2\pi}\int_{S_\theta} dp_z dz = \frac{1}{2\pi B_0 R_0^2}\int_{S_\theta}\frac{\partial A_z}{\partial R}dRdZ$$
$$= \int_{S_\theta}\nabla\times\mathbf{A}\cdot\mathbf{e}_\varphi dRdZ = \frac{1}{2\pi B_0 R_0^2}\int_{S_\theta} B_\varphi dRdZ. \tag{1.24}$$

Similarly, one can show that ψ_φ is a normalized magnetic flux of the poloidal field B_z through area S_φ enclosed by the closed toroidal contour C_φ (see Fig. 1.3), i.e., $\psi_p = -RA_\varphi/B_0R_0^2$,

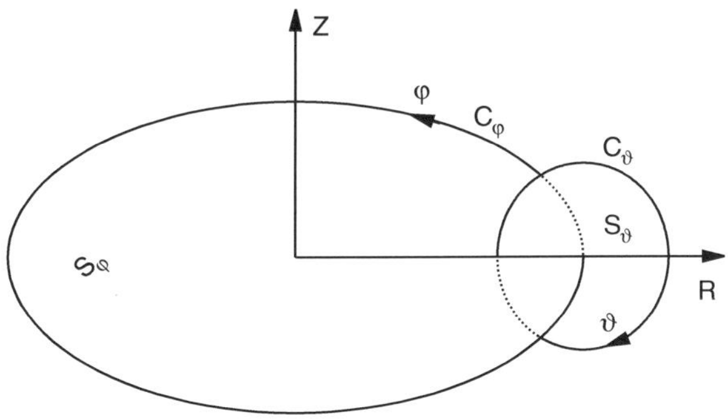

Fig. 1.3 Two basic irreducible closed contours C_θ and C_φ that wound torus along the poloidal angle ϑ (the short way around the torus) and the toroidal angle φ (the long way around the torus)

$$\psi_\varphi = -\frac{1}{2\pi}\oint_{C_\varphi} \frac{A_\varphi R d\varphi}{B_0 R_0^2} = \frac{1}{2\pi B_0 R_0^2}\int_{S_\varphi} \nabla \times \mathbf{A}\cdot \mathbf{e}_z R dR d\varphi$$
$$= -\frac{1}{2\pi B_0 R_0^2}\int_{S_\varphi} B_z dS. \tag{1.25}$$

The actions ψ_ϑ and ψ_φ are known as a *toroidal* and *poloidal* fluxes, respectively.[2]

The Eq. (1.22) also determine the relation between the flux coordinates $(\psi_\vartheta, \psi_\varphi, \vartheta, \varphi)$ and the quasitoroidal coordinates (r, θ, φ) (or the cylindrical coordinates (R, Z, φ)). Particularly, in equilibrium plasmas these geometrical coordinates are periodic functions of the angular variable ϑ,

$$R = \sum_m R_m(\psi)e^{im\vartheta}, \quad Z = \sum_m Z_m(\psi)e^{im\vartheta},$$
$$r = \sum_m r_m(\psi)e^{im\vartheta}, \quad \theta = \vartheta + \sum_m \alpha_m(\psi)e^{im\vartheta}, \tag{1.26}$$

where the Fourier coefficients $R_m(\psi)$, $Z_m(\psi)$, $r_m(\psi)$, and $\alpha_m(\psi)$ are functions of the flux coordinate ψ, (ψ_ϑ or ψ_φ). Typical dependencies of the poloidal angle ϑ on the geometrical angle θ is shown in Fig. 2.2.

Furthermore, we will notate the poloidal flux ψ_φ as ψ, and the toroidal flux ψ_ϑ as ψ_t.

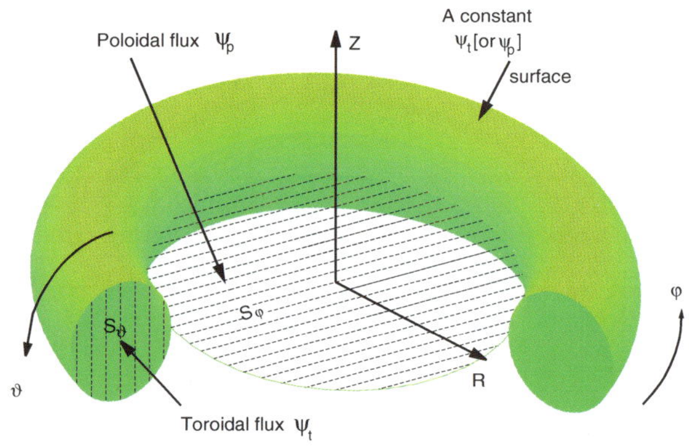

Fig. 1.4 Definitions of magnetic coordinates: the toroidal angle φ, the poloidal angle ϑ, the poloidal flux ψ_p, and the toroidal flux ψ_t

[2] The magnetic flux ψ cannot be arbitrary small. According to the quantization rule, $\oint p_z dz = hn$, $(n = 1, 2, \dots)$, $p_z = eA/c$ one has $\psi = \Phi_0 n$, where $\Phi_0 = hc/e$ is a quant of magnetic flux [h is the Planck's constant]. In fusion plasmas the magnetic flux is so large, that its discreteness does not play any role.

1.2.2 The Clebsch Form

The new variables $(\psi_t, \psi, \vartheta, \varphi)$ are called *magnetic coordinates* (or *Boozer coordinates*) (see Fig. 1.4). The position of field lines on the given magnetic surface is uniquely given by a *poloidal angle*, ϑ, (the short way around the torus) and *toroidal angle*, φ (the long way around the torus). In terms of toroidal and poloidal fluxes, ψ_t, ψ, poloidal and toroidal angles, ϑ, φ, the vector potential $\mathbf{A}$ and the divergence-free magnetic field $\mathbf{B}$ can be presented in the Clebsch form

$$\begin{aligned} \mathbf{A} &= B_0 R_0 \left(\nabla g + \psi_t \nabla\vartheta + \psi\nabla\varphi\right), \\ \mathbf{B} &= B_0 \left(\nabla\psi_t \times \nabla\vartheta - \nabla\varphi \times \nabla\psi\right). \end{aligned} \tag{1.27}$$

In find the equation of magnetic field lines we consider ψ as a function of ψ_t, ϑ, and φ, $\psi = \psi(\psi_t, \vartheta, \varphi)$. Then using

$$\nabla\psi = \frac{\partial\psi}{\partial\psi_\vartheta}\nabla\psi_\vartheta + \frac{\partial\psi}{\partial\vartheta}\nabla\vartheta + \frac{\partial\psi}{\partial\varphi}\nabla\varphi,$$

we obtain the magnetic field components in these coordinates are

$$\begin{aligned} B_\vartheta &= \mathbf{B}\cdot\nabla\vartheta = B_0 J\frac{\partial\psi}{\partial\psi_\vartheta}, \\ B_\psi &= \mathbf{B}\cdot\nabla\psi_t = -B_0 J\frac{\partial\psi}{\partial\vartheta}, \\ B_\varphi &= \mathbf{B}\cdot\nabla\varphi = J B_0, \end{aligned} \tag{1.28}$$

where $J = \nabla\psi_t \cdot (\nabla\vartheta \times \nabla\varphi)$ is the Jacobian of coordinate transformations. The equations for magnetic field lines take the Hamiltonian form

$$\frac{d\vartheta}{d\varphi} = \frac{B_\vartheta}{B_\varphi} = \frac{\partial\psi}{\partial\psi_t}, \qquad \frac{d\psi_t}{d\varphi} = \frac{B_\psi}{B_\varphi} = -\frac{\partial\psi}{\partial\vartheta}, \tag{1.29}$$

with ψ as a Hamiltonian function. This system is equivalent to the non-autonomous Hamiltonian system with 1+1/2-degree-of-freedom in which the toroidal angle φ plays a role of independent time-like variable, and (ϑ, ψ_t) as canonical-conjugated variables.

Field line equations can be also presented as the the autonomous two-degree-of-freedom Hamiltonian system similar to the system (1.12),

$$\begin{aligned} \frac{d\vartheta}{d\tau} &= \frac{\partial H}{\partial\psi_t}, \qquad \frac{d\psi_t}{d\tau} = -\frac{\partial H}{\partial\vartheta}, \\ \frac{d\varphi}{d\tau} &= \frac{\partial H}{\partial\psi}, \qquad \frac{d\psi}{d\tau} = -\frac{\partial H}{\partial\varphi}, \end{aligned} \tag{1.30}$$

with the Hamiltonian function

$$H \equiv H(\vartheta, \varphi, \psi_t, \psi) = \psi - \psi(\psi_t, \vartheta, \varphi) = 0, \tag{1.31}$$

and $(\vartheta, \varphi, \psi_t, \psi)$ as canonically- conjugated variables. The poloidal flux $\psi = \psi(\vartheta, \psi_t, \varphi)$ is a 2π—periodic function of ϑ, φ.

1.2.3 The Safety Factor

For the equilibrium magnetic field configuration with the nested magnetic surfaces, $\psi =$ constant, a field line on the magnetic surface $\psi =$ constant is a linear function of φ:

$$\vartheta = \frac{\varphi}{q(\psi)} + \vartheta_0 = \iota(\psi)\varphi + \vartheta_0, \tag{1.32}$$

where the quantities $q(\psi)$ and $\iota(\psi)$ are defined as

$$\frac{1}{q(\psi)} = \iota(\psi) = \frac{d\psi_p^{(0)}(\psi)}{d\psi_t}, \qquad \psi(\psi_t) = \int \frac{d\psi_t}{q(\psi)} = \int \iota(\psi) d\psi_t. \tag{1.33}$$

The $q(\psi)$ known a *safety factor* is used in a tokamak research, and $\iota(\psi)$ known a *rotational transform* (or a *winding number*) is used a stellarator research, respectively. The safety factor is equal increment of the toroidal angle, $\Delta\varphi$ of the magnetic surface per one full turn along the poloidal angle ϑ: $q(\psi) = \Delta\varphi/\Delta\vartheta = \Delta\varphi/2\pi$. The term a *safety factor* for $q(\psi)$ is introduced because its role in determining a stability of plasmas: higher values of q lead to greater stablity [see Wesson (2004)]. It also plays an important role in transport of heat and particles in a magnetically fusion plasmas.[3] The increment of the toroidal angle, $\Delta\varphi$ can be also calculated using Eq. (1.16). Then the safety factor is given by the integral

$$q(\psi) = \frac{\Delta\varphi}{2\pi} = \frac{1}{2\pi} \oint_C \frac{dz}{\partial\psi/\partial p_z}, \tag{1.34}$$

taken along the closed contour C of $\psi(z, p_z) =$ const.

One should note that the poloidal angle ϑ and the toroidal angle φ are not uniquely defined. By the transformation of angles (ϑ, φ) into new ones $(\bar{\vartheta}, \bar{\varphi})$

$$\begin{aligned} \bar{\vartheta} &= \vartheta + \iota(\psi) G(\psi, \vartheta, \varphi), \\ \bar{\varphi} &= \varphi + G(\psi, \vartheta, \varphi), \end{aligned} \tag{1.35}$$

[3] Particularly, as will be discussed Chap. 11 the barriers to a particle transport caused by a small scale turbulent field may be formed near the low–order rational values of q.

where $G(\psi, \vartheta, \varphi)$ is an arbitrary periodic function of (ϑ, φ), one can show that the magnetic field conserves the Clebsch form (1.27) also in the variables $(\bar{\vartheta}, \bar{\varphi})$.

Beside of this the magnetic field **B** is not changed if we add to ϑ (or φ) any arbitrary function $f(\psi)$ of ψ,

$$\Theta = \vartheta + f(\psi). \tag{1.36}$$

We will use this property later to define the poloidal angle ϑ which would be convenient to treat the field lines in a tokamak plasma with a magnetic separatrix (see Sect. 2.31).

1.3 Model of Single-Null Divertor Tokamak

In this section we consider the model of magnetic field which is topologically equivalent to the tokamak magnetic field with the separatrix. In this example we demonstrate the action-angle formalism of Hamiltonian dynamics to study the structure of field lines.

The poloidal flux of the magnetic field of the model is given by

$$\psi(z, p) = -\frac{\alpha^2}{2} z^2 \left(1 - \frac{z^2}{2z_a^2}\right) + \frac{\beta^2}{2} p^2, \tag{1.37}$$

where α, β, z_a are dimensionless parameters. The Hamiltonian equations of field lines are

$$\begin{aligned} \frac{dz}{d\varphi} &= \frac{\partial\psi(z, p)}{\partial p} = \beta^2 p, \\ \frac{dp}{d\varphi} &= -\frac{\partial\psi(z, p)}{\partial z} = \alpha^2 z \left(1 - \frac{z^2}{z_a^2}\right). \end{aligned} \tag{1.38}$$

The fixed points (z_s, p_s) of the system are determined by

$$\frac{dz}{d\varphi} = \frac{\partial\psi(z, p)}{\partial p} = 0, \qquad \frac{dp}{d\varphi} = -\frac{\partial\psi(z, p)}{\partial z} = 0. \tag{1.39}$$

According to (1.18) they correspond to the nulls of poloidal field B_R, B_z: $B_R(z_s, p_s) = 0$, $B_z(z_s, p_s) = 0$. The system (1.38) has two fixed points: one is a hyperbolic fixed point at $(z_s = 0, p_s = 0)$ which usually called the X-point and one is an elliptic fixed point at $(z_s = z_a, p_s = 0)$ corresponding to the magnetic axis (also called O-point) . The magnetic flux at the magnetic axis, corresponding to the minimum of $\psi(z, p)$ is equal to $\psi_a = -z_a^2\alpha^2/4$.

The magnetic field configuration of the system in the (z, p) plane is shown in Fig. 1.5a. In the plasma region the poloidal flux $\psi \equiv \psi(z, p) < 0$, and the field lines

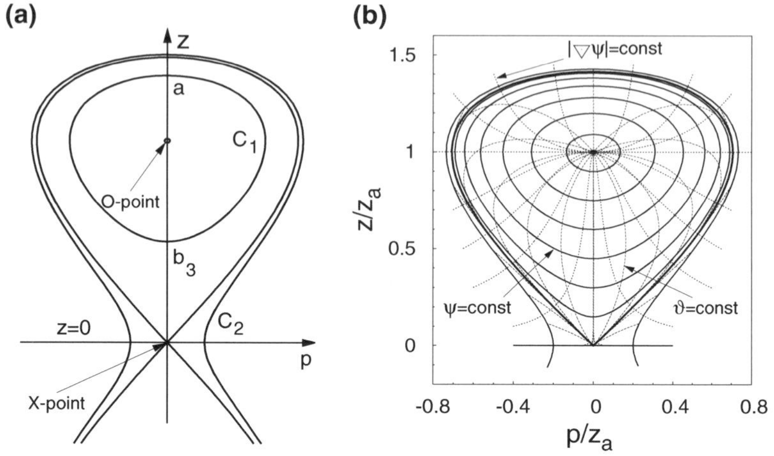

Fig. 1.5 **a** Magnetic configuration of the model in the (z, p) plane: *Curve* C_1 corresponds to field lines on the closed magnetic surface; *curve* C_2—an open field line; and *curve* 3—a magnetic separatrix. **b** Magnetic surfaces $\psi(z, p) = \text{const}$, isolines of $\vartheta(z, p) = \text{const}$, normals to magnetic surfaces $|\nabla\psi(z, p)| = \text{const}$, *thick curve* corresponds to the separatrix

lie on the magnetic surfaces $\psi(z, p) = \text{const}$ whose cross-section with the plane $\varphi = \text{const}$ is shown by the contour C_1. The open field lines lie on the magnetic surfaces with $\psi(z, p) > 0$ with the cross-section C_2. The magnetic separatrix is given by the surface $\psi(z, p) = 0$ (surface 3).

1.3.1 Field Lines in Action-Angle Variables

Below we determine the field lines using the action-angle variables (ψ_t, ϑ) which will be used later to study the effect of magnetic perturbations. Recall that in these variables field lines are straight lines, $\psi = \text{const}$, $\vartheta = \varphi/q(\psi) + \vartheta_0$.

1.3.1.1 Closed Field Lines

The action variable ψ_t for the closed field lines ($\psi < 0$) is defined as an integral along closed contour C_1 (see Fig. 1.5a),

$$\psi_t = \frac{1}{2\pi}\oint_{C_1} p(z;\psi)dz = \frac{1}{2\pi\beta}\oint_{C_1}\sqrt{2\left(\psi + \frac{\alpha^2 z^2}{2} - \frac{\alpha^2 z^4}{4z_a^2}\right)}dz$$
$$= \frac{\sqrt{2}\alpha}{2\pi z_a \beta}\int_b^a \sqrt{\left(a^2 - z^2\right)\left(z^2 - b^2\right)}dz,$$

where a and b are maximal and minimal vertical positions of field lines along the contour C_1, respectively,

$$(a, b) = z_a\sqrt{1 \mp \sqrt{\psi_N}}, \qquad \psi_N = 1 + \frac{\psi}{|\psi_a|}.$$

Here ψ_N is the normalized poloidal flux. Using the integral [Prudnikov et al. (1986)]

$$\int_b^a \sqrt{\left(a^2 - z^2\right)\left(z^2 - b^2\right)}dz = \frac{a}{3}\left[(a^2 + b^2)E(k) - 2b^2 K(k)\right],$$
$$k = \frac{\sqrt{a^2 - b^2}}{a},$$

where $K(k)$ and $E(k)$ are the complete elliptic integrals of the first kind and the second kind, respectively, with a module k, we obtain

$$\psi_t = \psi_t^{(s)}\frac{\sqrt{2(1 + \sqrt{\psi_N})}}{\pi}\left[E(k) - \left(1 - \sqrt{\psi_N}\right)K(k)\right], \tag{1.40}$$

with

$$k = \frac{\sqrt{2} \quad \psi_N^{1/4}}{\sqrt{1 + \psi_N^{1/2}}}. \tag{1.41}$$

Here $\psi_t^{(s)} = \alpha z_a^2/3\beta$ is the value of toroidal flux at the separatrix.

In order to introduce the angle variable ϑ we set that $\vartheta = 0$ at the highest point $z = a$ of the contour C_1, and $\vartheta = \mp\pi$ at the lowest point $z = b$. Then for the closed field lines ($\psi < 0$) the angle ϑ is introduced as the integral.

$$\vartheta(z) = \frac{\partial}{\partial \psi_t} \int_z^a p(z'; \psi) dz' = \frac{1}{\sqrt{2}\beta} \frac{d\psi}{d\psi_t} \int_z^a \frac{dz'}{\sqrt{\psi + \frac{\alpha^2 z^2}{2} - \frac{\alpha^2 z^4}{4 z_a^2}}}$$
$$= \frac{2 z_a}{\sqrt{2}\alpha\beta} \frac{d\psi}{d\psi_t} \int_z^a \frac{dz'}{\sqrt{\left(a^2 - z^2\right)\left(z^2 - b^2\right)}}$$
$$= \frac{\sqrt{2} z_a}{\alpha\beta a} \frac{d\psi}{d\psi_t} F\left(\arcsin \frac{\sqrt{a^2 - z^2}}{\sqrt{2} a \psi_N^{1/4}}, k\right), \tag{1.42}$$

where we have used the integral

$$\int_z^a \frac{dz}{\sqrt{\left(a^2 - z^2\right)\left(z^2 - b^2\right)}} = \frac{1}{a} F(\phi, k),$$
$$\phi = \arcsin \sqrt{\frac{a^2 - z^2}{a^2 - b^2}},$$

with the incomplete elliptic integral of the first kind $F(\phi, k)$. Using the condition $\vartheta(z = b) = \pi$ we find the derivative $d\psi/d\psi_t$ which is equal to the inverse safety factor,

$$q(\psi) = \left(\frac{d\psi}{d\psi_t}\right)^{-1} = \frac{\sqrt{2}}{\pi\alpha\beta\sqrt{1 + \sqrt{\psi_N}}} K(k). \tag{1.43}$$

Inverting the relation (1.42) with respect to the coordinate z and using the relation $p = \beta^{-2} dz/d\varphi = q^{-1}\beta^{-2} dz/d\vartheta$ which follows from (1.38) we obtain field line coordinates $z(\vartheta)$, $p(\vartheta)$ as a function of the angle variable ϑ:

$$z(\vartheta) = a\sqrt{1 - k^2 \operatorname{sn}^2\left(K(k)\vartheta/\pi, k\right)} = a \operatorname{dn}\left(K(k)\vartheta/\pi, k\right),$$
$$p(\vartheta) = -\frac{\sqrt{2} z_a \psi_N^{1/2} \alpha}{\beta} \operatorname{sn}\left(K(k)\vartheta/\pi, k\right) \operatorname{cn}\left(K(k)\vartheta/\pi, k\right), \tag{1.44}$$

where $\operatorname{sn}(u; k)$, $\operatorname{cn}(u; k)$, $\operatorname{dn}(u; k)$ are the Jacobi elliptic functions. The dependence of the field lines on the toroidal angle φ can be obtained by using the relation $\vartheta = \varphi/q(\psi)$. Field lines reach the lowest vertical point $z = b$ at $\varphi = \pm\pi q(\psi)$, and the highest vertical point $z = a$ at $\varphi = 0$.

1.3.1.2 Open Field Lines

We define the toroidal flux ψ_t for the open field lines $\psi_N > 1$ by requiring that it should be a continuous function of the poloidal flux ψ at the separatrix $\psi = 0$. This can be done by choosing the integration contour C_2 along the field line starting and ending at the p-axis ($z = 0$) after one full turn as shown in (see Fig. 1.5a), i.e.,

$$\psi_t = \frac{1}{2\pi}\oint_{C_2} p(z;\psi)dz = \frac{1}{\pi\beta}\int_0^a \sqrt{2\left(\psi + \frac{\alpha^2 z^2}{2} - \frac{\alpha^2 z^4}{4z_a^2}\right)}dz$$
$$= \frac{\sqrt{2}\alpha}{2\pi z_a \beta}\int_0^a \sqrt{\left(a^2 - z^2\right)\left(z^2 + b^2\right)}dz,$$

where $b^2 = z_a^2(\sqrt{\psi_N} - 1)$. Then using the integral

$$\int_0^a \sqrt{\left(a^2 - z^2\right)\left(z^2 + b^2\right)}dz = \frac{\sqrt{a^2 + b^2}}{3}\left[(a^2 - b^2)E\left(k^{-1}\right) + b^2 K\left(k^{-1}\right)\right],$$

where the module k is given by (1.41), one obtains

$$\psi_t = \psi_t^{(s)}\frac{2\psi_N^{1/4}}{\pi}\left[E\left(k^{-1}\right) + \frac{1}{2}\left(\sqrt{\psi_N} - 1\right)K\left(k^{-1}\right)\right]. \tag{1.45}$$

The angle variable ϑ is defined similar to (1.42) with the settings $\vartheta(z = a) = 0$ and $\vartheta(z = 0) = \pm\pi$. Then, it is not difficult to obtain

$$\vartheta(z) = \frac{1}{\alpha\beta\psi_N^{1/4}}\frac{d\psi}{d\psi_t}F\left(\arccos\frac{z}{a}, k^{-1}\right). \tag{1.46}$$

From the condition $\vartheta(z = 0) = \pi$ one obtains the safety factor,

$$q(\psi) = \frac{1}{\pi\alpha\beta\psi_N^{1/4}}K\left(k^{-1}\right). \tag{1.47}$$

Field line coordinates $z(\vartheta)$, $p(\vartheta)$ as a function of the angle variables ϑ are given by

$$z(\vartheta) = a\,\mathrm{cn}\left(\frac{K\left(k^{-1}\right)\vartheta}{\pi}, k^{-1}\right),$$
$$p(\vartheta) = -\frac{a\alpha\psi_N^{1/4}}{\beta}\,\mathrm{sn}\left(K(k^{-1})\vartheta/\pi, k^{-1}\right)\,\mathrm{dn}\left(K(k)\vartheta/\pi, k^{-1}\right). \tag{1.48}$$

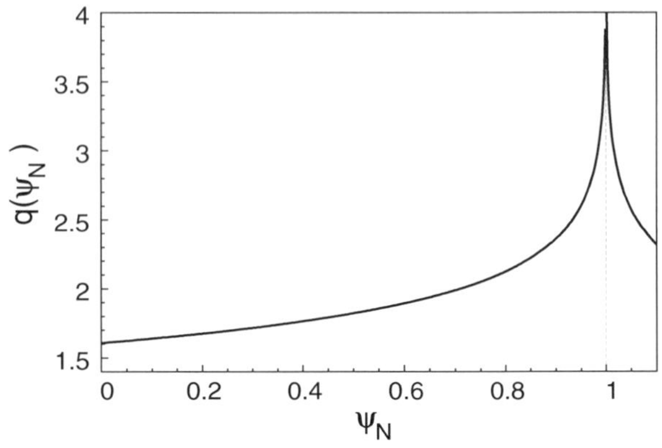

Fig. 1.6 Dependence of the safety factor $q(\psi)$ (1.43) and (1.47) on ψ_N. The values of the parameters $\gamma = 0.44$, $z_a = 1$, and $\alpha = 0.4553$. The value of q_{95} is 2.6

Figure 1.5b shows the isolines of $\vartheta(z, p) = \text{const}$, normals to magnetic surfaces $\nabla\psi(z, p) = \text{const}$, thick curve corresponds to the separatrix.

1.3.2 Magnetic Separatrix

The field line on the separatrix ($\psi = 0$) can be obtained from (1.44) or (1.48) by putting $\vartheta = \varphi/q(\psi)$ and taking the limit $\psi_N \to 1$, $k \to 1$ and using the relations $\mathrm{dn}(u, k = 1) = \mathrm{cn}(u, k = 1) = 1/\cosh(u)$, $\mathrm{sn}(u, k = 1) = \tanh(u)$, and $K(k)/\pi q(\psi) \to \gamma$. Therefore, we have

$$z_s(\varphi) = \frac{\sqrt{2}z_a}{\cosh(\gamma\varphi)}, \qquad p_s(\varphi) = -\frac{\sqrt{2}z_a\alpha\sinh(\gamma\varphi)}{\beta\cosh^2(\gamma\varphi)}. \tag{1.49}$$

The safety factor $q(\psi)$ (1.43) and (1.47) near the separatrix ($\psi \to \pm 0$, $\psi_N \to 1 \pm 0$) has the following asymptotics,

$$q(\psi) = \frac{1}{2\pi\gamma}\ln\frac{Q}{|\psi|} + O(\psi), \quad \psi \to \pm 0, \tag{1.50}$$

where $\gamma = \alpha\beta$ and $Q = 16z_a^2\alpha^2$. As we will see below, the asymptotical form (1.50) is a generic for the magnetic field configurations with the magnetic separatrix which has the so-called first order null of the poloidal magnetic field (see Sect. 2.4.1). Figure 1.6 shows the profile $q(\psi)$ for the values of the parameters $\gamma = 0.44$, $z_a = 1$, and $\alpha = 0.4553$. The number q_{95} is the value of the safety factor at the magnetic surface which has 95 % of the total poloidal flux through the cross section confined by the separatrix.

1.4 Bibliographic Comments

The Hamiltonian formulation of the equations of magnetic field lines has been already used in early studies of the problem of stability and destruction of magnetic surfaces in tokamaks and stellarators [Kerst (1962); Rosenbluth et al. (1966); Filonenko et al. (1967); Freis et al. (1973); Hamzeh (1974); Finn (1975), Matsuda and Yoshikawa (1975); Boozer and Rechester (1978)]. More general Hamiltonian formulation of magnetic field lines based on the variational principles of classical mechanics has been given by Cary and Littlejohn (1983). Boozer (1983) and White (2001) formulated the Hamiltonian equations of field lines in magnetic flux coordinates using the Clebsch form of magnetic field.

Historically, the Clebsch presentation of magnetic field has been first given by Kruskal and Kulsrud (1958); Hamada (1959, 1962); Greene and Johnson (1962) for equilibrium plasmas. The validity of this presentation for the arbitrary magnetic field was shown by Boozer (1983), and a rigorous mathematical proof that every solenoidal field in an arbitrary smooth simple toroidal domain can written in this form is given by Yoshida (1993). The detailed description of magnetic flux coordinates can be found, for example, in the books by Balescu (1988) and Dhaeseleer et al. (1991).

The formalism of action-angle variables is described in many textbooks on classical mechanics, for example, by Landau and Lifshits (1976) and Goldstein (1980). The general definition of action-angle variables in Hamiltonian systems with inseparable variables is given in a textbook by Arnold (1989).

Chapter 2
Magnetic Fields of Equilibrium Plasmas

In this chapter we describe the magnetic structure of equilibrium plasmas in tokamaks. First we briefly recall the Grad-Shafranov equation for the equilibrium magnetic configuration of plasmas. Then the magnetic configurations in the plasmas with the circular cross section will be considered. Particularly, the standard model of the magnetic field and the model of the equilibrium plasmas with the Shafranov shift will be presented. We describe also the analytical equilibria for the plasmas with the elongated shape, particularly, with the magnetic separatrix recently proposed in the literature. The generic properties of a magnetic field near the magnetic separatrix and the X-point will be studied. Finally, we consider also the so–called wire models of plasmas, which are often used to study the magnetic field of tokamak plasmas near the magnetic separatrix and the divertor. Particularly, we study the so-called magnetic field with the snowflake divertor configuration recently proposed to expand the heat load on the divertor plates.

2.1 The Grad–Shafranov Equation

In tokamaks the equilibrium magnetic field $\mathbf{B}(X, Y, Z)$ is created by external toroidal field B_φ and the poloidal field B_θ of the plasma current $\mathbf{j}(X, Y, Z)$. The magnetic field $\mathbf{B}$ confines the high temperature plasma consisting of fully ionized ions and electrons in order to isolate it from the walls of the vessel and to hold it in a bounded area. The equilibrium state of the plasma is maintained when the Ampere force $\mu_o \mathbf{j} \times \mathbf{B}$ acting on the plasma column is balanced with the pressure gradient ∇p of the plasma:

$$\mu_o \mathbf{j} \times \mathbf{B} = \nabla p, \tag{2.1}$$

where μ_o is the vacuum magnetic permeability. From (2.1) it follows the equations for the *equilibrium magnetic field*

S. Abdullaev, *Magnetic Stochasticity in Magnetically Confined Fusion Plasmas*, Springer Series on Atomic, Optical, and Plasma Physics 78,
DOI: 10.1007/978-3-319-01890-4_2,

$$\mathbf{B} \cdot \nabla p = 0, \qquad \mathbf{j} \cdot \nabla p = 0. \tag{2.2}$$

From Eq. (2.2) it follows that the pressure $p(X, Y, Z)$ is constant along the magnetic field $\mathbf{B}$ and the current $\mathbf{j}$. Since ∇p is perpendicular to the surface $p(X, Y, Y) = \text{const}$ the field lines lie on the surface of constant pressure $p(X, Y, Z) = \text{const}$. The surfaces of different constant pressure should not cross each other and they should be bounded in a finite domain. These conditions are satisfied if the magnetic surfaces are nested toroidal surfaces as was shown in Fig. 1.1.

The Eq. (2.1) for the equilibrium plasma can be reduced to the one for the poloidal flux $\Psi = B_0 R_0^2 \psi$ known the Grad–Shafranov equation (Grad and Rubin 1958, Shafranov 1958) [see, also Shafranov 1966; Wesson 2004]. In the cylindrical coordinate system the equation is given by

$$R\frac{\partial}{\partial R}\left(\frac{1}{R}\frac{\partial \Psi}{\partial R}\right) + \frac{\partial^2 \Psi}{\partial Z^2} = -R\mu_o j_\varphi,$$
$$j_\varphi = R\frac{dp}{d\Psi} + \frac{F}{R\mu_o}\frac{dF}{d\Psi}. \tag{2.3}$$

The plasma current j_φ, pressure p are functions of the poloidal flux Ψ, i.e., $j_\varphi = j_\varphi(\Psi)$, $p = p(\Psi)$. The flux function $F(\Psi)$ is related the toroidal field B_φ. The magnetic field $\mathbf{B}$ is determined via the flux functions Ψ and $F(\Psi)$,

$$\mathbf{B} = \frac{1}{R}\nabla\Psi \times \mathbf{e}_\varphi + \frac{F}{\mu_o R}\mathbf{e}_\varphi, \tag{2.4}$$

or

$$B_R = \frac{1}{R}\frac{\partial \Psi}{\partial Z}, \qquad B_Z = -\frac{1}{R}\frac{\partial \Psi}{\partial R}, \qquad B_\varphi = \frac{F(\psi)}{\mu_o R}. \tag{2.5}$$

For specific tokamak plasmas the Grad–Shafranov equation is solved numerically for the given plasma pressure $p(\Psi)$ and the current $j_\varphi(\Psi)$ profiles. The numerical codes have been developed for this purpose. For example, the computer code, EFIT (Equilibrium Fitting), solve the Grad–Shafranov equation using the available measurements as constraints on the toroidal current density Lao et al. (1985).

In plasmas with a circular cross sections the analytical solutions have been widely used in theoretical studies of stability, transport, and kinetic analysis. In the last decade analytical solutions to the Grad-Shafranov equations have been also proposed for tokamak plasmas with elongated cross sections (Zheng et al. 1996; Miller et al. 1998), and particularly, with magnetic separatrix and X-points (Srinivasan et al. 2001; Shi 2005; Srinivasan et al. 2010; Guazzotto and Freidberg 2007; Cerfon and Freidberg 2010). Below we consider some examples of analytical equilibria for the tokamak plasma with circular and elongated shapes of cross sections.

2.2 Equilibrium Plasmas with Circular Cross-Sections

2.2.1 *Standard Model of the Magnetic Field*

The magnetic field of plasmas modeled by

$$\mathbf{B}_0(r,\theta) = B_\varphi \mathbf{e}_\varphi + B_\theta \mathbf{e}_\theta = \frac{B_0}{1+\varepsilon\cos\theta}\left(\mathbf{e}_\varphi + \mathbf{e}_\theta \frac{r}{q(r)R_0}\right), \tag{2.6}$$

is known a *standard magnetic field* (Balescu 1988). In Eq. (2.6) B_0 is the strength of the toroidal magnetic field model at the magnetic axis $R = R_0$, and $q(r)$ is the safety factor. This model describes the main features of equilibrium magnetic field of the toroidal plasma, i.e., the radial decay of the toroidal field $B_\varphi \propto R^{-1}$, and nested, circular magnetic surfaces $r = \text{const}$.

The vector potential **A** corresponding to the magnetic field (2.6) is given by:

$$\begin{aligned} \mathbf{A}(r,\theta) &= \left(0, A_\varphi(R,Z), A_z(R,Z)\right), \\ A_\varphi(R,Z) &= \frac{R_0 B_0}{R}\int \frac{dr}{q(r)}, \qquad A_z(R) = -B_0 R_0 \ln R, \end{aligned} \tag{2.7}$$

where $r = \sqrt{(R-R_0)^2 + Z^2}$. The magnetic surfaces of the field (2.6) represent of circular surfaces $r = \text{const}$. According to (1.17) and (2.6) the canonical momentum p_z is

$$p_z = \ln(R/R_0) = \ln\left(1 + \sqrt{r^2/R_0^2 - z^2}\right), \tag{2.8}$$

Using Eq. (1.22) one can the relations between the action–angle (or the normalized toroidal flux and the intrinsic angle) variables (ψ_t, ϑ) and the geometrical coordinates (r, θ),

$$\begin{aligned} \psi_t &= 1 - \left(1 - r^2/R_0^2\right)^{1/2}, \\ \vartheta &= 2\arctan\left(\sqrt{\frac{1+r/R_0}{1-r/R_0}}\tan(\theta/2)\right). \end{aligned} \tag{2.9}$$

The poloidal flux $\psi(\psi_t) = -RA_\varphi/B_0R_0^2$ is determined by the integral

$$\psi(\psi_t) = \int \frac{d\psi_t}{q\left(r(\psi_t)\right)},$$

where $r(\psi_t) = R_0[1 - (1-\psi_t)^2]^{1/2}$.

For the large aspect ratio tokamaks, $\varepsilon = r/R_0 \ll 1$, one can approximate

$$\psi_t \approx \frac{r^2}{2R_0^2}, \qquad \vartheta = \theta - \varepsilon \sin\theta + O(\varepsilon^2). \tag{2.10}$$

In the standard model of magnetic field all circular magnetic surfaces $r = \text{const}$ have a common center $r = 0$.

2.2.2 *Magnetic Field of Plasmas with the Shafranov Shift*

Unlike the standard model in a real toroidal plasma the magnetic surfaces are shifted outwardly along the major radius R due to the plasma pressure and the electric current (Shafranov 1964). The geometry of such magnetic surfaces with circular cross sections is shown in Fig. 2.1. The quantity $R_p(\rho)$ is the radial position of the center of the magnetic surface of radius ρ. If a is the last magnetic surface, then the quantity $\Delta(\rho) = R_p(\rho) - R_p(a)$ describes the shift of magnetic surface of radius ρ with respect to the center of plasma center $R_p(a)$ which coincides with the torus center R_0. The shift is called the Shafranov shift .

The Shafranov shift $\Delta(\rho)$ can be found from the Grad–Shafranov Eq. (2.3) imposing the boundary conditions: $d\Delta(\rho)/d\rho = 0$ at $\rho = 0$ and $\Delta(a) = 0$ at $\rho = a$. For the large inverse aspect ratio tokamaks, $a/R_0 \ll 1$ one can obtain the following equation for $\Delta(\rho)$ (see, e.g., Kadomtsev 1988)

$$\frac{d\Delta(\rho)}{d\rho} = \frac{\rho}{R_0}(\Lambda - 1), \tag{2.11}$$

where $\Lambda = \beta_{pol} + l_i/2 - 1$. Here, β_{pol} is the ratio of the plasma pressure, $\langle p \rangle$ to the magnetic pressure, $B_\theta^2/8\pi$ of the poloidal field B_θ:

$$\beta_{pol} = \frac{8\pi(\langle p \rangle - p)}{B_\theta^2}, \qquad l_i = \frac{8\pi\langle B_\theta^2 \rangle}{\overline{B_\theta^2}}$$

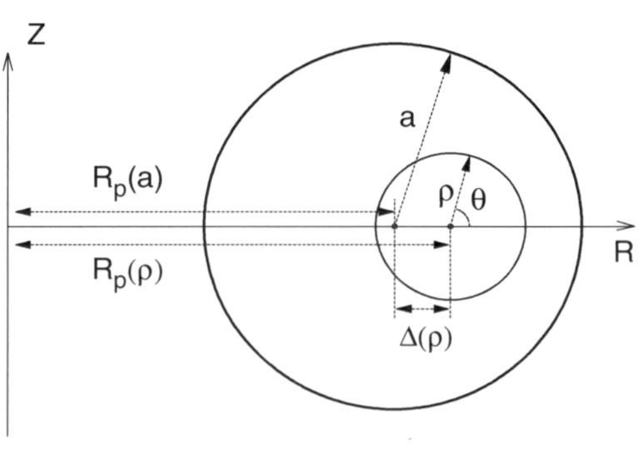

Fig. 2.1 Geometry of shifted magnetic flux surfaces in a toroidal system. Here $R_p(\rho)$ is a major radius of the center of magnetic surfaces of radius ρ, $\Delta(\rho) = R_p(\rho) - R_p(a)$ is the outward shift of magnetic surface from the center of the last magnetic surfaces of radius a

l_i is the internal inductance, and

$$\langle p \rangle = \frac{1}{\rho^2} \int_0^r 2p\rho d\rho, \qquad \langle B_\theta^2 \rangle = \frac{1}{\rho^2} \int_0^\rho 2B_{\theta 0}^2 \rho d\rho,$$

$$B_{\theta 0} = \frac{\mu_o I_p(\rho)}{2\pi\rho} \tag{2.12}$$

where $I_p(\rho)$ is the plasma current flowing through the section enclosed by the magnetic surface of radius ρ. The toroidal, B_φ, and poloidal, B_θ, fields on each magnetic surface of radius ρ are given by (see, Shafranov 1966; Kadomtsev 1988)

$$B_\varphi(R, Z) = B_0 \left(1 + \frac{\rho}{R_0} \cos\theta\right), \qquad B_0 = \frac{\mu_o I_\varphi}{2\pi R_0},$$

$$B_\theta(R, Z) = B_{\theta 0} \left(1 + \Lambda \frac{\rho}{R_0} \cos\theta\right), \tag{2.13}$$

where I_φ is the current of toroidal field, B_0 is the strength of toroidal field at the center $R = R_0$ of the last magnetic surface of radius $\rho = a$. The angle θ is defined as $\tan\theta = Z/(R - R_0)$ as shown in Fig. 2.1.

2.2.3 Model of a Magnetic Field with the Shafranov Shift

Below present another model of tokamak plasmas with nested, circular magnetic surfaces. This model describes the magnetic field of a toroidal plasma with the shifted center of the last magnetic surface along the major radius R. Let $R_p(\rho)$ be a major radius of the center of a magnetic surface of radius ρ. Unlike of the previous case in this model the major radius $R_p(a)$ of the last magnetic surface $\rho = a$ may not coincide with the torus center R_0.

The equilibrium magnetic field is given by the following vector potential,

$$\mathbf{A} = \left(0, A_\varphi(R, Z), A_z(R, Z)\right),$$

$$A_\varphi(R, z) = \frac{R_0 B_0}{R} \int \frac{d\rho}{q(\rho)},$$

$$A_z(R) = -B_0 R_0 \ln(R/R_0), \tag{2.14}$$

where $q(\rho)$ is a safety factor as a function of plasma radius ρ. We will use the following model for the Shafranov shift $\Delta(\rho)$ (Nguyen et al. 1995)

$$\Delta(\rho) = R_p(\rho) - R_p(a) = \left[R_p^2(a) + (\Lambda + 1)(a^2 - \rho^2)\right]^{1/2}. \tag{2.15}$$

The relation between the radius ρ of the magnetic surface and the cylindrical coordinates R, Z is $\rho = \sqrt{\left(R - R_p(a) - \Delta(\rho)\right)^2 + Z^2}$.

The relationship between the normalized toroidal flux ψ_t and the radius r of a magnetic surface:

$$\psi_t = \frac{R_p(\rho)}{R_p(a)} \left[1 - \left(1 - \frac{\rho^2}{R_p^2(\rho)} \right)^{1/2} \right]. \tag{2.16}$$

For the small Shafranov shift, $\Delta(\rho) \ll R_p(a)$, these formulas may be simplified to

$$\psi_t \approx 1 - \left(1 - \frac{\rho^2}{R_p^2(a)} \right)^{1/2},$$
$$\rho \approx R_p(a) \left[1 - (1 - \psi)^2 \right]^{1/2}. \tag{2.17}$$

The toroidal and the poloidal magnetic fields are

$$B_\varphi(R) = \frac{B_0 R_0}{R},$$
$$B_\theta(R, Z) = \frac{B_0}{qR} \frac{d\psi}{d\rho} \frac{1}{1 + \Delta'(\rho) \cos\bar{\theta}}, \tag{2.18}$$

where $\bar{\theta} = \arcsin(Z/\rho)$, and $\Delta'(\rho) = d\Delta(\rho)/d\rho$.

2.2.3.1 The Safety Factor Profile

In a cylindrical plasma the safety factor profile can be found by the density profile of the toroidal plasma current $j(\rho)$. The poloidal magnetic field B_θ is determined by according to the Ampère's law

$$B_\theta(R, Z) = \frac{\mu_o I(\rho)}{2\pi\rho} = \frac{\mu_o}{\rho} \int_0^r j(\rho')\rho' d\rho', \tag{2.19}$$

where $I(\rho)$ is the current flowing inside the magnetic surface of radius ρ. The safety factor $q(\rho)$ in the cylindrical geometry is defined as

$$q_{cyl}(\rho) = \frac{\rho B_0}{R B_\theta} = \frac{2\pi\rho^2 B_0}{\mu_o R_0 I(\rho)}. \tag{2.20}$$

The value of the safety factor at the plasma edge $\rho = a$ is determined by the total plasma current I_p ,

$$q_a = q_{cyl}(a) = \frac{2\pi a^2 B_0}{\mu_o R_0 I_p}. \tag{2.21}$$

Consider the model plasma current given by the density

$$j(\rho) = \frac{I_p(\nu+1)}{\pi a^2}\left(1-\rho^2/a^2\right)^\nu, \tag{2.22}$$

where ν in the constant parameter. The safety factor $q_{cyl}(r)$ corresponding to this plasma current distribution is given by

$$q_{cyl}(\rho) = q_a \frac{\rho^2/a^2}{1-\left(1-\rho^2/a^2\right)^{\nu+1}}. \tag{2.23}$$

The value of the safety factor at the magnetic axis $q(0)$ is determined by q_a and ν: $q(0) = q_a/(\nu+1)$.

In the toroidal geometry the safety factor $q(\rho)$ is changed due to bending the plasma column to the torus. The toroidal corrections to the safety factor can be found in the limit of large aspect ratio $R/r \gg 1$. The expansion of the safety factor in a series of powers $\varepsilon = \rho/R_p(\rho)$ is given by (see, Abdullaev et al. 1999),

$$q(\rho) = q_{cyl}(\rho)\frac{R_0^2}{R_0^2(\rho)}\left(1+\frac{1}{2}a_2\varepsilon^2+\frac{3}{8}a_4\varepsilon^4\right)+O(\varepsilon^8), \tag{2.24}$$

where $q_{cyl}(\rho)$ is the safety factor in the cylindrical geometry. The coefficients a_m are polynomial functions of the plasma parameter Λ:

$$a_m = (-1)^m \sum_{k=0}^{m}(m-k+1)\,\Lambda^k. \tag{2.25}$$

2.2.3.2 The Relation Between the Poloidal Angles θ and ϑ

The relation between the intrinsic poloidal angle ϑ and the geometrical one θ is rather complicated (Nguyen et al. 1995) and it is hard to find the inverse relationship $\theta = \theta(\vartheta, \rho)$. The relation between ϑ and θ can be presented by a series in powers of the inverse aspect ratio ε (see Abdullaev et al. 1999)

$$\vartheta(\theta,\rho) = \theta + \sum_{m=1}^{M}\alpha_m \sin m\theta + O(\varepsilon^{M+1}), \tag{2.26}$$

where the expansion coefficients α_m are series in powers of ε:

$$\alpha_m = \sum_{k=0}^{M} \alpha_m^{(k)} \varepsilon^{m+k} + O(\varepsilon^{M+1}).$$

The coefficients α_m for the case $M = 4$ are given by

$$\alpha_1 = a_1\varepsilon + \left(\frac{3a_3}{4} - \frac{a_1 a_2}{2}\right)\varepsilon^3 + O(\varepsilon^5),$$
$$\alpha_2 = \frac{\varepsilon^2}{4}\left[a_2 + \left(a_4 - \frac{a_2^2}{2}\right)\varepsilon^2\right] + O(\varepsilon^5),$$
$$\alpha_3 = \frac{1}{12}a_3\varepsilon^3 + O(\varepsilon^5), \qquad \alpha_4 = \frac{1}{32}a_4\varepsilon^4 + O(\varepsilon^5). \tag{2.27}$$

The relation (2.26), expressing the intrinsic poloidal angle ϑ via the geometrical poloidal angle θ, can be inverted to find θ in terms of ϑ,

$$\theta(\vartheta, \rho) = \vartheta + \sum_{m=1}^{M} \alpha_m^* \sin m\vartheta + O(\varepsilon^{M+1}). \tag{2.28}$$

The expansion coefficients α_m^* can be expressed in terms of coefficients α_m. A typical dependence ϑ on θ is shown in Fig. 2.2a at the magnetic surface $r = 0.43$ m, and the radial profile of the safety factor $q(\rho)$ is plotted in Fig. 2.2b. The plasma parameters are chosen: $\beta_{pol} = 1$, $l_i = 1.2$ and $R_p(a) = 1.75$ m. The radial profile of the plasma current density is chosen as $j(\rho) = (2I_p/\pi a^2)(1 - \rho^2/a^2)$ where I_p is the full plasma current. The toroidal field $B_\varphi = 2.25$ T.

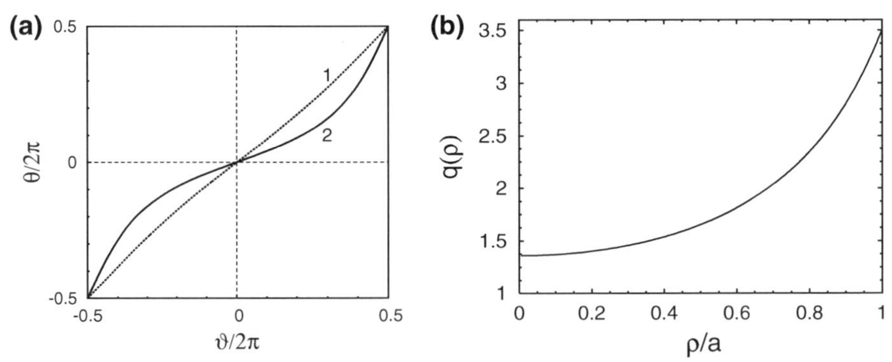

Fig. 2.2 **a** Typical dependence of the geometrical angle θ on the intrinsic angle ϑ at the two different magnetic surfaces close to the magnetic axis (*curve 1*) and close to the plasma edge (*curve 2*); **b** Typical radial profile of the safety factor $q(\rho)$ for the plasma with circular cross–section

2.3 Elongated Plasmas: Analytical Solutions

In order to obtain the analytical solutions of the Grad-Shafranov Eq. (2.3) one uses the simple profiles of pressure and poloidal current proposed by Solov'ev Solov'ev (1968) based on the assumptions (see, for example, Srinivasan et al. 2010; Cerfon and Freidberg 2010)

$$-\mu_o \frac{dp}{d\Psi} = A_1, \qquad -F\frac{dF}{d\Psi} = A_2, \tag{2.29}$$

where A_1 and A_2 are constants related to the plasma current and the plasma pressure. This assumption allows to obtain the full solution for Ψ in the form of linear combinations of polynomial functions. Below we describe the analytical solution obtained by Cerfon and Freidberg (2010).

Introducing the normalized coordinates $x = R/R_0$, $y = Z/R_0$, and $\psi = \Psi/\Psi_0$, the Grad-Shafranov Eq. (2.3) is presented as

$$\left[x\frac{\partial}{\partial x}\left(\frac{1}{x}\frac{\partial}{\partial x}\right) + \frac{\partial^2}{\partial y^2}\right]\psi = -\mu_o \frac{R_0^4}{\Psi_0^2}\frac{dp}{d\psi} - \frac{R_0^2 F}{\Psi_0^2}\frac{dF}{d\psi}, \tag{2.30}$$

with the following constraints,

$$-\mu_o \frac{R_0^4}{\Psi_0^2}\frac{dp}{d\psi} = A_1 = C, \qquad -\frac{R_0^2}{\Psi_0^2}F\frac{dF}{d\psi} = -A_2 = A. \tag{2.31}$$

The following condition $A+C = 1$ is imposed. Then the pressure p and the magnetic field are given by

$$p = -\frac{\Psi_0^2}{R_0^4 \mu_o}(1-A)\psi, \qquad B_\varphi^2(\psi) = \frac{R_0^2}{R^2}\left(B_0^2 - \frac{2\Psi_0^2}{R_0^4}A\psi\right),$$
$$B_R = \frac{\Psi_0}{R_0^2}\frac{1}{x}\frac{\partial\psi}{\partial y}, \qquad B_Z = -\frac{\Psi_0}{R_0^2}\frac{1}{x}\frac{\partial\psi}{\partial x}. \tag{2.32}$$

The solution of (2.30) is sought in the form

$$\psi(x, y) = \sum_{i=1}^{N} c_i \psi_i(x, y) + \frac{x^4}{8} + A\left(\frac{1}{2}x^2 \ln x - \frac{x^4}{8}\right), \tag{2.33}$$

where the functions $\psi_i(x, y)$, $(i = 1, \ldots, N)$ are polynomials of degree i. The functions $\psi_i(x, y)$, $(i = 1, 2, \ldots, N)$ are given in Table 2.1. The number $N = 7$ for the up/down symmetric plasmas with the even function $\psi_i(x, y)$ with respect to the vertical coordinate y: $\psi_i(x, -y) = \psi_i(x, y)$, $(i = 1, 2, \ldots, 7)$. For the up/down

Table 2.1 The functions $\psi_k(x, y$ in the solution (2.33) of the Grad–Shafranov equation

i	$\psi_i(x, y)$
1	1
2	x^2
3	$y^2 - x^2 \ln x$
4	$x^4 - 4x^2y^2$
5	$2y^4 - 9y^2x^2 + 3x^4 \ln x - 12x^2y^2 \ln x$
6	$x^6 - 12x^4y^2 + 8x^2y^4,$
7	$8y^6 - 140y^4x^2 + 75y^2x^4 - 15x^6 \ln x + 180x^4y^2 \ln x - 120x^2y^4 \ln x$
8	y
9	yx^2
10	$y^3 - 3yx^2 \ln x$
11	$3yx^4 - 4x^2y^3$
12	$8y^5 - 45yx^4 - 80y^3x^2 \ln x + 60yx^4 \ln x$

asymmetric plasmas the number of functions is $N = 12$ with the additional odd functions $\psi_i(x, y)$ with respect to y: $\psi_i(x, -y) = -\psi_i(x, y)$, $(i = 8, \ldots, 12)$.

The unknown coefficients c_k, A, and the constant Ψ_0 are found by imposing the boundary constraints on Ψ determined by the plasma current, pressure, shape. The total plasma current is determined by the integral,

$$I_p = \int j_\varphi(R, Z) dR dZ = -\frac{\Psi_0}{R_0 \mu_0} \int \left[\frac{A}{x} + (1 - A)x \right] dx dy. \tag{2.34}$$

The plasma position and its shape are characterized by the major radius R_0, the minor radius a, the elongation κ, and the triangularity δ. The plasma shape or the boundary of the upper cross section is given by the parametric equation

$$\begin{aligned} x(\tau) &= 1 + \varepsilon \cos(\tau + \alpha \sin \tau), \\ y(\tau) &= \varepsilon \kappa \sin \tau, \end{aligned} \tag{2.35}$$

where τ is a parameter covering the range: $0 \leq \tau \leq 2\pi$, $\alpha = \arcsin \delta$. The geometrical definitions of the elongation κ, and the triangularity δ are shown in Fig. 2.3. At the plasma boundary one sets $\psi(x, y) = 0$.

2.3.1 Poloidal Divertor Plasmas. Examples

Poloidal divertor tokamaks are important concept to control the plasma edge. The magnetic configuration of these tokamaks contains a magnetic surface called *magnetic separatrix* which sharply separates closed field lines on nested magnetic surfaces from the open field lines hitting the walls of fusion devices. It has one (or two)

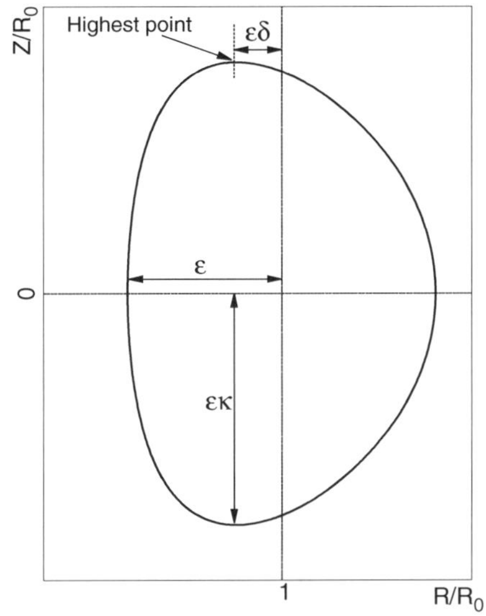

Fig. 2.3 Shape of the elongated plasma: $\varepsilon = a/R_0$ is the inverse aspect ratio, κ is the elongation, and δ is the triangularity

singular point(s), *X-point* , on the poloidal plane ($\varphi =$ const), where the poloidal magnetic field (B_R, B_Z) has zero(es). The so-called *single–null poloidal divertor* which has a separatrix with one X-point as in the example shown Fig. 2.4a. In *double–null divertor tokamaks* there are two X-points connected with two separatrices.

Below we give an example of analytical equilibrium obtained using the analytical solutions of the Grad–Shafranov equation given by Cerfon and Freidberg (2010). We consider the two typical single–null plasmas with large and small aspect ratios, the ITER–like plasma and the NSTX–like plasma. The latter is also known as a *spherical tokamak.* The plasma parameters and the corresponding coefficients c_i, A of the solutions (2.33) are given in Tables 2.2 and 2.3. Here Ψ_{pa} is the value of $\Psi(x, y)$ at the magnetic axis where it takes a minimal value.

Figure 2.4a shows the cross-sections of magnetic surfaces of the corresponding ITER-like plasma. The contours of $\psi(R, Z) =$ const are plotted by solid (black) curves, the isolines normal to the magnetic surfaces, i.e., the isolines of $\nabla\psi(R, Z) =$ const, are plotted by dashed (green) lines, and isolines of constant poloidal angles $\vartheta(R, Z) =$ const are given by dotted (blue) lines. As was noted before in Sect. 1.2.2 the choice of isolines $\vartheta(R, Z) =$ const is not unique. Using the property (1.36) the isolines $\vartheta(R, Z) =$ const have chosen such that it coincides with the isoline of $\nabla\psi(R, Z) =$ const which connects the X-point with the magnetic axis denoted as Σ_S as shown in Fig. 2.4a.

The section Σ_S (red curve in Fig. 2.4a) consists of two segments of the isolines $\vartheta(R, Z) =$ const which crosses the X-point: the one coming from the magnetic axis O in the plasma region and the one coming from outside the plasma region. It can be also presented by two isolines of $\nabla\psi(R, Z) =$ const which are infinitesimally close to Σ_S and connecting the magnetic axis with the outer plasma regions located

Table 2.2 Parameters of the ITER–like and the NSTX-like plasmas.

Parameters	ITER	NSTX
Plasma current I_p	14 MA	0.8 MA
Toroidal field B_t	5.3 T	0.45 T
Major radius R_0	6.2 m	0.85 m
Minor radius a	1.984 m	0.663 m
Plasma β_{pol}	1.365	1.2
Elongation κ	2.0	2.0
Triangularity δ	0.5	0.35
X-point (R_s, Z_s)	(5.456, −3.72) m	(0.595, −1.453) m
The value of q_{95}	3.71	12.2
The value of Ψ_0	2.0292×10^2	2.541×10^{-1}
$\Psi_{p0} = \Psi_0\psi_a$	−7.5257	-6.1625×10^{-2}

The quantity ψ_a is the value of the function $\psi(x, y)$ (2.33) at the magnetic axis

Table 2.3 Coefficients c_i of the solutions (2.33) and the parameter A for the ITER–like and the NSTX-like plasmas

Coefficients c_i	ITER	NSTX
c_1	1.00687012×10^{-1}	1.325526741×10^{-2}
c_2	4.16274456×10^{-1}	$-3.18336798 \times 10^{-2}$
c_3	$-6.53880989 \times 10^{-1}$	1.51573469×10^{-3}
c_4	$-2.95392244 \times 10^{-1}$	$-2.57246043 \times 10^{-2}$
c_5	4.40037966×10^{-1}	3.50996597×10^{-3}
c_6	$-4.01807386 \times 10^{-1}$	$-3.65210294 \times 10^{-3}$
c_7	$-1.66351381 \times 10^{-2}$	$-1.44205114 \times 10^{-4}$
c_8	1.92944621×10^{-1}	2.30001356×10^{-3}
c_9	8.36039453×10^{-1}	1.02822351×10^{-2}
c_{10}	$-5.30670714 \times 10^{-1}$	$-1.21901447 \times 10^{-2}$
c_{11}	$-1.26671504 \times 10^{-1}$	$-2.26823419 \times 10^{-3}$
c_{12}	1.47140977×10^{-2}	3.19619618×10^{-4}
A	−0.155	−0.05

on the left and right sides, respectively. We denote these branches as $\Sigma_S^{(\pm)}$. We set $\vartheta = -\pi$ on the right branch $\Sigma_S^{(+)}$, and $\vartheta = \pi$ on the left branch $\Sigma_S^{(-)}$, respectively. Inside the plasma region these branches coincide.

The introduction of the sections $\Sigma_S^{(\pm)}$ allows one to uniquely define the safety factor $q(\psi)$ not only inside the separatrix plasma, but also outside the separatrix where field lines are open. The safety factor $q(\psi)$ is then defined as the increment of the toroidal angle $\Delta\varphi$ per one poloidal turn from the section $\Sigma_S^{(-)}$ to the section $\Sigma_S^{(+)}$.

Figure 2.4b shows the safety factor profile $q(\psi)$ inside ($\psi_N < 1$) and outside ($\psi_N > 1$) the plasma calculated according to the given definition, where ψ_N is the normalized poloidal flux:

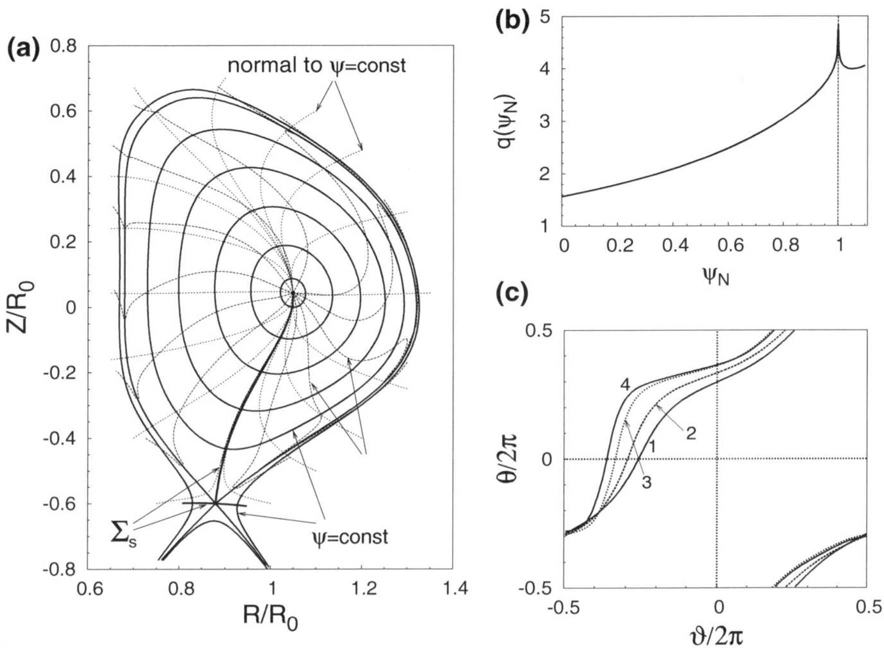

Fig. 2.4 Equilibrium configuration for the ITER like plasma: **a** Contours of the magnetic surfaces, $\psi(R; Z) = \text{const}$ solid lines, isolines of constant angle $\vartheta = \text{const}$, and lines normal to the magnetic surfaces; **b** the safety factor profiles $q(\psi)$ as a function of the normalized poloidal flux ψ_N. **c** Relation between geometrical poloidal angle θ on the poloidal angle ϑ at the different magnetic surfaces: *curve 1* corresponds to $\psi_N = 0.2$, *curve 2* $-\psi_N = 0.5$, *curve 3*—$\psi_N = 0.95$, and *curve 4*—$\psi_N = 1.05$

$$\psi_N = 1 - \frac{\Psi}{|\Psi_{pa}|} = 1 - \frac{\psi}{|\Psi_{pa}|/B_0 R_0^2}, \tag{2.36}$$

where Ψ_{pa} is the value of the poloidal flux at the magnetic axis. The relations between the geometrical (θ) and intrinsic (ϑ) poloidal angles at the different magnetic surfaces are shown in Fig. 2.4c.

The plasma configuration of the spherical tokamak, NSTX, the safety factor profile $q(\psi)$ and the relation between the angles θ and ϑ are presented in Fig. 2.5a–c. From the comparison of the latter with the corresponding parameters of the standard tokamak, the ITER, shown in Fig. 2.4b, c one can notice the following differences. The safety factor $q(\psi)$ of the spherical tokamak at the plasma plasma grows much faster than in the case of the standard tokamak. Its means the field lines have longer connection lengths $l_c = 2\pi q(\psi)$, i.e., they make more toroidal turns per single poloidal turn. Moreover, as seen from Fig. 2.5c, the field lines stay much longer on the high–field–side (HFS) ($\pi/2 < \theta < 3\pi/2$) of the torus than on the low–field–side (LFS) ($-\pi/2 < \theta < \pi/2$).

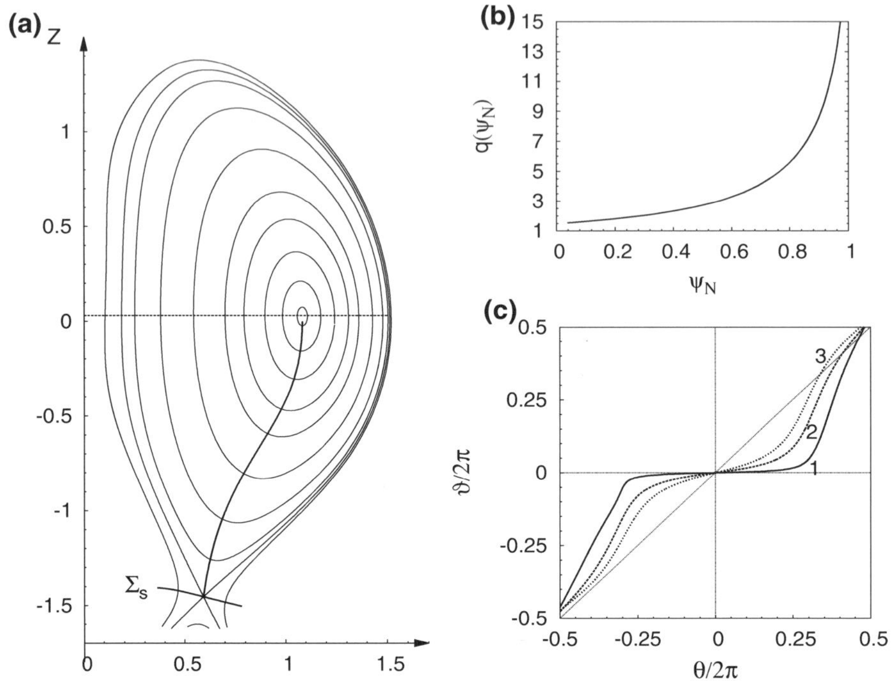

Fig. 2.5 Equilibrium configuration for the NSTX–like plasma: **a** Contours of the magnetic surfaces, $\psi(R; Z) = \text{const}$ *solid lines*; **b** the safety factor profiles $q(\psi)$ as a function of the normalized poloidal flux ψ_N. **c** Relation between geometrical poloidal angle θ on the poloidal angle ϑ at the different magnetic surfaces: *curve 1* corresponds to $\psi_N = 0.2$, *curve 2*—$\psi_N = 0.5$, *curve 3*— $\psi_N = 0.95$.

2.4 Field Lines Near the X-point

The behavior of field lines near the separatrix and X-point has a generic features independent on a specific device. Below we study these properties.

2.4.1 Standard X-point

Let $(z = z_s, p_z = p_s)$ be the coordinate of the X-point. The open and closed field lines, and the separatrix near the X-point are schematically plotted in Fig. 2.6. Consider first the case when the *poloidal magnetic field has the first order null.* The latter means that the expansion of the poloidal magnetic field (B_R, B_Z) near X-point contains non-zero coefficients before the first order terms, i.e., $B_R, B_Z \approx a_z(z - z_s) + a_p(p_z - p_s)$, where $a_z \neq 0, a_p \neq 0$. This case corresponds to the standard divertor configuration in tokamak plasmas.

According to (1.18) the X-point (R_s, Z_s), i.e., the null of the poloidal field (B_R, B_Z), on the poloidal section correspond to the hyperbolic fixed point (z_s, p_s) of the Hamiltonian system (1.12) determined by the equations,

$$\frac{dz}{d\varphi} = \frac{\partial\psi}{\partial p_z} = 0, \qquad \frac{dp_z}{d\varphi} = -\frac{\partial\psi}{\partial z} = 0. \tag{2.37}$$

Consider field lines near the X-point (see Fig. 2.6). The unperturbed Hamiltonian $H_0(z, p_z)$ can be expanded in a series of powers of $(z - z_s)$, $(p_z - p_s)$ near the X-points:

$$\begin{aligned}\psi(z, p_z) &= \psi(z_s, p_s) + \frac{1}{2}\psi_{zz}(z - z_s)^2 + \psi_{zp}(z - z_s)(p_z - p_s)\\ &\quad + \frac{1}{2}\psi_{pp}(p_z - p_s)^2 + O[(z - z_s)^3, (p_z - p_s)^3],\end{aligned} \tag{2.38}$$

where ψ_{zz}, ψ_{zp}, and ψ_{pp} are second derivatives of ψ, i.e., $\partial^2\psi/\partial z^2$, $\partial^2\psi/\partial z\partial p$, $\partial^2\psi/\partial p^2$, taken at the hyperbolic fixed point $z = z_s$, $p = p_s$. Then by the linear transformation of coordinates (z, p_z) to the coordinates (ξ, η) (see Fig. 2.6):

$$\begin{aligned}\xi &= (z - z_s)\cos\alpha_0 + (p_z - p_s)\sin\alpha_0,\\ \eta &= -(z - z_s)\sin\alpha_0 + (p_z - p_s)\cos\alpha_0,\end{aligned} \tag{2.39}$$

where α_0 is the angle between the z axis and the ξ axis, the Hamiltonian near the X-point can be presented in a normal form:

$$\psi(\xi, \eta) \equiv \psi(z, p_z) - \psi(z_s, p_s) = -\frac{|\lambda_1|}{2}\xi^2 + \frac{\lambda_2}{2}\eta^2 + O(\xi^3, \eta^3), \tag{2.40}$$

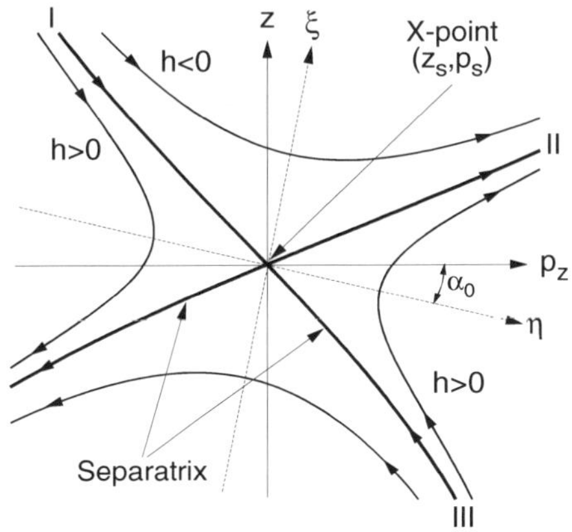

Fig. 2.6 Field lines near the X-point. Field lines along the segments *I* and *III* of the separatrix approach the X-point (z_s, p_s) at the limit $\varphi \to \infty$, ones along the segments *II* and *IV*—at the limit $\varphi \to -\infty$

Here $\psi(\xi, \eta)$ is the relative poloidal flux, the coefficients λ_1 and λ_2, $(\lambda_1 < 0, \lambda_2 > 0)$ given by

$$(\lambda_1, \lambda_2) = \frac{\psi_{zz} + \psi_{pp}}{2} \pm \sqrt{\frac{(\psi_{zz} - \psi_{pp})^2}{4} + \psi_{zp}^2}, \tag{2.41}$$

are the eigenvalues of the matrix,

$$\begin{pmatrix} \psi_{zz} & \psi_{zp} \\ \psi_{pz} & \psi_{pp} \end{pmatrix}. \tag{2.42}$$

The angle α_0 is determined by $\tan \alpha_0 = 2\psi_{zp}/(\psi_{pp} - \psi_{zz})$. Unperturbed field lines transversely cross the axis (ξ, η). The poloidal magnetic field components (B_R, B_Z) near the X- point are linear functions ξ, η, i.e., $B_R, B_Z \sim a_1\xi + a_2\eta$ with nonzero constant coefficients a_1, a_2.

The field line equation in the coordinates (ξ, η) are

$$\frac{d\xi}{d\varphi} = \frac{\partial\psi(\xi, \eta)}{\partial\eta} = \beta^2\eta, \qquad \frac{d\eta}{d\varphi} = -\frac{\partial\psi(\xi, \eta)}{\partial\xi} = \alpha^2\xi, \tag{2.43}$$

where $\alpha = \sqrt{|\lambda_1|}$, $\beta = \sqrt{\lambda_2}$. Field lines near the X-point are shown in Fig. 2.6.

Let ξ_0 and η_0 be field line coordinates at $\varphi = \varphi_0$. Field lines $(\xi(\varphi; h), \eta(\varphi; h))$ are given by

$$\xi(\varphi; \psi) = \xi_0 \cosh\left[\gamma\left(\varphi - \varphi_0\right)\right] + \frac{\alpha}{\beta}\eta_0 \sinh\left[\gamma\left(\varphi - \varphi_0\right)\right],$$
$$\eta(\varphi; \psi) = \eta_0 \cosh\left[\gamma\left(\varphi - \varphi_0\right)\right] + \frac{\beta}{\alpha}\xi_0 \sinh\left[\gamma\left(\varphi - \varphi_0\right)\right], \tag{2.44}$$

The parameter γ in (2.44) is defined by

$$\gamma = \alpha\beta = \sqrt{|\lambda_1\lambda_2|}. \tag{2.45}$$

In the coordinates,

$$x = \frac{1}{\sqrt{2\gamma}}\left(\alpha\xi + \beta\eta\right), \qquad y = \frac{1}{\sqrt{2\gamma}}\left(-\alpha\xi + \beta\eta\right), \tag{2.46}$$

the field lines can be presented by simple formulas,

$$x(\varphi; \psi) = x_0 e^{\gamma(\varphi - \varphi_0)}, \qquad y(\varphi; \psi) = y_0 e^{-\gamma(\varphi - \varphi_0)},$$
$$\psi(x, y) = \gamma xy = \gamma x_0 y_0. \tag{2.47}$$

At the separatrix, $\psi = 0$, field lines asymptotically approach the X-point $(\xi = 0, \eta = 0)$ at $\varphi \to \pm\infty$,

$$\xi(\varphi; 0) = \xi_0 e^{\mp\gamma(\varphi-\varphi_0)}, \qquad \eta(\varphi; 0) = \eta_0 e^{\mp\gamma(\varphi-\varphi_0)}, \tag{2.48}$$

where (ξ_0, η_0) is the field line coordinate at $\varphi = \varphi_0$ on the separatrix located to close to the X-point: $\xi_0 = \xi(\varphi; \varphi_0, h = 0)$, $\eta_0 = \xi_0|\lambda_1|/\lambda_2$. The upper signs $(-)$ correspond to field lines approaching the X-point at $\varphi \to \infty$ along the segments *I* and *III*, while the lower signs $(+)$ correspond to ones approaching the X-point along the segments *II* and *IV* at $\varphi \to -\infty$.

2.4.2 Asymptotics of the Safety Factor

Because of a generic behavior of field lines near the X-point the safety factor $q(\psi)$ has a generic asymptotics near the separatrix. To find this asymptotics we use Eq. (1.34). The latter we present the increment of the toroidal angle $\Delta\varphi$ along the closed contour C as a sum

$$\begin{aligned} &\Delta\varphi = \Delta\varphi_1 + \Delta\varphi_X, \\ &\Delta\varphi_1 = \int_{C_1} \frac{dz}{\partial\psi/\partial p_z}, \qquad \Delta\varphi_X = \int_{C_X} \frac{dz}{\partial\psi/\partial p_z}, \end{aligned} \tag{2.49}$$

where C_X is a segment of the closed contour C located near the X-point, and C_1 is the rest of C. Let (ξ_0, η_0) be a farthest coordinate of a field line on the segment C_x where the expansion (2.40) of the poloidal flux ψ in a series of powers ξ, η is valid. Then to find $\Delta\varphi_X$ one can use the solutions (2.44) and (2.47). Suppose that at $\varphi = \varphi_\Sigma$ the field line crosses the section Σ, i.e., the axes ξ or η. From (2.47) it follows that at the crossing points $|x_\Sigma| = \sqrt{|\psi|/\gamma}$, (or $|y_\Sigma| = \sqrt{|\psi|/\gamma}$) and

$$\Delta\varphi_X = 2(\varphi_\Sigma - \varphi_0) = -\frac{2}{\gamma}\ln\left|\frac{y_\Sigma}{y_0}\right| = -\frac{1}{\gamma}\ln\frac{|\psi|}{\gamma y_0^2}. \tag{2.50}$$

The increment $\Delta\varphi_1$ along the segment C_1 does not have a singularity at $\psi = 0$, and one can assume the asymptotics $\Delta\varphi_1 = Q_1 + O(\psi)$ at $|\psi| \to 0$, where Q_1 is finite constant. Finally, from Eqs. (1.34), (2.49) and (2.50) we obtain the following asymptotical formula for the safety factor

$$q(\psi) = \frac{1}{2\pi\gamma}\ln\frac{Q}{|\psi|} + O(\psi), \qquad |\psi| \to 0, \tag{2.51}$$

where $Q = \gamma y_0^2 \exp(\gamma Q_1/2\pi)$ is a positive constant depending on the specific magnetic field configuration. The safety factor near the separatrix is mainly determined by the parameter γ, while the dependence on Q is weak. The relative poloidal flux ψ is negative inside the plasma region ($\psi < 0$), positive outside the plasma region ($\psi > 0$) and $\psi = 0$ at the separatrix.

Near the separatrix the connection length $l_c = 2\pi R_0 q(\psi)$, i.e., the length of field line in one poloidal turn, grows with the distance ψ to the separatrix logarithmically weak,

$$l_c \sim \ln \frac{Q}{|\psi|}. \tag{2.52}$$

2.5 Wire Models of Plasmas

Behavior of magnetic field lines near the separatrix can be also conveniently studied using wire models of divertor plasmas. In such models a set of several straight and circular wires is used to simulate a magnetic field corresponding to the plasma and the divertor currents. In a number of works (see, e.g., Tomita et al. 1977; Boozer and Rechester 1978; Pomphrey and Reiman 1992; Abdullaev and Zaslavsky 1996; Reiman 1996; Abdullaev and Finken 1998; Morrison 2000) wire models of plasmas are used to study the onset of chaotic field lines near the separatrix.[1] However, the wire models are valid only at the regions close to the magnetic separatrix. Below we present general formulas for the magnetic field of wire–models, and consider the two-wire model in more details.

2.5.1 *Magnetic Field of a Set of Current Loops*

Consider a $M-$wire model consisting a set of M current loop (R_j, Z_j) carrying the current I_j, $(j = 1, 2, \dots, M)$. Figure 2.7 shows a two-wire model consisting of two circular currents loops.The toroidal component of the vector potential $A_\varphi(R, Z)$ of the magnetic field created by this set is given by (see, e.g., Morozov and Solov'ev 1966b; Jackson 1998)

$$A_\varphi(R, Z) = \sum_{j=1}^{M} \frac{\mu_0 I_j}{\pi k_j} \sqrt{\frac{R_j}{R}} \left[\left(1 - \frac{k_j^2}{2} \right) K(k_j) - E(k_j) \right], \tag{2.53}$$

where $K(k)$ and $E(k)$ are the complete elliptic integrals with the module

$$k_j^2 = \frac{4R_j R}{(R + R_j)^2 + (Z - Z_j)^2}.$$

[1] In recent papers by Aguirre and Peralta-Salas (2007); Aguirre et al. (2008); Hosoda et al. (2009); Miyaguchi et al. (2011) wire models are used to demonstrate the chaos of magnetic field lines, particularly, in modern electric circuits. Ram and Dasgupta (2010) used a wire model to study the dymanics of charged particles in a toroidal magnetic field.

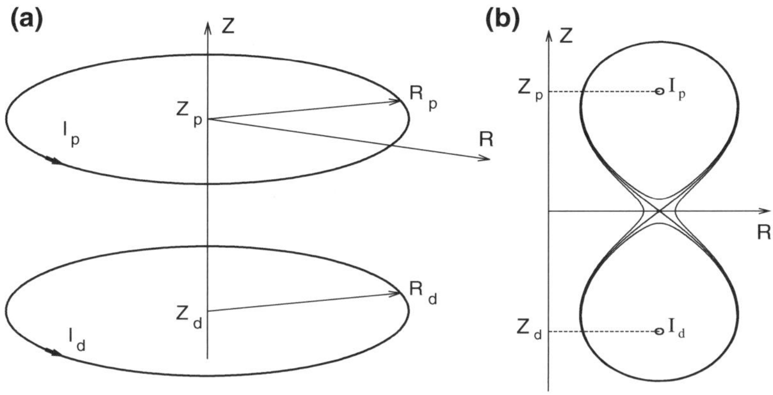

Fig. 2.7 **a** Two-wire model consisting of two circular current loops of radii R_p and R_d at the horizontal planes $Z = Z_p$ and $Z = Z_d$ carrying the currents I_p and I_d, respectively. **b** Magnetic configuration of the two wire model in the (R, Z)-plane

The toroidal magnetic field B_φ is supposed to be created by the straight wire located along the vertical axis Z, and given by $B_\varphi = B_0 R_0/R$.

For large aspect ratio tokamaks $R_j/a \gg 1$, $|1 - k_j^2| \ll 1$ and the normalized poloidal flux $\psi(z, p_z) = -RA_\varphi(R, Z)/B_0 R_0^2$ up to constant terms can be approximated by

$$\psi(z, p_z) = -R_0^{-1} \sum_{j=1}^{M} L_j \left(\ln \frac{64}{r_j^2} - 4 \right), \tag{2.54}$$

where

$$\begin{aligned} r_j^2 &= (x - x_j)^2 + (z - z_j)^2, & x &= R/R_0 - 1, \\ x_j &= R_j/R_0 - 1, & z_j &= Z_j/R_0, \\ L_j &= \mu_o I_j/4\pi B_0, & & (j = 1, 2, \ldots, M). \end{aligned} \tag{2.55}$$

The quantity L_j is the length scale.

According to (1.18) and (1.17) the poloidal components of the magnetic field of the wire–model are given by,

$$\begin{aligned} B_R &= -\frac{2B_0 R_0}{R^2} \sum_{j=1}^{M} L_j \frac{z - z_j}{r_j^2}, \\ B_Z &= \frac{2B_0 R_0}{R^2} \sum_{j=1}^{M} L_j \frac{x - x_j}{r_j^2} \frac{dx}{dp_z}. \end{aligned} \tag{2.56}$$

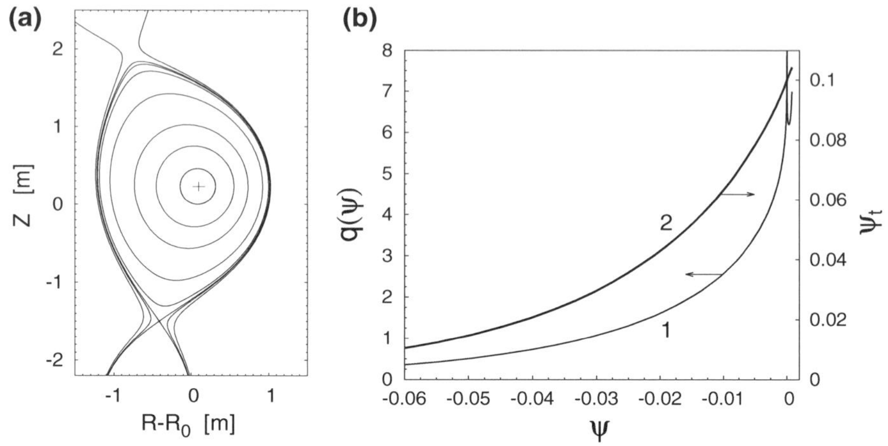

Fig. 2.8 Magnetic configuration of the three-wire system: **a** poloidal cross section; **b** the safety factor $q(\psi)$ (*curve 1*) and the toroidal flux ψ_t (*curve 2*) as functions of the relative poloidal flux ψ. Parameters: $B_0 = 2\,\mathrm{T}$, $R_0 = 3\,\mathrm{m}$, $I_p \equiv I_2 = 2\,\mathrm{MA}$, $I_1 = 2\,\mathrm{MA}$, $I_3 = 2.59\,\mathrm{MA}$. Coordinates of currents $(R_1, Z_1 = (2.4, -2.77)\,\mathrm{m}$, $(R_2, Z_2) = (3.1, 0.23)\,\mathrm{m}$, $(R_3, Z_3) = (1.6, 3.7)\,\mathrm{m}$. The lower X-point is located at $(X_s, Z_s) = (-0.4149, -1.5004)\,\mathrm{m}$

Using (1.17) one can show that the derivative $dx/dp_z = 1 + x$. In the cylindrical plasma $x = p_z$, and therefore $dx/dp_z = 1$.

Figure 2.8a, b show the magnetic configuration and the safety factor profile of the three-wire system simulating the plasma with the parameters close to the JET plasma: the major radius $R_0 = 3\,\mathrm{m}$, a toroidal field at the magnetic axis $B_0 = 2\,\mathrm{T}$. Magnetic axis is located at the position of the second wire $(R_2, Z_2) = (3.1, 0.23)\,\mathrm{m}$ with the plasma current $I_p \equiv I_2 = 2$ MA. For the positions of the first, $(R_1, Z_1 = (2.4, -2.77)\,\mathrm{m}$, and third, $(R_3, Z_3) = (1.6, 3.7)\,\mathrm{m}$, coils and their currents $I_1 = 2$ MA, and $I_3 = 2.59\,\mathrm{MA}$ are chosen such values in order to have a single null divertor configuration with the lower X-point at $(X_s, Z_s) = (-0.4149, -1.5004)\,\mathrm{m}$.

2.5.2 Two-Wire Model of the Plasma

The magnetic structure of magnetic field in a two-wire model can be studied analytically which allows one to obtain explicit formulas for field lines and the safety factor. The model consists of two current loops of radius $R_1 = R_2 = R_0$ located at the planes $Z_1 = Z_p = a$ and $Z_2 = Z_c = -a$ as shown in Fig. 2.7a. The first loop carries the plasma current I_p, and the second one describes the divertor current I_d. Such a model of the plasma well describes the single null divertor tokamak plasma configuration near the separatrix which does not depend on the radial profiles of plasma current I_p.

Using (2.54) the poloidal flux can be approximated by

$$\psi(z, p_z) = L_p R_0^{-1} \ln r_p^2 r_d^{2\delta}, \tag{2.57}$$

where $\delta = I_d/I_p$ is the ratio of the divertor current I_d to the plasma current I_p.

The Hamiltonian equations of field lines are

$$\begin{aligned}\frac{dz}{d\varphi} &= \frac{\partial \psi}{\partial p_z} = 2L_p R_0^{-1} x \left(r_p^{-2} + \delta r_c^{-2}\right) \exp(p_z), \\ \frac{dp_z}{d\varphi} &= -\frac{\partial \psi}{\partial z} = -2L_p R_0^{-1} \left(\frac{z - a/R_0}{r_p^2} + \delta \frac{z + a/R_0}{r_c^2}\right).\end{aligned} \tag{2.58}$$

The coordinates of the X-point is located at

$$(z_s, p_s) = \left(\frac{a}{R_0}\frac{1-\delta}{1+\delta}, 0\right).$$

One can show that the second derivatives $\psi_{zz} = -\psi_{pp}$ and the mixed derivative ψ_{zp} at this point vanishes. The eigenvalues λ_1 and λ_2 are given by

$$-\lambda_1 = \lambda_2 = \gamma = \frac{L_p R_0}{2a^2} \frac{(1+\delta)^3}{\delta}. \tag{2.59}$$

On the separatrix we have

$$\psi(z_s, p_s) = (1+\delta) L_p R_0^{-1} \ln \frac{2a^2\delta}{(1+\delta)R_0^2} - L_p R_0^{-1} \ln \delta. \tag{2.60}$$

Magnetic flux surfaces, $\psi(z, p_z(x)) = \text{const}$, are given by

$$\begin{aligned} r_p^2 r_c^{2\delta} &= \frac{1}{\delta}\left(\frac{2a^2\delta}{(1+\delta)R_0^2}\right)^{1+\delta} \Lambda, \\ \Lambda &= \exp(\psi R_0/L_p), \end{aligned} \tag{2.61}$$

where ψ is the relative poloidal flux $\psi = \psi(z, p) - \psi(z_s, p_s)$. Magnetic flux surfaces for $\delta = 1$ are shown in Fig. 2.7b. The width of the separatrix along R and Z axes are $\Delta R = a$ and $\Delta Z = \sqrt{2}a$, respectively. The width of the separatrix is maximum at the planes $Z = \pm\sqrt{3/2}a$. Elongation $\kappa = \Delta Z/\Delta R$, i.e., the ratio of the widths of the separatrix ΔZ and ΔR is equal to $\kappa = \sqrt{2}$.

Analytical integration of field line Eq. (2.58) in the toroidal system is rather difficult task. Further to simplify the problem we consider the case of large aspect ratio tokamak plasmas, $|x| \ll 1$, putting $p_z = \ln(1+x) \approx x$. Moreover, we suppose that the plasma, I_p, and coil, I_c, currents are equal, i.e., $\delta = 1$. Then, according to (1.34), (2.58) and (2.57), we have

$$q(\psi) = \frac{R_0}{4\pi L_p} \oint\limits_{C_p} \frac{dz}{x(z,h)(r_p^{-2} + r_c^{-2})}$$
$$= \frac{a^2 \Lambda}{4\pi L_p R_0} \begin{cases} K\left(\Lambda^{1/2}\right), & \text{for } \psi < 0, (\Lambda < 1) \\ \Lambda^{-1/2} K\left(\Lambda^{-1/2}\right), & \text{for } \psi > 0, (\Lambda > 1), \end{cases} \tag{2.62}$$

where integration is taken over closed magnetic surface C_p. Using the asymptotics of the complete elliptic integral $K(k)$ at $k \to 1$ ($|\psi| \to 0$), $K(k) \approx \ln(4/\sqrt{1-k^2})$ we obtain the asymptotic formula for the safety factor (2.51) with the constants

$$\gamma = \frac{4L_p R_0}{a^2}, \qquad Q = \frac{16 L_p}{R_0}.$$

Using the relation, $q(\psi) = d\psi_t/d\psi$, for the safety factor $q(\psi)$ one finds the toroidal flux ψ_t as a function of the relative poloidal flux ψ:

$$\psi_t = \int q(\psi) d\psi = \frac{a^2}{4\pi L_p R_0} \begin{cases} \int K\left(\Lambda^{1/2}\right) \Lambda d\psi, & \text{for } \psi < 0, \\ \int \Lambda^{1/2} K\left(\Lambda^{-1/2}\right) d\psi, & \text{for } \psi > 0, \end{cases}$$
$$= \frac{a^2}{2\pi R_0^2} \begin{cases} E\left(\Lambda^{1/2}\right) - (1-\Lambda) K\left(\Lambda^{1/2}\right), & \text{for } \psi < 0, \\ \Lambda^{1/2} E\left(\Lambda^{-1/2}\right), & \text{for } \psi > 0. \end{cases} \tag{2.63}$$

At the separatrix, $\psi = 0$ ($\Lambda = 1$), the toroidal flux ψ_t is equal to $\psi_a = a^2/(2\pi R_0^2)$.

The safety factor profile q (2.62) as a function of the toroidal flux ψ_t is plotted in Fig. 2.9. We have chosen the value of the length scale L_p equal to 0.0472 in order to fix the value $q_{95} \approx 3.395$, where q_{95} stands for the value of the safety factor at the magnetic surface which has 95 % of the total magnetic flux ψ_a, i.e., $\psi/\psi_a = 0.95$.

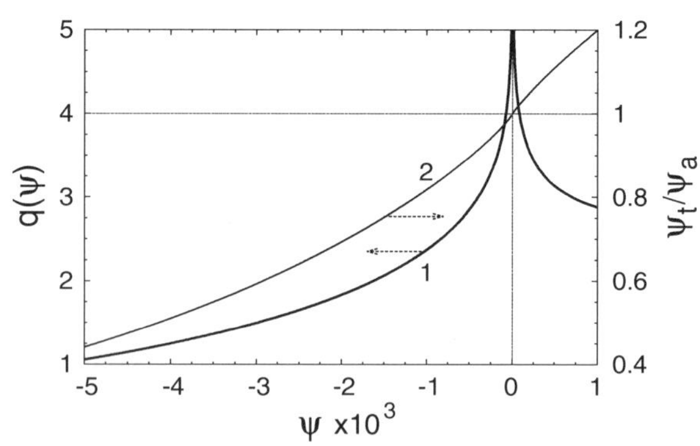

Fig. 2.9 Safety factor q (*curve 1*) [*left-hand side*] and the normalized toroidal flux ψ_t/ψ_a (*curve 2*) [*right-hand side*] as the functions of the relative poloidal flux ψ. The length scale L_p is chosen equal to 0.0472 to fix $q_{95} = 3.394585$ at $\psi/\psi_a = 0.95$. Other parameter are $R_0 = 6.2$ m, $a = 2$ m

The major radius is $R_0 = 6.2\,\text{m}$, and $a = 2\,\text{m}$. According to (2.58) the unperturbed field lines can be found by the integral

$$\varphi - \varphi_0 = \frac{R_0}{2L_p} \int_{z_0}^{z} \frac{dz}{x(z,h)(r_p^{-2} + r_c^{-2})}, \tag{2.64}$$

where z_0 is the highest z—coordinate of field line on the flux surface $f(x, z) \equiv H_0(z, p_z(x)) = \text{const}$. The integration gives

$$\varphi - \varphi_0 = \frac{a^2 \Lambda}{4L_p R_0} \begin{cases} F\left(\nu, \Lambda^{1/2}\right), & \text{for } \psi < 0, (\Lambda < 1) \\ \Lambda^{-1/2} F\left(\nu, \Lambda^{-1/2}\right), & \text{for } \psi > 0, (\Lambda > 1), \end{cases} \tag{2.65}$$

where $F(\nu, k)$ is the incomplete elliptic integral of the first kind. The argument ν is related to the coordinate $z = Z/R_0$:

$$\sin^2 \nu = \frac{2 + \Lambda^{1/2} - (\Lambda + 4Z^2/a^2)^{1/2}}{\Lambda^{1/2} + (\Lambda + 4Z^2/a^2)^{1/2}} \begin{cases} \Lambda^{-1/2}, & \text{for } \psi < 0, (\Lambda < 1) \\ \Lambda^{1/2}, & \text{for } \psi > 0, (\Lambda > 1). \end{cases} \tag{2.66}$$

The field line on the separatrix, $z_s(\varphi)$, $p_s(\varphi)$, is given by

$$\begin{aligned} z_s(\tau) &= \pm \frac{\sqrt{2}a}{R_0 \cosh(\gamma\tau)[1 + \tanh^2(\gamma\tau)]}, \\ p_s(\tau) \approx x_s(\tau) &= \mp \frac{\sqrt{2}a \tanh(\gamma\tau)}{R_0 \cosh(\gamma\tau)[1 + \tanh^2(\gamma\tau)]}, \end{aligned} \tag{2.67}$$

where $\tau = \varphi - \varphi_0$, upper signs stand for the plasma region, and the lower ones for the coil region. Note that when $\varphi = \varphi_0$ the field line coordinates are located at the farthest point from the X-point $z_s(\varphi_0) = \pm\sqrt{2}a/R_0$.

2.6 Snowflake Divertor Configuration

Up to now we have considered divertor tokamaks with the first order–null of the poloidal magnetic field at the X-point, where its first derivatives have finite values. By the appropriate choice of the divertor currents one can make the latter also vanish at the X-point. This case is called a *second order null of the poloidal magnetic field.* This kind of divertor tokamaks with the second order of null of the poloidal magnetic field have been recently proposed by Ryutov (2007); Ryutov et al. (2008). In these plasma configurations the flux of heat load is expanded over a much wider area than in the case of standard divertor discussed above. Such a divertor configuration called a *snowflake divertor* has been recently experimentally created on the TCV and the NSTX tokamaks (Pires et al. 2009; Soukhanovskii et al. 2011). Below we consider the simple example of the plasma configuration with the second order null divertor configuration consisting of three wires, and study the asymptotical behavior of the safety factor near the separatrix.

2.6.1 Three-Wire Model

Consider the cylindrical plasma in which one puts $x = p_z$, $R = R_0(1 + x)$. Let $(R_1 = R_0, Z_1 = R_0 a)$ be the coordinate of the plasma wire with the current I_p, and $\big(R_{2,3} = R_0(1 \pm d/2), Z_{2,3} = -R_0 b\big)$ be the coordinates of two divertor wires with the equal currents I_d located symmetrically with respect to the axis $R = R_0$ (or $x = 0$). The positions of coils are shown in Fig. 2.10. According to (2.56) the nulls (x_s, z_s) of the magnetic field (B_R, B_Z) are determined by the system

$$\frac{x}{\delta r_1^2} + \frac{x - d/2}{r_2^2} + \frac{x + d/2}{r_3^2} = 0,$$
$$\frac{z - a}{\delta r_1^2} + \frac{z + b}{r_2^2} + \frac{z + b}{r_3^2} = 0, \tag{2.68}$$

where $\delta = I_d/I_p$, and

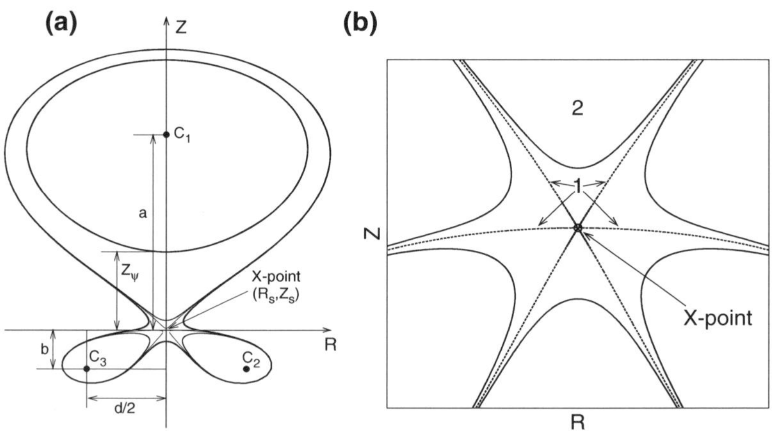

Fig. 2.10 **a** Magnetic configuration of the snowflake divertor obtained by three-wire model: C_1 is a coil for plasma current, $C_{2,3}$ are coils for the divertor currents. **b** Expanded view of the area near the X-point of the snowflake divertor. *Curves 1* corresponds to the separatrices, the region 2 corresponds to the plasma region

$$r_1^2 = x^2 + (z-a)^2, \qquad r_{2,3}^2 = (x \mp d/2)^2 + (z+b)^2.$$

From (2.68) it is not difficult to see that $x_s = 0$, and therefore the equation for the z—coordinate is reduced to

$$\frac{1}{\delta(z-a)} + \frac{2(z+b)}{(z+b)^2 + d^2/4} = 0. \tag{2.69}$$

The equations for the second order nulls are obtained by equating to zero the second derivatives from the poloidal flux $\psi(z, x)$. For the corresponding equation we have

$$\frac{1}{\delta(z-a)^2} + \frac{2[(z+b)^2 - d^2/4]}{[(z+b)^2 + d^2/4]^2} = 0. \tag{2.70}$$

From Eqs. (2.69) and (2.70) it follows that the second order null ($x_s = 0, z_s$) exists for the certain values of the ratio δ of the divertor I_d to the plasma currents I_p:

$$z_s = \frac{d^2}{4(a+b)}\left(\sqrt{1 + \frac{4(a+b)^2}{d^2}}\right) - b,$$
$$\delta = \frac{I_d}{I_p} = \frac{(z_s+b)^2 + d^2/4}{2a(z_s+b)}. \tag{2.71}$$

Particularly, the null is located at ($x_s = z_s = 0$) for the parameters,

$$d = 2b\sqrt{\frac{a+b}{a-b}}, \qquad \delta = \frac{I_d}{I_p} = \frac{b^2 + d^2/4}{2ab}. \tag{2.72}$$

The expansion of the poloidal flux near the second order null points ($z_s = 0, x_s = 0$) then contains the third order terms, and can be presented as

$$\psi(x, z) = -\frac{A}{6}\left(x^3 - 3xz^2\right), \tag{2.73}$$

where the constant A is given by

$$A = \frac{\mu_o I_d}{2\pi B_0 R_0} \frac{d^2}{b\left(b^2 + d^2/4\right)}.$$

The schematic view of field lines and the separatrices in the snowflake divertor is shown in Fig. 2.10 for the three-wire model. The plasma current $I_p = 1.5$ MA, the divertor current $I_2 = I_3 = 0.375$ MA, the positions of plasma wire ($R_1 = R_0 = 3$ m, $Z_p = 2$ m), and the divertor currents ($x_2 R_0 = -x_3 R_0 = 0.4899$ m, $Z_2 = Z_3 =$

$-0.4\,\mathrm{m}$). As seen the separatrices (curves 1) divide the area near the X-point into six zone, and the field line structure reminds the snowflakes.

2.6.2 The Safety Factor

The behavior of the safety factor near the separatrix with the second–order null can be found similar in the case of the standard divertor studied in Sect. 2.4.2. We present the toroidal angle increment $\Delta\varphi$ in (1.34) as the sum (2.49). The term $\Delta\varphi_X$ corresponding to the segment C_x located near the null ($x_s = 0, z_s = 0$) can be estimated using the expansion of the poloidal flux (2.73). Let z_ψ be the minimal vertical position of the magnetic surface in the plasma region (see Fig. 2.10a). Using (2.73) one obtains $z_\psi = (-6\psi/A)^{1/3}$, and the contour $\psi(z,x) = \mathrm{const}$ near the null point can be written as $x(z,\psi) = \sqrt{\left(z^3 - z_\psi^3\right)/3z}$. Then increment $\Delta\varphi_X$ is determined by the integral

$$\Delta\varphi_X = 2\int_{z_\psi}^{z_0} \frac{dz}{\partial\psi/\partial x} = -\frac{2}{A}\int_{z_\psi}^{z_0} \frac{dz}{x(z,\psi)z} = -\frac{2\sqrt{3}}{A}\int_{z_\psi}^{z_0} \frac{dz}{\sqrt{z\left(z^3 - z_\psi^3\right)}},$$

where z_0 is the highest vertical position of the magnetic surface in the segment C_x. Near the separatrix $\psi \to 0$ the ratio $z_0/z_\psi \gg 1$. Then $\Delta\varphi_X$ can be approximated by

$$|\Delta\varphi_X| = \frac{2\sqrt{3}}{Az_\psi}\int_1^{z_0/z_\psi} \frac{dy}{\sqrt{y\left(y^3-1\right)}} \approx \frac{2\sqrt{3}C}{6^{1/3}A^{2/3}|\psi|^{1/3}},$$

$$C = \int_1^{\infty} \frac{dy}{\sqrt{y\left(y^3-1\right)}}.$$

As in the case of the standard divertor the increment $|\Delta\varphi_1|$ along the contour C_1 does not have a singularity at $\psi = 0$. Therefore, the asymptotics of the safety factor near the separatrix has the form

$$q(\psi) \approx \frac{Q}{|\psi|^{1/3}} + O(\psi), \qquad \psi \to 0, \tag{2.74}$$

where Q is positive constant independent of ψ, i.e., $Q > 0$. The numerically calculated safety factor $q(\psi)$ as a function of the relative poloidal flux ψ is plotted in Fig. 2.11 (curve 1). Its fitting near the separatrix with the power–law function

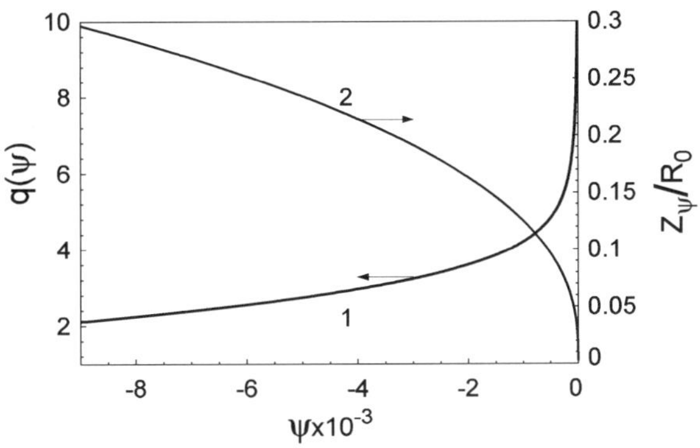

Fig. 2.11 Safety factor profile $q(\psi)$ (*curve 1*) [*left–hand side*] and the minimal vertical position of the field lines Z_ψ (*curve 2*) [*right–hand side*] as a function of the relative poloidal flux ψ for the snowflake plasma configuration in Fig. 2.10

$q(\psi) = Q/|\psi|^s + a\psi + b$ gives the value 0.327 of the exponent s which is close to 1/3 in the asymptotical formula (2.74).

From the power-law dependence (2.74) it follows that the safety factor grows much faster near the separatrix than in the case of the standard divertor where $q(\psi)$ grows weaker in the logarithmic–law (2.51). The connection length l_c grows with approaching the separatrix according to the power–law, $l_c \propto |\psi|^{-1/3}$, i.e., much steeper than in the standard divertor (2.52).

2.7 Summary

In this chapter we have presented the analytical models of magnetic field configurations of equilibrium plasmas. Particularly, these models describe magnetic fields of equilibrium toroidal plasmas with circular cross sections and elongated shapes possessing the magnetic separatrices. The plasmas with elongated shapes can be also described by simple wire models in which a plasma current is simulated by a single current flowing in a circular loop. The wire model is also used for description of the plasma with snowflake divertor configuration.

We analyzed a behavior of magnetic field lines near the magnetic separatrix and the X-point with the first and second order nulls of the poloidal magnetic field. The generic asymptotical formula for the safety factor near the separatrix has been found. Furthermore, it will be used to study magnetic field lines near the separatrix.

Chapter 3
Magnetic Field Perturbations

In real tokamak plasmas magnetic field deviates from the ideal equilibrium field. In typical situations the deviation of magnetic field from the equilibrium field is small. The magnetic perturbations may have a diverse nature: error fields produced by imperfect technical installation of poloidal and toroidal field coils, helical magnetic fields generated by MHD instabilities, magnetic fields created by external coils etc. The example of external coils used to create the resonant magnetic perturbations (RMPs) is illustrated in Fig. 3.6.

In this chapter we intend to formulate the Hamiltonian equations of field lines in the presence of magnetic perturbations. The aim is to present the perturbation Hamiltonian function in magnetic flux coordinates and to study the generic features of the poloidal and toroidal spectra of magnetic perturbations. First we obtain the asymptotical formulas for the poloidal spectra of the magnetic perturbations created by the helical coils. The generic features of the spectrum of the magnetic perturbations in the plasma configurations with the magnetic separatrix will be studied first in the analytical model of magnetic field studied in Sect. 1.3. Then it will be extended to the realistic magnetic fields in the tokamaks, the ITER like and the spherical tokamak, NSTX, with the installed sets of saddle coils. We establish the universal asymptotical spectra of the magnetic perturbations generated by these sets of saddle coils.

3.1 Hamiltonian Equations in the Presence of Magnetic Perturbations

The *non-axisymmetric magnetic perturbations* break the symmetry the equilibrium field along the toroidal, φ, and poloidal, θ, angles. The perturbation field can be given by the vector potential $\mathbf{A}^{(per)}(R, Z, \varphi)$. In general it has all three non–vanishing components, $\left(A_R^{(per)}, A_Z^{(per)}, A_\varphi^{(per)}\right)$.

S. Abdullaev, *Magnetic Stochasticity in Magnetically Confined Fusion Plasmas*, Springer Series on Atomic, Optical, and Plasma Physics 78,
DOI: 10.1007/978-3-319-01890-4_3,

3.1.1 Hamiltonian Equations in Flux Coordinates

From the Clebsch form (1.27) it follows that in the cylindrical coordinate system (R, Z, φ) the components of $\mathbf{A}$ are given by

$$A_R = B_0 R_0 \left(\frac{\partial g}{\partial R} + \psi_t \frac{\partial \vartheta}{\partial R} \right), \qquad A_Z = B_0 R_0 \left(\frac{\partial g}{\partial Z} + \psi_t \frac{\partial \vartheta}{\partial Z} \right),$$
$$A_\varphi = B_0 R_0 \left(\frac{1}{R} \frac{\partial g}{\partial \varphi} - \frac{1}{R} \psi \right). \tag{3.1}$$

The vector potential of the equilibrium field $\mathbf{A}^{(0)}$ can be chosen that

$$\psi_0 = -R A_\varphi^{(0)} / B_0 R_0^2, \qquad g = 0. \tag{3.2}$$

In the presence of magnetic perturbations we can choose such a gauge function g that the toroidal flux ψ_t and the poloidal angle ϑ in Eq. (1.27) are unchanged and the full poloidal flux ψ is determined by the sum of the unperturbed one ψ_0 (3.2) and the perturbation part $\psi^{(per)}$,

$$\psi = \psi_0(\psi) + \psi^{(pert)}(\psi, \vartheta, \varphi). \tag{3.3}$$

Then using the Clebsch presentation (1.27) one can obtain the following expression for the perturbation poloidal flux $\psi^{(pert)}(\psi, \vartheta, \varphi)$

$$\psi^{(pert)} = -f_\varphi^{(per)} + \frac{\partial g}{\partial \varphi}, \tag{3.4}$$

where the gauge function g is a solution of equation

$$\frac{\partial g}{\partial R} = R_0^{-1} f_R^{(per)}, \qquad \frac{\partial g}{\partial Z} = R_0^{-1} f_Z^{(per)}. \tag{3.5}$$

In (3.4) and (3.5), the functions $f_\varphi^{(per)}$, $f_R^{(per)}$, and $f_Z^{(per)}$ are defined as

$$f_\varphi^{(per)} = \frac{R A_\varphi^{(per)}}{B_0 R_0^2}, \qquad f_R^{(per)} = \frac{A_R^{(per)}}{B_0 R_0}, \qquad f_Z^{(per)} = \frac{A_Z^{(per)}}{B_0 R_0}. \tag{3.6}$$

The solution of Eq. (3.5) can be written as

$$g(R, Z, \varphi) = \frac{1}{R_0} \int \left(f_R^{(per)} dR + f_Z^{(per)} dZ \right). \tag{3.7}$$

On the given magnetic surface ψ_t = constant this solution can be rewritten via the poloidal angle ϑ:

$$g(R,Z,\varphi)=\int G\left(R(\vartheta,\psi_t),Z(\vartheta,\psi_t),\varphi\right)d\vartheta,$$
$$G(R,Z,\varphi)=\frac{1}{R_0}\left(f_R^{(1)}\frac{dR}{d\vartheta}+f_Z^{(1)}\frac{dZ}{d\vartheta}\right). \tag{3.8}$$

The Hamiltonian equation of perturbed field lines reads as

$$\frac{d\vartheta}{d\varphi}=\frac{\partial\psi}{\partial\psi_t},\quad \frac{d\psi_t}{d\varphi}=-\frac{\partial\psi}{\partial\vartheta}, \tag{3.9}$$

where the Hamiltonian function is presented as indexHamiltonian function!in flux coordinates

$$\psi=\psi_0\left(\psi_t\right)+\epsilon\psi_1\left(\vartheta,\psi_t,\varphi\right). \tag{3.10}$$

In (3.10) we have introduced a dimensionless perturbation parameter ϵ as $\psi^{(per)}=\epsilon\psi_1$, where the function $\psi_1\sim 1$, and ϵ stands for the relative strength of magnetic perturbations,

$$\epsilon=\frac{B_{pert}}{B_0}, \tag{3.11}$$

where B_{pert} is the characteristic magnitude of the perturbation magnetic field and B_0 is the magnitude of the toroidal magnetic field at the torus center. The perturbation poloidal flux $\psi_1(\vartheta,\psi,\varphi)\equiv\psi_1(z,p_z,\varphi)$ (3.4) can be presented as a Fourier series in ϑ and φ:

$$\psi_1(\vartheta,\psi_t,\varphi)=\sum_{m,n}H_{mn}(\psi_t)\exp\left(im\vartheta-in\varphi\right), \tag{3.12}$$
$$=\sum_{m,n}|H_{mn}(\psi_t)|\cos\left(m\vartheta-n\varphi+\chi_{mn}\right), \tag{3.13}$$

where $H_{mn}(\psi_t)=|H_{mn}(\psi_t)|e^{i\chi_{mn}}$ is the amplitude of Fourier harmonics corresponding the integer numbers m and n called the *poloidal* and *toroidal mode numbers*, $|H_{mn}(\psi_t)|$ and χ_{mn} represent their magnitudes and phases, respectively.

Using the relations (1.28) and the presentation (3.12) the magnetic field **B** can be also presented as a sum,

$$\mathbf{B}=\mathbf{B}^{(0)}+\mathbf{B}^{(1)}, \tag{3.14}$$

of the equilibrium magnetic field $\mathbf{B}^{(0)}$,

$$B_\vartheta^{(0)}=B_0J\frac{\partial\psi_0}{\partial\psi_t},\quad B_\psi^{(0)}=-B_0J\frac{\partial\psi_0}{\partial\vartheta},$$
$$B_\varphi^{(0)}=\mathbf{B}\cdot\nabla\varphi=JB_0, \tag{3.15}$$

and the perturbation magnetic field $\mathbf{B}^{(1)}$,

$$B_{\vartheta}^{(1)} = \epsilon B_0 J \frac{\partial \psi_1}{\partial \psi_t} = \epsilon B_{\varphi} \sum_{m,n} \frac{\partial |H_{mn}(\psi_t)|}{\partial \psi_t} \cos(m\vartheta - n\varphi + \chi_{mn}),$$

$$B_{\psi}^{(1)} = -\epsilon B_0 J \frac{\partial \psi_1}{\partial \vartheta} = \epsilon B_{\varphi} \sum_{m,n} b_{mn} \sin(m\vartheta - n\varphi + \chi_{mn}), \tag{3.16}$$

where

$$b_{mn} = m|H_{mn}(\psi_t)| \tag{3.17}$$

are the normalized Fourier coefficients of the component B_{ψ} of the perturbation magnetic field normal to the magnetic surface $\psi = \text{const}$.

The poloidal and toroidal spectra $H_{mn}(\psi_t)$ are the important characteristics determining the effect of RMPs on the plasma. There exists a number of numerical codes for the calculations of the spectra for a specific plasma and a given coil configuration. For example, the Gourdon code (Kaleck et al. (1997)) has been used to calculate the magnetic perturbation spectrum of the TEXTOR-DED, the MASTOC code (Nguyen et al. (1995)) is developed for the Ergodic Divertor (ED) of the Tore Supra tokamak, and the TRIPND and TRIP3D codes (Evans, et al. (2002)) are created to study the magnetic field perturbations of the internal coils of the DIII-D.

However, the numerically calculated spectra of the RMPs are not convenient to analyze the generic features of the spectra of RMPs for typical tokamak plasmas and external coils, like their dependence on the plasma parameters, a coil configurations, etc. To study these features of the RMPs it is useful to obtain analytical formulas for the spectra H_{mn}. In the next sections we shall derive the asymptotical formulas for the spectra of the RMPs created by two type of external coils, a set of helical coils and a set of saddle coils. The details of analytical calculations of the RMPs are given in Appendix A for the set of helical coils and Appendix B for the set of saddle coils.

3.2 Asymptotic Behavior of Spectra of Helical Magnetic Perturbations

The magnetic perturbations created by the set helical coils can be presented as a sum of helical magnetic perturbations,

$$f_{\varphi}^{(per)}(r, \theta, \varphi) = \epsilon \sum_{mn} a_{mn}(r) \cos(m\theta - n\varphi + \chi_{mn}). \tag{3.18}$$

Here the Fourier components $a_{mn}(r)$ represent the spectra of the perturbation field in the quasitoroidal coordinates (r, θ, φ). Such a form is convenient for the RMPs created by a set of helical coils, for example, in ergodic divertors. They will be used in Chap. 9 to study the magnetic field configuration of the ergodic divertor tokamaks.

The relation between the spectra $a_{mn}(r)$ and the corresponding one $H_{mn}(\psi)$ in the magnetic coordinates $(\psi, \vartheta, \varphi)$ is given by the integral

$$\epsilon H_{mn}(\psi) = \frac{1}{(2\pi)^2}\mathrm{Re} \int\int_0^{2\pi} f_{\varphi}^{(per)}(r, \theta, \varphi)$$
$$\times e^{-i(m\vartheta - n\varphi)} d\vartheta d\varphi = \sum_{m'} S_{mm'}(\psi) a_{m'n}(\psi), \qquad (3.19)$$

where the $S_{mm'}(\psi)$ is the transformation matrix determined by the integral

$$S_{mm'}(\psi) \approx \frac{1}{2\pi} \int_0^{2\pi} e^{-i(m\vartheta - m'\theta(\vartheta,\psi))} d\vartheta. \qquad (3.20)$$

The Fourier integrals (3.20) depend on the relation between the poloidal angles θ and ϑ given by Eq. (1.26). For the large aspect ratio tokamaks, $\varepsilon = r/R_0 \ll 1$, one expects that the main contribution to H_{mn} comes from the harmonics a_{mn} of the magnetic field with the same poloidal mode m, while the contribution to this poloidal mode m from the sideband modes $a_{m\pm k,n}$ decreases as ε^k,

$$H_{mn}(\psi) \approx a_{mn} + \sum_{k=1} \varepsilon^k c_k a_{m\pm k,n}, \qquad (3.21)$$

where the coefficients $c_k \sim 1$.

However, this feature of mode transformation is valid only for small poloidal mode numbers m, m' (see Abdullaev et al. 1999). For large mode numbers m, m', $(m, m' \gg 1)$, the matrix $S_{mm'}$ may grow with m' at the fixed m and the main contributions to H_{mn} may come from the poloidal mode m' located far from m. Below we recall these features of the mode transformation matrix $S_{mm'}$, based on the asymptotic analysis of the integral (3.20) (see Abdullaev et al. (1999); Abdullaev (2006) for details).

The method of stationary phase approximation which usually employed for the asymptotical estimations of integrals from the fast varying functions is not applicable for the integral (3.20). It is related with the behavior of the phase function,

$$F(\vartheta) = m\vartheta - m'\theta(\vartheta, \psi), \qquad (3.22)$$

in (3.20) near the critical points $\vartheta = 0$ and $\vartheta = \pi$. The second derivatives of $F(\vartheta)$ vanish at this critical points. The asymptotical estimations of such integrals can be done by reducing to them the special integrals (see, e.g., Fedoryuk (1989)).

From the typical dependence θ of ϑ shown in Fig. 2.2a it follows that the function $\theta = \theta(\vartheta)$ has two critical points, $\theta(\vartheta = 0) = 0$, i.e., on the high field side (HFS), and $\theta(\vartheta = \pi) = \pi$, on the low field side (LFS), where the second derivatives $d^2\theta/d\vartheta^2$ vanish. For the fixed values of m the main contribution to the integral (3.20) comes from two intervals of m'. For small values of $m' \lessapprox m/\gamma_1$, the integral $S_{mm'}(\psi)$ has the following asymptotics

$$S_{mm'}(\psi) \approx \left(\frac{2}{|\gamma_3|m'}\right)^{1/3} \mathrm{Ai}\left(-\frac{\gamma_1 m' - m}{(|\gamma_3|m'/2)^{1/3}}\right), \tag{3.23}$$

where $\mathrm{Ai}(z)$ is the Airy function. This function oscillates for $x < 0$ and exponentially decays for $x > 0$. It has a local maximum at $x_c \approx -1$. In Eq. (3.23) the parameters γ_1 and γ_3 are the values of the derivatives at the points $\vartheta = 0$ and $\vartheta = \pi$:

$$\gamma_1 = \left.\frac{d\theta}{d\vartheta}\right|_{\theta=0}, \qquad \gamma_3 = \left.\frac{d^3\theta}{d\vartheta^3}\right|_{\theta=0}, \tag{3.24}$$

Similarly, for the values of $m' \gtrapprox m/\beta_1$, we have the following asymptotics

$$S_{mm'}(\psi) \approx (-1)^{m+m'}\left(\frac{2}{\beta_3 m'}\right)^{1/3} \mathrm{Ai}\left(\frac{\beta_1 m' - m}{(\beta_3 m'/2)^{1/3}}\right), \tag{3.25}$$

where

$$\beta_1 = \left.\frac{d\theta}{d\vartheta}\right|_{\theta=\pi}, \qquad \beta_3 = \left.\frac{d^3\theta}{d\vartheta^3}\right|_{\theta=\pi}. \tag{3.26}$$

For the intermediate values of m': $m/\gamma_1 < m' < m/\beta_1$ the integral (3.20) is proportional to $1/\sqrt{m'}$. Contributions from these terms may be neglected due to the rapid oscillations of $S_{mm'}$ with m'.

The asymptotics of the integrals (3.23) and (3.25) are mainly determined by the values γ_1, β_1, γ_3, β_3, of derivatives $d\theta/d\vartheta$ and $d^3\theta/d\vartheta^3$ taken on the LFS $[\vartheta = 0]$ and HFS $[\vartheta = \pi]$, respectively. In general they satisfy the following inequalities: $\gamma_1 > 1$, $\gamma_3 < 0$, $0 < \beta_1 < 1$, and $\beta_3 > 0$. [1] Particularly, for the standard model of the magnetic field (see Sect. 2.2.1) one can obtain the analytical formulas,

$$\gamma_1 = A = \sqrt{\frac{1 + r/R_0}{1 - r/R_0}}, \quad \beta_1 = \frac{1}{A}, \tag{3.27}$$

$$\gamma_3 = -A(A^2 - 1)/2, \qquad \beta_3 = \frac{A^2 - 1}{2A^3}. \tag{3.28}$$

The typical radial profiles of these parameters are shown in Fig. 3.1.

An example of the dependence of the transformation matrix $S_{mm'}$ on m' at the fixed value $m = 12$ is shown in Fig. 3.2a shows $S_{mm'}$ itself, and Fig. 3.2b shows it after multiplication by $(-1)^{m'}$. The solid curves correspond to the exact numerical evaluation the integral (3.20), and the dashed curves describe their asymptotics by the Airy functions. As seen from Fig. 3.2b the asymptotic formulas well describe the transformation matrix $S_{mm'}$ near the values $m' \approx m/\gamma_1$ and $m' \approx m/\beta_1$.

---

[1] These inequalities can change the signs in plasmas with the Shafranov shift for magnetic surfaces whose radii ρ is less that the Shafranov shift $\Delta(\rho)$.

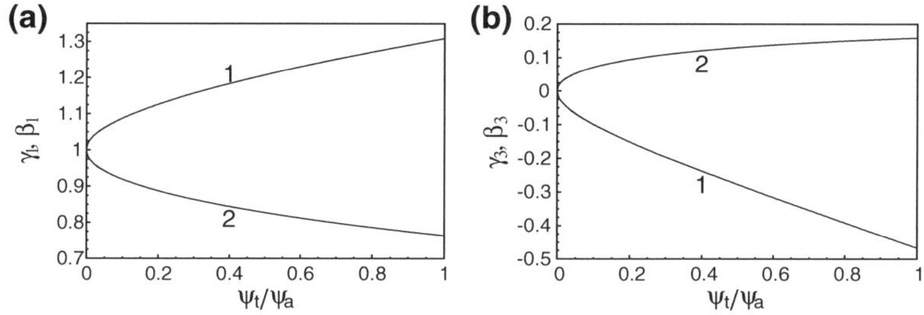

Fig. 3.1 **a** Dependencies of γ_1 (*curve 1*) and β_1 (*curve 2*) on the normalized toroidal flux ψ_t/ψ_a for the standard model of magnetic field; **b** The same as in (**a**) but for γ_3 (*curve 1*) β_3 (*curve 2*). The plasma parameters: the major radius $R_0 = 1.75$ m, the minor radius $a = 0.46$ m; $\psi_a = 1 - \sqrt{1 - a^2/R_0^2}$ is the toroidal flux at the last magnetic surface

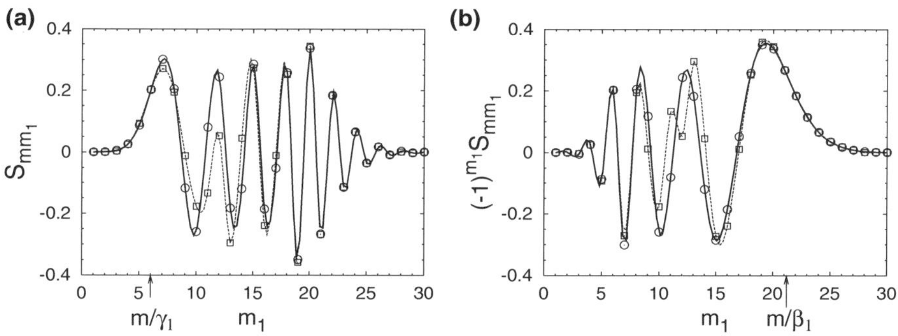

Fig. 3.2 Typical behavior of the integrals S_{mm_1} as a function of m_1 for the fixed $m = 12$: **a** S_{mm_1}; **b** $(-1)^{m_1} S_{mm_1}$. *Solid curve* describes the exact numerical integration, *dashed curve* corresponds to the asymptotics given by (3.23)– (3.25)

From Fig. 3.2a, b also follows that the main contribution to $H_{m,n}$ with the fixed m comes from a few magnetic perturbation modes $A_{m',n}$ located near $m' \approx m/\gamma_1$ or $m' \approx m/\beta_1$. The contributions from other modes are negligible because of rapid oscillations of the transformation matrix $S_{m,m'}$.

First consider the case when a_{mn} changes smoothly with m and the perturbation field is localized on the LFS. Then using (3.23) the sum in (3.19) can be approximated by the integral

$$H_{mn}(\psi) \approx \int\limits_{-\infty}^{\infty} \left(\frac{2}{m'\gamma_3}\right)^{1/3} \mathrm{Ai}\left(\frac{\gamma_1 m' - m}{(m'\gamma_3)^{1/3}}\right) a_{m'n}(\psi) dm'$$

$$= C \int\limits_{-\infty}^{\infty} \mathrm{Ai}(\xi) a_n \left(\frac{m + \xi\left(m'\gamma_3/2\gamma_1\right)^{1/3}}{\gamma_1}\right) d\xi,$$

where $a_n(m) \equiv a_{m,n}$. Integrating the latter integral by the Laplace method we obtain the following asymptotical formula

$$H_{mn}(\psi) \approx C a_n\left(m'\right), \tag{3.29}$$

where

$$\begin{aligned} m' &= \frac{m + \xi_c (m\gamma_3/2\gamma_1)^{1/3}}{\gamma_1} \approx \frac{m}{\gamma_1}, \\ C &= \frac{Bm}{m\gamma_1 + \xi_c (m\gamma_3/2\gamma_1)^{1/3}} \approx \frac{B}{\gamma_1}. \end{aligned} \tag{3.30}$$

Here $B \approx \sqrt{2\pi/|\xi_c|}\mathrm{Ai}(\xi_c)$, and $\xi_c \approx -1$ is the local maximum of the Airy function, and $a_n(m) \equiv a_{m,n}$.

The similar formula can be obianed in the case when the perturbation field is localized on the HFS. The corresponding asymptotic formula for $H_{m,n}$ is given by

$$\begin{aligned} H_{mn}(\psi) &\approx (-1)^m \beta_1^{-1} B\, \tilde{a}_n\left(m'\right), \\ m' &= \frac{m + \xi_c (m\beta_3/2\beta_1)^{1/3}}{\beta_1} \approx \frac{m}{\beta_1}, \end{aligned} \tag{3.31}$$

where $\tilde{a}_n(m) \equiv (-1)^m a_{m,n}$.

Asymptotic formulas (3.29) and (3.31) describe the main features of transformations of poloidal mode spectra of magnetic perturbations in toroidal plasmas. To illustrate the mode transformation consider the following example with the magnetic perturbation modes, a_{mn}, is given by

$$a_{mn}(\psi) = C_m \frac{\sin(m - m_0)\theta_c}{\pi m (m - m_0)}. \tag{3.32}$$

We consider two case type of coefficients C_m: (i) $C_m = (-1)^m$ and (ii) $C_m = 1$. The first case corresponds to the magnetic perturbation localized on the HFS side of the torus, and the second case to the one localized on the LFS. The mode spectrum is localized near the central mode m_0 with a width $\Delta m = 2\pi/\theta_c$.

For the spectrum a_{mn} (3.32) the formulas (3.29) and (3.31) are reduced to

$$H_{mn}(\psi) \approx C_m B \frac{\sin(m/\delta - m_0)\theta_c}{\pi m (m/\delta - m_0)}, \tag{3.33}$$

where $\delta = \beta_1$ for the case (i) and $\delta = \gamma_1$ for the case (ii). As seen from (3.33) and Fig. 3.3a in the first case (i) the center of the mode distribution m_0 of magnetic mode perturbations, A_{mn}, (solid curve 1) is shifted to the lower mode $m_0^* \approx m_0\beta_1$ (solid curve 2 and dashed curve 3). The width of the distribution $\Delta m = 2\pi/\theta_c$ of magnetic perturbations became narrow: $\Delta m^* \approx \beta_1 2\pi/\theta_c$. In the second case (ii) (Fig. 3.3b)

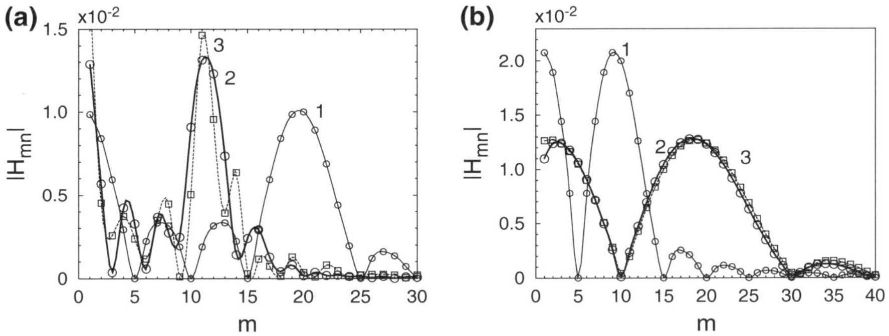

Fig. 3.3 Spectrum of magnetic a_{mn} (*curve 1*) and Hamiltonian perturbations $H_{m,n}$ obtained by numerical integration (*solid curve 2*) and using the asymptotics (3.31) or (3.29) (*dashed curve 3*): **a** describes the case $C_m = (-1)^m$, $m_0 = 20, \theta_c = \pi/5$. **b** the case (ii) $C_m = 1, m_0 = 10, \theta_c = \pi/5$

the central mode number m_0 of magnetic perturbations is shifted to higher mode $m_0^* \approx m_0\gamma_1$ and the distribution became wider: $\Delta m^* \approx \gamma_1 2\pi/\theta_c$. The asymptotical formulae (3.29) and (3.31) qualitatively correct describe the transformation of magnetic perturbations into Hamiltonian perturbations.

The described asymptotic method of the calculation of the spectra of magnetic perturbations can be also used to estimate the spectra of perturbation Hamiltonian for particle drift motion. In this case one obtains the formulas similar to the asymptotical formulas of type (3.29) and (3.31). In Sect. 11.2 we will use these formulas for the spectra of perturbation Hamiltonian of runaway electrons in a turbulent magnetic field.

3.3 Magnetic Perturbations in Tokamaks with X-Point. Model

In this section we consider the effect of magnetic perturbations in tokamaks with the magnetic separatrix on the example of the model of magnetic system considered in Sect. 1.3. This model allows one to study analytically the spectrum of magnetic perturbations which turns out to be a generic for a realistic magnetic field configurations.

We choose the perturbation magnetic flux in the form

$$\begin{aligned} \epsilon\psi_1(z,p,\varphi) &= \epsilon\psi_n(z,p)\sin(n\varphi), \\ \psi_n(z,p) &= ([z-b]\cos\alpha_0 + p\sin\alpha_0)\,, \end{aligned} \tag{3.34}$$

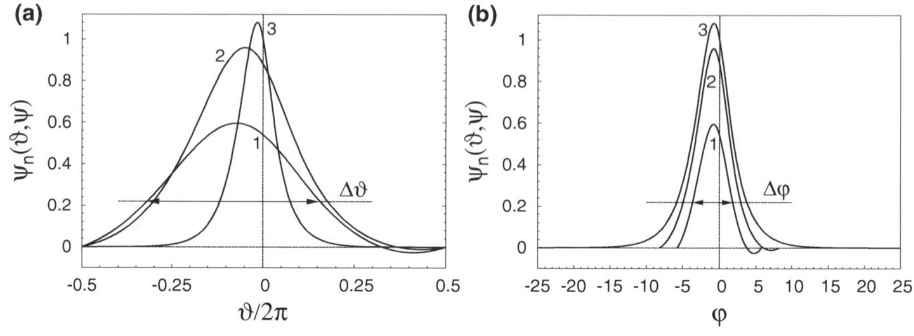

Fig. 3.4 Dependencies of the perturbation function $\psi_n(\vartheta, \psi)$ (3.34) on the angle variable ϑ (**a**) and on the toroidal angle $\varphi = \vartheta q(\psi)$ (**b**) at the several magnetic surfaces: *curve 1* corresponds to $\psi_N = 0.5$, *curve 2* corresponds to $\psi_N = 0.95$, and *curve 3* corresponds to $\psi_N = 1 - 10^{-6}$. The values of the model parameters are $\alpha = 0.44$, $\beta = 1$, $z_a = 1$, and $\alpha_0 = \pi/4$

where α_0 is a constant coefficient. The dependencies of the perturbation function $\psi_n(\vartheta, \psi) \equiv \psi_n(z(\vartheta, \psi), p(\vartheta, \psi))$ on the poloidal angle ϑ and on the toroidal angle φ at the several magnetic surfaces is shown in Fig. 3.4.

One can notice that the perturbation $\psi_n(\vartheta, \psi)$ behaves as a pulse—like function of the poloidal, ϑ, and the toroidal, φ, angles. The width $\Delta\varphi$ of "pulse" along the toroidal angle φ is weakly changed with the magnetic surface ψ, while the width $\Delta\vartheta$ along the poloidal angle ϑ decreases with growing ψ_N. Using (1.49) we have $\Delta\varphi \approx 2/\gamma$ and $\Delta\vartheta \approx \Delta\varphi/q(\psi) = 2/\gamma q(\psi) \sim 1/\ln|\psi|$ near the magnetic separatrix $\psi_N \to 1$, $(\psi \to 0)$.

Using the Fourier expansion of the Jacobi elliptic function $\mathrm{dn}(u, k)$, (Abramowitz and Stegun 1965),

$$\mathrm{dn}\,(u, k) = \frac{\pi}{2K(k)} + \frac{2\pi}{K(k)} \sum_{m=1}^{\infty} \frac{q^m}{1+q^{2m}} \cos\left(\frac{m\pi u}{K(k)}\right),$$
$$q = \exp\left[-\pi K\left(\sqrt{1-k^2}\right)/K(k)\right],$$

and Eq. (1.44), the perturbation Hamiltonian $\psi_1(z, p, \varphi)$ in Eq. (1.37) can be expanded in a Fourier series in the angle variable ϑ. For the closed field lines $(\psi < 0)$ one has

$$\psi_n(z, p) = \sum_{m=1}^{\infty} \left[H_m^{(c)}(\psi)\cos(m\vartheta) + H_m^{(s)}(\psi)\sin(m\vartheta)\right], \tag{3.35}$$

where

$$H_m^{(c)}(\psi) = H_m(\psi)\cos\alpha_0, \qquad H_m^{(s)}(\psi) = -\frac{m}{\beta^2 q(\psi)} H_m(\psi)\sin\alpha_0,$$

$$H_m(\psi) = \frac{2\sqrt{2}z_a}{\gamma} \frac{1}{q(\psi)\cosh\left[mC(\psi)/q(\psi)\right]}, \tag{3.36}$$

where

$$C(\psi) = \frac{\sqrt{2}K\left(\sqrt{1-k^2}\right)}{\gamma\sqrt{1+\psi_N^{1/2}}}.$$

At the limit $|\psi| \to 0$ or for large mode numbers $m \gg 1$, the Fourier components $H_m(\psi)$ have the asymptotics

$$H_m(\psi) \approx \frac{2\sqrt{2}}{\gamma q(\psi)} \exp\left(-\frac{mC(\psi)}{q(\psi)}\right). \tag{3.37}$$

The function $C(\psi)$ in the exponent of (3.37) has the following asymptotics near the magnetic axis and the separatrix

$$C(\psi) \approx \begin{cases} \sqrt{2}/2\gamma \ln\left(8/\psi_N^{1/2}\right), & \text{for} \quad \psi_N \to 0, \\ \pi/2\gamma, & \text{for} \quad \psi_N \to 1. \end{cases} \tag{3.38}$$

Its dependence on the normalized poloidal flux ψ_N is plotted in Fig. 3.5. From Eq. (3.38) it follows that near the magnetic axis $H_m(\psi)$ has the following asymptotics

$$H_{mn}(\psi) \propto \psi_N^{m/2}, \quad \psi_N \to 0. \tag{3.39}$$

The considered analytical model of magnetic field allows one to predict generic features of more realistic magnetic fields in tokamaks where the exact analytical calculations are impossible. Particularly, the asymptotical behavior of the poloidal spectrum of magnetic perturbations found for this model turns out to be generic for toroidal magnetic configurations.

Fig. 3.5 Profile of the function $C(\psi)$ in (3.36): The value of the model parameters are the same as in Fig. 3.4

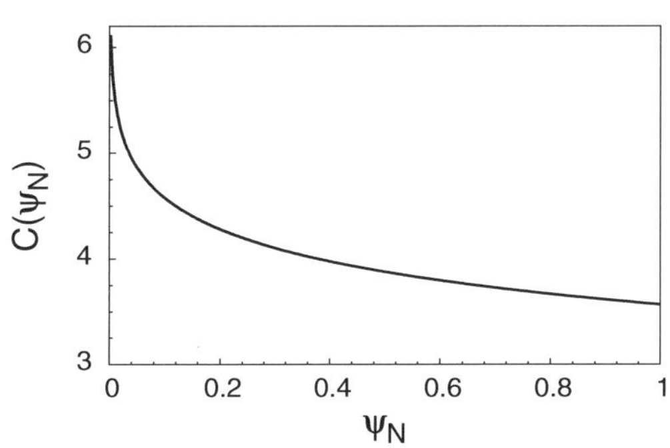

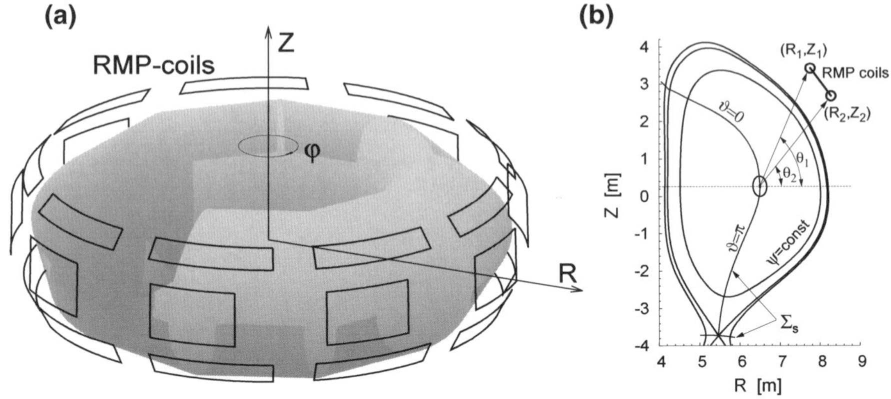

Fig. 3.6 Plasma configurations and RMP coils for the ITER-like plasma: **a** 3D sketch of the devices; **b** Poloidal section in the (R, Z) plane. Only the RMP coils at the upper equatorial plane are shown. The coil positions are taken at $(R_1, Z_1) = (7.75345, 3.42)$ m, $(R_2, Z_2) = (8.27958, 2.67661)$ m. The sections where $\vartheta(R, Z) = 0$ and $\vartheta(R, Z) = \pi$ are shown by *solid lines*. The geometrical angle θ is counted with respect to the outer equatorial plane $Z_a = \text{const}$ crossing the magnetic axis (R_a, Z_a)

3.4 Universal Asymptotics of Poloidal Spectra of the RMPs Created by a Set of Saddle Coils

In this section we consider the spectra of the RMPs created by a set of saddle coils. Typically, they are installed in tokamaks with the elongated shape of plasmas and magnetic separatrices. The examples of such coil configurations are shown in Fig. 3.6a which foreseen for the ITER plasma and Fig. 3.7a for the spherical tokamak NSTX. It turns out that the poloidal mode spectra of perturbation magnetic field generated by a set of saddle coils can be presented by a linear combination of three universal asymptotical formulas similar to (3.37) (Abdullaev 2011b, 2012). They are determined only by the safety factor of plasmas and the geometrical positions coils. The nature of this universality is related to the generic behavior of magnetic field lines due to the toroidicity of system and the presence of the magnetic separatrix.

Below we study the poloidal spectra of magnetic perturbations created by a set of saddle coils. Specifically we consider the two type of tokamaks with a large aspect ratio, the ITER–like plasma and a small aspect-ratio tokamak, the NSTX. Their configuration and plasma parameters were given in Table 2.2, and shown in Figs. 2.4 and 2.5.

3.4.1 *Saddle Coils and Their Magnetic Fields*

For simplicity we consider a single set of saddle coils located upper equatorial plane of the tokamak as shown in Fig. 3.6a, b for the ITER like plasma device, and Fig. 3.7a, b

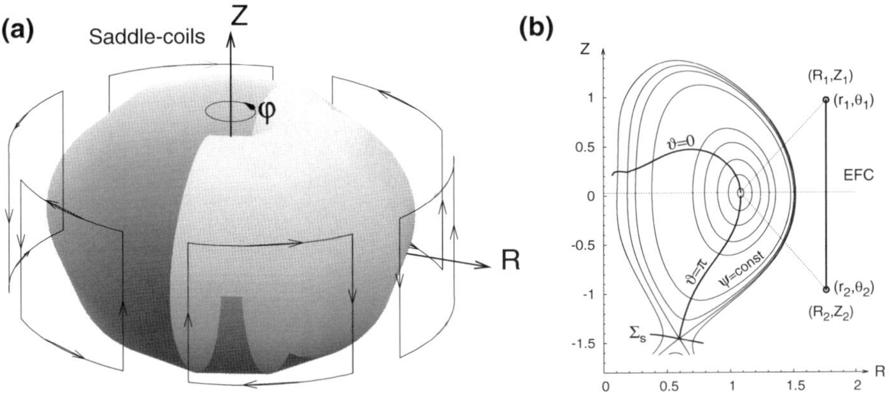

Fig. 3.7 The same as in Fig. 3.6 but for the NSTX like plasma with the Error Field Coils (EFC): **a** 3D sketch of the devices; **b** Poloidal section in the (R, Z) plane. The coil positions are taken at $(R_1 = R_2) = 1.761$ m, $(Z_1 = -Z_2) = 0.965$ m

for the NSTX like plasma device, respectively. Suppose that the set consists of N of saddle coils whose the upper and lower horizontal segments lie on the circular loops of radii R_1 and R_2 located at the horizontal planes $Z = Z_1$ and $Z = Z_2$, respectively, as shown in Fig. 3.8 (see also Figs. 3.6 and 3.7a, b). The values of $(R_j, Z_j), (j = 1, 2)$ for the are given in the caption of Figs. 3.6 and 3.7. We separately consider magnetic fields created by the segments of saddle coils on the horizontal circular loops on planes $Z = Z_j, (j = 1, 2)$ and the segments on the vertical planes (R, Z).

The vector potential $\mathbf{A}(R, Z, \varphi)$ of the magnetic field at the point $\mathbf{r} = (R, Z, \varphi)$ generated by the saddle coils can be found by the Biot–Savart law (Jackson (1998)),

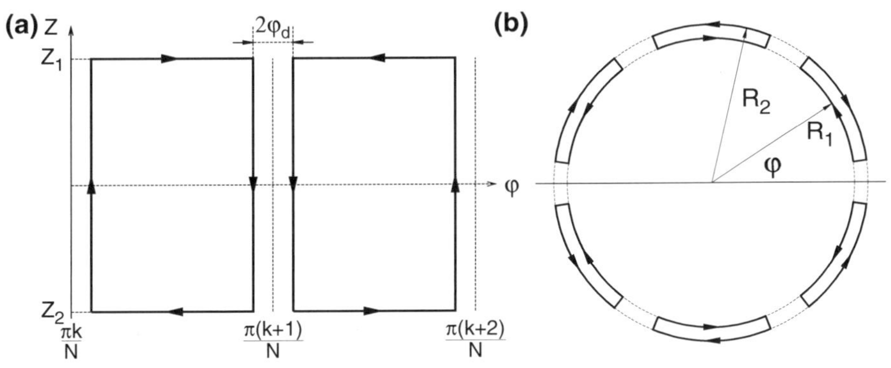

Fig. 3.8 **a** Geometry of one pair of saddle coils in the (φ, Z) plane. **b** Projection of set of saddle coils into (R, φ) plane

$$\mathbf{A}(R, Z, \varphi) = \frac{\mu_o}{4\pi} \int_G \frac{\mathbf{j}(R_0, Z_0, \varphi_0) dV_0}{|\mathbf{r} - \mathbf{r}_0|}, \tag{3.40}$$

where $\mathbf{j}(R_0, Z_0, \varphi_0)$ is the current density, $dV = RdRdZd\varphi$ is a volume element, $\mathbf{r}_0 = (R_0, Z_0, \varphi_0)$ are coordinates in the area of coils G, and

$$|\mathbf{r} - \mathbf{r}_0| = \sqrt{R^2 + R_0^2 + (Z - Z_0)^2 - 2RR_0 \cos(\varphi - \varphi_0)}.$$

3.4.2 Magnetic Field of Horizontal Segments

The current $I^{(j)}(\varphi)$, $(j = 1, 2)$ flowing in the j-th circular loop can be presented as a Fourier series in the toroidal angle φ:

$$I_c^{(j)}(\varphi) = I_c \sum_n i_n \sin(n\varphi), \tag{3.41}$$

where I_c is the current magnitude, i_n is the Fourier coefficient corresponding to the toroidal mode number n. For the set consisting of $2N$ saddle coils the toroidal mode number n takes odd numbers $n = (2s + 1)N$, $(s = 0, 1, 2, \dots)$. For the geometry of saddle coils shown in Fig. 3.8 the Fourier coefficients i_n are given by (see Sect. B.1),

$$i_n = \frac{4}{\pi} \frac{\cos[(2s + 1)N\varphi_d]}{2s + 1},$$

where φ_d is the half–distance between neighboring saddle coils along the toroidal angle φ, as it schematically shown in Fig. 3.8a.

The current loops on the horizontal planes contribute only to the toroidal component of the vector potential, $A_\varphi(R, Z, \varphi)$. The calculations provided in Sect. B.1 of Appendix B show that the integral (3.40) can be reduced to

$$A_\varphi^{(j)}(R, Z, \varphi) = \frac{\mu_o R_j}{4\pi} \int_0^{2\pi} \frac{I_c^{(j)}(\varphi_j) \cos(\varphi_j - \varphi) d\varphi_j}{|\mathbf{r} - \mathbf{r}_j|}. \tag{3.42}$$

Using the relation (3.4) the perturbation poloidal flux $\psi^{(pert)}$ in (3.10) generated by the current (3.41) on the j-th loop can be presented as

$$\psi^{(pert)}(R, Z, \varphi) = \epsilon \sum_{n=N}^{\infty} \psi_n^{(j)}(R, Z) \sin(n\varphi), \tag{3.43}$$

where $\psi_n^{(j)}(R, Z)$ corresponding to the n-th toroidal mode is given by the following integral,

$$\psi_n^{(j)}(R, Z) = -\frac{i_n D_j}{R_0}\left[\left(1 - \frac{k_j^2}{2}\right) K_n(k_j) - E_n(k_j)\right]. \tag{3.44}$$

Here $K_n(k)$ and $E_n(k)$ are the generalized complete elliptic integrals defined by

$$K_n(k) = \int_0^{\pi/2} \frac{\cos(2n\phi)\, d\phi}{\sqrt{1 - k^2 \sin^2 \phi}},$$

$$E_n(k) = \int_0^{\pi/2} \cos(2n\phi) \sqrt{1 - k^2 \sin^2 \phi}\, d\phi. \tag{3.45}$$

The module k_j in (3.44) is given by

$$k_j^2 = 4RR_j / D_j^2, \qquad D_j = \sqrt{(R + R_j)^2 + (Z - Z_j)^2}.$$

In (3.43) the dimensionless perturbation parameter ϵ is defined as

$$\epsilon = \frac{\mu_o I_c}{4\pi B_0 R_0}. \tag{3.46}$$

On the given magnetic surface ψ =const the function $\psi_n(R, Z)$ is localized function of geometrical poloidal angle θ and the intrinsic poloidal angle ϑ with extrema at the angles θ_j and ϑ_j, $(j = 1, 2)$. Particularly, θ_j are the geometrical angles between the straight line connecting the magnetic axis of the plasma with coil position and the equatorial plane as shown in Figs. 3.6b and 3.7b. The dependencies of these functions on the poloidal angles θ and the intrinsic poloidal angle ϑ at the magnetic surface $\psi_N = 0.95$ for the toroidal mode $n = 3$ are shown in Fig. 3.9a and b for the RMP coils in the ITER and the EFCs in the NSTX. The values of the plasma and the RMP coils are presented in Figs. 3.6, 3.7 and in Table 2.2. The current amplitude I_c is taken equal to $I_c = 1$ kA.

As seen from Fig. 3.9 the functions $\psi_n(R, Z)$ corresponding to the single horizontal segments of saddle coils have a pulse like form with the extremal (maximal or minimal) values at θ_j (or ϑ_j) $(j = 1, 2)$, where the magnetic surface is the closest distance to the corresponding horizontal segment of saddle coils as shown in Fig. 3.6b. They have a "pulse"–like dependence on the poloidal angle ϑ. The pulses are much sharper in the NSTX plasma. It can be explained with the fact that in the small aspect ratio tokamaks the length of a field line on the low–field side where the perturbation coils are located is much shorter than on the high–field side. It is related with the large value of the safety factor for the NSTX tokamak than the one for the ITER tokamak (see Table 2.2 and Figs. 2.4b and 2.5b), since as we have pointed out in Sect. 3.3 the width of pulses is inversely proportional to the safety factor.

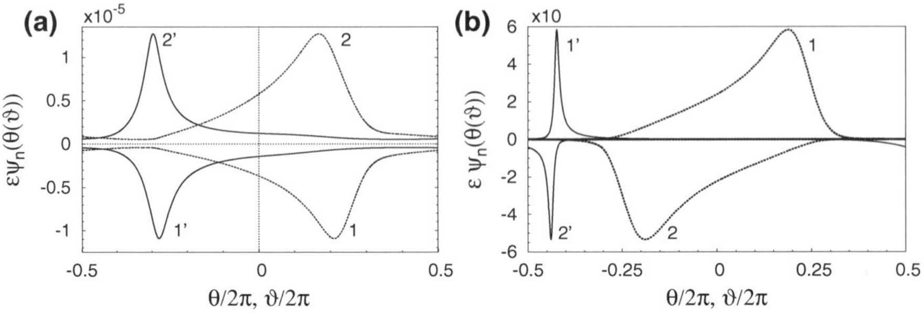

Fig. 3.9 Dependencies of the perturbation poloidal flux $\psi_n^{(j)}(R, Z)$ (3.44) created by the horizontal segments of the set of saddle coils at the $\psi_N = 0.95$ magnetic surface on the geometric, θ, (*curves 1* and *2*) and the intrinsic, ϑ (*curves 1′* and *2′*) poloidal angles for the ITER–like plasma (**a**) and for the NSTX like plasma (**b**). *Curves 1* and *1′* correspond to the upper horizontal segment, *curves 2* and *2′* correspond to the lower horizontal segment. The plasma parameters and the RMP coils positions are given in Figs. 3.6, 3.7 and in Table 2.2. The perturbation current $I_c = 1$ kA and the perturbation parameter ϵ (3.46) is 3.043×10^{-6} for the ITER plasma and 2.614×10^{-4} for the NSTX plasma, respectively

The poloidal spectrum of the perturbation flux $\psi_n^{(j)}(R, Z)$ is given by the coefficients $H_{mn}^{(j)}(\psi)$ of the Fourier series in ϑ,

$$\psi_n^{(j)}(R, Z) = \sum_{m=-\infty}^{\infty} H_{mn}^{(j)}(\psi) e^{im\vartheta}$$
$$= H_{0n}^{(j)}(\psi) + 2 \sum_{m=1}^{\infty} |H_{mn}^{(j)}(\psi)| \cos(m\vartheta + \chi_{mn}^{(j)}), \quad (3.47)$$

where $|H_{mn}^{(j)}(\psi)|$ and $\chi_{mn}^{(j)}$ are the magnitude and phase of the poloidal mode m:

$$H_{mn}^{(j)}(\psi) = |H_{mn}^{(j)}(\psi)| \exp\left(i\chi_{mn}^{(j)}\right).$$

3.4.3 Magnetic Field of Vertical Segments

Similar asymptotical formulas can be established for the vertical segments of the set saddle coils (see Abdullaev (2012) for details). The current density $I^{(v)}(\varphi)$ flowing in the vertical segments of saddle coils can be expanded into the following Fourier series in the toroidal angle φ:

$$\mathbf{I}^{(v)}(\varphi) = (I_R, I_Z) = \sum_{n=0}^{\infty} \mathbf{i}_n \cos(n\varphi), \quad (3.48)$$

where the vector $\mathbf{i}_n$ is a some function of (R, Z). The vector potential components (A_R, A_Z) of the magnetic field generated by the current (3.48) are found by the Biot-Savart law (3.40). The details of calculation are given in Sect. B.2 of Appendix B. Using Eqs. (3.4)–(3.8) the perturbation poloidal flux $\psi^{(pert)}$ generated by the current (3.48) can be reduced to

$$\psi^{(pert)}(R, Z, \varphi) = \epsilon \sum_{n=N}^{\infty} \psi_n^{(3)}(R, Z) \sin(n\varphi), \tag{3.49}$$

where the perturbation parameter ϵ is defined by Eq. (3.46), $\psi_n^{(3)}(R, Z)$ is determined by the integral similar to (3.44). Its Fourier expansion is given by

$$\begin{aligned} \psi_n^{(3)}(R, Z) &= 2 \sum_{m=1}^{\infty} \left| H_{mn}^{(3)}(\psi) \right| \cos\left(m\vartheta + \chi_{mn}^{(3)} \right), \\ H_{mn}^{(3)}(\psi) &= -\frac{n}{m} G_{mn}(\psi), \quad \chi_{mn}^{(3)} = \phi_{mn} + \frac{\pi}{2}, \end{aligned} \tag{3.50}$$

where $G_{mn}(\psi)$ are the coefficients in the Fourier expansion of the function $G(R, Z, \varphi)$ in Eq. (3.8),

$$\begin{aligned} G(R, Z, \varphi) &= \sum_{s=0}^{\infty} G_{n_s}(R, Z) \cos(n_s \varphi), \\ G_n(R, Z) &= \sum_{m=0}^{\infty} |G_{mn}(\psi)| \cos(m\vartheta + \phi_{mn}). \end{aligned} \tag{3.51}$$

The details of the numerical calculations of the poloidal mode amplitudes $H_{mn}^{(j)}(\psi)$, $(j = 1, 2, 3)$, in Eqs. (3.47) and (3.50) are presented in Sect. B.3 of Appendix B.

The typical angular dependence of the function $G_n(R, Z)$ in (3.7) on the given magnetic surface $\psi = \text{const}$ is shown in Fig. 3.10. For the considered saddle coil the magnitude of $G_n(R, Z)$ corresponding to the vertical segments is of order of one smaller than one $\psi_n^{(j)}(R, Z)$ corresponding to the horizontal segments.

3.4.4 Asymptotics of Poloidal Spectra

The analytical calculations made for the model magnetic system (1.3) have shown the poloidal spectrum H_{mn} given by Eq. (3.36). It turns out that the asymptotic behavior of H_{mn} described by (3.37) is a generic feature for magnetic field configurations with the separatrices (Abdullaev (2011b, 2012)). In general, the poloidal spectrum $H_{mn}^{(j)}(\psi)$ of

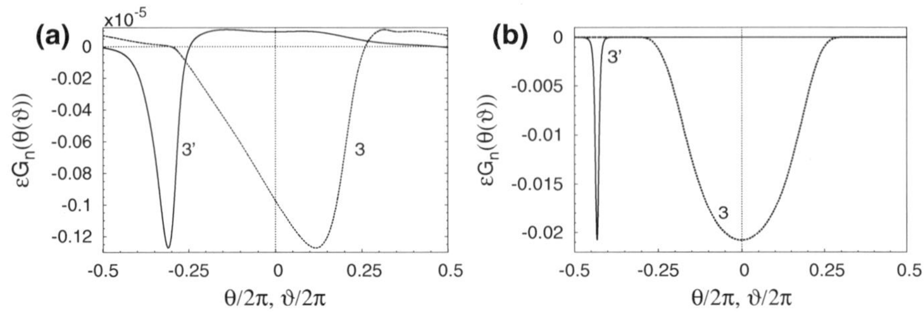

Fig. 3.10 Dependencies of the perturbation poloidal flux $G_n(R, Z)$ (3.7) created by the vertical segments of the set of saddle coils at the $\psi_N = 0.95$ magnetic surface on the geometric, θ, (*curve 3*) and the intrinsic, ϑ (*curve 3'*) poloidal angles for the ITER–like plasma (**a**) and for the NSTX like plasma (**b**). The parameters are the same as in Fig. 3.9

the localized magnetic perturbation $\psi_n^{(j)}(R, Z)$ with a pulse-like angular dependence has universal asymptotical behavior. To find this one can use the following arguments.

The dependence of $\psi_n^{(j)}(R, Z)$ on the angle ϑ near its peaks can be approximated by the Lorentzian pulse function

$$\psi_L(\vartheta) = \frac{a\sigma^2}{(\vartheta - \vartheta_0)^2 + \sigma^2}, \tag{3.52}$$

where a, ϑ_0, and σ are the amplitude, the width, and the center of the pulse, respectively. The Fourier transform H_{mn} of $\psi_L(\vartheta)$ has the exponential dependence on m similar to the one (3.37),

$$H_{mn} = \frac{1}{2\pi} \int_0^{2\pi} e^{-im\vartheta} \psi_L(\vartheta) d\vartheta = \frac{a\sigma}{2} e^{-im\vartheta_0 - m\sigma}. \tag{3.53}$$

From the numerical calculations show that the width σ of the pulse-like dependence of $\psi_n^{(j)}(R, Z)$ on ϑ on the given magnetic surface decrease with the increase of its safety factor, i.e., $\sigma \propto 1/q(\psi)$.

Using these observations one can propose the following asymptotical formulas for H_{mn} in the complex form,

$$H_{mn}^{(j)}(\psi) = \frac{A_j(\psi)}{2q(\psi)} \exp\left(-\frac{mC_j(\psi)}{q(\psi)} + i\chi_m^{(j)} \right), \tag{3.54}$$

where $A_j(\psi)$ is amplitude, $C_j(\psi)$ is the exponent, and $\chi_{mn}^{(j)}$ is the phase as a linear function of mode m

$$\chi_m^{(j)} = \chi^{(j)} - m\vartheta_j, \qquad (j = 1, 2). \tag{3.55}$$

The phase $\chi^{(j)}$ takes values 0 and π for the upper and lower horizontal segments, respectively.

The spectrum is determined by the safety factor of the plasma $q(\psi)$ and the three functions A_j, C_j, and ϑ_j $(j = 1, 2)$ of the flux variables ψ. The functions A_j, C_j have finite values at the separatrix $\psi_N = 1$,

$$C_j(\psi_N = 1) \neq 0, \qquad A_j(\psi_N = 1) \neq 0. \tag{3.56}$$

One expects that the asymptotics (3.54) are valid for large mode numbers m and its accuracy of these formulas improves with increasing of the toroidal mode n.

The asymptotical formula for the poloidal spectrum $H_{mn}^{(3)}(\psi)$ corresponding to the vertical segments of saddle coils can be found by the following arguments. The function $G_n(R, Z)$ in (3.51) has a pulse–like dependence on the poloidal angle ϑ. Therefore, one can expect the following asymptotical formula for its Fourier components $G_{mn}(\psi)$,

$$G_{mn}(\psi) = \frac{1}{2} A_3(\psi) \exp\left(-\frac{mC_3(\psi)}{q(\psi)}\right) e^{-im\vartheta_3}. \tag{3.57}$$

The functions $A_3(\psi)$, $C_3(\psi)$, and ϑ_3 are similar to the functions A_j, C_j, $\vartheta_{mn}^{(j)}$ in (3.54 and 3.55). Then from the relation (3.50) between $H_{mn}^{(3)}(\psi)$ and $G_{mn}(\psi)$ one obtain the following formula

$$H_{mn}^{(3)}(\psi) = \frac{nA_3(\psi)}{2m} \exp\left(-\frac{mC_3(\psi)}{q(\psi)}\right) e^{i\chi_m^{(3)}},$$
$$\chi_m^{(3)} = \pi/2 - m\vartheta_3. \tag{3.58}$$

One should note that $H_{mn}^{(3)}(\psi)$ can be obtained from $H_{mn}^{(j)}(\psi)$ (3.54) by replacing the safety factor $q(\psi)$ in the pre-exponent is replaced by the ratio m/n of the mode numbers.

The example of the numerically calculated dependencies of the magnitudes $|H_{mn}^{(j)}(\psi)|$ and the phases $\chi_{mn}^{(j)}(\psi)$, $(j = 1, 2, 3)$, on the poloidal mode number m are shown in Fig. 3.11a, b, respectively, at the $\psi_N = 0.95$ magnetic surface for the toroidal mode $n = 3$. Dots with the symbols $\circ$, $\square$, and $\triangle$ correspond to the upper horizontal, lower horizontal, and vertical segments of a set of coils, respectively. Solid curves 1, 2, and 3 describe the asymptotical formulas for $|H_{mn}^{(j)}(\psi)|$, $(j = 1, 2, 3)$, given by Eqs. (3.54 and 3.58). The coefficients C_j, A_j, $(j = 1, 2, 3)$ are found by fitting those formulas with the numerical calculated ones.

The phases $\chi_{mn}^{(j)}(\psi)$ are also well presented by the linear functions of the poloidal mode m (3.55). However, their values depend on the calibration of the poloidal angle ϑ. For the calibration $\vartheta = -\pi$ at the section Σ_s as shown in Figs. 3.6b and 3.7b

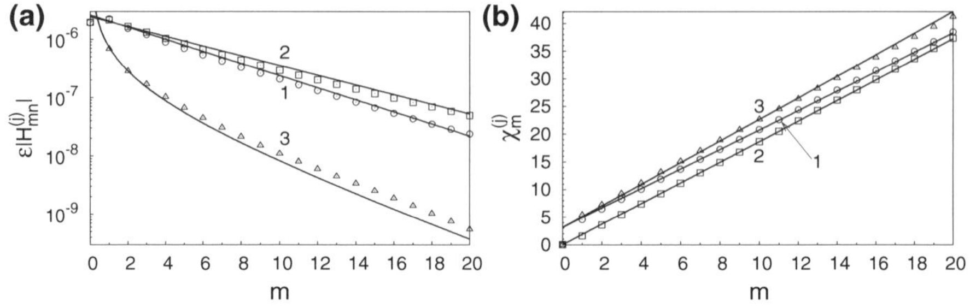

Fig. 3.11 Magnitudes of Fourier components $H_{mn}(\psi)$ (**a**) and phases χ_{mn} (**b**) versus the poloidal mode number m versus the poloidal mode number m on the magnetic surface $\psi_N = 0.95$: Symbols o, □, and △ correspond to the upper horizontal, lower horizontal, and vertical segments of a set of coils, respectively. *Solid curves 1, 2, 3* describe to the corresponding asymptotical formulas (3.54) with the fitted parameters C_j, A_j, $(j = 1, 2, 3)$. The plasma and the RMP–coils parameters are the same as in Fig. 3.6, and the toroidal mode $n = 3$

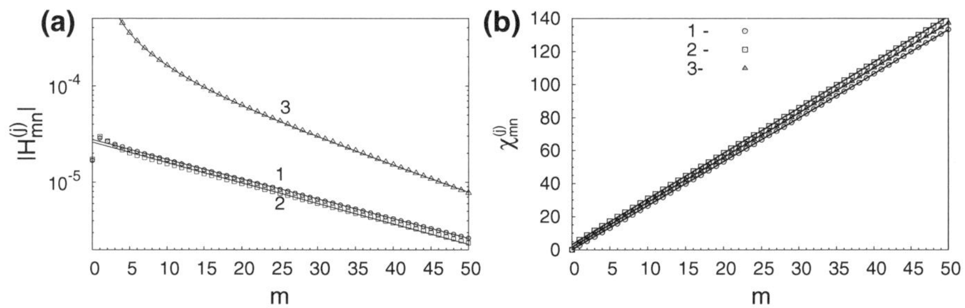

Fig. 3.12 The same as in Fig. 3.11 but for the NSTX like plasma with the EFC. The plasma and the EFCs parameters are the same as in Fig. 3.7, and the toroidal mode $n = 3$

the dependencies $\chi_{mn}^{(j)}(\psi)$ on m are shown in Fig. 3.11b for the ITER-like plasma and Fig. 3.12b for the NSTX like plasma, respectively. The solid curves 1–3 corresponding to the upper and lower horizontal, and the vertical segments, respectively. As seen the phase $\chi_{mn}^{(j)}(\psi)$ are well described by the linear functions of the mode number m.

We should note that that the accuracy of the asymptotical formulae (3.54) and (3.58) depends on the aspect ratio of tokamaks. Particularly, in a spherical tokamak they describe the poloidal spectra with higher accuracy than in the standard tokamaks (Abdullaev 2011b, 2012).

3.4.5 Determination of Coefficients C_j, A_j, ϑ_j

In order to use the formulae (3.54) and (3.58) to study the structure of magnetic field lines one should determine the coefficients C_j, A_j,and ϑ_j, $(j = 1, 2, 3)$ with a possible high accuracy. To find these coefficients we use the following criteria. In the formation of chaotic layer the crucial role plays the resonant interaction of field lines with magnetic perturbation which is predominantly determined by the amplitude of poloidal modes H_{mn} at the resonant magnetic surfaces $\psi = \psi_{mn}$. For this reason the above coefficients should have such values that the formulae (3.54) and (3.58) are most close the numerical ones near the resonant surfaces. To satisfy this criteria we will calculate the above coefficients at the given magnetic surface ψ by the fitting of the formulae (3.54) and (3.58) with numerically calculated spectrum H_{mn} in the interval of modes m located around the resonant mode number $m_0 = nq(\psi)$: $m_0 - \delta m \le m \le m_0 + \delta m$.

The another, more simplest way to calculate these coefficients is to fit the calculated dependence of the perturbation poloidal flux $\psi_n^{(3)}(R, Z)$ on the poloidal angle ϑ near its peaks with the Lorentzian function $\psi_L(\vartheta)$ (3.52). The fitting coefficients a, σ and ϑ_0 of the latter determine the coefficients C_j, A_j,and ϑ_j:

$$C_j = q(\psi)\sigma, \quad A = q(\psi)a\sigma, \quad \vartheta_j = \vartheta_0. \tag{3.59}$$

3.4.5.1 Coefficients C_j, A_j

Below we describe the radial profiles of these functions found using the above criteria. The functions C_j, $(j = 1, 2, 3)$, in the exponents of Eq. (3.54) have similar properties as the function $C(\psi)$ in (3.36) in the model system (1.3). They are smooth functions of ψ with the finite values at the separatrix, $C_j \to C_j(1) \neq 0$ at $\psi_N \to 1$ as shown in Fig. 3.13a. Their values do not depend on the magnitude of the perturbation current I_0, but they depend on the mutual position of the plasma and coils.

Similar to (3.38) and (3.39) near the magnetic axis $\psi_N = 0$ the functions C_j have the logarithmic dependence, $C_j \approx -q(0)/2 \ln \psi_N$ on ψ_N, which corresponds to the power-law dependence the amplitudes $H_{mn}(\psi)$,

$$|H_{mn}^{(j)}(\psi)| \propto \psi_N^{m/2}. \tag{3.60}$$

In the plasma region and near the separatrix the functions C_j, $(j = 1, 2, 3)$ can be well approximated by series,

$$C_j(\psi_N) = \sum_{s=0}^{M} c_s \psi^s - (q(0)/2 + d_1\psi) \ln \psi_N, \qquad \psi = 1 - \psi_N, \tag{3.61}$$

where the constant coefficients $c_s, s = 0, \dots, M, d_1$ can be found by the fitting these functions with the numerically calculated ones.

The functions C_j corresponding to the horizontal $(j = 1, 2)$ and the vertical segments of the saddle coil are plotted in Fig. 3.13a by solid curves. The numerical estimations confirm the asymptotics of $|H_{mn}^{(j)}(\psi)|$ (3.60) near the magnetic axis for the all three functions $C_j(\psi_N)$, $(j = 1, 2, 3)$. As it was shown in Sect. 1.3 such an

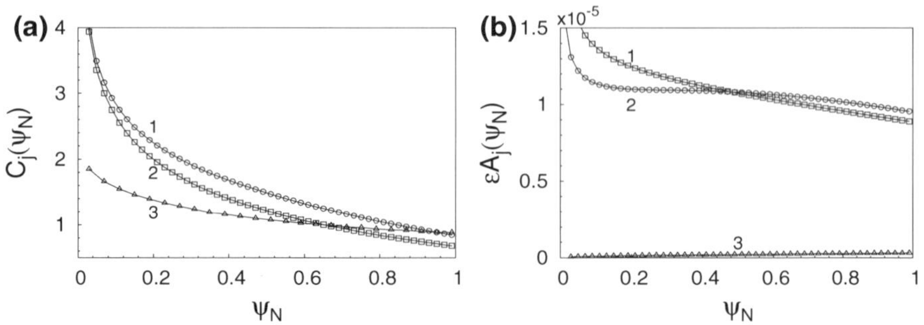

Fig. 3.13 **a** Calculated exponents $C_j(\psi_N)$, $(j = 1, 2, 3)$, in Fourier components $H_{mn}^{(j)}(\psi)$ Eq.(3.54). Solid curves describes the fitting functions (3.61). *Curve 1* and symbol ∘ correspond to the upper horizontal segment, *curve 2* and symbol □—to the lower horizontal segment, *curve 3* and symbol △—to the vertical segments of saddle coils. **b** Radial profiles of the amplitudes A_j. *Curve 1* corresponds to the upper horizontal segment, *curve 2*—to the lower horizontal segment, *curve 3*—to the vertical segments of saddle coils

asymptotics is also confirmed by the analytical calculations for the analytical model of a magnetic field (see, Eq. (3.39)).

The radial profiles if the amplitudes A_j, $(j = 1, 2, 3)$ are shown in Fig. 3.13b. The functions A_j weakly change along the radial coordinate ψ in the whole plasma region. One should note for the considered set of coils the magnitude of A_3 is almost one order smaller than ones for the the horizontal segments, i.e., $A_{1,2} \gg A_3$. The amplitudes A_j linearly depend on the magnitude of the perturbation current I_c.

However, the magnitude of A_3 may become comparable to $A_{1,2}$ for the larger length of the vertical segments of saddle coils. In the NSTX tokamak the vertical segments of saddle coils have a length $\Delta Z = |Z_1 - Z_2|$ close to its major radius R_j. In this case the amplitude A_3 corresponding to the vertical segments is the same order as ones $A_{1,2}$ corresponding to the horizontal ones (see Abdullaev 2011b, 2012). The latter are shown in Fig. 3.14b.

In the region $0.5 < \psi_N < 1.1$ the amplitudes A_j, $(j = 1, 2)$ can be also approximated by series,

$$A_j(\psi_N) = \sum_{s=0}^{M} a_s \psi^s, \quad \psi = 1 - \psi_N, \tag{3.62}$$

with the constant coefficients a_s, $s = 0, \ldots, M$.

The amplitude A_3 can be approximated by the $A_3(\psi_N) = a_0 + a_1 m$ with the fitting coefficients a_0, a_1. In this case the function a_0 tends to the finite non-zero value $a_0(\psi_N) \to a_0(1) \neq 1$, while a_1 goes to zero, $a_1(\psi_N) \to 0$, at the separatrix $\psi_N \to 1$ (see Abdullaev (2012)).

The dependencies of $C(n) \equiv C_j(1)$, and $A(n) \equiv A_j(1)$ for the upper horizontal segment $j = 1$ on the toroidal mode n are shown in Fig. 3.15a, b. For the small values of $n \lesssim 6$ the exponent $C_j(1)$ decays fast with n, and for $n \gtrsim 6$ it slowly

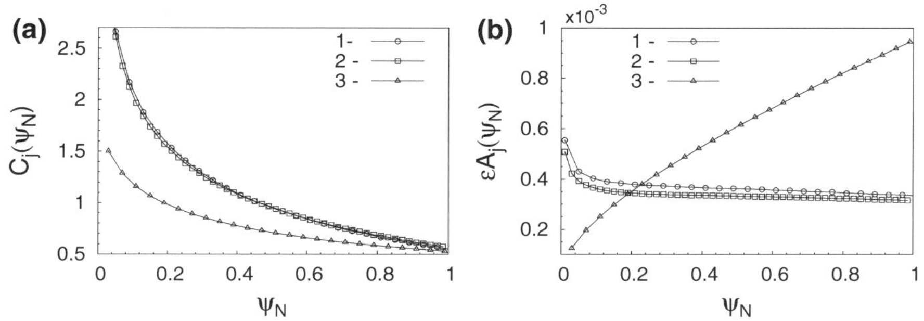

Fig. 3.14 The same as in Fig. 3.13a, b but for the EFC in the NSTX like plasma: **a** The profiles of $C_j(\psi_N)$; **b** The amplitudes $A_1(\psi_N)$, $A_2(\psi_N)$, $A_3(\psi_N)$ in Fourier components $H_{mn}^{(j)}(\psi)$ (see Eqs. (3.54) and (3.58)) for the NSTX like plasma with the EFC coils. *Curve 1* corresponds to the upper horizontal segment, *curve 2* corresponds to the lower horizontal segment, *curve 3* corresponds to the vertical segments of saddle coils, respectively

tends to the non-zero constant value C_∞. The numerically calculated $C(n)$ can be well approximated by the function $C(n) = C_\infty + bn^{-\beta}$ with the fitting coefficients $C_\infty = 0.4534$, $b = 1.1284$, and $\beta = 0.9197$ as shown by solid curve in Fig. 3.15a.

On the other hand the amplitude $A(n)$ decays with increasing n. For large n, $n \gtrsim 10$, it decays according to the exponential law, $A_j \sim \exp(-Bn)$, where B is a non-zero constant, $B \neq 0$ as shown in Fig. 3.15b by solid straight curve 2.

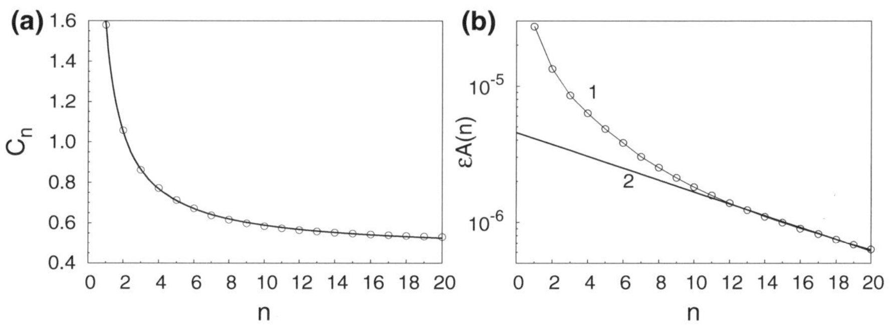

Fig. 3.15 **a** Dependence of the exponent $C(1)$ near the separatrix $\psi_N = 1$ on the toroidal mode number n for the upper horizontal coil $j = 1$. The dependence C on n can be approximated by the power–law function $C(n) = C_\infty + bn^{-\beta}$ (*solid curve*) with the fitting coefficients $C_\infty = 0.4534$, $b = 1.1284$, and $\beta = 0.9197$. **b** Dependence of the amplitude $A(1)$ near the separatrix $\psi_N = 1$ on the toroidal mode number n for the upper horizontal coil $j = 1$. *Curve 1* corresponds to the dependence A on n, *curve 2* (*solid*) corresponds to its exponential asymptotics at the large n, $n \gtrsim 10$, i.e., $A = a\exp(-Bn)$ with the fitting coefficients $a = 0.9252$ and $B = 0.1$

3.4.5.2 Angles $\vartheta_j(\psi)$

The angles $\vartheta_j(\psi)$ determine the angular locations of poloidal flux extrema created by the corresponding segments of saddle coils. (Recall that we have calibrated the angle ϑ_j to take the value $-\pi$ at the section Σ_S as shown in Figs. 3.6 and 3.7.) They are smooth functions of magnetic surface ψ and slowly increases near the separatrix. On the other hand the locations $\Theta_j(\psi)$ of poloidal flux extrema in the gauge $\Theta = 0$ at the equatorial plane $\theta = 0$ goes to zero near the separatrix (the corresponding angle in this gauge is notated as $\Theta(\psi)$). The relation between the gauges ϑ and Θ is given by Eq. (1.36) where the function $f(\psi)$ is determined by the safety factor $q(\psi)$ and the distance $\Delta\varphi_s(\psi)$ between the sections $\vartheta = 0$ and $\Theta = 0$ along the toroidal angle φ on the magnetic surface ψ:

$$\Theta = \vartheta + f(\psi), \quad f(\psi) = \Delta\varphi_s/q(\psi). \tag{3.63}$$

The product of $\Theta_j(\psi)$ to the safety factor $q(\psi)$ has a finite value at the limit $\psi_N \to 1$, i.e. at the separatrix. The radial profiles of this product is shown in Fig. 3.16 a, b for the RMPs in the ITER like and the NSTX plasmas, respectively. Similar to the approximations of the functions $C_j(\psi_N)$, $A_j(\psi_N)$ by the series of type (3.61 and 3.62) the function $\Theta_j(\psi)$ can be also presented as

$$\Theta_j(\psi_N) = \sum_{s=0}^{M} \theta_s \psi^s, \quad \psi = 1 - \psi_N, \tag{3.64}$$

with the constant coefficients θ_s, $s = 0, 1, \ldots, M$.

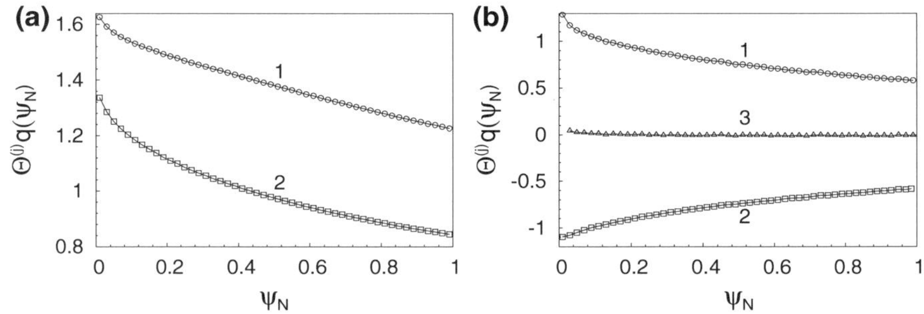

Fig. 3.16 Radial profiles of the product $\Theta_j(\psi)q(\psi)$ for the ITER plasma (**a**) and for the NSTX plasma (**b**). *Curve 1* corresponds to the upper horizontal segment, *curve 2* corresponds to the lower horizontal segment, *curve 3* corresponds to the vertical segments of the saddle coils

3.4.6 Poloidal Fluxes

The analytical formulas for $|H_{mn}^{(j)}(\psi)|$, $(j = 1, 2, 3)$ given by Eqs. (3.54) and (3.58) allows one to obtain a compact analytical formulas for the perturbation fluxes $\psi_n^{(j)}(R, Z), (j = 1, 2, 3)$ (3.47).

First consider the poloidal fluxes corresponding to the horizontal segments $j = 1, 2$. Using the spectrum H_{mn} (3.54) the poloidal flux (3.47) can be reduced to

$$
\begin{aligned}
\psi_n^{(j)}(\psi, \vartheta) &= H_{0n}^{(j)}(\psi) + 2\sum_{m=1}^{\infty}\left|H_{mn}^{(j)}(\psi)\right| \cos\left(m[\vartheta - \vartheta_0^{(j)}] + \chi^{(j)}\right) \\
&= \frac{A_j(\psi)}{2q(\psi)}\mathrm{Re}\left\{e^{i\chi_j}\left(-1 + 2\sum_{m=0}^{\infty} r^m\right)\right\} = \frac{A_j(\psi)}{q(\psi)}\mathrm{Re}\left\{e^{i\chi_j}\frac{1+r}{1-r}\right\} \\
&= \frac{A_j(\psi)}{2q(\psi)\Lambda_j}\mathrm{Re}\left\{e^{i\chi_j}\left(1 + 2i\sin\left(\vartheta - \vartheta_0^{(j)}\right) - e^{-2b_j}\right)\right\}, \qquad (3.65)
\end{aligned}
$$

where the following notations are used

$$
\begin{aligned}
r &= \exp\left[i(\vartheta - \vartheta_0^{(j)}) - b_j\right], \\
\Lambda_j &= 1 + e^{-2b_j} - 2e^{-b_j}\cos\left(\vartheta - \vartheta_0^{(j)}\right), \qquad b_j = \frac{C_j(\psi)}{q(\psi)}. \qquad (3.66)
\end{aligned}
$$

Recalling that $\chi_j = 0$ for $j = 1$ and $\chi_j = \pi$ for $j = 2$ the poloidal flux $\psi_n(\psi, \vartheta)$ (3.65) for each segments $j = 1, 2$ can be presented as

$$
\psi_n^{(j)}(\psi, \vartheta) = \pm\frac{A_j}{2q(\psi)\Lambda_j}\left(1 - e^{-2b_j}\right), \qquad (j = 1, 2). \qquad (3.67)
$$

The perturbation flux $\psi_n^{(3)}(\psi, \vartheta)$ (3.50) corresponding to the vertical segments of saddle coils can be calculated similar to (3.65). Putting Eq. (3.58) in Eq. (3.50) we have

$$
\begin{aligned}
\psi_n^{(3)}(\psi, \vartheta) &= 2\sum_{m=1}^{\infty}\left|H_{mn}^{(3)}(\psi)\right| \cos\left(m[\vartheta - \vartheta_0^{(3)}] + \pi/2\right) \\
&= nA_3(\psi)\mathrm{Re}\left\{e^{i\pi/2}\sum_{m=1}^{\infty} e^{-m\alpha_3}\frac{\cos(mx) + i\sin(mx)}{m}\right\}, \qquad (3.68)
\end{aligned}
$$

where $x = \vartheta - \vartheta_0^{(3)}$. Using the sum (see Prudnikov et al. (1986)),

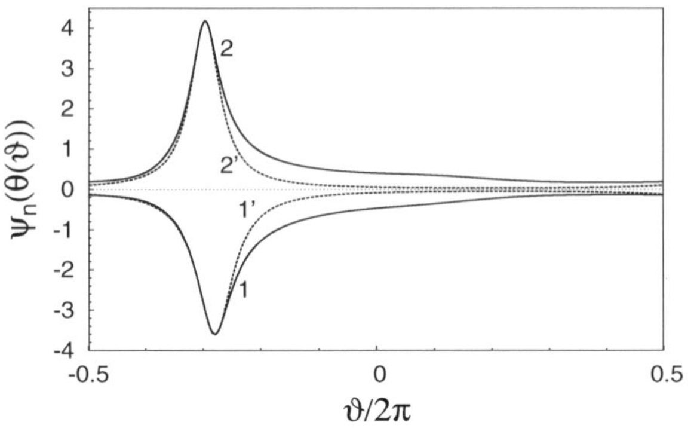

Fig. 3.17 Comparison of angular dependencies of the poloidal fluxes $\epsilon\psi_j(\vartheta)$ created by the horizontal segments $j = 1, 2$ obtained by the numerical calculations (*curves 1* and *2*) and the analytical formula (*curves 1′, 2′*) given (3.67). *Curves 1* and *1′* corresponds to the upper horizontal segment 1, *curve 2* corresponds to the lower horizontal segment 2 of saddle coils. The magnetic surface $\psi_N = 0.95$

$$\sum_{m=1}^{\infty} m^{-1} e^{-m\alpha} \sin(mx) = -\arctan\left(\frac{\sin x}{\cos x - e^{\alpha}}\right),$$

we obtain

$$\psi_n^{(3)}(\psi, \vartheta) = -nA_3 \arctan\left(\frac{\sin(\vartheta - \vartheta_3)}{\cos(\vartheta - \vartheta_3) - e^{\alpha_3}}\right), \tag{3.69}$$

where Λ_3, α_3 are defined similar to ones in Eq. (3.66).

The comparison of the analytical formula (3.67) for the perturbation poloidal fluxes $\psi^{(j)}(\psi, \vartheta)$ $(j = 1, 2)$ with the numerically calculated ones are shown in Fig. 3.17 where the angular dependencies of $\psi_j(\psi, \vartheta)$ on the given magnetic surface $\psi_N = 0.95$ are plotted. It shows that the simple analytical formulae sufficiently well describe the perturbation poloidal fluxes, particularly, the locations of extremal points.

3.5 Summary

In this chapter we have studied the generic features of poloidal spectra of magnetic perturbations in typical tokamaks. The two types of magnetic perturbations created by a set of helical coils and a set of saddle coils are considered. It was found that the poloidal spectra of these magnetic perturbations has a generic behavior independent on the specific parameters of plasmas. It is mainly due to the toroidicity of tokamak plasmas.

Particularly, as was shown in Sect. (3.2) the poloidal spectra of magnetic perturbations generated by a set of helical coils are mainly determined by a few parameters. Namely, the spectrum transformation matrix (3.20) asymptotically depends only on the values of the derivatives (3.24) and (3.26) at the HFS ($\theta = \pi$) and the LFS ($\theta = 0$), respectively.

The asymptotical formulas (3.54) and (3.58) for magnetic perturbations created by a set of saddle coils are also determined by a few parameters. Their accuracy improves with increasing the inverse aspect ratio $\varepsilon = a/R$ of tokamaks. Particularly, in spherical tokamaks the poloidal mode spectra of RMPs is described by these formulas with a higher accuracy than in low aspect ratio tokamaks.

The universal features of poloidal spectra of magnetic perturbations allows one to construct a realistic model of the RMPs in tokamaks. These models would be useful to study the onset of chaos of magnetic field lines and their properties. In Chap. 9 we will construct such models of magnetic perturbations in the ergodic divertors and poloidal divertor tokamaks.

The universal asymptotical presentation of the RMPs spectra may be useful for the theoretical interpretations of the plasma response to external magnetic perturbations created by a set of saddle coils. Asymptotical formulas can be also employed in a technical design of saddle coils in modern tokamaks to obtain a desired spectrum of external magnetic perturbations. In Sect. 9.2 we consider the application of the asymptotical formulas to analyze some features of poloidal mode spectra in the even/odd parity coil configurations in the DIII-D like plasmas.

Chapter 4
Hamilton Equations for Guiding Center Motion of Particles

The description of charged particle motion in a magnetic field is more complicated than magnetic field lines. It is related with the fact that a charged particle orbit gyrates around the magnetic field while moving freely close its direction as illustrated in Fig. 4.1. The gyration period is much shorter than the characteristic period of particle motion around the torus. The existence of two different time scales makes a difficult to study the particle motion in an inhomogeneous magnetic field in long time intervals. To avoid this problem the equations of particle motion in an inhomogeneous electromagnetic magnetic field are reduced to ones for the guiding–center of gyrating orbits (see, e.g., Northrop 1963; Morozov and Solov'ev 1966a). The guiding–center approximations are based on the existence of an adiabatic invariant, a magnetic moment $\mu = m_0 v_\perp^2/2B$, where m_0 is a particle mass, $v_\perp$ is the velocity component perpendicular to magnetic field $\mathbf{B}$, and $B = |\mathbf{B}|$.

The Hamiltonian approach to study this problem been subject to many studies since early 1960s. The equations of guiding center motion in Hamiltonian from have several advantages in the study of particle motion in magnetic confinement devices. It allows one to use the methods of the Hamiltonian theory, like methods of canonical transformations, the Hamilton-Jacobi method, Liouville's invariants, perturbation theory and the symplectic maps to efficiently solve the problem.

Several Hamiltonian formulations of the guiding center motion have been proposed since the early 1980s by Littlejohn (1981, 1983), White et al. (1982), Boozer (1983), White and Chance (1984), White (1990), Meiss and Hazeltine (1990), Wobig and Pfirsch (2001), White and Zakharov (2003) (see also a book by White (2001)). These approaches use magnetic coordinates as canonical variables and therefore would require the existence of magnetic surfaces. More general approaches have been proposed by Hazeltine (1990), Wobig and Pfirsch (2001) in which one should find a coordinate system in which one of the covariant components of the magnetic vector $\mathbf{B}$ and the vector potential $\mathbf{A}$ vanish, e.g., $A_R = B_R = 0$.

In this chapter we present an alternative Hamiltonian formulation of guiding-center motion of charged particles in a toroidal system. It has been proposed by Abdullaev and Finken (2002), Abdullaev et al. (2006a), Wingen et al. (2006).

S. Abdullaev, *Magnetic Stochasticity in Magnetically Confined Fusion Plasmas*, Springer Series on Atomic, Optical, and Plasma Physics 78,
DOI: 10.1007/978-3-319-01890-4_4, © Springer International Publishing Switzerland 2014

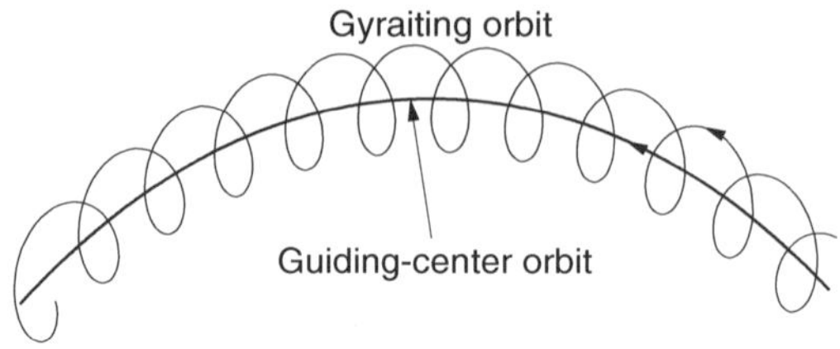

Fig. 4.1 Schematic presentation of a gyrating particle orbit and its guiding center orbit

The idea is similar to the one proposed in 1980s by Mynick (1979, 1980a, b) for the mirror and field–reversed magnetic configurations where the particle gyroradius is comparable to the system size.

The main idea of our formulation is the following. The Hamiltonian of the three-dimensional (3D) gyrating motion of a particle in magnetic and electric fields may be considered as a Hamiltonian, $H = P_R^2/2 + U$, for 1D motion along the radial coordinate R in the effective potential well $U = U(R; Z, P_z, \varphi, P_\varphi, T)$ which is a steep function of R and slowly–varying along the other coordinates (Z, ϕ), the momentum (P_z, P_φ) and time T. For a strong magnetic field the motion along the radial coordinate R is highly oscillatory near the minimum of $U(R)$ with the amplitude ρ_R and the frequency ω_R being of the order of the gyro-radius $\rho_c = ev_\perp B/mc$ and the gyro-frequency $\omega_c = eB/mc$, respectively. Introducing action–angle variables (I_R, ϑ_R) to describe such a motion by means of the canonical transformations of variables $(R, Z, \varphi, P_R, P_z, P_\varphi)$ to the new variables $(\vartheta_R, Z_c, \varphi_c, I_R, P_{zc}, P_{\varphi c})$ allows one to derive the system of Hamiltonian equations for the action–angle variables (ϑ_R, I_R) and the slowly varying guiding–center variables $(Z_c, \varphi_c, P_{zc}, P_{\varphi c})$. One can show that in typical tokamak plasmas the action I_R is an adiabatic invariant, and therefore one consider it as a constant motion. Thus we obtain a 4D system of Hamiltonian equations for guiding–center motion.

Below we present a systematic derivation of Hamiltonian equations for guiding–center motion based on this idea. Keeping in mind the application to the energetic relativistic particles, for instance, runaway electrons, we derive the relativistic Hamiltonian equations for guiding-center motion. In Sect. 4.1 we write down the Hamiltonian equations of charged particles in a cylindrical coordinate system and analyze the radial gyro–oscillations. The introduction of the radial action–angle variables via a canonical change of variables and the corresponding Hamiltonian for guiding–center variables are presented in Sect. 4.2. The simplified version of the equations of guiding–center motion is derived in Sect. 4.3.

4.1 Hamiltonian Equations in a Cylindrical Coordinate System

Let us consider a particle with a mass m_0 and charge $q_a = Z_q e$, where e is the elementary charge, particularly, $Z_q = 1$ for the protons and $Z_q = -1$ for the electrons. The relativistic motion of a particle in an electromagnetic field is governed by the Hamiltonian system

$$\frac{d\mathbf{Q}}{dT} = \frac{\partial H}{\partial \mathbf{P}}, \quad \frac{d\mathbf{P}}{dT} = -\frac{\partial H}{\partial \mathbf{Q}}, \tag{4.1}$$

with the Hamiltonian function

$$H = \left[m_0^2 c^4 + c^2 \left(\mathbf{P} - \frac{q_a}{c} \mathbf{A} \right)^2 \right]^{1/2} + q_a \Phi, \tag{4.2}$$

where ($\mathbf{Q} = (X, Y, Z), \mathbf{P} = (P_x, P_y, P_z)$) are the canonical coordinates and momenta, T is a time, $\mathbf{A} = \mathbf{A}(\mathbf{Q}, T)$ and $\Phi = \Phi(\mathbf{Q}, T)$ are the vector and scalar potentials of the electromagnetic field, respectively. It is convenient to expand the six-dimensional phase-space ($\mathbf{Q}, \mathbf{P}$) of the Hamiltonian system to the eight-dimensional one (Q_i, P_i), including time T and the energy H as additional canonical variables, $Q_4 = T$, $P_4 \equiv P_t = -H$. Then the dynamical system can be presented as an autonomous Hamiltonian system,

$$\begin{aligned} \frac{dQ_i}{d\tau} &= \frac{\partial \mathcal{H}}{\partial P_i}, \quad \frac{dP_i}{d\tau} = -\frac{\partial \mathcal{H}}{\partial Q_i}, \quad (i = 1, 2, 3, 4), \\ \mathcal{H} &= \frac{1}{2m_0} \left[m_0^2 c^2 + \left(\mathbf{P} - \frac{q_a}{c} \mathbf{A} \right)^2 - \frac{1}{c^2} \left(P_t + q_a \Phi \right)^2 \right], \end{aligned} \tag{4.3}$$

where τ is the time–like independent variables, coinciding with the relativistic time within the co-moving frame of reference.

In the cylindrical coordinate system $\mathbf{Q} = (R, \varphi, Z)$ the Hamiltonian $\mathcal{H}$ can be reduced to the following form

$$\mathcal{H} = \frac{1}{2} m_0 c^2 + \frac{1}{2m_0} P_R^2 + U = 0, \tag{4.4}$$

where we have introduced the effective potential $U \equiv U(R, \varphi, Z, T, P_\varphi, P_Z, P_t)$,

$$U = \frac{1}{2m_0} \left[\frac{1}{R^2} \left(P_\varphi - \frac{q_a}{c} R A_\varphi \right)^2 + \left(P_Z - \frac{q_a}{c} A_Z \right)^2 - c^{-2} \left(P_t + q_a \Phi \right)^2 \right]. \tag{4.5}$$

In (4.5) we assumed that the radial component of the vector potential A_R is zero, i.e., $A_R = 0$, due to the gauge invariance of an electromagnetic field.

4.1.1 Radial Gyro–Oscillations

The system (4.3) with the Hamiltonian $\mathcal{H}$ (4.4) effectively describes the one-dimensional motion along the radial coordinate R in the potential field

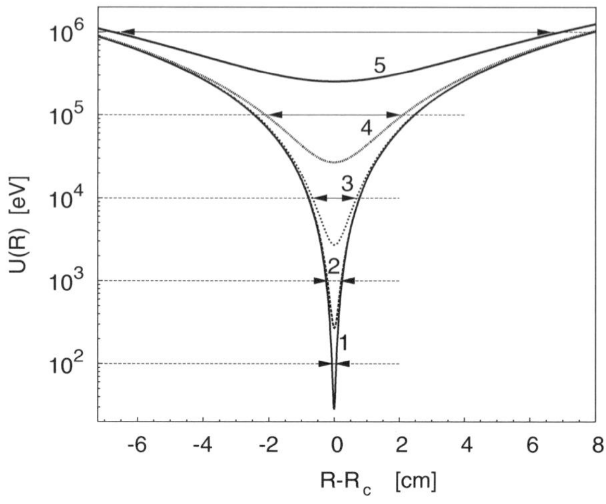

Fig. 4.2 Radial dependencies of the effective potential $U_{eff}(R)$ for different values of the hydrogen ion (H^+) energy E. The tokamak magnetic field is described by the standard equilibrium magnetic field (2.2.1). *Curve 1* corresponds to the energy $E = 100\,\text{eV}$, *2*—$E = 1\,\text{keV}$, *3*—$E = 10\,\text{keV}$, *4*—$E = 100\,\text{keV}$, and *5*—$E = 1\,\text{MeV}$. The corresponding *horizontal arrows* show the radial extend of particle oscillations. The parameters of the equilibrium plasma are $B_0 = 2.25\,\text{T}$, $I_p = 0.4\,\text{MA}$, $R_0 = 175\,\text{cm}$, $a = 46\,\text{cm}$

$U_{eff}(R) = U(R, \varphi, Z, T, P_\varphi, P_Z, P_t)$. In the strong toroidal magnetic field B_t the effective potential U has a minimum value at $R = R_c$ where $\partial U(R_c)/\partial R = 0$ and sharply grows with increasing distance $|R - R_c|$. The typical shape of the potential U along the radial coordinate $r = R - R_c$ is shown in Fig. 4.2 for the different values of particle energy (the hydrogen ion H^+) and for the typical tokamak parameters: $B_0 = 2.25\,\text{T}$, $R_0 = 175\,\text{cm}$, $a = 46\,\text{cm}$.

Particles are trapped inside the potential–well and oscillate along the radial axis about their minimum R_c. The latter and the oscillation frequency ω_R can be determined by expanding the potential function $U_{eff}(R)$ in powers of $R - R_c$.

Introducing the relativistic factor

$$\gamma_t = \frac{-P_t - q_a \Phi}{m_0 c^2}, \tag{4.6}$$

and the toroidal and the z—components of the particle's velocity,

$$v_\varphi = \frac{P_\varphi - (q_a/c) R A_\varphi}{m_0 R \gamma_t}, \qquad v_Z = \frac{P_Z - (q_a/c) A_Z}{m_0 \gamma_t},$$

one can find the equation for R_c, $\partial U(R)/\partial R = 0$,

$$\frac{\partial U}{\partial R} = -\frac{m_0 v_\varphi^2 \gamma_t^2}{R} - \frac{q_a v_\varphi}{cR}\frac{\partial (RA_\varphi)}{\partial R} - \frac{q_a v_z}{c}\frac{\partial A_z}{\partial R} - q_a \gamma_t \frac{\partial \Phi}{\partial R}$$
$$= -\frac{m_0 v_\varphi^2 \gamma_t^2}{R} - \frac{q_a \gamma_t}{c}\left(v_\varphi B_Z - v_z B_\varphi\right) + \gamma_t E_R = 0, \tag{4.7}$$

where B_φ, B_Z are the toroidal and z-components of the magnetic field (1.18), $E_R = -\partial \Phi / \partial R$ is the radial component of the electric field. Equation (4.7) describes the balance of the centripetal force $F_c = m v_\varphi^2 \gamma_t / R$ and the radial component of the Lorentz force $\mathbf{F} = q_a \mathbf{E} + (q_a/c)\mathbf{v} \times \mathbf{B}$,

$$\frac{m_0 v_\varphi^2 \gamma_t}{R} = q_a E_R + \frac{q_a}{c}\left(v_z B_\varphi - v_\varphi B_Z\right). \tag{4.8}$$

Keeping only nonzero quadratic terms in the expansion the potential U in a series of powers $(R - R_c)$ near its minimum at R_c,

$$U = U_0 + \frac{1}{2}\frac{\partial^2 U(R_c)}{\partial R^2}(R - R_c)^2 + O\left((R - R_c)^3\right), \quad U_0 = U(R_c), \tag{4.9}$$

the Hamiltonian $\mathcal{H}$ (4.4) is reduced to the one of the harmonic oscillator,

$$\mathcal{H} \approx \frac{1}{2} m_0 c^2 + U_0 + \frac{1}{2m_0}\left[P_R^2 + m_0^2 \omega_R^2 (R - R_c)^2\right], \tag{4.10}$$

where ω_R is the frequency of small amplitude radial oscillations,

$$\omega_R = \left(\frac{1}{m_0}\frac{\partial^2 U}{\partial R^2}\bigg|_{R=R_c}\right)^{1/2}. \tag{4.11}$$

It is not difficult to obtain the expression for ω_R,

$$\omega_R = Z_q \omega_0 (1 + f)^{1/2}, \quad \omega_0 = \omega_c \frac{B}{B_0}, \quad B = \left(B_\varphi^2 + B_Z^2\right)^{1/2}, \tag{4.12}$$

where $\omega_c = eB_0/m_0 c$ in the reference gyrofrequency in the magnetic field B_0, and the coefficient f is given by

$$f = \frac{3 v_\varphi^2 \gamma_t^2}{\omega_c^2 R^2} + \frac{3 v_\varphi \gamma_t B_z}{\omega_c R B} + \frac{\gamma_t v_\varphi}{\omega_c B}\left(\frac{\partial B_\varphi}{\partial R} - \frac{\partial B_z}{\partial R}\right)$$
$$+ \frac{q_a \gamma_t}{m_0 \omega_c^2}\left(\frac{\partial E_R}{\partial R} + \frac{1}{m_0 c^2 \gamma_t} E_R^2\right). \tag{4.13}$$

For the typical values of tokamak magnetic field $B_0 \sim 1$ T, the coefficient $f \ll 1$, and the radial gyrofrequency ω_R is of order of ω_0.

4.2 Hamiltonian Guiding–Center Equations

The minimum R_c of the potential U along the radial coordinate R slowly changes in time T and space Z, φ. It coincides with the radial coordinate of the guiding center orbit. In order to obtain the equation for the guiding center orbit we use the following observations.

The Hamiltonian system (4.10) is similar to a one-dimensional oscillator with a slowly varying frequency of oscillations ω_R and the equilibrium point R_c in space and time. For small amplitude of the radial oscillations, where one can neglect the expansion terms higher than $(R - R_c)^2$, the problem is similar to the well-known one of the harmonic oscillator under slow changes in the frequency (see, e.g., Kulsrud 1957; von Hertweck and Schlüter 1957; Meyer 1973). If the gyration period $2\pi/\omega_R$ is much smaller than the time scale t_B of a change of ω_R and R_c and the transition time of the system L/v_T, where L is a characteristic spatial scale of system, and v_T is a particle's velocity, then the action variable I_R defined as $I_R = [P_R^2/\omega_R + \omega_R(R - R_c)^2]/2$ is the adiabatic invariant. The existence of the adiabatic invariant significantly simplifies the study of particle motion in an inhomogeneous magnetic field.

Consider a mathematically rigorous approach of the introduction of slowly varying guiding center variables using the adiabatic invariant I_R. Let ϑ_R be the canonical variable (or a gyrophase of radial oscillations) conjugated to I_R. The pair of variables (ϑ_R, I_R) describes the gyrating components of the particle orbit. Let $(\mathbf{Q}_c, \mathbf{P}_c) \equiv (Z_c, \varphi_c, T_c, P_{cZ}, P_{c\varphi}, P_{ct})$ be the phase–space coordinates of the guiding-center orbit. The Hamiltonian equations for these variables can be derived by the transformation of the gyrating particle coordinates $(\mathbf{Q}, \mathbf{P}) = (Z, \varphi, t, P_z, P_\varphi, P_t)$ to the guiding–center coordinates $(\mathbf{Q}_c, \mathbf{P}_c)$. Such a transformation is a canonical if it is determined by the generating function $F = F(R, \mathbf{Q}, I_R, \mathbf{P}_c)$ of mixed variables, old coordinates $\mathbf{Q}$ and new momenta $\mathbf{P}_c$). Then the old momenta $\mathbf{P}$ and the new coordinates $\mathbf{Q}_c$ are determined by

$$P_R = \frac{\partial F(R, \mathbf{Q}, I_R \mathbf{P}_c)}{\partial R} \quad \vartheta_R = \frac{\partial F(R, \mathbf{Q}, I_R \mathbf{P}_c)}{\partial I_R},$$
$$\mathbf{P} = \frac{\partial F(R, \mathbf{Q}, I_R \mathbf{P}_c)}{\partial \mathbf{Q}}, \quad \mathbf{Q}_c = \frac{\partial F(R, \mathbf{Q}, I_R, \mathbf{P}_c)}{\partial \mathbf{P}_c}. \tag{4.14}$$

The new Hamiltonian function $\mathcal{H}_c$ in the guiding–center variables $\mathbf{Q}_c, \mathbf{P}_c$ is given by

$$\mathcal{H}_c(\vartheta_R, I_R, \mathbf{Q}_c, \mathbf{P}_c) = \mathcal{H} + \frac{\partial F(R, \mathbf{Q}, I_R \mathbf{P}_c)}{\partial \tau} = \mathcal{H}(R, P_z, \mathbf{Q}, \mathbf{P}), \tag{4.15}$$

since the generating function F does not explicitly depend on the time-like variable τ. In Eq. (4.15) the arguments $(R, P_z, \mathbf{Q}, \mathbf{P})$ of the Hamiltonian $\mathcal{H}$ are assumed as functions of the new variables $(\vartheta_R, I_R, \mathbf{Q}_c, \mathbf{P}_c)$.

The generating function F is sought in the form

$$F = \mathbf{Q} \cdot \mathbf{P}_c + S\left(R, I_R, \varepsilon\mathbf{Q}, \varepsilon\mathbf{P}_c\right), \tag{4.16}$$

where ε is a small parameter given by the ratio ρ_R/L, of the radial gyro–radius to the characteristic scale L. In (4.16) the function S is a slowly varying function of the variables $(\mathbf{Q}, \mathbf{P}_c)$. Using Eqs. (4.14), (4.10) and (4.16) one finds the function S

$$\begin{aligned} S_R\left(R, I_R, \varepsilon\mathbf{Q}, \varepsilon\mathbf{P}_c\right) &= \int^R P_R\left(R', I_R, \mathbf{Q}, \mathbf{P}_c\right) dR', \\ P_R\left(R', I_R, \mathbf{Q}, \mathbf{P}_c\right) &= \sqrt{2m_0}\left[\bar{\mathcal{H}} - m_0\omega_R^2(R - R_c)^2/2\right]^{1/2}, \\ \bar{\mathcal{H}} &= \mathcal{H} - m_0c^2/2 - U(R_c), \end{aligned} \tag{4.17}$$

where the action variable I_R is defined by the integral

$$I_R = \frac{1}{2\pi}\oint_C P_R\left(R', I_R, \mathbf{Q}, \mathbf{P}_c\right) dR', \tag{4.18}$$

taken over along the closed contour C of constant $\bar{\mathcal{H}}$ in the $(R,\ P_R)$ plane. Performing the integrations in (4.17) and (4.18) one obtains

$$\begin{aligned} \bar{\mathcal{H}}_c &= \omega_R I_R, \\ S_R\left(R, I_R, \varepsilon\mathbf{Q}, \varepsilon\mathbf{P}_c\right) &= I_R\left(\arcsin y + y\sqrt{1 - y^2}\right), \\ y = \frac{R - R_c}{\rho_R}, &\quad \rho_R = \sqrt{\frac{2I_R}{m_0\omega_R}}. \end{aligned} \tag{4.19}$$

The quantity ρ_R is the amplitude of gyro–oscillations along the radial coordinate R, and can be called *a radial gyroradius*. The frequency of radial oscillations ω_R, the radial coordinate of guiding center R_c, are slowly varying functions of coordinates $(\mathbf{Q}, \mathbf{P}_c)$.

Using (4.14), (4.16) and (4.19) one can obtain the relations between the gyrating particle's coordinates $(R_R, P_R, \mathbf{Q}_c, \mathbf{P}_c)$ and the guiding–center coordinates $(\vartheta_R, I_R, \mathbf{Q}_c, \mathbf{P}_c)$,

$$\begin{aligned} R &= R_c + \rho_R \sin\vartheta_R, \\ P_R &= \sqrt{2I_R\omega_R m_0}\cos\vartheta_R, \\ \mathbf{Q} &= \mathbf{Q}_c - \frac{\partial S}{\partial \mathbf{P}_c} = \mathbf{Q}_c - 2I_R\cos\vartheta_R\frac{\partial y}{\partial \mathbf{P}_c}, \\ \mathbf{P} &= \mathbf{P}_c + \frac{\partial S}{\partial \mathbf{Q}} = \mathbf{P}_c + 2I_R\cos\vartheta_R\frac{\partial y}{\partial \mathbf{Q}}, \end{aligned} \tag{4.20}$$

where

$$\varepsilon\frac{\partial y}{\partial \xi} = \frac{1}{2}\sin\vartheta_R \frac{\partial \ln \omega_R}{\partial \xi} - \frac{1}{\sqrt{I_R/m_0\omega_R}}\frac{\partial R_c}{\partial \xi}, \qquad \xi = (\mathbf{Q}, \mathbf{P}_c). \tag{4.21}$$

The slow–varying guiding center variables $(\mathbf{Q}, \mathbf{P})$ and (ϑ_R, I_R) are governed by the Hamiltonian equations

$$\begin{aligned} \frac{d\vartheta_R}{d\tau} &= \frac{\partial \mathcal{H}_c}{\partial I_R}, & \frac{dI_R}{d\tau} &= -\frac{\partial \mathcal{H}_c}{\partial \vartheta_R}, \\ \frac{d\mathbf{Q}_c}{d\tau} &= \frac{\partial \mathcal{H}_c}{\partial \mathbf{P}_c}, & \frac{d\mathbf{P}_c}{d\tau} &= -\frac{\partial \mathcal{H}_c}{\partial \mathbf{Q}_c}. \end{aligned} \tag{4.22}$$

with the Hamiltonian function

$$\begin{aligned} &\mathcal{H}_c \equiv \mathcal{H}_c(\vartheta_R, I_R, \mathbf{Q}_c, \mathbf{P}_c) = m_0 c^2/2 + U_0 + \omega_R I_R = 0, \\ &U_0 \equiv U_0\left(\mathbf{Q}_c - \varepsilon\frac{\partial S}{\partial \varepsilon \mathbf{P}_c}, \mathbf{P}_c + \varepsilon\frac{\partial S}{\partial \varepsilon \mathbf{Q}}\right). \end{aligned} \tag{4.23}$$

The Hamiltonian function $\mathcal{H}_c$ is exact and no approximations were made in its derivation, except the expansion of U (4.9). It depends on the gyro–phase ϑ_R through the derivatives $\partial S/\partial \mathbf{Q}$, $\partial S/\partial \mathbf{P}_c$ given by (4.20).

4.2.1 6D System of Guiding–Center Motion Equations

The autonomous system (4.22) is actually equivalent to the non-autonomous system in the 6D phase space $(\vartheta_R, Z, \varphi, I_R, P_Z, P_\varphi)$. Considering time t as independent variable instead τ and $P_t = H_c$ as the Hamiltonian instead of $\mathcal{H}_c$ we have

$$\begin{aligned} \frac{d\vartheta_R}{dt} &= \frac{\partial \tilde{H}_c}{\partial I_R}, & \frac{dI_R}{dt} &= -\frac{\partial \tilde{H}_c}{\partial \vartheta_R}, \\ \frac{dZ}{dt} &= \frac{\partial \tilde{H}_c}{\partial P_z}, & \frac{dP_z}{dt} &= -\frac{\partial \tilde{H}_c}{\partial Z}, \\ \frac{d\varphi}{dt} &= \frac{\partial \tilde{H}_c}{\partial P_\varphi}, & \frac{dP_\varphi}{dt} &= -\frac{\partial \tilde{H}_c}{\partial \varphi}. \end{aligned} \tag{4.24}$$

where we have omitted the subscript "c" from the variables $Z, \varphi, P_z, P_\varphi$ of the guiding–center orbit. Using (4.23) and (4.5) one finds H_c

$$\tilde{H}_c = \tilde{H}_c(\vartheta_R, Z, \varphi, I_R, P_Z, P_\varphi, t)$$
$$= H_c\Bigg(Z - \varepsilon\frac{\partial S}{\partial \varepsilon P_z}, \varphi - \varepsilon\frac{\partial S}{\partial \varepsilon P_\varphi},$$
$$I_R, P_Z + \varepsilon\frac{\partial S}{\partial \varepsilon Z}, P_\varphi + \varepsilon\frac{\partial S}{\partial \varepsilon\varphi}\Bigg) + \varepsilon\frac{\partial S}{\partial \varepsilon t}, \tag{4.25}$$

where

$$H_c\left(Z, \varphi, I_R, P_Z, P_\varphi\right) = \Bigg[m_0 c^2 + 2\omega_R I_R + c^2\left(P_Z - \frac{q_a}{c}A_z\right)^2$$
$$+ \frac{c^2}{R_c^2}\left(P_\varphi - \frac{q_a}{c}A_\varphi\right)^2\Bigg]^{1/2} + q_a\Phi(R_c, Z, \varphi, t). \tag{4.26}$$

The gyro-frequency ω_R defined by Eq. (4.11), as well as the vector potential A_z, A_φ, and the radial coordinate R_c are considered as functions of variables $Z, \varphi, I_R, P_Z, P_\varphi, t$, i.e., $\omega_R = \omega_R\left(Z, \varphi, I_R, P_Z, P_\varphi, t\right), A_i = A_i\left(Z, \varphi, I_R, P_Z, P_\varphi, t\right)$, $(i = z, \varphi)$, and $R_c = R_c\left(Z, \varphi, I_R, P_Z, P_\varphi, t\right)$.

4.2.2 Adiabatic Invariant and 4D System of Guiding–Center Motion Equations

The action variable I_R (4.18) is introduced in the Hamiltonian system (4.10) for the harmonic oscillator with slowly changing variables. Using the methods of KAM theory one can show that I_R is a perpetually conserved adiabatic invariant similar to the magnetic moment $\mu = mv_\perp^2/2B$, (see Sect. 7.1.3).[1] The numerical calculations done by Abdullaev and Finken (2002) also show that I_R is conserved during the slow drift motion of a particle with sufficiently high accuracy. The existence of the adiabatic invariant I_R allows one to simplify the equations of motion by reducing the 8D autonomous system (4.22) to the 6D autonomous system of Hamiltonian equations.

Using the smallness of the ratio ε one can expand the Hamiltonian (4.23) in powers of ε. Neglecting the small terms of order of ε^2 and higher, we can present the Hamiltonian in the form

$$\mathcal{H}_c(\vartheta_R, I_R, \mathbf{Q}_c, \mathbf{P}_c) = \mathcal{H}_{c0}(I_R, \mathbf{Q}_c, \mathbf{P}_c) + \varepsilon\mathcal{H}_{c1}(\vartheta_R, I_R, \mathbf{Q}_c, \mathbf{P}_c), \tag{4.27}$$

where the first term $\mathcal{H}_{c0}$ on the right hand side describes the part of the Hamiltonian which does not depend on the gyro-phase ϑ_R:

---

[1] A more detailed discussion of this problem is given by Arnold (1963b) and in the book by Arnold et al. (2006) (Sect. 6.4, page 314).

$$\mathcal{H}_{c0}(\vartheta_R, I_R, \mathbf{Q}_c, \mathbf{P}_c) = \omega_R I_R + U_0(\mathbf{Q}_c, \mathbf{P}_c) = 0. \tag{4.28}$$

To eliminate the gyro–phase ϑ_R from the equations of motion one should neglect the Hamiltonian $\mathcal{H}_{c1}$ depending on ϑ_R. Then the 8D-system of the guiding center Eq. (4.22) is reduced to the 6D autonomous system for its guiding center variables $(\mathbf{Q}_c, \mathbf{P}_c)$,

$$\frac{d\mathbf{Q}_c}{d\tau} = \frac{\partial \mathcal{H}_{c0}}{\partial \mathbf{P}_c}, \qquad \frac{d\mathbf{P}_c}{d\tau} = -\frac{\partial \mathcal{H}_{c0}}{\partial \mathbf{Q}_c}. \tag{4.29}$$

The gyrophase ϑ_R is given by a linear function of time τ,

$$\vartheta_R = \int \omega_R d\tau + \vartheta_{R0}. \tag{4.30}$$

The system (4.29) can be also presented as the 4D non-autonomous system of Hamiltonian equations,

$$\begin{aligned} \frac{dZ}{dt} &= \frac{\partial H_c}{\partial P_z}, & \frac{dP_z}{dt} &= -\frac{\partial H_c}{\partial Z}, \\ \frac{d\varphi}{dt} &= \frac{\partial H_c}{\partial P_\varphi}, & \frac{dP_\varphi}{dt} &= -\frac{\partial H_c}{\partial \varphi}, \end{aligned} \tag{4.31}$$

with the Hamiltonian function

$$\begin{aligned} H_c\left(Z, \varphi, I_R, P_Z, P_\varphi, t\right) = \Big[m_0 c^2 + 2\omega_R I_R + c^2 \left(P_Z - \frac{q_a}{c} A_z\right)^2 \\ + \frac{c^2}{R_c^2}\left(P_\varphi - \frac{q_a}{c} A_\varphi\right)^2 \Big]^{1/2} + q_a \Phi(R_c, Z, \varphi). \end{aligned} \tag{4.32}$$

4.3 Simplified Guiding Center Equations

Furthermore we rewrite the equations of motion in normalized variables. We introduce a reference time $T_{ref} = 1/\omega_c$, reference energy $E_{ref} = m_0 \omega_c^2 R_0^2$, a reference velocity $v_{ref} = \omega_c R_0$, and a reference momentum $p_{ref} = m_0 \omega_c R_0$.

Using the normalized quantities (1.14) and normalized time $t = t/T_{ref} = \omega_c T$, and we introduce new ones

$$\begin{aligned} b_z &= \frac{B_z}{B_0}, \quad b_\varphi = \frac{B_\varphi}{B_0}, \\ p_\varphi &= \frac{P_{c\varphi}}{p_{ref} R_0}, \quad p_z = \frac{P_{cz}}{p_{ref}}, \\ u_\varphi &= \frac{p_\varphi + Z_q \psi}{x}, \quad u_z = p_z - Z_q \psi_z, \end{aligned}$$

$$\varepsilon_0 = \frac{m_0 c^2}{E_{ref}} = \left(\frac{c}{\omega_c R_0}\right)^2, \quad \phi = \frac{e\Phi}{E_{ref}}. \tag{4.33}$$

The quantity ε_0 is the normalized energy of a particle at rest, and ϕ is the normalized electric field potential.

The equations of motion may significantly be simplified using the smallness of the poloidal magnetic field B_Z in comparison with the toroidal field B_φ, and the velocity of particles v_φ, $v_z \ll v_{ref}$.

4.3.1 Radial Guiding–Center Coordinate R_c

In the normalized variables (4.33) the equation for the radial coordinate $R_c = R_0 x_c$ of the guiding center (4.8) can be written as

$$\frac{u_\varphi^2}{x Z_q} = u_\varphi b_z - u_z b_\varphi + \gamma_t E_x, \tag{4.34}$$

or

$$p_z - Z_q \psi_z = \frac{1}{b_\varphi}\left[u_\varphi b_z + \gamma_t E_x - u_\varphi^2 / \left(x Z_q\right)\right], \tag{4.35}$$

where $E_x = -\partial\phi/\partial x$ is the normalized radial electric field.

For the typical tokamak parameters ($R_0 \approx 1-6\,\text{m}$, $B_0 = 2-6\,\text{T}$) the normalized ion and electron velocities v_φ are much smaller than $\omega_c R_0$: $u_\varphi \ll 1$. Even for the relativistic electron $v_\varphi \lesssim c$, $u_\varphi \sim 10^{-2}$. On the other hand the normalized poloidal field $b_z = B_Z/B_0$ at the plasma edge is of order of 10^{-1}. Therefore, in the zero approximation one can neglect the terms proportional to $\mu \simeq u_\varphi^2 \sim |b_z u_\varphi| \ll 1$ in the r.h.s. of the Eq. (4.35) for the guiding center coordinate x_c. In typical tokamaks the term in (4.35) containing the electric field E_x is also small. Let $x_c^{(0)}$ be the solution of the equation

$$p_z - Z_q \psi_z = 0. \tag{4.36}$$

Then according to (4.35) the radial guiding center coordinate x_c and the momentum p_z up to the first order of μ are given by

$$\begin{aligned} x_c &= x_c^{(0)} + \Delta x_c + \mathcal{O}\left(\mu^2\right), \\ p_z &= Z_q \psi_z(x_c^{(0)}) + \Delta p_z + \mathcal{O}\left(\mu^2\right), \end{aligned} \tag{4.37}$$

where Δx_c and Δp_z are the correction to x_c and p_z in the first order of μ,

$$\Delta p_z = \frac{1}{b_\varphi}\left(u_\varphi b_z + \gamma_t E_x - u_\varphi^2/\left(x Z_q\right)\right)\Big|_{x=x_c^{(0)}},$$
$$\Delta x_c = \Delta p_z / b_\varphi. \tag{4.38}$$

The radial gyro–frequency ω_R (4.12) can be reduced to

$$\omega_R = Z_q \frac{\omega_c}{x}\left(1 - x E_x - x^2 \partial E_x/\partial x\right)^{1/2}\Big|_{x=x_c^{(0)}} + \mathcal{O}\left(\mu^2\right). \tag{4.39}$$

Particularly, for the toroidal magnetic field $B_\varphi = B_0/R$, $A_z = B_0 R_0 \ln R/R_0$, and in the absence of the electric field $E_x \equiv 0$ the solution $x_c^{(0)}$ of the Eq. (4.36) and the corresponding frequency ω_R are given by

$$x_c^{(0)} = e^{-p_z/Z_q},$$
$$\omega_R = Z_q \frac{\omega_c}{x_c^{(0)}} = Z_q \omega_c e^{p_z/Z_q}. \tag{4.40}$$

The relation for ω_R in (4.40) can be also obtained from Eq. (4.12) by neglecting the poloidal field B_Z in comparison of B_φ. Then the relations (4.20) between the particle and guiding-center coordinates are simplified,

$$x = x_c + \sqrt{\frac{2I_x}{\omega_x}} \sin\vartheta_x, \quad p_x = \sqrt{2I_x\omega_x}\cos\vartheta_x,$$
$$z = z_c + \sqrt{\frac{2I_x}{\omega_x}} \cos\vartheta_x \quad p_z = p_{cz},$$
$$\varphi = \varphi_c, \quad p_\varphi = p_{c\varphi}, \quad t = t_c, \quad p_t = p_{ct}, \tag{4.41}$$

where $I_x = I_R/m_0\omega_c R_0^2$ and $\omega_x = \omega_R/\omega_c$ are the normalized radial action and gyro–frequency, respectively.

4.3.2 Equations of Motion

Furthermore, we will omit the subscript "c" in the guiding–center variables. Then normalized Hamiltonian $h = \mathcal{H}_0/m_0\omega_c^2 R_0^2$ (4.28) reads

$$h = \frac{\varepsilon_0}{2} + \omega_x(p_z) I_x + U + \mathcal{O}(\mu^2) = 0,$$
$$U = \frac{1}{2}\left[\frac{(p_\varphi + Z_q\psi)^2}{x_c^2} - \frac{(-p_t - Z_q\phi)^2}{\varepsilon_0}\right]. \tag{4.42}$$

The Hamiltonian equations of motion (4.29) with the canonical coordinates $\mathbf{q} = (z, \varphi, t)$ and the canonical momenta $\mathbf{p} = (p_z, p_\varphi, p_t)$ can be presented as,

$$\begin{aligned} \frac{dz}{d\tau} &= \frac{\partial h}{\partial p_z} = \frac{\partial \omega_x}{\partial p_z} I_x + \frac{\partial U}{\partial x_c}\frac{\partial x_c}{\partial p_z}, \\ \frac{dp_z}{d\tau} &= -\frac{\partial h}{\partial z} = -\frac{\partial U}{\partial z}, \\ \frac{d\varphi}{d\tau} &= \frac{\partial h}{\partial p_\varphi} = \frac{u_\varphi}{x_c}, \quad \frac{dp_\varphi}{d\tau} = -\frac{\partial h}{\partial \varphi} = -\frac{\partial U}{\partial \varphi}, \\ \frac{dt}{d\tau} &= \frac{\partial h}{\partial p_t} = \gamma_t, \quad \frac{dp_t}{d\tau} = -\frac{\partial h}{\partial t} = -\frac{\partial U}{\partial t}, \end{aligned} \tag{4.43}$$

where the relativistic factor $\gamma_t = (-p_t - Z_q\phi)/\varepsilon_0$, and u_φ is defined in Eq. (4.33). The radial guiding center x_c is assumed to be the function of p_z. The derivatives $\partial U/\partial x$ and so on are given by

$$\frac{\partial U}{\partial x} = -\frac{u_\varphi^2}{x} + Z_q \frac{u_\varphi}{x}\frac{\partial \psi}{\partial x} + Z_q\gamma_t \frac{\partial \phi}{\partial x}, \tag{4.44}$$

$$\frac{\partial U}{\partial q_i} = \frac{u_\varphi}{x}\frac{\partial \psi}{\partial q_i} + Z_q\gamma_t \frac{\partial \phi}{\partial q_i}, \tag{4.45}$$

$$(i = 1, 2, 3), \quad (q_1, q_2, q_3) = (z, \varphi, t).$$

The action variables I_x is considered as an adiabatic invariant, and from the equation for the qyrophase ϑ_x,

$$\frac{d\vartheta_x}{d\tau} = \frac{\partial h}{\partial I_x} = \omega_x(p_z), \tag{4.46}$$

we have

$$\vartheta_x = \int^{\tau} \omega_x\left(p_z(\tau')\right) d\tau' + \vartheta_{x0} = \int^{t} \frac{\omega_x\left(p_z(t')\right)}{\gamma_t(t')} dt' + \vartheta_{x0}, \tag{4.47}$$

where ϑ_{x0} is the initial gyrophase, $\vartheta_{x0} = \vartheta_x(0)$.

The equations of the guiding center motion can be also presented as the Hamiltonian system with the independent time variable t and the canonical variables $(z, \varphi, p_z, p_\varphi)$,

$$\begin{aligned} \frac{dz}{dt} &= \frac{\partial h_c}{\partial p_z} = \frac{1}{\gamma_t}\left[\frac{\partial \omega_x}{\partial p_z} I_x + \frac{\partial U}{\partial x_c}\frac{\partial x_c}{\partial p_z}\right], \quad \frac{dp_z}{dt} = -\frac{\partial h_c}{\partial z} = -\frac{1}{\gamma_t}\frac{\partial U}{\partial z}, \\ \frac{d\varphi}{dt} &= \frac{\partial h_c}{\partial p_\varphi} = \frac{u_\varphi}{x_c\gamma_t}, \quad \frac{dp_\varphi}{dt} = -\frac{\partial h_c}{\partial \varphi} = -\frac{1}{\gamma_t}\frac{\partial U}{\partial \varphi}, \end{aligned} \tag{4.48}$$

with the Hamiltonian function $h_c = h_c(z, \varphi, p_z, p_\varphi, t)$,

$$h_c = \varepsilon_0 \left[1 + \frac{2\omega_x I_x}{\varepsilon_0} + \frac{\left(p_\varphi + Z_q \psi\right)^2}{\varepsilon_0 x_c^2} \right]^{1/2} + Z_q \phi. \tag{4.49}$$

The relativistic factor γ_t is expressed through the canonical variables $(z, \varphi, p_z, p_\varphi)$,

$$\gamma_t^2 = 1 + \frac{2}{\varepsilon_0} \left(\omega_x(p_z) I_x + \frac{\left(p_\varphi + Z_q \psi\right)^2}{2x_c^2} \right). \tag{4.50}$$

The system of Eq. (4.48) should be supplemented by the initial conditions: by the values $(z_0, \varphi_0, p_{z0}, p_{\varphi 0})$ of the variables $(z, \varphi, p_z, p_\varphi)$ at the initial time $t = t_0$ and the action variable I_0.

For the low energy particles whose velocity v is much smaller than the speed of light c the relativistic factor γ_t is close to the unity, i.e., $\gamma_t - 1 \ll 1$. Then, in Eq. (4.48) one can put $\gamma_t = 1$ and the Hamiltonian (4.49) can be approximated by

$$h_c \approx \varepsilon_0 + \omega_x(p_z) I_x + \frac{\left(p_\varphi + Z_q \psi\right)^2}{2x_c^2} + Z_q \phi. \tag{4.51}$$

The quantity $\tilde{h} = h_c - \varepsilon_0$ corresponds to the non–relativistic particle energy.

4.3.3 Hamiltonian Equations for Gyrating Particles

For the comparison of the guiding–center orbits with the exact gyrating orbits we explicitly write down the Hamiltonian Eqs. (4.1), (4.2) for the particle orbits in the normalized variables introduced in Eq. (4.33). The Hamiltonian function (4.2) in the normalized variables takes the form

$$h = \left[\varepsilon_0^2 + \varepsilon_0 \left(p_x^2 + u_z^2 + u_\varphi^2 \right) \right]^{1/2} + Z_a \phi. \tag{4.52}$$

The corresponding equations of motion are given by

$$\begin{aligned}
\frac{dx}{dt} &= \frac{\partial h}{\partial p_x} = \frac{p_x}{\gamma_t}, \\
\frac{dp_x}{dt} &= -\frac{\partial h}{\partial x} = -\frac{1}{\gamma_t} \left[\frac{u_\varphi^2}{x} - \frac{Z_q u_\varphi}{x} \frac{\partial \psi}{\partial x} + Z_q u_z \frac{\partial \psi_z}{\partial x} \right] - Z_q \frac{\partial \phi}{\partial x}, \\
\frac{dz}{dt} &= \frac{\partial h}{\partial p_z} = \frac{u_z}{\gamma_t}, \\
\frac{dp_z}{dt} &= -\frac{\partial h}{\partial z} = \frac{Z_q}{\gamma_t} \left[-\frac{u_\varphi}{x} \frac{\partial \psi}{\partial z} + u_z \frac{\partial \psi_z}{\partial z} \right] - Z_q \frac{\partial \phi}{\partial z},
\end{aligned}$$

$$\frac{d\varphi}{dt} = \frac{\partial h}{\partial p_\varphi} = \frac{u_\varphi}{\gamma_t x},$$
$$\frac{dp_\varphi}{dt} = -\frac{\partial h}{\partial \varphi} = \frac{Z_q}{\gamma_t}\left[-\frac{u_\varphi}{x}\frac{\partial \psi}{\partial z} + u_z \frac{\partial \psi_z}{\partial \varphi}\right] - Z_q \frac{\partial \phi}{\partial \varphi}, \tag{4.53}$$

where the relativistic factor γ_t is given by

$$\gamma_t = \sqrt{1 + \varepsilon_0 \left(p_x^2 + u_z^2 + u_\varphi^2\right)}.$$

4.3.4 Hamiltonian Equations for Passing Particles

The particles which do not change their direction of motion along the toroidal angle φ are called *passing particles*. The toroidal component of their velocity u_φ does not change the sign. For such passing particles, the Hamiltonian drift equations can be reformulated by introducing the toroidal angle φ as the independent, time-like variable and the corresponding canonical momentum p_φ as a new Hamiltonian $K = -p_\varphi$. Then the variable (z, t, p_z, p_t) would be the canonically–conjugated coordinates and momenta. From Eq. (4.42) we find $K = K(z, t, p_z, p_t, \varphi)$,

$$K = \psi - \sigma x_c \left[\varepsilon_0(\gamma_t^2 - 1) - 2\omega_x I_x\right]^{1/2}. \tag{4.54}$$

Here we introduced the new parameter $\sigma = u_\varphi/|u_\varphi| = \pm 1$, which determines the direction of motion relatively to the field lines. For $\sigma = 1$ the particles are moving in the positive direction of the toroidal angle (*co-passing particles*). For $\sigma = -1$ the particles are moving in the opposite direction, the so called *counter-passing particles*. The dynamics is given by the Hamiltonian equations of motion, now with respect to the new Hamiltonian K,

$$\frac{dz}{d\varphi} = \frac{\partial K}{\partial p_z} = \frac{x_c}{Z_q}\left[\frac{\partial \psi}{\partial x_c} + \sigma \Gamma \left(\varepsilon_0(\gamma_t^2 - 1) - \omega_x I_x - x_c \gamma_t \frac{\partial \phi}{\partial x_c}\right)\right],$$
$$\frac{dp_z}{d\varphi} = -\frac{\partial K}{\partial z} = \frac{\partial \psi}{\partial z} - \sigma x_c \gamma_t \Gamma \frac{\partial \phi}{\partial z},$$
$$\frac{dt}{d\varphi} = \frac{\partial K}{\partial p_t} = \sigma x_c \gamma_t \Gamma,$$
$$\frac{dp_t}{d\varphi} = -\frac{\partial K}{\partial t} = \frac{\partial \psi}{\partial t} - \sigma x_c \gamma_t \Gamma \frac{\partial \phi}{\partial t}, \tag{4.55}$$

where

$$\Gamma = 1\Big/\sqrt{\varepsilon_0(\gamma_t^2 - 1) - 2\omega_x I_x}.$$

For the co-passing particles ($\sigma = +1$) the set of Hamiltonian Eq. (4.55) should be integrated along the positive direction of the toroidal angle φ, and vice versa for counter–passing particles ($\sigma = -1$).

It is not difficult to see that the Hamiltonian equation for the field lines (1.16) is particularly case of the Eq. (4.55). Putting $\Gamma = 0$ in (1.16) and using the relation (4.40) between x_c and the canonical momentum p_z we recover the equation for the field lines.

4.3.5 Summary

The Hamiltonian formulation of guiding–center motion equation presented in this chapter uses a cylindrical coordinate system rather than a complex magnetic coordinate system. It makes these equations much simpler as compared to corresponding guiding center equations in a magnetic coordinate system. More important, that the use of the action variable I_R as an adiabatic invariant makes more sense than using the magnetic moment μ, this is because the action I_R is related to the invariant J_R of the Hamiltonian system whereas the magnetic moment $\mu = mv_\perp^2/2B$ is related to the magnetic field only (see Sect. 7.1.3). Therefore, for instance, in the presence of the electric field the magnetic moment may be conserved with less accuracy than without. As shown by Dirickx and Weyssow (1998) one needs to modify the magnetic moment μ in a strong sheared electric field in order to make it an adiabatic invariant. In our approach the action variable I_x may be considered as an adiabatic invariant no matter whether the electric field is present or not. From the very beginning it is introduced for the whole system which may include not only the effect of the magnetic field, but also the electric field within the plasma as well.

4.4 Bibliographic Notes

The standard theory of guiding–center motion of charged particles in a magnetic and electric fields is described in many textbooks and monographs on plasma physics (see, for example, Goldston and Rutherford (1995), Miyamoto, (2005), Wesson (2004)). The detailed derivation of the equations of guiding-center motion and their discussions theory are reviewed by Northrop (1963) and by Morozov and Solov'ev (1966a).

Since early 1960s the Hamiltonian approach to study the charged particle motion in magnetic field has been subject many studies by Gardner (1959, 1966), Weitzner (1981), Littlejohn (1981, 1983). The formulation of Hamiltonian equations for the guiding–center motion in magnetic flux coordinates has been developed at the beginning of 1980s by White et al. (1982), Boozer and White (1982), Boozer (1983, 1984), White and Chance (1984). Using magnetic flux coordinates as canonical variables requires the existence of magnetic surface. The case of destroyed magnetic surfaces are treated by assuming a special form of magnetic field perturbations. To avoid this problem more general approaches has been proposed by Meiss and Hazeltine (1990) and later by Wobig and Pfirsch (2001). However, they also require a special conditions on the vector potential **A** and magnetic field **B**. The Hamiltonian theory of guiding–center motion is described in the book by White (2001).

Further development of the Hamiltonian theory in magnetic flux coordinates can be found in the works by White and Zakharov (2003), Wang (2006), Cooper et al. (2007, 2011) (see also references therein). The recent review by Cary and Brizard (2009) gives the comprehensive presentation of the Hamiltonian theory of guiding–center motion.

These approaches based on the existence of an adiabatic invariant, a magnetic moment $\mu = m_0 v_\perp^2/2B$, where m_0 is a particle mass, $v_\perp$ is the velocity component perpendicular to magnetic field **B**, and $B = |\mathbf{B}|$. However, in some cases, for instance, in spherical tokamaks with a weak but strongly spatially varying magnetic field or in the presence of an electric field with strong gradients the magnetic moment μ may break down (see Carlsson 2001, Dirickx and Weyssow 1998). Under such circumstances the above methods may not work well. Moreover, as shown in the works by Weitzner (1995), Weitzner and Pfirsch (1999) that even when the magnetic moment μ is a good adiabatic invariant in general magnetic configurations its expansion series is not a single-valued function of position **r**. This makes it difficult to use the magnetic moment as an invariant in the solutions of the gyro-kinetic equations.

At the beginning of 1980s Mynick (1979, 1980a, b) has developed the Hamiltonian theory of guiding-center motion in a magnetic system using a cylindrical coordinate system instead of magnetic flux coordinates. This approach has been applied to study a charged particle dynamics in mirror and field–reversed configurations in which a gyroradius comparable to the system size. Using the similar idea Abdullaev and Finken (2002) have derived the Hamiltonian guiding–center equations in a toroidal magnetic system, particularly, in tokamaks with arbitrary values of a gyroradius. This theory has been applied by Abdullaev et al. (2006a) to study the chaotic dynamics of particles in a stochastic magnetic field. The relativistic version of this theory has been employed to study the dynamics of runaway electrons in tokamaks by Finken et al. (2006, 2007), Wingen et al. (2006), Abdullaev et al. (2010, 2012).

Chapter 5
Drift Orbits in Equilibrium Plasmas

In this chapter we study the guiding center orbits in equilibrium plasmas using the Hamiltonian equations for the guiding-center motion given in the previous chapter. Particularly, we define drift surfaces, and study their structures for passing and trapped particles, estimate their widths. They will be illustrates with a number of examples in Sect. 5.1.1. In Sect. 5.2 we introduce the action-angle variables for guiding-center motion, study the toroidal procession of orbits, and formulate the Hamiltonian equations in the presence of perturbation magnetic and electric fields. In Sect. 5.3 the behavior of guiding-center orbits near hyperbolic fixed points, particularly, fixed points near the magnetic X-point and the banana orbit tips will be analyzed. In the final Sect. 5.4 we study a generic behavior of transit times of orbits.

5.1 Drift Surfaces

The trajectory of the guiding-center of a particle is called a *drift orbit*. In the equilibrium plasmas with the axisymmetric magnetic field and the electric field, $\psi = \psi(x, z)$ and $\phi = \phi(x, z)$ do not depend on the toroidal angle φ, and therefore $\partial U/\partial\varphi \equiv 0$. Then from Eq. (4.48) it follows that the toroidal momentum p_φ is the constant of motion, p_φ =const. Since the particle energy $\varepsilon = -p_t$ is also the constant of motion, the relation (4.50) determines the surface on which the drift orbits lie. This surface called a *drift surface*. The latter, according to (4.42) and (4.39), can be presented as

$$F(x, z) \equiv -p_\varphi Z_q^{-1} = \psi(x, z) - \frac{\sigma}{Z_q} x|u_\varphi| = \text{const}, \tag{5.1}$$

where

$$u_\varphi^2 = \frac{(\varepsilon - Z_q\phi(x, z))^2}{\varepsilon_0} - \varepsilon_0 - 2\omega_x I_x = 2\left[\varepsilon_k - \omega_x I_x\right]. \tag{5.2}$$

S. Abdullaev, *Magnetic Stochasticity in Magnetically Confined Fusion Plasmas*, Springer Series on Atomic, Optical, and Plasma Physics 78,
DOI: 10.1007/978-3-319-01890-4_5,

Here ε_k is related the kinetic energy $T_k = \varepsilon_0 (\gamma_t - 1)$:

$$\varepsilon_k = \frac{1}{2}\varepsilon_0 \left(\gamma_t^2 - 1\right) = \frac{1}{2}\varepsilon_0 \left[(1 + T_k/\varepsilon_0)^2 - 1\right].$$

For the nonrelativistic particles $\gamma_t - 1 \ll 1$, $\varepsilon_k \approx T_k$. The quantity $\sigma = u_\varphi/|u_\varphi|$ determines the direction of particle motion: $\sigma = 1$ for a co-passing particle, and $\sigma = -1$ for a counter-passing particle. If $u_\varphi = 0$ at the certain positions in the (x, z)-plane, then the particle changes its direction without making the full poloidal turn around the magnetic axis. Such a particle is called a *trapped particle*.

It is convenient to present the drift surface $F(x, z) \equiv -p_\varphi$ in the parametric form

$$x(\vartheta) \equiv x(\vartheta; \varepsilon, p_\varphi), \quad z(\vartheta) \equiv z(\vartheta; \varepsilon, p_\varphi),$$

where the parameter ϑ changes in the interval $[0, 2\pi]$. [1] In the plasmas with the up-down symmetry with respect to the equatorial plane $z = 0$ one can set $\vartheta = 0, \pi$ at $z = 0$. For the passing particles which make a full poloidal turn it coincides with the poloidal angle (see Fig. 5.1). For the trapped particles the poloidal velocity $u_\varphi(\vartheta)$ as a function of ϑ vanishes at the certain angle ϑ_t, $u_\varphi(\vartheta_t) = 0$, where it changes the direction.

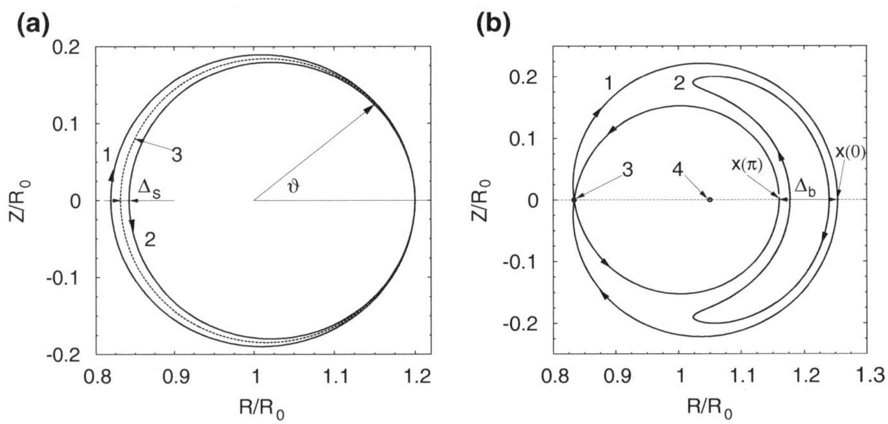

Fig. 5.1 **a** Guiding center orbits of the passing ion (proton H^+) in the plasma with the circular cross section: *curve 1* corresponds to the co-passing ion, *curve 2* corresponds counter-passing particle, *curve 3* corresponds to the magnetic surface. The ion's kinetic energy is $E_k = E_{ref} T_k = 5$ keV and the ratio $\lambda_l = 0.5$. **b** Guiding center orbits of the trapped ion of the kinetic energy $E_k = 50$ keV: *curve 1* corresponds to the nearly trapped ion with $\lambda_l = 0.853738$ near the separatrix, *curve 2* corresponds to $\lambda_l = 0.9$, point 3 corresponds to the banana orbit tip, point 4 corresponds to the stagnation orbit

[1] As we will see in Sect. 5.2.1 the angle variable can be taken as the parameter ϑ.

As seen from (5.1) that the deviation of the drift surfaces from the magnetic surfaces $\psi(x, z) =$ constant depends not only the full energy ε and the ratio of the action of the radial gyro-oscillations I_x to the kinetic energy T_k,

$$\lambda_I = \frac{I_x}{T_k}, \tag{5.3}$$

but also on the direction of the particle σ.

The dependence of the drift surfaces on σ has a simple explanation. It is known that the drift velocity $\mathbf{u}_d$ due to $\nabla\mathbf{B}$ and the curvature drifts does not depend on the directions of particle velocity, i.e. $\mathbf{u}_d \propto v_{\parallel}^2 + v_{\perp}^2/2$ (see, e.g., Morozov and Solove'v (1966a); Wesson (2004)). The total velocity of particle's guiding center, $\mathbf{v}_c$, is then given by the sum of $\mathbf{u}_d$ and the parallel velocity $\mathbf{v}_{\parallel}$, i.e., $\mathbf{v}_c = \mathbf{v}_{\parallel} + \mathbf{u}_d$. When the direction of particle motion is changed to the opposite, $\sigma \leftrightarrow -\sigma$, the velocity $\mathbf{v}_c^{(\sigma)}$ is changed to $\mathbf{v}_c^{(-\sigma)} = -\mathbf{v}_{\parallel} + \mathbf{u}_d$, i.e., $\mathbf{v}_c^{(\sigma)} \neq -\mathbf{v}_c^{(-\sigma)}$. Therefore, the guiding center orbits of particles moving in opposite directions (at a given energy) do not coincide.

5.1.1 Drift Orbits: Supra-Thermal Ions

Below, we present some typical examples of the drift orbits and the gyrating orbits of ions and electrons in the plasmas with the circular cross-section and with the elongated shapes. They are obtained by the integration of the guiding-center equations (4.48) and the Hamiltonian equations (4.3) using the equilibrium plasma configurations described in Sects. 2.2.2 and 2.3. For the plasma with the circular cross-section we have used the TEXTOR plasma parameters, the major and minor radii $R_0 = 1.75$ m and $a = 0.46$ m, the toroidal field $B_0 = 2.2$ T, the plasma current $I_p = 0.4$ MA, the ratio $\beta_{pol} = 0.3$, the plasma inductance $l_i = 1.2$. Note that in the TEXTOR plasma the direction of the plasma current is opposite to the direction of the toroidal field B_φ. The latter is considered as the positive direction of the toroidal angle φ.

For the plasma with the elongated shape the ITER-like plasma with the following parameters are chosen: the major and minor radii $R_0 = 6.2$ m and $a = 1.98$ m, the elongation $\kappa = 1.7$, triangularity $\delta = 0.33$, the toroidal field $B_0 = 5.3$ T, the plasma current $I_p = 15$ MA, the ratio $\beta_{pol} = 0.05$. The current direction is along the toroidal magnetic field.

The drift orbits of the passing and the trapped ion (a proton H^+) in the (R, Z)-plane in the plasma with the circular cross-section are given in Fig. 5.1a, b, respectively. In (a) the particle energy is $E_k = E_{ref} T_k = 5$ keV and the ratio $\lambda_I = 0.5$, curve 1 corresponds to the co-passing ($\sigma = +1$) particle and curve 2 corresponds to the counter-passing ($\sigma = -1$) particle. Curve 3 describes the magnetic surface $\psi(x, z) =$ const which has a common point with the drift surfaces at the outer equatorial plane $z = 0$.

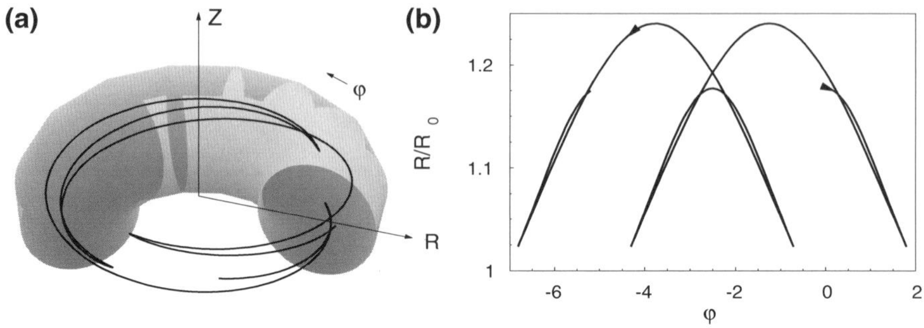

Fig. 5.2 Toroidal procession of the trapped ion's guiding-center orbit shown by *curve 2* in Fig. 5.1b: **a** in the 3D geometry; **b** in the (φ, R)-plane. The ion's kinetic energy is $E_k = 50\,\mathrm{keV}$ and the ratio λ_I is 0.9. The plasma parameters are the same as in the Fig. 5.1

Figure 5.1b shows the guiding-center motion the trapped ion of the energy $E_k = 50\,\mathrm{keV}$ and the ratio $\lambda_I = 0.853738$ for the barely trapped particle (curve 1) and $\lambda_I = 0.95$ for the trapped particle (curve 2), respectively. Because of their resemblance to the banana shape they are called *banana orbits*. The orbit of the barely trapped motion forms the separatrix. Point 3 corresponds to the X-point of the banana orbit separatrix and called a *banana tip*. Point 4 stands for the projection of the stagnation passing orbit into the (R, Z)-plane.

After one full poloidal turn the orbit advances along the toroidal direction φ. This phenomenon is known as a *toroidal procession*. The toroidal procession of trapped motion is shown Fig. 5.2 in the 3D toroidal geometry (a) and and the (φ, R)-plane (b). This phenomenon will be analyzed in Sect. 5.2.3.

5.1.2 Drift Orbits: High Energetic Particles

The deviation of the drift surface of the electron from the magnetic surface is of order of its gyroradius, and much smaller than the one of the ions. The drift becomes significant for the energetic electrons with the energies of order of 1 MeV and higher. Figure 5.3a shows the drift orbits and the gyrating orbits of such electrons in the plasmas with the circular cross-section: curves 1–3 correspond to the co-passing electrons ($\sigma = +1$) of the energies 1 MeV, 10 MeV, and 20 MeV, respectively, the ratio $\lambda_I = 0.1$. Such electrons correspond to the runaway electrons generated in tokamak plasma discharges during the start-up, low density operation, or plasma disruptions. The direction of a runaway electron motion is opposite to the plasma current.

The drift orbits of the electrons with the kinetic energy 100 MeV and $\lambda_I = 0.1$ in the ITER-like plasma shown in Fig. 5.3b by curve 1 for counter-passing electrons and by curve 2 for the co-passing electrons. In this case the counter-passing electrons would corresponds to the runaway electrons.

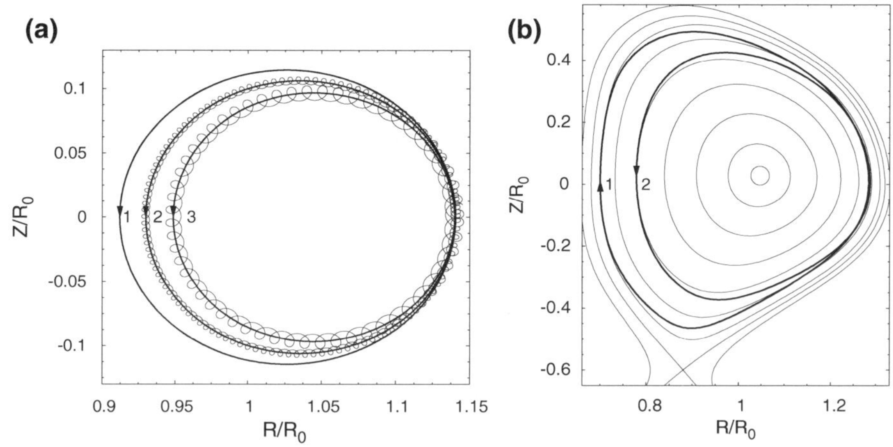

Fig. 5.3 **a** Guiding center and gyrating orbits of the energetic electrons in the plasma with circular cross-section: *curve 1* corresponds to the electron kinetic energy $E_k = 1\,\text{MeV}$, *curve 2* corresponds to $E_k = 10\,\text{MeV}$, and *curve 3* corresponds to $E_k = 20\,\text{MeV}$. The value of λ_I is 0.1. The plasma parameters are the same as in Fig. 5.1. **b** Electron guiding center orbits in the ITER-like plasma: *curve 1* corresponds to the counter-passing electron and *curve 2* corresponds to the co-passing electron with the energy of 100 MeV

The confinement region of runaway electrons shrinks with increasing their energy E_k or decreasing the plasma current I_p. The typical gyrating orbits and guiding-center orbits of such runaway electrons shown in Fig. 5.4 for the TEXTOR plasma with the plasma current $I_p = 100\,\text{kA}$. The electron energy is taken $E = 30\,\text{MeV}$, the toroidal field $B_t = 2.4\,\text{T}$, and the plasma radius $a = 30\,\text{cm}$.

As seen from Fig. 5.4a there are two types of unconfined orbits (curves 1 and 2) and the confined orbits (curve 3) are located inside the separatrix (curve 4). The appearance of the separatrix has been predicted by Zehrfeld et al. (1981). The area S_{conf} of confined orbits depends on the electron energy E_k and the plasma current I_p. At the given current the confined particle's area shrinks with increasing the energy E_k. This tendency is shown in Fig. 5.4b where the separatrices of the guiding-center orbits are plotted for the four different values of the electron energy: curves 1–4 correspond to the energies $E_k = 10$, 20, 30, and 40 MeV, respectively. The particles with the energy E exceeding the certain critical value E_{cr} are not confined in the plasma. The value E_{cr} depends on the plasma current I_p.

The numerical calculations show that I_p grows quadratically with E_{cr}. The dependence of E_{cr} on the square root of the current I_p is plotted in Fig. 5.5. This dependence can be fitted by the linear function (solid curve) $E_{cr} = a + bI_p^{1/2}$ with the constant coefficients a and b. Such a dependence is an agreement with the qualitative theoretical estimation given by Knoepfel and Spong (1979).

Finally, the orbits of the energetic (or fast) ions are displayed in Fig. 5.6a shows the drift orbits of the ion of the energy $E_k = 30\,\text{MeV}$ and the ratio $\lambda_I = 1$, and Fig. 5.6b shows the corresponding orbits of the α-particle of the energy $E_k = 5\,\text{MeV}$

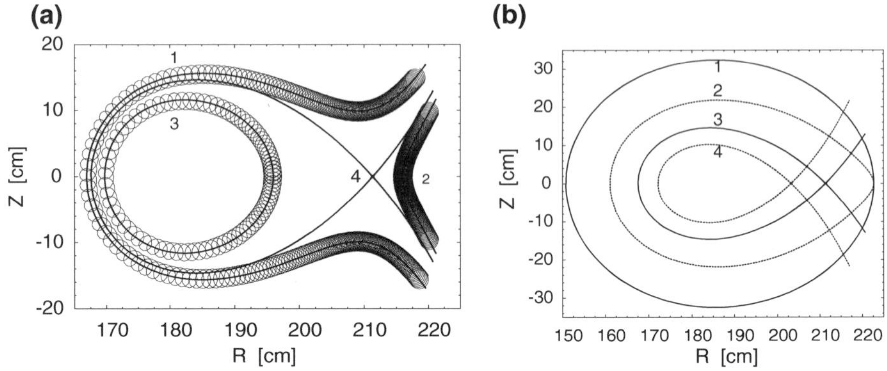

Fig. 5.4 **a** Runaway electron orbits in the (R, Z) plane for $E_k = 30$ MeV for the plasma current $I_p = 100$ kA, the toroidal field $B_t = 2.4$ T, the minor radius $a = 30$ cm. *Curves 1* and *2* correspond to unconfined electrons, *curve 3* corresponds to the confined electrons, *curve 4* corresponds to the curve (the separatrix) separating the confined and unconfined orbits. **b** Separatrices of runaway electron orbits for the four values of the electron energy: *curves 1–4* correspond to $E_k = 10$, 20, 30, and 40 MeV, respectively. The plasma parameters are the same as in Fig. 5.4

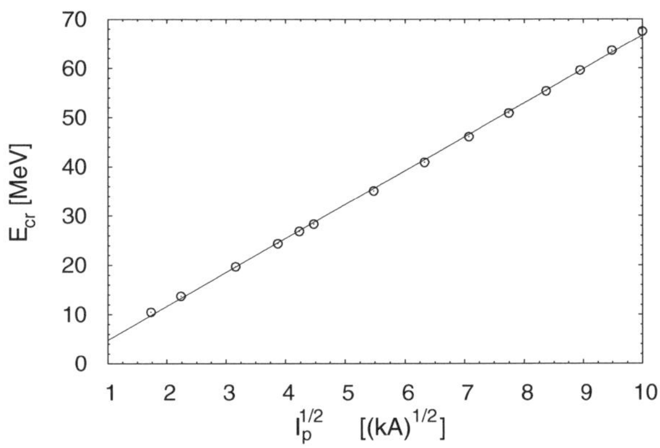

Fig. 5.5 Critical energy E_{cr} versus the plasma current I_p: The symbols $\odot$ correspond to the numerical calculations, the *solid line* corresponds to the fitting by the straight line $E_{cr} = a + bI_p^{1/2}$ ($a = -2.1285$, $b = 6.88258$)

and $\lambda_I = 0.1$ for the counter-passing (curve 1) and the co-passing (curve 2) motions. These orbits lie on the drift surface which are close to the magnetic surfaces. However, at large values of λ_I the orbits may strongly deviate from the magnetic surfaces have more complicated topology. Changing of the orbits topology with p_φ is shown Fig. 5.6a. Curve 1 describe the passing orbit not encircling the axis, curve 2 describes the *limiting* orbit with the X-point, curve 3 corresponds passing orbit, and curves 4 and 5 corresponds barely trapped and trapped orbits, respectively. The limiting orbit (curve 2) is called a *pinch orbit*.

Near the magnetic axis fast ions may have orbits much broader than the banana orbits which are called *potato orbits*. Such an orbit of α-particle produced in the core of the plasma is shown by curve 3 in Fig. 5.6.

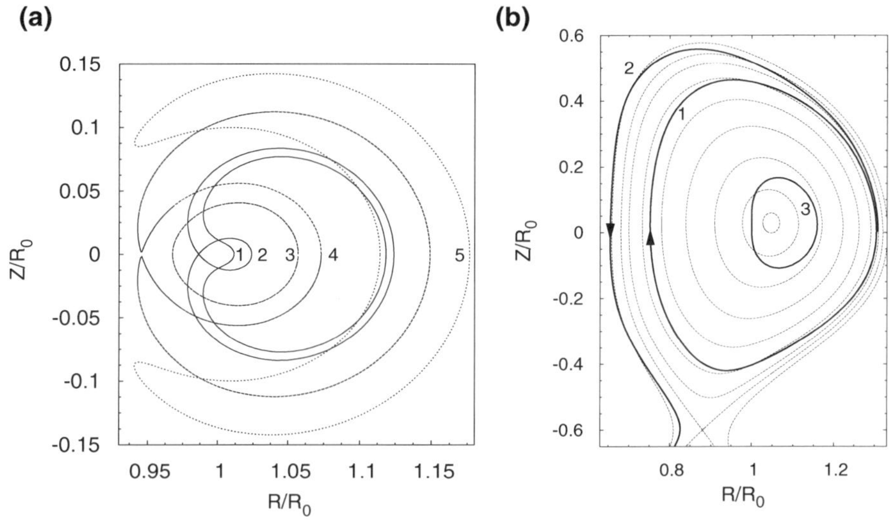

Fig. 5.6 **a** Guiding center orbits of the energetic ion (H^+) of the energy $E_k = 30\,\text{MeV}$ and the ratio $\lambda_I = 1$ in the plasma with circular cross-section. The value of the momentum p_φ decreases with the curves number. *Curve 2* corresponds to the pinch orbit. The plasma parameters are the same as in Fig. 5.1. **b** The same as in (**a**) but for the α-particle of the energy $E_k = 5\,\text{MeV}$ and $\lambda_I = 0.1$ (*curves 1* and *2*) and $E_k = 3.5\,\text{MeV}$ and $\lambda_I = 1$ (*curve 3*) in the ITER-like plasma with the parameters as in Fig. 5.3b. *Curve 1* corresponds to the counter-passing α-particle, *curve 2* corresponds to the co-passing α-particle, *curve 3* corresponds to the potato orbit formed near the magnetic axis

5.1.3 A Deviation of Drift Surfaces of Passing Particles from Magnetic Surfaces

Below we estimate the deviation of the drift surfaces $F(x, z)$ =const from the magnetic ones $\psi(x, z) = \text{const}$. Let $(x_M(\vartheta), z_m(\vartheta))$ be the parametric form of the magnetic surfaces. Assume that $\vartheta = 0$ at the low-field side and $\vartheta = \pi$ at the high-field side. Consider the magnetic surface which crosses the point $(x(0), z(0))$, i.e., $(x_M(0), z_M(0)) = (x(0), z(0))$, as shown in Fig. 5.1a. Since, on the magnetic surface $\psi(x_M(0), z_M(0)) = \psi(x(0), z(0)) = \psi(x_M(\vartheta), z_M(\vartheta))$, the Eq. (5.1) can be rewritten as

$$\psi\left(x(\vartheta), z(\vartheta)\right) - \psi\left(x_M(\vartheta), z_M(\vartheta)\right) = \frac{1}{Z_q}\left[x u_\varphi(\vartheta) - x(0)u_\varphi(0)\right],$$
$$u_\varphi(\vartheta) = \sigma\sqrt{2\varepsilon_k}\left[1 - \omega_x(x(\vartheta))\lambda_I\right]^{1/2}. \quad (5.4)$$

Assume that the deviation of the drift surface from the magnetic surfaces is small. Then the l.h.s of Eq. (5.4) can be presented as

$$\psi\left(x(\vartheta), z(\vartheta)\right) - \psi\left(x_M(\vartheta), z_M(\vartheta)\right) \approx \frac{\partial \psi}{\partial x}\left(x - x_M\right) + \frac{\partial \psi}{\partial z}\left(z - z_M\right).$$

Consider the plasma with the circular cross section in which there is the relation $\partial\psi/\partial x \approx s(x-1)/q(\psi) = sr/[R_0 q(\psi)]$, where $q(\psi)$ is the safety factor and s stands for the direction of the plasma current: $s = 1$ if the current flows in the positive direction of the toroidal angle φ, and $s = -1$ if the current flows in the opposite direction. Then one obtains the following formula for the shift $\Delta_p = x(\pi) - x_M(\pi)$ of the drift surface at the high-field side $\vartheta = \pi$ of the equatorial plane $z = 0$,

$$\Delta_p = s\frac{q(\psi)}{Z_q(x-1)}\left[x(\pi)u_\varphi(\pi) - x(0)u_\varphi(0)\right]. \tag{5.5}$$

In the non-normalized variables we have

$$R_0\Delta_p = s\frac{q(\psi)c}{Z_q reB_0}\left[\tilde{p}_\varphi(\pi) - \tilde{p}_\varphi(0)\right], \tag{5.6}$$

where $\tilde{p}_\varphi = P_{c\varphi} - Z_q eRA_\varphi$ is the toroidal momentum of the particle.

Since $\tilde{p}_\varphi(\pi) < \tilde{p}_\varphi(0)$, for the ions, $Z_q > 0$, the drift surface is shifted inwardly, ($\Delta_p < 0$), for the co-passing motion ($\tilde{p}_\varphi(0) > 0$), and outwardly, ($\Delta_p > 0$), for the counter-passing motion ($\tilde{p}_\varphi(0) < 0$).

For the electrons, $Z_q = -1$, the situation is opposite: the co-passing electron orbits are shifted outwardly, and the counter-passing electron orbits are shifted inwardly, as shown in Fig. 5.3a.

The maximum deviation occurs for the barely passing particles with $\tilde{p}_\varphi(\pi) \approx 0$ corresponding to $\lambda_I = 1/\omega_x(x(\pi))$,

$$R_0\Delta_p = \frac{qc}{Z_q reB_0}\left|\tilde{p}_\varphi(0)\right| = \frac{q\left|\tilde{p}_\varphi(0)\right|}{rm_0 Z_q\omega_c}. \tag{5.7}$$

In the absence of the electric field, $\phi \equiv 0$, using the relation (4.39) the shift Δ_p (5.5) can be reduced to

$$R_0\Delta_p = \frac{q(\psi)\rho_c}{r}\left|R(\pi)\sqrt{1 - \frac{R_0\lambda_I}{R(\pi)}} - R(0)\sqrt{1 - \frac{R_0\lambda_I}{R(0)}}\right|, \tag{5.8}$$

where

$$\rho_c = \frac{\sqrt{2E_k/m_0}}{(Z_q\omega_c)} \tag{5.9}$$

is the gyroradius. For the barely passing particle with $\lambda_I = R(\pi)/R_0$, the shift Δ_p reaches its maximal value,

$$R_0\Delta_p = \frac{q(\psi)\rho_c}{r}R(0)\sqrt{1-\frac{R_0\lambda_I}{R(0)}}$$
$$\approx \sqrt{2}q(\psi)\rho_c\left(\frac{R_0+r}{r}\right)^{1/2}, \quad (5.10)$$

where we have used the relations $R(0)-R(\pi)\approx 2r$, $R(0)=R_0+r$. From (5.8) it follows that the drift orbits of the ions are shifted from the magnetic surfaces much larger than the ones for the electrons with the same energy. The ratio of the corresponding shifts $\Delta_p^{(i)}$ and $\Delta_p^{(e)}$ is proportional to the square root of their mass ratio, i.e., $\sqrt{m_i/m_e}$.

5.1.4 The Width of Banana Orbits

Consider now the trapped particles whose guiding center orbits have the banana shape as was shown in Fig. 5.1b. Using the equation for the drift surface (5.1) one can estimate the width of the banana orbit Δ_b. Let $x(0)$ and $x(\pi)$ be the radial coordinates of the outer and inner crossing points of the orbit the equatorial plane $z=0$. Since they lie on the same drift surface and the directions of motion at these points are opposite, $\sigma(0)=-\sigma(\pi)$, we have

$$|\psi\,(x(\pi),0)-\psi\,(x(0),0)| = \frac{1}{Z_q}\left[x(\pi)|u_\varphi\,(\pi)\,|+x(0)|u_\varphi(0)|\right]. \quad (5.11)$$

From (5.11) it follows that the width of the banana orbit $R_0\Delta_b = R_0|x(0)-x(\pi)|$ is proportional to the momentum $\tilde{p}_\varphi$, similar to the shift $R_0\Delta_p$ of the drift surfaces of the passing particles from the magnetic ones (5.6).

Consider the banana orbits with the small widths. Then in the absence of the electric field and for the plasma with the circular cross section one can get

$$R_0\Delta_t = \frac{q(\psi)\rho_c}{r}\left[R(\pi)\sqrt{1-\frac{R_0\lambda_I}{R(\pi)}}+R(0)\sqrt{1-\frac{R_0\lambda_I}{R(0)}}\right], \quad (5.12)$$

and the barely trapped particles,

$$R_0\Delta_t \approx 2\frac{q(\psi)\rho_c}{r}R(0)\sqrt{1-\frac{R_0\lambda_I}{R(0)}} = 2\sqrt{2}q(\psi)\rho_c\left(\frac{R_0+r}{r}\right)^{1/2}. \quad (5.13)$$

From the comparison of this formula with (5.10) one can notice that the maximum banana orbit width $R_0\Delta_b$ is approximately two times larger than the shift of the drift surface $R_0\Delta_p$ of the barely passing particle.

5.2 Equations of Drift Motion in Action-Angle Variables

In this section we formulate the equations of motion (4.48) for the guiding center in the action-angle variables (I, ϑ) which can be introduced similar to the ones for the field lines (see in Sect. 1.2.1). Using these variables we describe the toroidal procession of guiding center orbits.

5.2.1 Action-Angle Variables

Consider the axisymmetric plasma in which there exist the drift surfaces for the confined particles. Let C_z and C_φ be the closed contours formed by the projection of the drift orbit in the the (z, p_z) and (φ, p_φ)-planes of the phase space, respectively. The contour C_z for the passing and trapped particle motion is shown in Fig. 5.7a, b respectively.

The action variable J is given by the integrals

$$J = \frac{1}{2\pi} \oint_{C_z} p_z(z; h_c, p_\varphi) dz, \tag{5.14}$$

taken along the contour C_z. The action J is known also as the *integral invariant*. The angle variable ϑ conjugated to the action variable J is introduced by the canonical transformation of variables $(z, \varphi, p_z, p_\varphi)$ to the action-angle variables $(\vartheta, \vartheta_\varphi, J, I_\varphi = p_\varphi)$,

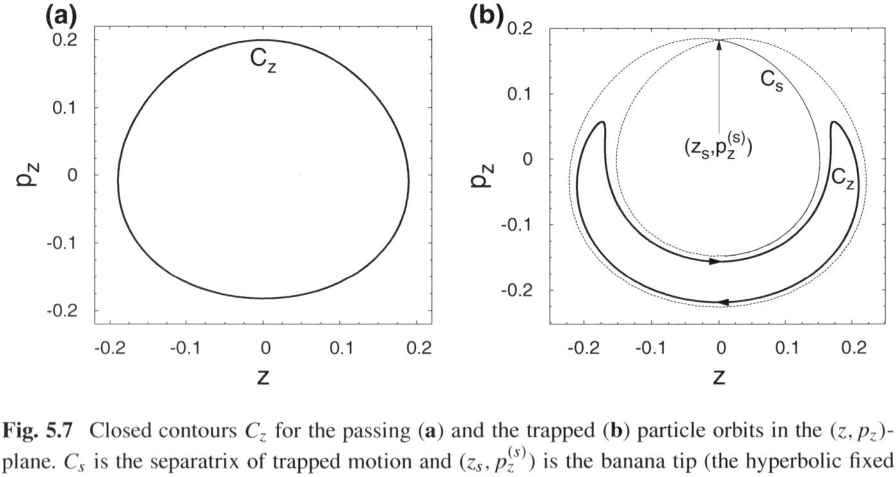

Fig. 5.7 Closed contours C_z for the passing (**a**) and the trapped (**b**) particle orbits in the (z, p_z)-plane. C_s is the separatrix of trapped motion and $(z_s, p_z^{(s)})$ is the banana tip (the hyperbolic fixed point)

$$\vartheta = \frac{\partial S}{\partial J}, \qquad \vartheta_\varphi = \frac{\partial S}{\partial I_\varphi}, \tag{5.15}$$

determined by the generating function,

$$S = S(z, \varphi, J, I_\varphi) = \varphi I_\varphi + \int^z p_z(z'; h_c, I_\varphi)dz'. \tag{5.16}$$

In the action-angle variables the Hamiltonian h_c depends on the action variables (J, I_φ), $h_c(J, I_\varphi)$ and the equation of motion (4.48) reads as

$$\begin{aligned} \frac{d\vartheta}{dt} &= \frac{\partial h_c}{\partial J} = \omega(J, I_\varphi), & \frac{dJ}{dt} &= -\frac{\partial h_c}{\partial \vartheta} = 0, \\ \frac{d\vartheta_\varphi}{dt} &= \frac{\partial h_c}{\partial I_\varphi} = \omega_\varphi(J, I_\varphi), & \frac{dI_\varphi}{dt} &= -\frac{\partial h_c}{\partial \vartheta_\varphi} = 0. \end{aligned} \tag{5.17}$$

The unperturbed motion is characterized by the poloidal, $\omega(J, I_\varphi)$, and the toroidal, $\omega_\varphi(J, I_\varphi)$, frequencies of the conditionally-periodic motion on the drift surface. These oscillations along the poloidal and toroidal axes are called *transit oscillations* (or *bounce oscillations*). The quantities ω and ω_φ are called *transit frequencies* (or *bounce frequencies*), while $T = 2\pi/\omega$ is the *transit time* (or *bounce time*) to make full poloidal turn. Furthermore we apply the terminology "transit" for all type of orbits, particularly for passing and trapped ones.[2]

The angle variables ϑ and ϑ_φ determining the position of the guiding-center on the drift surface, are the linear functions of time:

$$\begin{aligned} \vartheta &= \omega(J, I_\varphi)t + \vartheta_{z0}, \\ \vartheta_\varphi &= \omega_\varphi(J, I_\varphi)t + \vartheta_{\varphi 0}. \end{aligned} \tag{5.18}$$

From (5.15) we obtain the relation between the old variables $(z, \varphi, p_z, p_\varphi)$ and the new ones $(\vartheta, \vartheta_\varphi, J, I_\varphi = p_\varphi)$,

$$\vartheta = \frac{\partial}{\partial J} \int^z p_z(z', h_c, p_\varphi)dz', \tag{5.19}$$

$$\vartheta_\varphi = \varphi + \frac{\partial}{\partial I_\varphi} \int^z p_z(z', h_c, p_\varphi)dz'. \tag{5.20}$$

The variables (z, p_z) are the 2π–periodic functions of the angle variable ϑ,

$$z(\vartheta; J, I_\varphi) = z(\vartheta + 2\pi; J, I_\varphi), \qquad p_z(\vartheta; J, I_\varphi) = p_z(\vartheta + 2\pi; J, I_\varphi). \tag{5.21}$$

The relation between the toroidal angle φ and the angle variables ϑ, ϑ_φ comes from (5.20)

---

[2] Usually the term "bounce" is applied for the trapped orbits, while "transit" for the passing ones.

$$\varphi(\vartheta, \vartheta_\varphi; J, I_\varphi) = \vartheta_\varphi - G(\vartheta; J, I_\varphi), \tag{5.22}$$

where the function $G(\vartheta; J, I_\varphi)$ is a 2π–periodic in ϑ and determined by the second term on the r.h.s. of (5.20). Therefore, in the action-angle variables $(\vartheta, \vartheta_\varphi, J, I_\varphi)$ the guiding-center orbit (z, p_z, φ) can be presented by the Fourier series in ϑ,

$$\begin{aligned} z(\vartheta; J, I_\varphi) &= \sum_m z_m(J, I_\varphi) e^{im\vartheta}, \\ p_z(\vartheta; J, I_\varphi) &= \sum_m p_m(J, I_\varphi) e^{im\vartheta}, \\ \varphi(\vartheta, \vartheta_\varphi; J, I_\varphi) &= \vartheta_\varphi - \sum_m g_m(J, I_\varphi) e^{im\vartheta}, \end{aligned} \tag{5.23}$$

with the Fourier expansion coefficients $z_m(J, I_\varphi), p_m(J, I_\varphi), g_m(J, I_\varphi)$.

Using (5.18) one can obtain the time-dependencies of orbits. The typical time-dependencies of the toroidal angle φ of the guiding-center orbits are shown in Fig. 5.8 for the several passing and trapped motions.

It is not difficult to see that the action variable J in the zero order in μ is proportional to the toroidal magnetic flux ψ_t (1.24) of the toroidal field B_φ through the area enclosed by the contour C_z. Indeed, using the definition (1.14), the Eq. (4.36) for the guiding-center coordinate x_c, and the calibration $A_R = 0$ we have

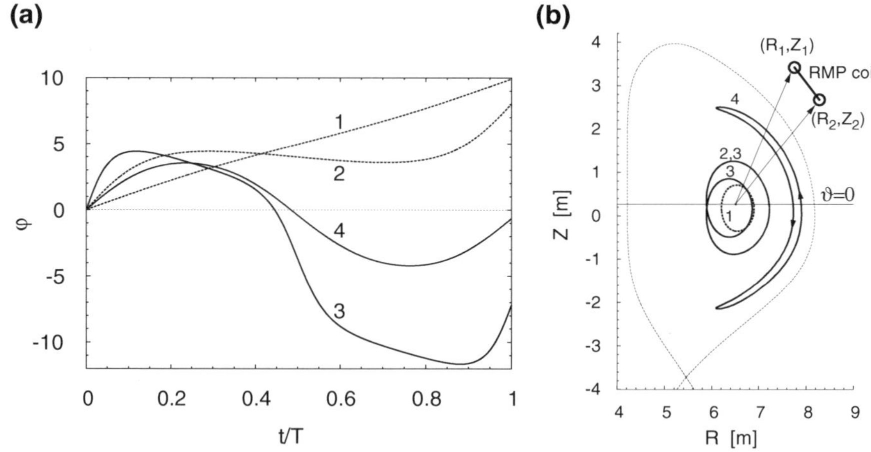

Fig. 5.8 **a** Dependencies of the toroidal angle φ on time t (normalized to the transit time $T = T(J, I_\varphi) = 2\pi/\omega(J, I_\varphi)$) for the several guiding-center orbits of the α-particle of energy $E = 1.0\,\text{MeV}$: *curve 1* corresponds to a passing particle, *curve 2*—to a barely passing particle, *curve 3*—to a trapped particle, and *curve 4*—a barely trapped particle. **b** Corresponding guiding center orbits in the (R, Z)-plane; The plasma parameters are the same as in Fig. 5.6b

$$
\begin{aligned}
J &= \frac{Z_q}{2\pi} \oint_{C_z} \psi_z dz = \frac{Z_q}{2\pi B_0 R_0^2} \oint_{C_z} \mathbf{A} \cdot d\mathbf{r}_\perp \\
&= \frac{Z_q}{2\pi B_0 R_0^2} \int_{S_z} \nabla \times \mathbf{A} \cdot \mathbf{e}_\varphi dR dZ = Z_q \psi_t,
\end{aligned} \tag{5.24}
$$

where $d\mathbf{r}_\perp = (dR, dZ)$, S_z is the area enclosed by the contour C_z.

The Hamiltonian system (5.17) can be reformulated by presenting the action J as a Hamiltonian $J = J(h_c, p_\varphi)$ and ϑ as a "time"-like variable. The Hamiltonian system with the canonically conjugated variables $(\vartheta_\varphi, t, p_\varphi, h_c)$ reads as

$$
\frac{d\vartheta_\varphi}{d\vartheta} = \frac{\partial J}{\partial I_\varphi} = \frac{\omega_\varphi(J, I_\varphi)}{\omega(J, I_\varphi)}, \quad \frac{dI_\varphi}{d\vartheta} = 0, \tag{5.25}
$$

$$
\frac{dt}{d\vartheta} = \frac{\partial J}{\partial h_c} = \frac{1}{\omega(J, I_\varphi)}, \quad \frac{dh_c}{d\vartheta} = 0. \tag{5.26}
$$

The first system of Eq. (5.25) can be considered as a one-degree-of-freedom Hamiltonian system with the canonical variables $(\vartheta_\varphi, I_\varphi)$ and the Hamiltonian $J = J(I_\varphi; h_c)$.

As we have seen that the canonical transformation of the gyrating particle variables $(x, z, \varphi, p_x, p_z, p_\varphi)$ to the guiding-center ones $(z, \varphi, p_z, p_\varphi)$ reduces the three-degrees-of-freedom Hamiltonian system to the two-degrees-of-freedom system. This is due to the guiding magnetic field which contracts the degree of freedom of motion. The next transformation of the guiding-center variables $(z, \varphi, p_z, p_\varphi)$ to the action-angle variable (ϑ, J) reduces the system further to the one-degree-of-freedom Hamiltonian system. In the latter case the action variable J is equivalent to the Hamiltonian function (see White (2001).

5.2.2 The Effective Safety Factor

The guiding-center orbits of particles on the drift surfaces p_φ =const can be also characterized by the quantity similar to the safety factor $q(\psi)$ of the magnetic field lines. We notate this quantity as $q_p(J, E)$ and refer as an *effective safety factor* . It is defined as the ratio of the increments of the toroidal φ and the poloidal ϑ angles of the orbit to make a full poloidal turn, i.e., $q_p(J, E) = \Delta\varphi/\Delta\vartheta = \Delta\varphi/2\pi$. From Eq. (5.25) it follows that it is determined by the ratio

$$
q_p(J, E) = \frac{\omega_\varphi(J, E)}{\omega(J, E)}, \tag{5.27}
$$

of the transit frequencies $\omega_\varphi(J, E), \omega(J, E)$ along the poloidal and the toroidal angles, respectively.

Since, the deviation of the drift surface from the magnetic ones is of order of the gyroradius, ρ_c, the difference between $q_p(J,E)$ and $q(\psi)$ is of order of $\rho_c/R_0 \ll 1$,

$$q_p(J,E) = q(\psi) + \sigma \mathcal{O}\left(\rho_c/R_0\right). \tag{5.28}$$

One should also note that $q_p(J,E)$ depends not only on the particle energy E but also on the direction of motion $\sigma = u_\varphi/|u_\varphi|$, $q_p(J,E) = q_p(J,E,\sigma)$.

5.2.3 Toroidal Procession

Consider a toroidal procession of orbits shown in Fig. 5.2. According to (5.22) the toroidal angle φ is nor a periodic function of time t neither a linear in t (or angle variable ϑ). As can see from Figs. 5.2 and 5.8 the trapped particle orbit advances along the toroidal angle after one poloidal turn.

Let $\Delta\varphi$ be a toroidal procession angle of an orbit along the toroidal angle after one full poloidal turn. The general formula for $\Delta\varphi$ can be found from Eq. (5.18). Since $\Delta\varphi$ is equal to $\omega_\varphi T$, we have

$$\vartheta_\varphi = \omega_\varphi T = \Delta\varphi + \frac{\partial}{\partial I_\varphi}\oint_{C_z} p_z(z', h_c, p_\varphi)dz' = \Delta\varphi + 2\pi\frac{\partial J(h_c, p_\varphi)}{\partial I_\varphi},$$

Using the definition of J (5.14) and $T = 2\pi/\omega$ one finds $\Delta\varphi$ through the derivative,[3]

$$\begin{aligned}\Delta\varphi &= \omega_\varphi T - 2\pi\frac{\partial J(h_c, p_\varphi)}{\partial I_\varphi}\\ &= 2\pi\frac{\partial h_c}{\partial I_\varphi}\frac{\partial J}{\partial h_c} - 2\pi\left(\frac{\partial J}{\partial p_\varphi}\frac{\partial p_\varphi}{\partial I_\varphi} + \frac{\partial J}{\partial h_c}\frac{\partial h_c}{\partial I_\varphi}\right) = -2\pi\frac{\partial J}{\partial I_\varphi}.\end{aligned} \tag{5.29}$$

From the latter it follows the averaged toroidal procession velocity $v_{dr} = \Delta\varphi/T = \omega_\varphi$,

$$\frac{\Delta\varphi}{T} = -\frac{\partial J/\partial I_\varphi}{\partial J/\partial h_c}. \tag{5.30}$$

It describes the toroidal drift of guiding center orbits. Eq. (5.29) can be considered as an analog of the relation $T = 2\pi\partial J/\partial h_c$ between the period of motion and the Hamiltonian function $h_c = h_c(J)$.

---

[3] In general, as was shown by Perkins and Bellan (2010) in the systems containing one fast-oscillating coordinate the net displacement other coordinate over single period is given by the differentiation of the action variable associated oscillating coordinate.

5.2.4 Equations of Motion in the Presence of Perturbations

Let $\psi^{(per)}(x, z, \varphi, t)$ and $\phi^{(per)}(x, z, \varphi, t)$ be the perturbations of the poloidal flux and the electric field potential with respect to their equilibrium quantities $\psi_0(x, z)$ and $\phi_0(x, z)$, i.e.,

$$\psi \equiv \psi(x, z, \varphi, t) = \psi_0(x, z) + \psi^{(per)}(x, z, \varphi, t),$$
$$\phi \equiv \phi(x, z, \varphi, t) = \phi_0(x, z) + \phi^{(per)}(x, z, \varphi, t). \tag{5.31}$$

Then the Hamiltonian functions (4.42) and (4.49) can be presented as sums

$$h = h_0(z, p_z, p_\varphi, p_t) + \epsilon h_1(z, \varphi, t, p_z, p_\varphi, p_t), \tag{5.32}$$
$$h_c = h_c^{(0)}(z, p_z, p_\varphi) + \epsilon h_c^{(1)}(z, \varphi, t, p_z, p_\varphi), \tag{5.33}$$

where the perturbation parts of the Hamiltonians are given by

$$\epsilon h_1(z, \varphi, t, p_z, p_\varphi, p_t) \approx Z_q \frac{u_\varphi \psi^{(pert)}}{x_c} + Z_q \phi_\varphi^{(pert)}, \tag{5.34}$$
$$\epsilon h_c^{(1)}(z, \varphi, t, p_z, p_\varphi) \approx Z_q \frac{u_\varphi \psi^{(pert)}}{\gamma_t x_c} + Z_q \phi_\varphi^{(pert)}. \tag{5.35}$$

Here ϵ stands for the amplitude of the perturbation fields, $\epsilon \sim \max|\psi^{(per)}|/|\psi_0|$, $u_\varphi = \left(p_\varphi + Z_q \psi_0\right)/x_c$, $\gamma_t = \left(-p_t - Z_q \phi_0\right)/\varepsilon_0$. In (5.34) and (5.35) we have neglected the terms of order of ϵ^2.

Introducing the action-angle variables $(\vartheta, \vartheta_\varphi, J, I_\varphi)$ for the unperturbed Hamiltonian system with h_c the equation of motion (4.48) can be presented as

$$\frac{d\vartheta}{dt} = \frac{\partial h_c}{\partial J} = \omega(J, I_\varphi) + \epsilon \frac{\partial h_c^{(1)}}{\partial J}, \qquad \frac{dJ}{dt} = -\frac{\partial h_c}{\partial \vartheta} = -\epsilon \frac{\partial h_c^{(1)}}{\partial \vartheta},$$
$$\frac{d\vartheta_\varphi}{dt} = \frac{\partial h_c}{\partial I_\varphi} = \omega_\varphi(J, I_\varphi) + \epsilon \frac{\partial h_c^{(1)}}{\partial I_\varphi}, \qquad \frac{dI_\varphi}{dt} = -\frac{\partial h_c}{\partial \vartheta_\varphi} = -\epsilon \frac{\partial h_c^{(1)}}{\partial \vartheta_\varphi}, \tag{5.36}$$

where

$$h_c = h_c^{(0)}(J, I_\varphi) + \epsilon h_c^{(1)}(\vartheta, \vartheta_\varphi, t, J, I_\varphi),$$
$$\epsilon h_c^{(1)}\left(\vartheta, \vartheta_\varphi, t, J, I_\varphi\right) \equiv \epsilon h_c^{(1)}\left(z(\vartheta, J, I_\varphi), \varphi(\vartheta_\varphi), t, p_z(\vartheta, J, I_\varphi), I_\varphi\right) \tag{5.37}$$

is the Hamiltonian in the action-angle variables $(\vartheta, \vartheta_\varphi, J, I_\varphi)$ (5.14) and (5.15). We suppose the perturbation is a periodic in time with the frequency Ω. Since the perturbation Hamiltonian $h_c^{(1)}(\vartheta, \vartheta_\varphi, t, J, I_\varphi)$ is a 2π–periodic function of $(\vartheta, \vartheta_\varphi)$ it can be expanded into a Fourier series,

$$h_c^{(1)}\left(\vartheta,\vartheta_\varphi,t,J,I_\varphi\right)=\sum_{m,n,s}H_{mns}(J,I_\varphi)e^{im\vartheta+in\vartheta_\varphi-is\Omega t},\tag{5.38}$$

where $H_{mn}(J,I_\varphi)$ are the Fourier coefficients corresponding to the perturbation mode (m,n).

The typical behavior of perturbation Hamiltonian (5.38) corresponding to the magnetic perturbations created by a set of saddle coils is studied in Appendix A.

5.3 Fixed Points of Drift Orbits

Analysis of the fixed points of the Hamiltonian equations (4.48) for guiding center orbits in the phase space $(z,\varphi,p_z,p_\varphi)$ allows one to classify a different types orbits and to study a generic behavior of orbits and their transit times near the hyperbolic fixed points. Below we consider the standard passing and banana orbits which may possess hyperbolic fixed points. Specifically, we study guiding-center orbits (i) in magnetic configurations possessing the magnetic separatrix and the X-point; (ii) banana orbits. Below we analyze the appearance of such fixed points and study a behavior of the guiding-center orbits near them.

For simplicity we consider the non-relativistic case described by the Hamiltonian (4.51),

$$\bar{h}_c(z,p_z,p_\varphi)=\omega_x(p_z)I_x+\frac{1}{2}u_\varphi^2+Z_q\phi(x,z),\tag{5.39}$$

where

$$u_\varphi=\frac{p_\varphi+Z_q\psi(x,z)}{x},\quad x\equiv x(p_z)=e^{-p_z/Z_q}.$$

Let $(z_s,p_z^{(s)},p_s)$ be a hyperbolic fixed point in the (z,p_z,p_φ) space. It is illustrated in Fig. 5.10a for the motion in the magnetic field with the magnetic separatrix. The fixed point and the corresponding separatrix the (z,p_z) plane for the banana orbit are shown in Fig. 5.7b. The fixed points are determined by the condition

$$\frac{\partial\bar{h}_c(z,p_z,p_\varphi)}{\partial z}=0,\quad\frac{\partial\bar{h}_c(z,p_z,p_\varphi)}{\partial p_z}=0,$$
$$\bar{h}_c(z,p_z,p_\varphi)=E,$$

where E is the full particle energy. Using Eqs. (4.39) and (4.40), the system of equations for $(z_s,p^{(s)},p_s)$ can be reduced to

$$
\begin{aligned}
f_1(z,p_z,p_\varphi) &= u_\varphi b_R + \frac{\partial\phi}{\partial z} = 0,\\
f_2(z,p_z,p_\varphi) &= \omega_x(p_z)I_x + u_\varphi^2 - Z_q u_\varphi b_z - Z_q x\frac{\partial\phi}{\partial x} = 0,\\
f_3(z,p_z,p_\varphi) &= \bar{h}_c(z,p_z,p_\varphi) = E,
\end{aligned}
\tag{5.40}
$$

where

$$
b_R = \frac{B_R}{B_0} = \frac{1}{x}\frac{\partial\psi}{\partial z}, \qquad b_z = \frac{B_Z}{B_0} = -\frac{1}{x}\frac{\partial\psi}{\partial x}.
$$

In the absence of the electric field, $\phi \equiv 0$, the first two equations in (5.40) take the following form

$$
\begin{aligned}
f_1(z,p_z,p_\varphi) &= b_R(z,p_z) = 0,\\
f_2(z,p_z,p_\varphi) &= \omega_x(p_z)I_x + u_\varphi^2 - Z_q u_\varphi b_z = 0.
\end{aligned}
\tag{5.41}
$$

Notice that the radial component of the poloidal magnetic field b_R vanishes at the fixed points. There are two solutions of the second equation in (5.41) with respect to u_φ:

$$
u_\varphi^{(1,2)} = \frac{\sigma}{2}Z_q b_z\left(1 \pm \sqrt{1 - \frac{4\omega_x I_x}{Z_q^2 b_z^2}}\right).
\tag{5.42}
$$

[We recall that $\sigma = \pm 1$ determines the direction of motion.] By requiring that there are no complex solutions $u_\varphi^{(1,2)}$ (5.42) one finds that the orbits can be hold by plasma at the following condition,

$$
I_x < I_c = \frac{Z_q^2 b_z^2}{4\omega_x}, \quad \text{or} \quad \lambda_I < \lambda_{Ic} = \frac{Z_q^2 b_z^2}{4\omega_x T_k}.
\tag{5.43}
$$

At $\lambda_I > \lambda_{Ic}$ no orbits can be hold by the plasma. The critical action I_c (or the ratio λ_{Ic}) depends on the particle kinetic energy $T_k = \varepsilon_0(\gamma_t - 1)$, on the plasma current I_p (through the magnetic field b_z), and a particle charge Z_q.

5.3.1 Orbits Near the Fixed Points

Furthermore, we consider the case in the absence of the electric field, $\phi \equiv 0$. The expansion of the Hamiltonian near the fixed point $(z_s, p_z^{(s)}, p_s)$ into the series of powers of $(z - z_s)$, $\left(p_z - p_z^{(s)}\right)$, and $\left(p_\varphi - p_s\right)$ reads as

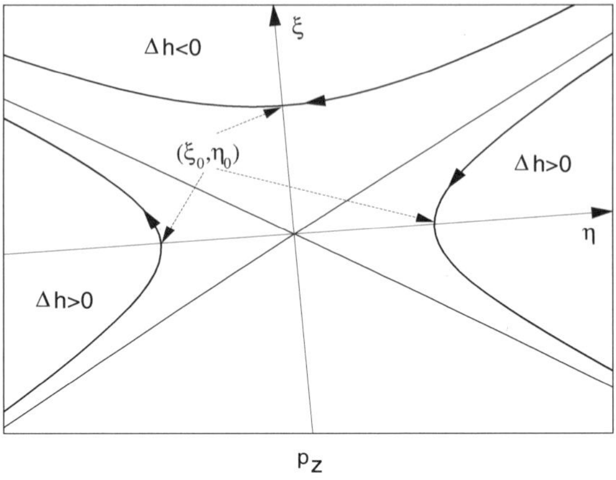

Fig. 5.9 Guiding center orbit of the counter-passing particle near the X-point in the (z, p_z) plane

$$\bar{h}_c(z, p_z, p_\varphi) = h_s + \frac{u_s}{x_s}(p_\varphi - p_s) + A(z - z_s)(p_\varphi - p_s) + B\left(p_z - p_z^{(s)}\right)(p_\varphi - p_s)$$
$$+ \frac{h_{zz}}{2}(z - z_s)^2 + h_{zp}(z - z_s)\left(p_z - p_z^{(s)}\right) + \frac{h_{pp}}{2}\left(p_z - p_z^{(s)}\right)^2 + \cdots, \quad (5.44)$$

where $h_s \equiv \bar{h}_c(z_s, p_z^{(s)}, p_\varphi)$, $u_s = u_\varphi(x_s)$. In (5.44) the following notations are introduced

$$A = \frac{\partial^2 \bar{h}_c}{\partial p_\varphi \partial z} = Z_q b_R = 0, \qquad B = \frac{\partial^2 \bar{h}_c}{\partial p_\varphi \partial p_z} = \frac{u_s}{Z_q x_s} - \frac{Z_q b_z}{x_s},$$

and

$$h_{zz} = \frac{\partial^2 \bar{h}_c}{\partial z^2} = Z_q^2 b_R^2 + Z_q u_s \frac{\partial b_R}{\partial z} = Z_q u_s \frac{\partial b_R}{\partial z} = Z_q \frac{u_s}{x_c}\frac{\partial^2 \psi}{\partial z^2},$$
$$h_{zp} = \frac{\partial^2 \bar{h}_c}{\partial z \partial p_z} = x_s u_s b_R - x_s^2 b_R b_z - Z_q x_s u_s \frac{\partial b_z}{\partial x} = Z_q \frac{u_s}{x_c}\frac{\partial^2 \psi}{\partial z \partial p_z},$$
$$h_{pp} = \frac{\partial^2 \bar{h}_c}{\partial p_z^2} = Z_q^{-2}\left(\omega_x(p_z) I_x + 2u_s^2\right) - 2u_s b_z + b_z^2 + Z_q \frac{u_s}{x_c}\frac{\partial^2 \psi}{\partial p_z^2}. \quad (5.45)$$

The rotational transformation of coordinates (z, p_z) around the fixed point $(z_s, p_z^{(s)})$ to the new coordinates (ξ, η) shown in Fig. 5.9 (see also (2.39) and Fig. 2.6):

$$\xi = (z - z_s)\cos\alpha_0 + \left(p_z - p_z^{(s)}\right)\sin\alpha_0,$$
$$\eta = -(z - z_s)\sin\alpha_0 + \left(p_z - p_z^{(s)}\right)\cos\alpha_0, \quad (5.46)$$

one can reduce the Hamiltonian h_c to

$$\bar{h}_c(z, p_z, p_\varphi) = h_s + \frac{u_s}{x_s}(p_\varphi - p_s) + \left[\tilde{A}\xi + \tilde{B}\eta\right](p_\varphi - p_s) + \frac{\lambda_1}{2}\xi^2 + \frac{\lambda_2}{2}\eta^2 + \cdots, \tag{5.47}$$

where

$$\tilde{A} = B \sin \alpha_0, \quad \tilde{B} = B \cos \alpha_0.$$

The coefficients λ_1 and λ_2 are the eigenvalues of the matrix,

$$\begin{pmatrix} h_{zz} & h_{zp} \\ h_{pz} & h_{pp} \end{pmatrix}. \tag{5.48}$$

and given by

$$(\lambda_1, \lambda_2) = \frac{h_{zz} + h_{pp}}{2} \pm \sqrt{\frac{(h_{zz} - h_{pp})^2}{4} + h_{zp}^2}, \tag{5.49}$$

The angle α_0 is determined by $\tan \alpha_0 = 2h_{zp}/(h_{pp} - h_{zz})$.

For the trapped particles the coefficient $h_{zp} = 0$ at the banana tip and therefore $\alpha_0 = 0$. The eigenvalues (λ_1, λ_2) are given by $\lambda_1 = h_{zz} < 0$, $\lambda_2 = h_{pp} > 0$.

On the other hand for the passing particles near the magnetic separatrix we have $(\lambda_1 > 0, \lambda_2 < 0)$ for the co-passing motion $(u_s > 0)$, and $(\lambda_1 < 0, \lambda_2 > 0)$ for the counter-passing motion $(u_s < 0)$.

Using the expansion (5.44) the equations of motion (4.48) can be reduced to

$$\frac{d\xi}{dt} = \lambda_2\eta + \tilde{B}(p_\varphi - p_s), \quad \frac{d\eta}{dt} = -\lambda_1\xi - \tilde{A}(p_\varphi - p_s),$$
$$\frac{d\varphi}{dt} = \frac{u_\varphi^{(s)}}{x_s} + \tilde{A}\xi + \tilde{B}\eta, \quad \frac{dp_\varphi}{dt} = 0. \tag{5.50}$$

Consider the case of the counter-passing motion, $u_s < 0$, $(\lambda_1 < 0, \lambda_2 > 0)$. Let (ξ_0, η_0) be the coordinate of the closest point of the orbit $(\xi(t), \eta(t))$ to the X-point. From the Hamiltonian (5.47) one can obtain

$$\left(\xi_0 = \pm\sqrt{2|\Delta h|/|\lambda_1|}, \eta_0 = 0\right), \quad \text{for } \Delta h < 0,$$
$$\left(\xi_0 = 0, \eta_0 = \pm\sqrt{2|\Delta h|/\lambda_2}\right), \quad \text{for } \Delta h > 0, \tag{5.51}$$

where

$$\Delta h = h_c(z, p_z, p_\varphi) - h_s - \frac{u_s}{x_s}(p_\varphi - p_s). \tag{5.52}$$

Then the solutions $(\xi(t), \eta(t), \varphi(t))$ of Eq. (5.50) near the fixed point $(z_s, p_z^{(s)}, p_s)$ satisfying the initial condition $(\xi_0, \eta_0, 0)$ at the time instant $t = 0$ are given by

$$\xi = \tilde{B}\left(p_\varphi - p_s\right)t + \xi_0 \begin{cases} -\sinh\left(\gamma_d t\right), & \text{for } \Delta h > 0, \\ \cosh\left(\gamma_d t\right), & \text{for } \Delta h < 0, \end{cases}$$
$$\eta = -\tilde{A}\left(p_\varphi - p_s\right)t + \eta_0 \begin{cases} \cosh\left(\gamma_d t\right), & \text{for } \Delta h > 0, \\ -\sinh\left(\gamma_d t\right), & \text{for } \Delta h < 0, \end{cases} \tag{5.53}$$

and

$$\varphi(t) = \frac{u_\varphi^{(s)} t}{x_s} + \int^t \left(\tilde{A}\xi(t') + \tilde{B}\eta(t')\right) dt' = \frac{u_\varphi^{(s)} t}{x_s}$$
$$+ \frac{1}{\gamma_d} \begin{cases} -\tilde{A}\xi_0 \cosh\left(\gamma_d t\right) + \tilde{B}/\eta_0 \sinh\left(\gamma_d t\right), & \text{for } \Delta h > 0, \\ \tilde{A}\xi_0 \sinh\left(\gamma_d t\right) + -\tilde{B}\eta_0 \cosh\left(\gamma_d t\right), & \text{for } \Delta h < 0, \end{cases} \tag{5.54}$$

where the parameter γ_d is defined as

$$\gamma_d = \sqrt{|\lambda_1 \lambda_2|}. \tag{5.55}$$

Similar formulas can be obtained for the orbits of co-passing motion near the X-point and the banana tips.

Since the particle energy is a constant, $h_c(z, p_z, p_\varphi) = h_s$, then $\Delta h = -(u_\varphi/x_s)$ $(p_\varphi - p_s)$. The orbit at the fixed point $(z_s, p_z^{(s)}, p_s)$ uniformly moves along the toroidal angle φ,

$$\varphi_s(t) = (u_s/x_s)t. \tag{5.56}$$

For the passing motion from Eq. (5.45) it is not difficult to show that the coefficient γ_d near the X-point of the drift separatrix is proportional to the parameter γ (2.45) for the magnetic separatrix,

$$\gamma_d = \frac{|Z_q u_s|}{x_s}\gamma\left[1 + \mathcal{O}\left(\rho_c/R_0\right)\right]. \tag{5.57}$$

Using the procedure similar to the one in Sect. 2.4.2 one can show that the transit time T, i.e., the time spent by a particle to make a full poloidal turn, has the following asymptotics near the banana separatrix,

$$T = \frac{s}{\gamma_d} \ln \frac{Q}{|\Delta h|} + \mathcal{O}(\Delta h), \quad \text{at } \Delta h \to 0, \tag{5.58}$$

where $s = 2$ for the trapped motion and $s = 1$ for the passing motion, and Q is a constant. Such an asymptotics is also valid for the passing particles near the magnetic separatrix.

5.3.2 Hyperbolic Fixed Points Near the Magnetic X-point

Now we consider a magnetic field with the magnetic separatrix and the X-point where the poloidal field vanish, $b_R = b_z = 0$. The typical orbits near the hyperbolic fixed point $(z_s, p_z^{(s)}, p_s)$ in the (z, p_z, p_φ) space is shown in Fig. 5.10a For the passing motion ($u_\varphi^{(s)} \equiv u_\varphi(x_s) \neq 0$) the fixed points are shifted to respect the magnetic X-point due to a drift motion and the electric field, $(\partial\phi/\partial x, \partial\phi/\partial z)$ as shown in Fig. 5.10b.

In the absence of the latter from (5.40) it follows that the radial component of the poloidal magnetic field vanish, $b_R = 0$. From the second equation in (5.40) it follows that the product $Z_q u_\varphi b_z$ should be positive, $Z_q u_\varphi^{(s)} b_z > 0$. Let $(x_s^{(M)}, z_s^{(M)})$ be the coordinates of the magnetic X-point. To find the shift of the X-point of the drift orbit from the magnetic X-point, i.e., $\Delta x = x_s - x_s^{(M)}$, $\Delta z = z_s - z_s^{(M)}$ we expand the system (5.40) near the magnetic X-point $(x_s^{(M)}, z_s^{(M)})$. Taking into account that

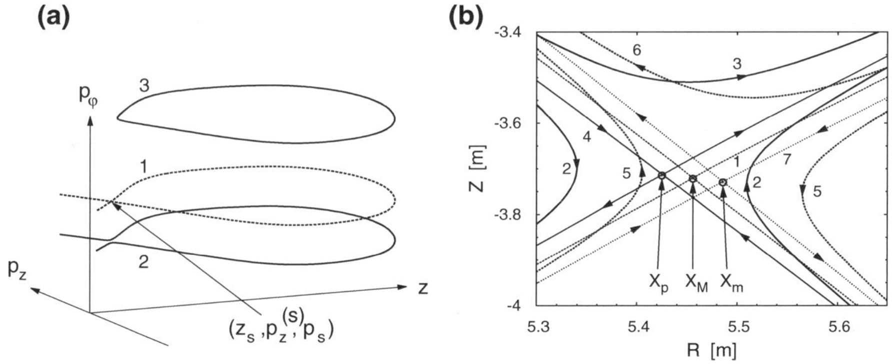

Fig. 5.10 **a** Guiding center orbits the α particles near the separatrix in the (z, p_z, p_φ) space: *curve 1* is the separatrix, *curve 2* corresponds the de-confined orbit, and *curve 3* corresponds confined orbit; **b** Orbits near the hyperbolic fixed points in the (R, Z) plane. *Curve 1* corresponds to the magnetic separatrix and X_M is the magnetic X-point, *curves 2–4* correspond to the unconfined orbit, confined orbit, and the separatrix of the co-passing motion $u_\varphi > 0$, respectively, and X_p is the corresponding hyperbolic fixed point; *curves 5–7* correspond to similar types of orbits of the counter-passing motion $u_\varphi < 0$, and X_m is the corresponding hyperbolic fixed point. The energy of the α particle $E_k = 3.5$ MeV and the ratio $\lambda_I = 0.1$. The parameters of the ITER like plasma are the same in Fig. 5.6b

$$b_R(x_s^{(M)}, z_s^{(M)}) = 0, \quad b_z(x_s^{(M)}, z_s^{(M)}) = 0,$$

one obtains

$$x_s = x_s^{(M)} + \Delta x, \quad z_s = z_s^{(M)} + \Delta z, \tag{5.59}$$

where

$$\Delta x = \frac{u_\varphi D_x}{Z_q}\left(1 + \frac{\omega_x I_x}{u_\varphi^2}\right),$$
$$\Delta z = -\frac{u_\varphi D_z}{Z_q}\left(1 + \frac{\omega_x I_x}{u_\varphi^2}\right), \tag{5.60}$$

$$D_x = \frac{\partial b_R}{\partial z}\left[\frac{\partial b_z}{\partial x}\frac{\partial b_R}{\partial z} - \frac{\partial b_z}{\partial z}\frac{\partial b_R}{\partial x}\right]^{-1},$$
$$D_z = \frac{\partial b_R}{\partial x}\left[\frac{\partial b_z}{\partial x}\frac{\partial b_R}{\partial z} - \frac{\partial b_z}{\partial z}\frac{\partial b_R}{\partial x}\right]^{-1}.$$

For the passing particle whose parallel energy is much larger than the perpendicular one, $\omega_x I_x / u_\varphi^2 \ll 1$, the shift of the X-point similar to the deviation $R_0 \Delta_p$ of the drift surface from the magnetic one (5.8) is of order of the gyroradius ρ_c (5.9), i.e.,

$$d = R_0\sqrt{(\Delta x)^2 + (\Delta z)^2} = \rho_c x_s D\sqrt{1 - \frac{\lambda_I}{x_s}}, \tag{5.61}$$

where $D = \sqrt{D_x^2 + D_z^2}$.

5.3.3 Banana Orbit Tips

For the trapped motion there exists the reversal point where $u_\varphi = 0$. The fixed point or the banana tip (see Fig. 5.7b) does not coincide with the reversal point and the toroidal velocity u_φ has a finite nonzero value $u_\varphi^{(s)} \neq 0$ there.

In the absence of the electric field the banana tip is determined by the smallest in magnitude solution $u_\varphi^{(2)}$ (5.42) of Eq. (5.41). Since the product $Z_q u_\varphi b_z > 0$ the direction of the toroidal velocity $u_\varphi^{(s)}$ at the banana tip is determined by the sign of the vertical component of the poloidal field b_z at the high field side and the charge Z_q.

The typical behavior of orbits near the banana tips is shown in Fig. 5.11 in the (z, p_z) plane. The curves at the upper part correspond to the orbits with the negative

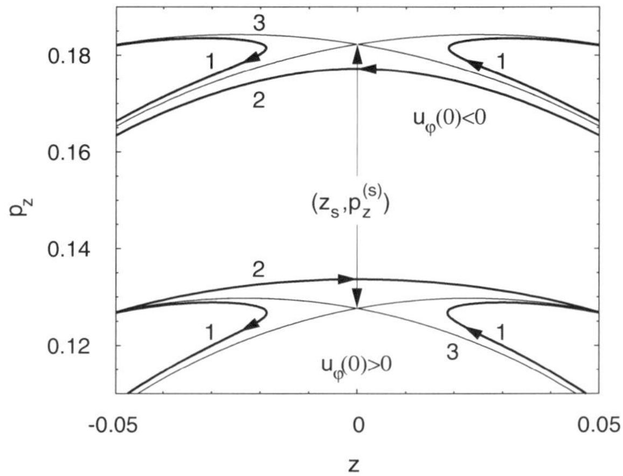

Fig. 5.11 Typical behavior of guiding center orbits the ion (H^+) near the banana tips in the (z, p_z) plane. Curves at the top correspond to the case with $u_\varphi(0) < 0$, and at the bottom correspond to the case $u_\varphi(0) > 0$. *Curves 1* correspond to the trapped motion with the momentum $p_\varphi < p_s$, *curves 2* correspond to the passing motion with the momentum $p_\varphi > p_s$, *curves 3* describe the separatrix. The energy of ions is $E_k = 50\,\mathrm{keV}$ and the ratio $\lambda_I = 0.853738$. The parameters of the plasma with the circular cross-section are the same in Fig. 5.1

value of the toroidal velocity at the low field side, $u_\varphi(0) < 0$, and the curve at the lower part correspond to the orbits with $u_\varphi(0) > 0$. Each of curves lie at the different planes $p_\varphi = \text{const}$ in the (z, p_z, p_φ) space. The banana orbits appear at the values of the toroidal momentum p_φ lower than the some critical value p_s depending on the energy E_k and the parameter λ_I, $p_\varphi < p_s$.

5.4 Numerical Examples

In this section we present and anylyze the typical profiles of the action variables, the transit times, and the toroidal drifts of passing and trapped particles calculated numerically.

Figures 5.12a, b illustrate typical dependencies of the action J and the transit times T on the momentum p_φ for the confined ion motions with the fixed energy E_k at the different values of λ_I in the plasma with the circular section. The ions of energy E_k is taken equal 1 keV, curves 1, 2, 3, and 4 correspond to $\lambda_I = 0.1, 0.5, 0.7$, and 0.9, respectively. Similar dependencies of the α-particles in the ITER-like plasma are shown in Fig. 5.13. The energy of the α-particle is 3.5 MeV, and the values of λ_I are taken equal to 0.5 (curves 1 and 1′), 0.7 (curves 2 and 2′), and 0.9, (curves 3 and 3′) respectively.

At $\lambda_I = 0.1$ all confined ions are passing, i.e., there are no trapped ones. With increasing λ_I, i.e., increasing the radial gyro-oscillation energy $\omega_R I_R$, particles trap

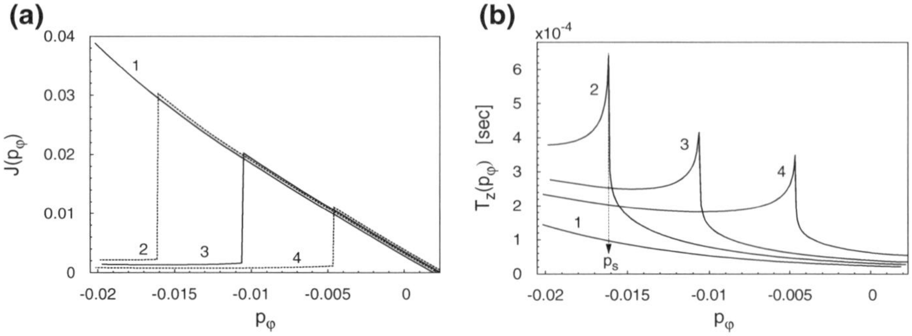

Fig. 5.12 Dependencies of the action variable J (**a**) and the transit time T (in a real unit) (**b**) on the momentum p_φ for the ions of the energy $E_k = 1$ keV at different values of λ_I in the plasma with the circular cross-section: *curve 1* corresponds to $\lambda_I = 0.1$, *curve 2*—$\lambda_I = 0.5$, curve 3—$\lambda_I = 0.7$, and *curve 4*—$\lambda_I = 0.9$. The plasma parameters are the same in Fig. 5.1 and $\omega_c = 2.1076 \times 10^8$ s^{-1}

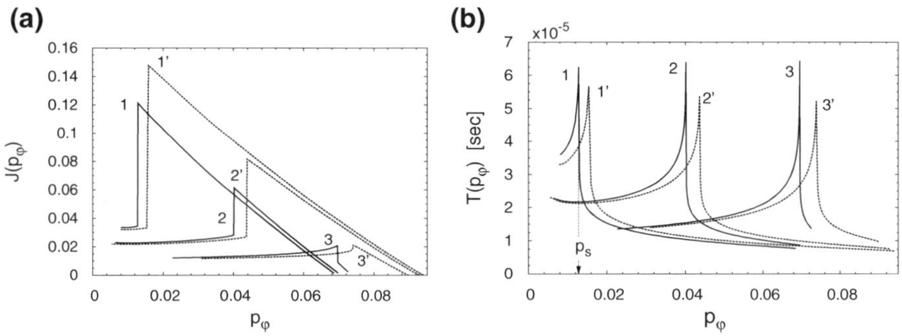

Fig. 5.13 The same as in Fig. 5.12 but for the α-particles of the energy $E_k = 3.5$ MeV in the ITER-like plasma: (**a**) and (**b**) the action variable J and the transit time T (in a real unit) versus p_φ; *Curves 1* and *1'* correspond to $\lambda_I = 0.5$, *curves 2* and *2'*—$\lambda_I = 0.6$, *curves 3* and *3'*—$\lambda_I = 0.8$. *Solid curves 1, 2,* and *3* correspond to the counter passing particles with $u_\varphi < 0$ at the outer equatorial plane, and the *dashed curves1', 2'*, and *3'* to the co-passing particles with $u_\varphi > 0$ at the outer equatorial plane. The plasma parameters are the same in Fig. 5.6b, and $\omega_c = 1.2781 \times 10^8$ s^{-1}

at the values of the momentum p_φ lower than a certain critical value p_s, $p_\varphi < p_s$ (see Figs. 5.12 and 5.13). The action J takes the zero value for the stagnation point and grows monotonically with decreasing p_φ. But J is changed discontinuously when p_φ crosses p_s when the transition from the passing orbits to the trapped ones occurs (see Fig. 5.1). That is the momentum p_φ is the *multivalued function* of the action variable J. Therefore the drift surfaces cannot be labeled by the action J while the analogous quantity, the toroidal flux ψ_t does label magnetic surfaces. This property of the Hamiltonian system of guiding-center motion is a manifestation of a phenomenon known as a *Hamiltonian monodromy*. It was introduced by Duistermaat (1980) a topological obstruction to the existence of global action-angle variables [see also Cushman and Bates (1997) and references therein].

The dependence of J on p_φ can be approximated by the following formula containing the logarithmic and polynomial functions,

$$J(p_\varphi) = \begin{cases} J_0 x \ln|x| + p_1(x), & \text{for trapped motion, } x < 0, \\ p_2(x), & \text{for the passing motion, } x > 0, \end{cases} \tag{5.62}$$

where $x = p_\varphi - p_s$, and J_0 is a constant, p_i, $(i = 1, 2, \dots)$ are polynomial functions whose coefficients depending on the energy E_k and λ_I.

According to (5.24) the action J is mainly proportional to the magnetic flux through the area S_z enclosed by contour C_z, that is the action J is determined by the area S_z. Therefore, for the passing particles the action J weakly depends on the particle energy E_k and the radial energy $I_R = \lambda_I E_k$ since their guiding center orbit is only slightly modified under of these parameters. In contrary, for the trapped particles the action J strongly depends on E_k and λ_I because the area enclosed by the banana orbits is determined by their width Δ_b. Since Δ_b grows with the gyroradius ρ_c (5.9) it follows that $J \propto \sqrt{2E_k}$ for the trapped particles.

5.4.1 Transit Times

The transit times T also change smoothly with p_φ except the point $p_\varphi = p_s$ where they have singularities because of the appearance of the separatrix of the banana orbit. The singularities of T are similar to the those of the safety factor $q(h)$ near the separatrix (see Sect. 2.4.2 and Eq. (2.51)), i.e., the time period T grows to infinity according to the logarithmic law $T \propto -\ln|p_\varphi - p_s|$. The dependence of the transit time T on p_φ can be well approximated by the following formulas,

$$T(p_\varphi) = \begin{cases} 2\gamma_d^{-1} \ln(Q_t/|x|) + p_t(x), & \text{for the trapped motion, } x < 0, \\ \gamma_d^{-1} \ln\left(Q_p/|x|\right) + p_p(x), & \text{for the passing motion, } x > 0, \end{cases} \tag{5.63}$$

where $x = p_\varphi - p_s$, $Q_{(t,p)}$, are constants depending on the energy E_k and λ_I, $p_t(x)$ and $p_p(x)$ are the polynomial functions of x. The parameter γ_d in (5.63) is determined by the expansion of the Hamiltonian (4.49) near the hyperbolic fixed point $(z_s, p_z^{(s)}, p_s)$ of the banana separatrix, i.e., Eq. (5.55).

Figure 5.14 shows the fitting of the numerically calculated transit time T by the formula (5.63) with the fitting parameters $Q_{t,p}$, and $a_{t,p}$, $b_{t,p}$ in the polynomial functions $p_{p,t}(x) = a_{t,p}x + b_{t,p}x^2$. The parameter γ_d found from the expansion coefficients α and β (5.45).

As one expects the transit $T(p_\varphi)$ is inverse proportional the velocity u_φ. The numerical analysis shows that the transit $T(p_\varphi)$ as a function of p_φ or J can be presented as

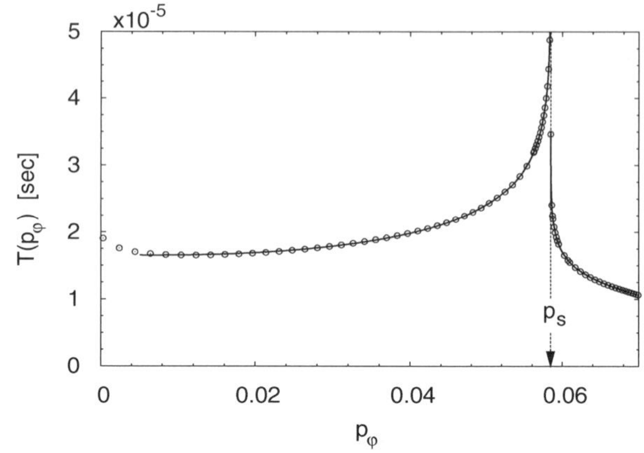

Fig. 5.14 Transit time T (in a real unit) versus p_φ (*dots*) and its fitting by by Eq. (5.63) (*solid curve*). The value $\lambda_I = 0.7$. The plasma parameters are the same in Fig. 5.6b . The parameter $\gamma_d\omega_c = 3.2464 \times 10^5\ \mathrm{s}^{-1}$

$$T \approx \frac{1}{u_\varphi(0)} \begin{cases} f(J), & \text{for the passing particles,} \\ g(p_\varphi), & \text{for the trapped particles,} \end{cases} \tag{5.64}$$

where $u_\varphi(0)$ is the toroidal velocity at the outer equatorial plane, $f(J)$ and $g(p_\varphi)$ are the functions depending only on J and p_φ, respectively. Moreover, one can show that for the passing particles the ratio $l_c/v_\varphi(0)t_c$, where $l_c = R_0\Delta\varphi$ is the connection length along the guiding-center orbit, $\Delta\varphi$ is the increment of the toroidal angle during one full poloidal turn, $v_\varphi(0) = R_0\omega_c u_\varphi(0)$, and $t_c = T/\omega_c$, can be also presented as

$$\frac{l_c}{v_\varphi(0)t_c} = \frac{\Delta\varphi}{u_\varphi(0)T} = f_t(J) + \mathcal{O}(\rho_c), \tag{5.65}$$

where $f_t(J)$ is the function depending only on J and $\mathcal{O}(\rho_c)$ is the corrections of order of the gyroradius.

5.4.2 Toroidal Drift Velocity

The dependencies of the angular toroidal drift velocity $\Delta\varphi/t_c = \omega_c\Delta\varphi/T$ on the toroidal momentum p_φ for the trapped motion are presented in Fig. 5.15a shows $\Delta\varphi/T$ versus p_φ at the different energies of ion with the fixed $\lambda_I = 1.0$ in the circular cross-section plasma; Fig. 5.15b for the trapped α-particles of energy $E_k = 3.5$ MeV and $\lambda_I = 0.8$ in the ITER-like plasma: curves 1 and 2 correspond to the particles with $u_\varphi < 0$ and $u_\varphi > 0$, respectively, at the outer equatorial plane.

Since the drift velocity due to curvature is $v_d \propto v_\varphi^2/\omega_c R_0$, one expects the quadratic dependence of the angular toroidal procession velocity $\Delta\varphi/t_c$ on u_φ. The numerical analysis show that the ratio $\Delta\varphi/Tu_\varphi^2(0)$ weakly depends of the energy E_k as shown in Fig. 5.16. Therefore the angular toroidal drift velocity $u_d = \Delta\varphi/T$ can be presented

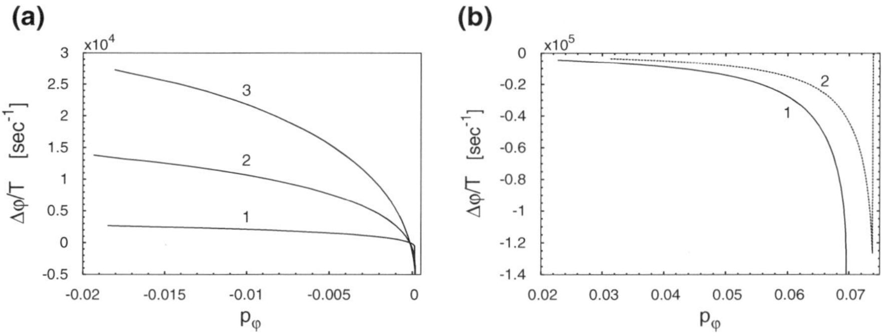

Fig. 5.15 **a** Averaged angular toroidal procession velocity $\Delta\varphi/t_c$ (in real units) versus the momentum p_φ for the trapped ions at different energies in the plasma with the circular cross section: *curve 1* corresponds to the energy $E_k = 1\,\text{keV}$, *curve 2*—$E_k = 5\,\text{keV}$, and *curve 2*—$E_k = 10\,\text{keV}$. The parameter $\lambda_I = 1.0$; **b** The same as in (**a**) but for the trapped α-particles in the ITER-like plasma. The energy $E_k = 3.5\,\text{MeV}$ and $\lambda_I = 0.8$: *curves 1* and *2* correspond to the particles with $u_\varphi < 0$ and $u_\varphi > 0$, respectively, at the outer equatorial plane

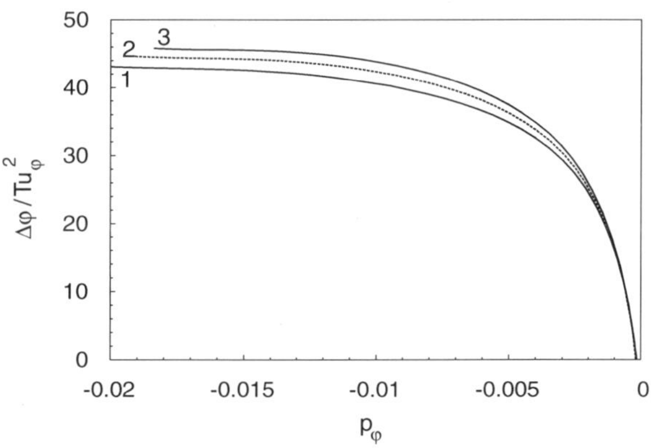

Fig. 5.16 Ratio $\Delta\varphi/u_\varphi^2 T$ versus p_φ for the trapped ions shown in Fig. 5.15a: *curve 1* corresponds to the energy $E_k = 1\,\text{keV}$, *curve 2*—$E_k = 5\,\text{keV}$, and *curve 2*—$E_k = 10\,\text{keV}$

by the following function of the momentum,

$$u_d = \frac{\Delta\varphi}{T} = u_\varphi^2(0) f_d(p_\varphi) + \mathcal{O}\left(\sqrt{\rho_c}\right), \tag{5.66}$$

where $f_d(p_\varphi)$ is the function of p_φ alone.

When one approaches the separatrix, i.e., at the limit $p_\varphi \to p_s$, the velocity u_d tends to the finite value $u_d^{(s)}$. From (5.56) it follows that $u_d^{(s)} = u_s/x_s$ which is in agreement with the numerical results.

Finally, we consider also the example illustrating typical dependencies of the effective safety factor q_p (5.27) on the particle energy. Figure 5.17 show the dependencies of the effective safety factors $q_p(J, E)$ on the action variable J for the passing ions (H^+) in the circular plasma (a) and for the α-particles in the

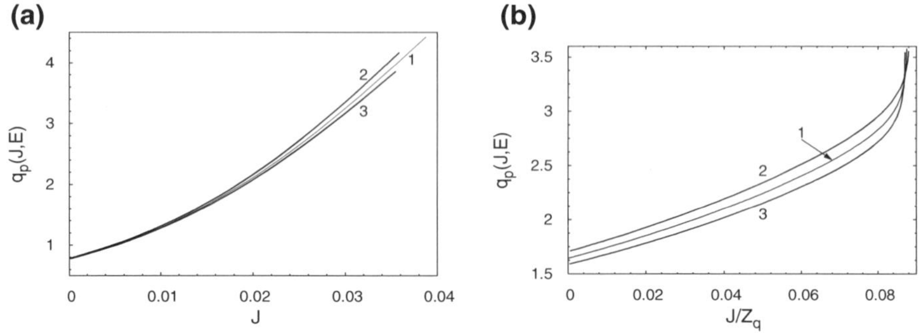

Fig. 5.17 Effective safety factor $q_p(J, E)$ versus the action variable J for the passing particles: (**a**) the ions of energy $E_k = 10\,\text{keV}$ in the plasma with the circular cross-section; (**b**) the α-particles of the energy $E_k = 3.5\,\text{MeV}$ in the ITER-like plasma. *Curve 1* corresponds the safety factor of the equilibrium magnetic field, *curve 2* corresponds to the counter-passing particles, and *curve 3*—the co-passing particles

ITER-like plasma (b). As seen from Fig. 5.17a, b at the given value of J the effective safety factor $q_p(J, E)$ is smaller than the magnetic safety factor $q(\psi)$ for the co-passing passing particles, is larger than $q(\psi)$ for the counter-passing passing particles. On the other hand the quantity $q_p(J, E)$ takes the same values for the co-passing and counter passing particles at the different values of $J = J_+$, $J = J_-$, i.e., $q_p(J_-, E, \sigma = -1) = q_p(J_+, E, \sigma = +1)$, and $J_+ > J_-$. The latter can be seen from Fig. 5.1: the co-passing orbit (curve 1) has a cross-section area than the counter-passing orbit (curve 2).

5.5 Bibliographic Notes

The analyses of guiding-center orbits in tokamaks are given in many textbooks on plasma physics, for example, by White (2001), Miyamoto (2005). The guiding-center orbits of energetic particles, especially fast ions for example α-particles, may have more complicated topology than low energy particles. The detailed classification and the topology of fast ion orbits in magnetically fusion plasmas can be found in reviews by Rome and Peng (1979); Eriksson and Porcelli (2001) (see also Egedal (2000)). The compact analytical formulas for the action vartiable J and the frequencies of trapped and passing guiding-center orbits in large aspect-ratio plasmas with circular cross sections can be found in Brizard (2011) (see also White (2001)).

Chapter 6
Methods to Study the Hamiltonian Systems

The methods developed to study Hamiltonian dynamical systems are also applicable to study magnetic field lines. In this chapter we present the classical methods and new recently developed advanced methods to study chaotic motion in Hamiltonian systems. We will mostly concentrate on the analytical and semi-analytical mappings and describe the technique to derive these mappings from Hamiltonian systems.

The important tool in the study of dynamical systems is the Poincaré mapping. First we give the definition of this mapping for magnetic field lines. In Sect. 6.2 we formulate the fundamental problem of dynamics as the study Hamiltonian systems in the presence of perturbations whose particular case is the Hamiltonian system (3.9) and (3.10) for magnetic field lines in tokamaks the presence of non–axisymmetric magnetic perturbations. Then we describe the method to integrate such Hamiltonian systems which via the canonical transformation of variables reduces to the time–evolution continuous system of equations to the discrete one, i.e., to the canonical mapping. The generating function associated with the mapping is determined by the solutions of the Hamilton–Jacobi equation in a finite time intervals whose solutions are found using the classical perturbation theory. The simplified forms of the generating function are derived in Sect. 6.3.

In Sect. 6.4 we derive the full-turn transfer mapping describing, particularly, the evolution of field lines and particles in one full poloidal turn. These mappings will be used not only for the fast integration of Hamiltonian systems, but also to analyze the onset of chaos near the separatrix and derive the analytical formulas for the spatial forms of stable and unstable manifolds etc. The properties of the Poincaré and the Melnikov integrals associated with the full-turn transfer mappings are studied in Sect. 6.5. The simple example of the application of the full–turn transfer mapping to study the field lines is presented in Sect. 6.6. Particularly, the construction of the Poincaré sections, field line behavior near the X-point, and the determination of magnetic footprints on the divertor plates are discussed.

S. Abdullaev, *Magnetic Stochasticity in Magnetically Confined Fusion Plasmas*, Springer Series on Atomic, Optical, and Plasma Physics 78,
DOI: 10.1007/978-3-319-01890-4_6, © Springer International Publishing Switzerland 2014

In the final Sect. 6.9 we give a brief overview of the mapping models proposed in the literature to study chaotic field lines and particles in fusion devices. Such models include the standard mapping, the Wobig–Mendonça mapping, the tokamap and other.

6.1 Poincaré Mapping

Henri Poincaré introduced a powerful tool to study dynamical systems known now as Poincaré mapping, which replaces the continuous system with the discrete one. This mapping has important advantages, which allows, first (i) to reduce dimensions of the system at least by one, (ii) to visualize the dynamics of the system at certain sections of phase space called Poincaré sections, and thereby display the global behavior of the system. Below we recall a definition of the Poincaré map.

Let P_k be a crossing point of a field line with the poloidal section $\varphi = \text{const}$ shown in Fig. 6.1. The mapping of the crossing point P_k to the next one P_{k+1},

$$P_{k+1} = \hat{M} P_k, \tag{6.1}$$

is Poincaré mapping. A plot of crossing points at the given section $\varphi = \text{const}$ is known as *Poincaré section*. As a Hamiltonian system Poincaré maps for field lines should be *symplectic*. The latter means the property of the system in the conservation of certain invariants in phase space. The flux-preserving property of magnetic field lines is manifested in this property.

The traditional method to construct Poincaré sections of field lines for specific plasmas is based on field line tracing codes, like the FLOC code (Fowler et al. 1978), the Gourdon code (Kaleck et al. 1997), the MASTOC code (Nguyen et al. 1995), the TRIPND and TRIP3D codes (Evans et al. 2002) which are widely used in many applications. The codes integrate the field line Eq. (1.19) in the cylindrical coordinate system (R, Z, φ) using the standard integration schemes, like the Runge–Kutta

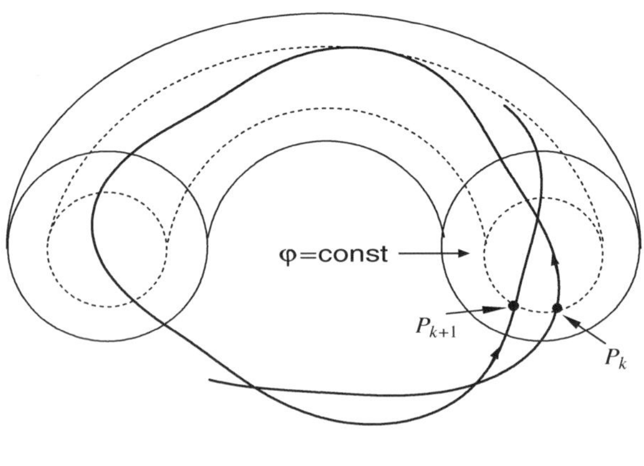

Fig. 6.1 Poincaré mapping (6.1): P_k and P_{k+1} are the consecutive crossing points of the section $\varphi = \text{const}$

integrator. Although these schemes are not symplectic they give a sufficiently accurate results for the practical applications in tokamaks and stellarators.

In the next sections we present a general method to integrate Hamiltonian systems of type given by Eq. (3.12). It is based on the construction of the symplectic mappings for Hamiltonian systems. The detailed description of the method are given in Abdullaev (1999, 2002, 2006).

6.2 Mapping Method to Integrate Hamiltonian Equations

In this section we present a general method to integrate Hamiltonian equations based on the canonical transformation of variables and the classical perturbation theory in a finite time. The main idea of the method is the presentation of the time evolution of the system by the symplectic mapping. First we formulate this method for the fundamental problem of dynamics and then consider its application to magnetic field lines and the particle dynamics.

Consider a Hamiltonian system with N degrees of freedom with (q_i, p_i), $(i = 1, \dots, N)$ being the canonically conjugated coordinates and momenta. The time-evolution of the system in the $2N-$ dimensional phase space $(q_1, \dots, q_N, p_1, \dots, p_N)$ is governed by the Hamiltonian system

$$\frac{dq_i}{dt} = \frac{\partial H}{\partial p_i}, \qquad \frac{dp_i}{dt} = -\frac{\partial H}{\partial q_i} \qquad i = 1, \dots, N+1, \tag{6.2}$$

determined with a Hamiltonian function $H(q, p)$. There exists a unique solution of (6.2)

$$q_i(t) = q_i\left(t, q^{(0)}, p^{(0)}\right), \qquad p_i(t) = p_i\left(t, q^{(0)}, p^{(0)}\right), \tag{6.3}$$

with given initial conditions $q(t_0) \equiv q^{(0)} = (q_1(t_0), \dots, q_N(t_0))$, $p(t_0) \equiv p^{(0)} = (p_1(t_0), \dots, p_N(t_0))$, at the time instant $t = t_0$. The trajectories (6.3) is often considered as a *flow* (or *a Hamiltonian flow*) of $2N-$ dimensional fluid in phase space (q, p). This "fluid" is incompressible, i.e., any closed volume of phase space is conserved under Hamiltonian flow. This means that the Jacobian of the transformation from initial state coordinates $q(t_0)$, $p(t_0)$ to final state coordinates $q(t)$, $p(t)$ equals one,

$$\frac{\partial\,(q(t), p(t))}{\partial\,(q(t_0), p(t_0))} = 1. \tag{6.4}$$

Such a transformation $(q(t_0), p(t_0)) \to (q(t), p(t))$ is called a *volume-preserving*.

6.2.1 Fundamental Problem of Dynamics

Suppose the Hamiltonian system can be presented as a sum of unperturbed and perturbation parts. Assume that unperturbed system is completely integrable. [1] The latter means that there exist as many as independent constants (also called integrals) of motions $F_i(q, p) = \text{const}$, $(i = 1, \ldots, N)$ as degrees of freedom. Then one can introduce action-angle variables (I, ϑ): $(I_1, \ldots, I_N, \vartheta_1, \ldots, \vartheta_N)$ to replace the canonical variables (q, p).[2] The time t and the energy H can be also considered as a pair of canonically–conjugated variables $(\vartheta_{N+1} = t, I_{N+1} = H)$. Then one should introduce the independent time–like variable which will be notated τ.

The unperturbed Hamiltonian H_0 then depends only on the action variables, $H_0 = H_0(I)$, and the unperturbed equations of motion, $\dot{I}_i = 0$, $\dot{\vartheta}_i = \partial H_0/\partial I_i$, are integrable

$$I_i = \text{const}, \qquad \vartheta_i(\tau) = \omega_i(I)\tau + \vartheta_{i0}, \qquad i = 1, \ldots, N+1, \tag{6.5}$$

where

$$\omega_i(I_1, \cdots, I_N) = \frac{\partial H_0}{\partial I_i}, \qquad i = 1, \ldots, N+1, \tag{6.6}$$

are the unperturbed frequencies of motion. Often, especially in a mathematical literature it is referred that these orbits of the unperturbed system describe the motion on the N-dimensional tori determined $I_i = \text{const}$ of the $2(N+1)$-dimensional phase space (see, e.g., Arnold 1989).[3]

Suppose that the system is subjected to the perturbation, and described by the Hamiltonian,

$$H(I, \vartheta, \epsilon) = H_0(I) + \epsilon H_1(I, \vartheta, \epsilon), \tag{6.7}$$

where the dimensionless parameter ϵ stands for the perturbation strength. The perturbation Hamiltonian $H_1(I, \vartheta, \epsilon)$ is a 2π−periodic function in ϑ. In many cases one supposes that the perturbation is small, $\epsilon \ll 1$. However, we will not require this condition in the derivation of mapping.

The set of perturbed Hamiltonian equations can be written as

$$\frac{d\vartheta_i}{d\tau} = \frac{\partial H}{\partial I_i} = \omega_i(I) + \epsilon \frac{\partial H_1}{\partial I_i}, \qquad \frac{dI_i}{d\tau} = -\epsilon \frac{\partial H_1}{\partial I_i}. \tag{6.8}$$

Furthermore, we present the perturbation Hamiltonian $H_1(I, \vartheta, \epsilon)$ as a Fourier series in angles ϑ:

---

[1] The definition of integrable of Hamiltonian systems is given by the Liouville's theorem. Its modern formulation can be found in Arnold (1989).

[2] As was noted in Sec. 5.4 in some Hamiltonian systems with degrees of freedom $N \geq 2$ one cannot introduce global action–angle variables in whole phase space. Such a phenomenon is known as Hamiltonian monodromy (Duistermaat (1980)).

[3] Coincidently, the magnetic field of the toroidally confined plasma is a real example of the tori.

$$H_1(I, \vartheta) = \sum_m H_m(I) e^{im\cdot\vartheta} = \sum_m |H_m(I)| \cos(m \cdot \vartheta + \chi_m), \tag{6.9}$$

where $H_m(I) = |H_m(I)| \exp(i\chi_m), m = (m_1, m_2, \ldots, m_N, m_{N+1}), m\cdot\vartheta = m_1\vartheta_1 + m_2\vartheta_2 + \cdots + m_N\vartheta_N + m_{N+1}\vartheta_{N+1}$. It is assumed that the perturbation is a time–periodic with the frequency $\omega_{N+1} = \Omega$.

The Hamiltonian problem (6.8) and (6.9) describing the dynamics of integrable system affected by small perturbation has numerous applications in different areas of physics, mechanics and astronomy. Poincaré (1892/1893/1899) called this problem a *fundamental problem of dynamics*.

6.2.2 Canonical Transformation and Mapping

Our task is to describe the time evolution of the system (6.8) and (6.28) by the mapping connecting the variables $\vartheta_k = \vartheta(\tau_k)$, $I_k = I(\tau_k)$ at the sequence of time instants τ_k $(k = 0, \pm 1, \pm 2, \ldots)$,

$$(\vartheta_k, I_k) \to (\vartheta_{k+1}, I_{k+1}). \tag{6.10}$$

The mapping should be area-preserving or symplectic,[4]

$$\left|\frac{\partial(\vartheta_{k+1}, I_{k+1})}{\partial(\vartheta_k, I_k)}\right| = \left|\frac{\partial\vartheta_{k+1}}{\partial\vartheta_k}\frac{\partial I_{k+1}}{\partial I_k} - \frac{\partial\vartheta_{k+1}}{\partial I_k}\frac{\partial I_{k+1}}{\partial\vartheta_k}\right| = 1. \tag{6.11}$$

The mapping $(\vartheta_k, I_k) \to (\vartheta_{k+1}, I_{k+1})$ (6.10) can be presented in the following symplectic form[5]

$$J_k = I_k - \epsilon\frac{\partial S_k}{\partial\vartheta_k}, \qquad \psi_k = \vartheta_k + \epsilon\frac{\partial S_k}{\partial J_k}, \tag{6.12}$$

$$\bar{\psi}_k = \psi_k + w(J_k, \epsilon)(\tau_{k+1} - \tau_k), \tag{6.13}$$

$$I_{k+1} = J_k + \epsilon\frac{\partial S_{k+1}}{\partial\vartheta_{k+1}}, \qquad \vartheta_{k+1} = \bar{\psi}_k - \epsilon\frac{\partial S_{k+1}}{\partial J_k}, \tag{6.14}$$

where $S_k \equiv S(\vartheta_k, J_k, \tau_k, \tau_0, \epsilon)$, $S_{k+1} \equiv S(\vartheta_{k+1}, J_k, \tau_{k+1}, \tau_0, \epsilon)$. The generating function $S(\vartheta, J, t, \tau_0, \epsilon)$ is the solution of the Hamilton-Jacobi equation

---

[4] The word "symplectic" comes from Greek meaning "twining or plaiting together", since the canonical coordinates q_i and momenta p_i are intertwined by symplectic 2-form given as a sum of wedge products of dp_i with dq_i, $\omega^2 = \sum_{i=1}^N dq_i \wedge dp_i = \sum_{i=1}^N dI_i \wedge d\vartheta_i$. This structure is invariant under the Hamiltonian flow given by Eq. (6.8) (see, e.g. Arnold 1989). The wedge product $dq_i \wedge dp_i$ is equal to an area of a parallelogram with sides dq_i and dp_i.

[5] Symplectic mapping method for integrating the Hamiltonian system (6.7) and (6.8) based on the method of canonical transformation of variables and the classical perturbation theory has been developed by Abdullaev et al. (1999); Abdullaev (1999, 2002, 2004a) (see also a book by Abdullaev (2006)).

$$H_0\left(J+\epsilon\frac{\partial S}{\partial\vartheta}\right)+\epsilon H_1\left(J+\epsilon\frac{\partial S}{\partial\vartheta},\vartheta,\epsilon\right)+\epsilon\frac{\partial S}{\partial\tau}=\mathcal{H}(J,\epsilon), \tag{6.15}$$

satisfying the initial condition

$$S(J,\vartheta,\tau,\tau_0,\epsilon)\Big|_{\tau=\tau_0}=0. \tag{6.16}$$

The quantity $w(J,\epsilon)$ is the perturbed frequency $w(J,\epsilon)$ of the system determined by the new Hamiltonian $\mathcal{H}(J,\epsilon)$,

$$w(J,\epsilon)=\frac{\partial\mathcal{H}(J,\epsilon)}{\partial J}. \tag{6.17}$$

6.2.3 Generating Function

For small perturbation parameter, ϵ, we can seek the generating function $S(\theta, J, t; \epsilon)$ as series in powers of ϵ:

$$S(J,\vartheta,\tau,\tau_0,\epsilon)=S_1(J,\vartheta,\tau,\tau_0)+\epsilon S_2(J,\vartheta,\tau,\tau_0)+\cdots. \tag{6.18}$$

We also expand the old, $H_1(I,\vartheta,\tau,\epsilon)$ and new, $\mathcal{H}(J;\epsilon)$, Hamiltonians in a similar series

$$\begin{aligned}\epsilon H_1(I,\vartheta,\epsilon)&=\epsilon H_1(I,\vartheta)+\epsilon^2H_2(I,\vartheta)+\cdots,\\ \mathcal{H}(J,\epsilon)&=\mathcal{H}_0(J)+\epsilon\mathcal{H}_1(J)+\epsilon^2\mathcal{H}_2(J)+\cdots.\end{aligned} \tag{6.19}$$

Expanding the Hamilton–Jacobi Eq. (6.15) in a series in powers of ϵ and equating the terms with the same power in ϵ we obtain the relation

$$\mathcal{H}_0(J)=H_0(J), \tag{6.20}$$

and the equations for the expansion coefficients $S_i\equiv S_i(\vartheta,J,\tau)$ of the generating function $S(\vartheta,J,\tau;\epsilon)$:

$$\frac{\partial S_1}{\partial\tau}+\frac{\partial H_0}{\partial J}\cdot\frac{\partial S_1}{\partial\vartheta}=\mathcal{H}_1(J)-H_1(\vartheta,J), \tag{6.21}$$

$$\frac{\partial S_j}{\partial\tau}+\frac{\partial H_0}{\partial J}\cdot\frac{\partial S_j}{\partial\vartheta}=\mathcal{H}_j(J)-F_j(\vartheta,J,\tau),\qquad j\geq 2, \tag{6.22}$$

where $F_j(\vartheta,J,\tau)$ are the polynomial functions of derivatives $\partial S_1/\partial\vartheta,\ldots,\partial S_{j-1}/\partial\vartheta$. Particularly, for $F_2(\vartheta,J,\tau)$ we have

$$F_2(\vartheta, J, \tau) = H_2(\vartheta, J, \tau) + \frac{1}{2}\frac{\partial S_1}{\partial \vartheta} \cdot \frac{\partial^2 H_0}{\partial J \partial J} \cdot \frac{\partial S_1}{\partial \vartheta} + \frac{\partial H_1}{\partial J} \cdot \frac{\partial S_1}{\partial \vartheta}. \tag{6.23}$$

In Eqs. (6.21), (6.22) and (6.23) the notations $a \cdot b$, $a \cdot c \cdot b$ stand for $a \cdot b = \sum_{i=1}^{N} a_i b_i$ and $a \cdot c \cdot b = \sum_{i,j=1}^{N} a_i c_{ij} b_j$, respectively.

The equations for S_i are first order partial differential equations and can be solved by the method of characteristic equations. The left hand sides of Eqs. (6.21) and (6.22) may be written as total time derivatives of $S_i \equiv S_i(\vartheta, J, \tau, \tau_0)$ taken along the unperturbed trajectory $(\vartheta(\tau), J(\tau))$ of the Hamiltonian $H_0(J)$ and satisfying the condition $\vartheta(\tau = \tau_0) = \vartheta_0$, i.e., $\vartheta(\tau) = \vartheta_0 + \omega(J)(\tau - \tau_0)$, $J(\tau) = \text{const}$. The solution of (6.21) can be written as the integral

$$S_1(\vartheta, J, \tau, \tau_0) = \mathcal{H}_1(J)(\tau - \tau_0) - \int_{\tau_0}^{\tau} H_1(\vartheta_0(\tau'), J) d\tau', \tag{6.24}$$

where integration is taken along the unperturbed trajectory $\vartheta_0(\tau') = \vartheta + \omega(J)$ $(\tau - \tau')$. The latter it can be rewritten in the form of Fourier series

$$\begin{aligned} S_1(\vartheta, J, \tau, \tau_0) &= -(\tau - \tau_0) \sum_m c(x_m) H_{mn}(J) e^{im\cdot\vartheta} \\ &= -(\tau - \tau_0) \sum_m |H_m(J)| [a(x_m) \sin(m \cdot \vartheta + \chi_m) \\ &\quad + b(x_m) \cos(m \cdot \vartheta + \chi_m)], \end{aligned} \tag{6.25}$$

where $c(x) = b(x) - ia(x)$ is the complex function with real and imaginary parts:

$$a(x) = \frac{1 - \cos x}{x}, \quad b(x) = \frac{\sin x}{x}. \tag{6.26}$$

Here the argument x_{mn} of these functions is given by

$$x_m = [m \cdot \omega(J)](\tau - \tau_0).$$

The higher order generating functions S_i $(i \geq 2)$ are determined by the integrals

$$S_i(\vartheta, J, \tau, \tau_0) = \mathcal{H}_i(J)(\tau - \tau_0) - \int_{\tau_0}^{\tau} F_i(\vartheta_0(\tau'), J, \tau', \tau_0) d\tau', \quad (i \geq 2). \tag{6.27}$$

If the functions $F_j(\vartheta, J, \tau, \tau_0)$ are presented as a Fourier series in angles ϑ,

$$F_j(\vartheta, J, \tau, \tau_0) = \sum_k F_k^{(j)}(J)e^{ik\cdot\vartheta}$$
$$= \sum_k |F_k^{(j)}(J)| \cos\left(k \cdot \vartheta + \chi_k^{(j)}\right), \tag{6.28}$$

where $F_k^{(j)}(I) = |F_k^{(j)}(I)| \exp(i\chi_k^{(j)})$, the generating functions S_i ($i \geq 2$) can be also reduced to the form like Eq. (6.25).

Finally, according to the expansion (6.19) the perturbed frequency $w(J, \epsilon)$ (6.17) can be determined by the series in powers of ϵ,

$$w(J, \epsilon) = \omega(J) + \epsilon \frac{\partial \mathcal{H}_1(J)}{\partial J} + \cdots . \tag{6.29}$$

The mapping (6.12)–(6.14) can be also considered as the alternative method of the symplectic integration of Hamiltonian systems (6.8) and (6.28) (see Abdullaev (2002)). This mapping with large integration steps, $\Delta\tau = \tau_{k+1} - \tau_k$, comparable with the characteristic periods of the system (e.g., a perturbation period) has the same accuracy as the conventional symplectic integrators (see, for example, a review Sanz-Serna 1991) with integration steps two or three orders smaller. The mapping can be applied to Hamiltonian systems with moderately large perturbation $\epsilon \sim 1$ by taking the mapping step $\Delta\tau$ sufficiently small. The details of the computational aspects of the mapping (6.12)–(6.14) can be found in Abdullaev (2002, 2006).

6.2.4 Mapping of Magnetic Field Lines

Consider the application of the mapping procedure for the Hamiltonian system (6.8) and (6.28) to the magnetic field lines. The corresponding equations for the field lines are given by Eqs. (3.9) and (3.10). The toroidal angle φ corresponds to the independent time–like variable τ, the toroidal flux ψ_t and the poloidal angle ϑ are the action–angle variables (ϑ, I), and the poloidal flux ψ is the Hamiltonian.

Let us introduce the poloidal sections $\varphi = \varphi_k = 2\pi k/N$, $(k = 1, \ldots, N)$ $(N \geq 1)$. Let (ϑ_k, ψ_k) be the poloidal angle and toroidal flux at the poloidal sections $\varphi = \varphi_k$, $(\vartheta_k = \vartheta(\varphi_k), \psi_k = \psi_t(\varphi_k)$. The mapping scheme is illustrated in Fig. 6.2a, b for the case $N = 6$. We define the mapping of field lines,

$$(\vartheta_{k+1}, \psi_{k+1}) = \hat{M}(\vartheta_k, \psi_k), \tag{6.30}$$

which relates the variables (ϑ, ψ) at the successive sections φ_k and φ_{k+1}. Then the mapping $(\hat{M})^N$ defines Poincaré mapping. The area-preserving property (6.11) corresponds to the flux-preserving property

$$\left| \frac{\partial(\vartheta_{k+1}, \psi_{k+1})}{\partial(\vartheta_k, \psi_k)} \right| = 1. \tag{6.31}$$

The general form of the mapping (6.12)–(6.14) for field lines takes the form

$$\Psi_k = \psi_k - \epsilon \frac{\partial S^{(k)}}{\partial \vartheta_k}, \quad \Theta_k = \vartheta_k + \epsilon \frac{\partial S^{(k)}}{\partial \Psi_k}, \tag{6.32}$$

$$\Psi_{k+1} = \Psi_k, \quad \bar{\Theta}_k = \Theta_k + w(\Psi_k, \epsilon)(\varphi_{k+1} - \varphi_k), \tag{6.33}$$

$$\psi_{k+1} = \Psi_{k+1} + \epsilon \frac{\partial S^{(k+1)}}{\partial \vartheta_{k+1}}, \quad \vartheta_{k+1} = \bar{\Theta}_k - \epsilon \frac{\partial S^{(k+1)}}{\partial \Psi_{k+1}}, \tag{6.34}$$

where $w(\Psi, \epsilon) = \partial \mathcal{H}/\partial \Psi$ is the rotational transform (inverse of the safety factor) of perturbed equilibrium magnetic field. The generating function $S^{(k)} \equiv S(\vartheta_k, \Psi_k)$ is the value of the generating function $S(\vartheta, \Psi, \varphi, \varphi_0; \epsilon)$ taken at sections $\varphi = \varphi_k$, and according to (6.24) and (6.27) is given by the expansion

$$S(\vartheta, \Psi, \varphi, \varphi_0; \epsilon) = S_1(\vartheta, \Psi, \varphi, \varphi_0) + \epsilon S_2(\vartheta, \Psi, \varphi, \varphi_0) + \cdots . \tag{6.35}$$

In the first order of ϵ the frequency $w(\Psi, \epsilon)$ is determined by the inverse safety factor $q(\Psi)$:

$$w(\Psi, \epsilon) = 1/q(\Psi), \tag{6.36}$$

and the generating function $S(\vartheta, \Psi, \varphi, \varphi_0)$ in the finite interval $\varphi_{k+1} < \varphi < \varphi_k$ is given by

$$S_1(\vartheta, \Psi, \varphi, \varphi_0) = - \int_{\varphi_0}^{\varphi} \psi_1 \left(\vartheta(\varphi'), \Psi, \varphi' \right) d\varphi'. \tag{6.37}$$

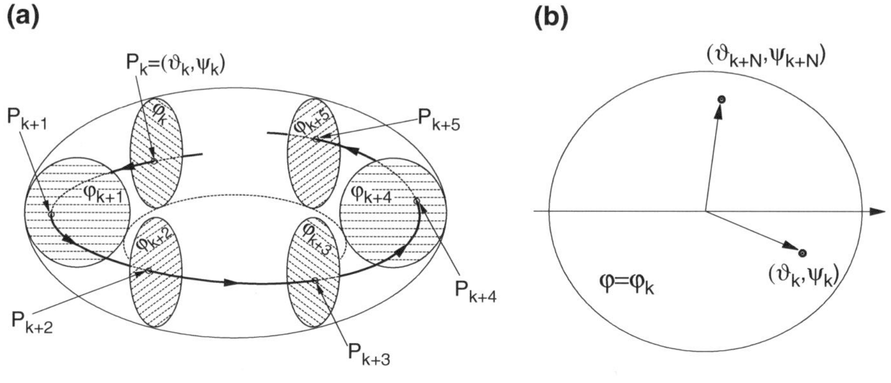

Fig. 6.2 Geometry of the mapping in a toroidal system: **a** Scheme of the mapping along the toroidal angle for the case $N = 6$; **b** Poincaré mapping of field lines in the poloidal plane $\varphi = \varphi_k$

The integration in (6.37) is taken along the unperturbed orbit $\vartheta(\varphi') = \vartheta(\varphi) + (\varphi' - \varphi)/q(\Psi)$. Similar expressions we obtain for the higher–order generating functions $S_j, (j \geq 2)$,

$$S_j(\vartheta, \Psi, \varphi, \varphi_0) = -\int\limits_{\varphi_0}^{\varphi} F_j\left(\vartheta(\varphi'), \Psi, \varphi'\right) d\varphi'. \tag{6.38}$$

Using the Fourier expansions (3.12) and (3.13) for the perturbation function ψ_1 $(\vartheta, \psi, \varphi)$ one obtains

$$S_1(\vartheta, \Psi, \varphi, \varphi_0) = -(\varphi - \varphi_0) \sum_{m,n} H_{mn}(\Psi)\big[a(x_{mn}) \sin(m\vartheta - n\varphi + \chi_{mn}) + b(x_{mn}) \cos(m\vartheta - n\varphi + \chi_{mn})\big], \tag{6.39}$$

where the arguments x_{mn} of the functions $a(x)$ and $b(x)$ (6.26) are given by

$$x_{mn} = \left(\frac{m}{q(\Psi)} - n\right)(\varphi - \varphi_0).$$

The free parameter φ_0 lies in the interval $\varphi_k \leq \varphi_0 \leq \varphi_{k+1}$. The mapping (6.32) is called *symmetric* if the free parameter is taken exactly in the middle of the interval $[\varphi_k, \varphi_{k+1}]$, i.e., $\varphi_0 = (\varphi_k + \varphi_{k+1})/2$.

6.3 Simplified Forms of the Mapping

Below we consider the approximation which simplifies the form of the generating functions S_k, S_{k+1} of the mapping (6.12)–(6.14). To be specific we identify the independent time–like variable τ with the time t. The corresponding frequency ω_N and the mode number m_N are notated as Ω and s.

6.3.1 Mappings for the Broad Toroidal Perturbation Mode Spectrum

Consider the time–periodic perturbation with broad spectrum in s. The corresponding perturbation Hamiltonian $H_1(\vartheta, I, t)$ can be presented in the form

$$H_1(\vartheta, I, t) = \sum_m \sum_{s=-M}^{M} H_m(\psi) \cos(m\vartheta - s\Omega t + \chi_m), \tag{6.40}$$

with a finite but a large number of harmonics M ($M \gg 1$) of the fundamental mode of the frequency Ω. In many works starting from the pioneering papers by Zaslavsky and Chirikov (1971), Chirikov (1979) the such Hamiltonian systems are simplified using the Poisson summation rule,

$$\sum_{s=-\infty}^{\infty} \cos(s\Omega t) = \frac{2\pi}{\Omega} \sum_{k=-\infty}^{\infty} \delta\left(t - \frac{k2\pi}{\Omega}\right), \tag{6.41}$$

where $\delta(x)$ is the Dirac's delta function. Then the Hamiltonian (6.40) is replaced by the delta-kicked one

$$\begin{aligned} H_1(\vartheta, I, t) &= \frac{2\pi}{\Omega} V(I, \vartheta) \sum_{k=-\infty}^{\infty} \delta\left(t - \frac{k2\pi}{\Omega}\right), \\ V(\psi, \vartheta) &= \sum_m H_m(\psi) \cos(m\vartheta + \chi_m). \end{aligned} \tag{6.42}$$

It is supposed that the Hamiltonian Eq. (6.8) with this Hamiltonian function is easy to integrate. The uncritical use of integration along the delta function yields the symplectic mapping in the form

$$\begin{aligned} I_{k+1} &= I_k + \epsilon \frac{\partial G(\vartheta_{k+1}, I_k)}{\partial \vartheta_{k+1}}, \\ \vartheta_{k+1} &= \vartheta_k + \frac{2\pi\omega(I_{k+1})}{\Omega} - \epsilon \frac{\partial G(\vartheta_{k+1}, I_{k+1})}{\partial I_{k+1}}, \end{aligned} \tag{6.43}$$

with the generating function

$$G(\vartheta, I) = \frac{2\pi}{\Omega} V(I, \vartheta). \tag{6.44}$$

However, this method of delta functions to construct mappings encounters the fundamental mathematical difficulty (see Abdullaev (2006), (2007) and references therein) related with the undefined procedure of the integration of dynamical equations along the δ functions. The mathematically correct way to avoid this difficulty is follow. First one should construct the symplectic maps (6.12)–(6.14) for these systems by the canonical change of variables and then take the limit $M \to \infty$ in the generating function. According to (6.24) and (6.42) one obtains

$$S_1(\vartheta, J, t, t_0) = -\frac{2\pi}{\Omega} \int_{t_0}^{t} V\left(J, \vartheta(t')\right) \sum_{k=-\infty}^{\infty} \delta\left(t' - \frac{k2\pi}{\Omega}\right) dt'. \tag{6.45}$$

From (6.45) it follows that the function $S_1(\vartheta, J, t, t_0)$ vanishes in the time intervals $t_k < t < t_{k+1}$, $t_k = (2\pi/\Omega)k$, $t_k < t_0 < t_{k+1}$. But at the time instants $t = t_k + 0$

$t = t_{k+1} - 0$ it takes the finite values,

$$S_k = S_1(\vartheta, J, t_k, t_0) = \frac{\pi}{n} V(J, \vartheta_k),$$
$$S_{k+1} = S_1(\vartheta, J, t_{k+1}, t_0) = -\frac{\pi}{\Omega} V(J, \vartheta_{k+1}). \qquad (6.46)$$

With these generating functions the mapping (6.12)–(6.14) is reduced to

$$\begin{aligned} J_k &= I_k - \epsilon \frac{\partial S(\vartheta_k, J_k)}{\partial \vartheta_k}, \\ \Theta_k &= \vartheta_k + \epsilon \frac{\partial S(\vartheta_k, J_k)}{\partial J_k}, \\ \bar{\Theta}_k &= \Theta_k + \frac{2\pi\omega(J_k)}{\Omega}, \\ I_{k+1} &= J_k - \epsilon \frac{\partial S(\vartheta_{k+1}, J_k)}{\partial \vartheta_{k+1}}, \\ \vartheta_{k+1} &= \bar{\Theta}_k + \epsilon \frac{\partial S(\vartheta_{k+1}, J_k)}{\partial J_k}, \end{aligned} \qquad (6.47)$$

with the generating function $S(\vartheta, J) = G(\vartheta, J)/2$, where $G(\vartheta, J)$ is given by (6.44).

As was shown by Abdullaev (2004a) that it is *impossible to obtain the mapping in the form* (6.43) with the generating function (6.44). If one attempts to construct the mapping (6.43) its generating function would have a complicated form

$$G(\vartheta, J) = \frac{2\pi}{\Omega} \sum_m H_m(J) \cos\left[\frac{\pi m \omega(J)}{\Omega}\right] \cos\left(m\vartheta + \chi_m + \frac{\pi m \omega(J)}{\Omega}\right). \qquad (6.48)$$

In the case when the perturbation harmonics $H_m(J)$ does not depend on the action variable J (or the flux variable Ψ) the mapping (6.47) can be reduced to the form (6.43), but for the intermediate variables (Θ, J), i.e.,

$$\begin{aligned} J_k &= J_{k-1} - \epsilon \frac{\partial S(\Theta_k)}{\partial \Theta_k}, \\ \Theta_{k+1} &= \Theta_k + \frac{2\pi\omega(J_k)}{\Omega}. \end{aligned} \qquad (6.49)$$

However, the action variable J in the mapping (6.49) does not coincide with the corresponding variable of continuous Hamiltonian system.

The nonsymmetric mapping of the type (6.43) for the intermediate variables $(\bar{\Theta}, J)$, i.e., $(\bar{\Theta}_k, J_k) \to (\bar{\Theta}_{k+1}, J_{k+1})$, can be approximately obtained from the symmetric mapping (6.47) for the small perturbation parameter ϵ.

6.3.2 The Simple Form of the Generating Function

The above procedure to obtain the mapping (6.47) give an idea to get a more simple form of the generating functions S_k, S_{k+1} for the mapping (6.12)–(6.14). It is based on the canonical change of variables which reduces the perturbation Hamiltonian to the form (6.42). The details of this procedure to obtain the generating functions are given in Sect. D. The simplified form of the generating functions $S^{(k)}$, $S^{(k+1)}$ of the mappings (6.12)–(6.14) and (6.32) are given by

$$\begin{aligned} S^{(k)} &= S(\vartheta_k, J_k, t_k), \\ S^{(k+1)} &= -S(\vartheta_{k+1}, J_{k+1}, t_{k+1}), \\ S(\vartheta, J, \varphi) &= \frac{\Delta t}{2} H_1(J, \vartheta, t). \end{aligned} \tag{6.50}$$

The above mappings with these generating functions requires less computational times than with the one given by (6.25) and (6.39).

6.4 Full-Turn Transfer Mapping: The Separatrix Mapping

In this section we construct the Poincaré mapping (6.1) for the general Hamiltonian system (6.8) and (6.9), and particularly, for field lines and guiding-center orbits to the certain sections of the torus. Following the terminology used in the accelerator physics we will call this mapping a *full–turn transfer mapping*. Figures 6.3a, b show the geometrical scheme of such mappings: (a) the orbits are mapped to the outer equatorial plane $z = 0$; (b) the orbits are mapped to the section Σ_S in the toroidal plasmas with the magnetic separatrix. In the latter case the Poincaré mapping describes the behavior of field lines and guiding–center orbits near the separatrix and the divertor plates, and it is often called the *separatrix mapping*.

In the next chapters we will use the full–turn transfer mapping an efficient computational and analytical tools to study magnetic field lines and particle dynamics. Further we will use the term "separatrix mapping" when the mapping is applied to the dynamics of system near the separatrix.

6.4.1 Definition of the Full-Turn Transfer Mapping

The geometries of the full–turn transfer mapping are shown in Figs. 6.3a, b. Let P_k be the kth crossing point of a perturbed field line with the section Σ_s where the angles ϑ are set $\pm\pi$. In plasmas with the circular cross section this section can be taken at the $\vartheta = 0$ (Fig. 6.3a). In plasmas with the magnetic separatrix it is convenient to take the section Σ_s consisting of two mutual perpendicular planes (ξ, φ) and (η, φ) as shown in Fig. 6.3b. The full–turn transfer mapping is the Poincaré mapping which

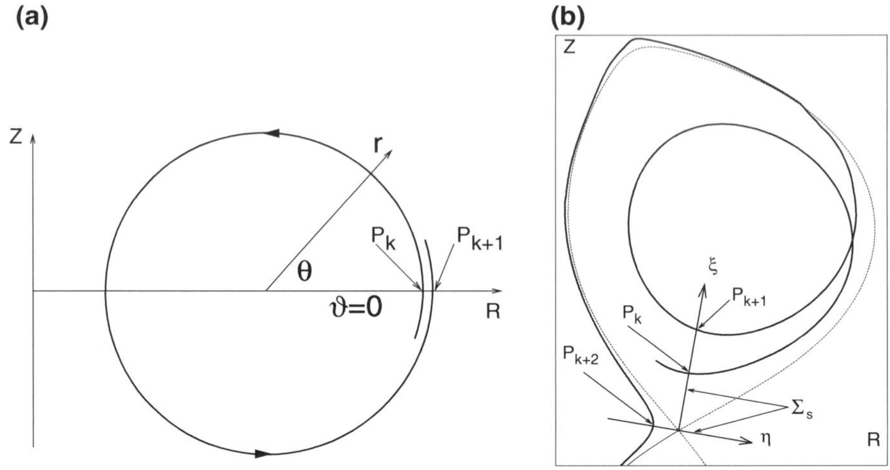

Fig. 6.3 Schematic view of the full-turn transfer mapping $P_k \rightarrow P_{k+1}$ of field lines on the $\theta = 0$-plane (**a**) and near the separatrix on the section Σ_s [exaggerated] (**b**). *Dashed curve* describe the separatrix, *solid curve* corresponds to perturbed field lines; the section Σ_s consists of two segments of the ξ- and η-axes. The unperturbed field lines transversely cross the axes (ξ, η)

relates the point P_k with the next crossing point P_{k+1}: $P_{k+1} = \hat{M} P_k$. For the full description one should introduce a *forward* and *backward* mappings which map the orbit along the positive and negative direction of the toroidal axis φ, respectively. Particularly, for field lines the crossing point P_k is defined by the canonical variables: the toroidal angle, φ_k, and the relative poloidal flux, ψ_k.

The canonical forms of the such mapping for the Hamiltonian systems with the $1 + \frac{1}{2}$–degrees–of–freedom are constructed by Abdullaev (2004a, 2005). Below first we construct the mapping for the general many–dimensional Hamiltonian systems (6.8) and (6.9), and then write down its explicit form for the guiding center orbits in the presence of the non-axisymmetric magnetic and electric perturbations (5.31). The mapping for the field lines will be a particular case of this mapping.

6.4.2 The General Form the Mapping

Consider Hamiltonian systems (6.8) and (6.9). Let $\vartheta \equiv \vartheta_1$ be the one of the angle variables $(\vartheta_1, \ldots, \vartheta_N, \vartheta_{N+1})$ which takes constant value $\vartheta_\Sigma = \pm\pi$ by mod2π at the Poincaré section Σ as was shown in Fig. 6.1. We notate the other angular variables as $\Theta = (\vartheta_2, \ldots, \vartheta_N, \vartheta_{N+1} = \tau)$. Let (Θ_k, I_k) be the values of the angle variables Θ and the action $I = (I_1, \ldots, I_N, I_{N+1} = H)$ at the section Σ after the kth full–turn of the orbit the along angle ϑ, i.e., $\vartheta_{k+1} = \vartheta_k + 2\pi$. Below we will use the short notation $(\vartheta_0(\tau), Z_0(\tau, I))$ for the unperturbed orbit, where $Z_0(\tau, I) = (I(\tau), \Theta_0(\tau))$. We also introduce the notations $\omega(I) \equiv \omega_1(I, H)$ and $\Omega(I) = (\omega_2, \ldots, \omega_N, 1)$ for the unperturbed frequencies of motion (6.6).

Using the general form of the mapping (6.12)–(6.14), and the smallness of perturbation, $\epsilon \ll 1$, one can derive the full–turn transfer mapping. The details of the derivation of the mapping is given in Appendix G. The mapping $P_{k+1} = \hat{M} P_k$ has the following form,

$$\begin{aligned} I_{k+1} &= I_k - \epsilon \frac{\partial G(\Theta_k, J_k)}{\partial \Theta_k}, \\ \Theta_{k+1} &= \Theta_k + \frac{2\pi \Omega(J_k)}{\omega(J_k)} + \epsilon \frac{\partial G(\Theta_k, J_k)}{\partial J_k}, \end{aligned} \tag{6.51}$$

where $G(\Theta_k, J_k)$ is the generating function

$$G(\Theta_k, J_k) = P\left(\Theta_k + \frac{\pi \Omega(J_k)}{\omega(J_k)}, J_k\right), \tag{6.52}$$

defined by the integral

$$P(\Theta(\tau_0), J) = \int_{-\pi/\omega(J)}^{\pi/\omega(J)} H_1(\vartheta_0(\tau), Z_0(\tau + \tau_0, J))\, d\tau, \tag{6.53}$$

taken over the perturbation Hamiltonian $H_1(I, \vartheta)$ (6.9) along the unperturbed orbit $Z_0(t, J)$. In (6.53) the quantity τ_0 is the time instant when the orbit crosses the section $\vartheta = 0$. We will call the function $P(\Theta; I)$ as the *Poincaré integral.*

The mapping (6.51) contains the intermediate action variables J which are found by the canonical transformations of the initial variables Θ_k and I_k,

$$J_k = I_k - \epsilon \frac{\partial S^{(-)}(\Theta_k, J_k)}{\partial \Theta_k}, \tag{6.54}$$

with the generating function $S^{(-)}(\Theta, J)$,

$$S^{(-)}(\Theta(\tau_0), J) = \int_{-\pi/\omega(J)}^{0} H_1(\vartheta_0(\tau), Z_0(\tau + \tau_0, J))\, d\tau. \tag{6.55}$$

The full–turn transfer mapping $(\Theta_k, I_k) \rightarrow (\Theta_{k+1}, I_{k+1})$ given by Eq. (6.51) is an invariant with respect to time–reverse $\tau \rightarrow -\tau$, i.e., $k \leftrightarrow k+1$. It is also the area–preserving and satisfies the condition (6.11).

The full-turn transfer mapping (6.51) can be simplified by transforming it into the form containing only the variables (I_k, I_{k+1}). As was shown in Appendix G such a form of the mapping which has a time–reversal symmetry and an area preserving reads,

$$
\begin{aligned}
I_{k+1} &= I_k - \epsilon \frac{\partial G(\Theta_k, I_{k+1})}{\partial \Theta_k}, \\
\Theta_{k+1} &= \Theta_k + \frac{\pi \Omega(I_k)}{\omega(I_k)} + \frac{\pi \Omega(I_{k+1})}{\omega(I_{k+1})} + \epsilon \frac{\partial G(\Theta_k, I_{k+1})}{\partial I_{k+1}},
\end{aligned} \tag{6.56}
$$

where

$$
G(\Theta_k, I_{k+1}) = P\left(\Theta_k + \frac{\pi \Omega(I_k)}{\omega(I_k)}, I_{k+1}\right). \tag{6.57}
$$

6.4.3 The Mapping of Guiding Center Orbits

The dynamics of the guiding center orbit is governed by Eqs. (5.32) or (4.48) with the Hamiltonian (5.32) and (5.33). In the action–angle variables $(\vartheta, \vartheta_\varphi, J, I_\varphi = p_\varphi)$ the equations of motion in these variables are given by Eq. (5.36) with the Hamiltonian functions (5.37) and (5.38). We assume that the amplitude of the perturbations are small, $\epsilon \ll 1$.

Let $(\vartheta_k, \vartheta_{k\varphi}, t_k, J_k, p_k, h_k)$ be the values of the angle variables $\vartheta, \vartheta_\varphi$, the time instant t, the action J, the toroidal momentum p_φ, and the energy h_c at the kth crossing point P_k of the orbit the section Σ_s. According to the relation (5.22) between the toroidal angle φ and the angles $(\vartheta, \vartheta_\varphi)$ one can set $\varphi_k = \vartheta_{k\varphi}$. The transit frequency ω of an orbit stands the frequency of motion along the angle ϑ. The frequency $\Omega(p, h) = (\omega_\varphi, 1)$ where ω_φ is the frequency of toroidal motion. According to (5.27) the ratio ω_φ/ω is equal to the effective safety factor $q_p(p, h)$. Then the mapping (6.56) for the guiding center orbits can be reduced to

$$
\begin{aligned}
h_{k+1} &= h_k - \epsilon \frac{\partial G_k}{\partial t_k}, \\
t_{k+1} &= t_k + \frac{\pi}{\omega(p_k, h_k)} + \frac{\pi}{\omega(p_{k+1}, h_{k+1})} + \epsilon \frac{\partial G_k}{\partial h_{k+1}},
\end{aligned} \tag{6.58}
$$

$$
\begin{aligned}
p_{k+1} &= p_k - \epsilon \frac{\partial G_k}{\partial \varphi_k}, \\
\varphi_{k+1} &= \varphi_k + \pi \left[q_p(p_k, h_k) + q_p(p_{k+1}, h_{k+1}) \right] + \epsilon \frac{\partial G_k}{\partial p_{k+1}},
\end{aligned} \tag{6.59}
$$

with the Poincare integral P_k is given by

$$
G_k = P\left(t_k + \frac{\pi}{\omega(p_k, h_k)}, \varphi_k + \pi q_p(p_k, h_k), p_{k+1}, h_{k+1}\right). \tag{6.60}
$$

The Poincaré integral $P\,(t, \varphi; p, h)$ defined by (6.53) is taken over the perturbation Hamiltonian $h_1(Z_0)$ (5.34) along the unperturbed guiding center orbit $Z_0(\tau; J) =$

$(z_0(\tau), \varphi_0(\tau), t_0(\tau), p_{z0}(\tau), p_{0\varphi}(\tau), h_0(\tau))$ in the 6-D phase–space $(z, \varphi, t, p_z, p_\varphi, h)$.

6.4.4 The Full-Turn Transfer Mapping for Field Lines

From the mapping (6.59) one can readily obtain the mapping of field lines. For field lines the toroidal momentum p_φ coincides with the poloidal flux ψ. Then omitting the time t, energy h variables, and replacing the effective safety factor $q_p(p_\varphi, h)$ by the safety factor $q(\psi)$ of the magnetic field, we obtain

$$\psi_{k+1} = \psi_k - \epsilon\frac{\partial P\,(\varphi_k + \pi q(\psi_k); \psi_{k+1})}{\partial \varphi_k},$$
$$\varphi_{k+1} = \varphi_k + \pi[q(\psi_k) + q(\psi_{k+1})] + \epsilon\frac{\partial P\,(\varphi_k + \pi q(\psi_k); \psi_{k+1})}{\partial \psi_{k+1}}, \tag{6.61}$$

where $\psi_{k+1} = \psi(\tau_k)$ is the value of the poloidal flux at the kth crossing point P_k of the field line the section Σ_s. The Poincaré integral $P\,(\varphi, \psi)$ for field lines,

$$P(\varphi; \psi) = \int_{-\pi q(\psi)}^{\pi q(\psi)} \psi_1\left(z(\varphi'; \psi), p_z(\varphi'; \psi), \varphi' + \varphi\right) d\varphi', \tag{6.62}$$

is obtained from (6.53) by changing the integration variable τ to the toroidal angle $\varphi = \tau\omega_\varphi$ and replacing the perturbation Hamiltonian h_1 by the perturbed poloidal flux $\psi_1\,(z, p_z, \varphi)$ given by Eq. (3.12).

The mapping (6.61) is the *forward mapping* of field lines, i.e., it maps field line along the positive direction of the toroidal angle φ. The *backward mapping* can be obtained from (6.59) similar to the forward mapping but changing the integration variable τ to $\varphi = -\tau\omega_\varphi$. It gives

$$\psi_{k+1} = \psi_k + \epsilon\frac{\partial P\,(\varphi_k - \pi q(\psi_k); \psi_{k+1})}{\partial \varphi_k},$$
$$\varphi_{k+1} = \varphi_k - \pi[q(\psi_k) + q(\psi_{k+1})] - \epsilon\frac{\partial P\,(\varphi_k - \pi q(\psi_k); \psi_{k+1})}{\partial \psi_{k+1}}. \tag{6.63}$$

It is easy to show that the mappings (6.61) and (6.63) are a flux–preserving, i.e.,

$$\left|\frac{\partial(\psi_{k+1}, \varphi_{k+1})}{\partial(\psi_k, \varphi_k)}\right| = \left|\frac{\partial\psi_{k+1}}{\partial\psi_k}\frac{\partial\varphi_{k+1}}{\partial\varphi_k} - \frac{\partial\psi_{k+1}}{\partial\varphi_k}\frac{\partial\varphi_{k+1}}{\partial\psi_k}\right| = 1. \tag{6.64}$$

The following rules are applied to the mapping (6.61) and (6.63). For the simplicity set $\psi = 0$ at the separatrix. The field line is inside the plasma region when the relative poloidal flux ψ is negative at the crossing point of a field line with the upper branch of the ξ- axis, $\psi_k < 0$ (see Fig. 6.3. The field line is supposed to be left the plasma region if $\psi_k > 0$, i.e., when it crosses the η-axis. This assumption works practically well for the sufficiently small values of the perturbation parameter ϵ, which is confirmed with the numerical computations (Abdullaev (2005)).

6.5 Poincaré and Melnikov Integrals

In this section we study the property of the Poincaré integral (6.53) introduced in Sect. 6.4.2 to construct the full–turn transfer mappings. We shall also introduce the so-called *Melnikov integral*. These integrals are the generalization of the corresponding integrals introduced by Poincaré (1892/1893/1899) and Melnikov (1963b) to study the dynamics of systems near the separatrices, in particular, for the quantitative estimations of the splittings of separatrices perturbed dynamical systems (Poincaré (1892/1893/1899); Melnikov 1963b; Arnold et al. 2006, see also Sect. 7.5).

6.5.1 General Definitions

Consider the Hamiltonian system (6.8) and (6.9) and the integral (6.62) over the perturbation Hamiltonian $H_1(I, \vartheta)$ taken along the unperturbed orbit $Z(t)$ on the given tori $I = \text{const}$ for one full turn along the angle ϑ.

The integral $P(\Theta; I)$ (6.53) is the generalization of the integral $P(t) = P(t)$ introduced by Poincaré to study the splitting of the asymptotic surfaces (separatrices). The latter is defined as the integral taken over the perturbation Hamiltonian $H_1(Z)$ along the unperturbed orbits on the separatrices $Z_{sep}(t)$ (see, e.g., Arnold et al. 2006), i.e.,

$$P(t) = \int_{-\infty}^{\infty} H_1 \left[Z_{sep}(\tau + t), \tau\right] d\tau. \tag{6.65}$$

The similar integral known as the *Melnikov integral* has been introduced by Melnikov (1963b)[6]

$$M(t) = \int_{-\infty}^{\infty} \{H_0, H_1\} \left(Z_{sep}(\tau + t; I, H, \tau\right) d\tau, \tag{6.66}$$

where $\{(\cdots)\}$ stands for the Poisson bracket,

[6] Interestingly that this integral has been introduced to study the magnetic field lines produced by the helical current flowing the surface of a torus (Melnikov (1963a)).

$$\{f, g\} \equiv \sum_{i=1}^{N} \left(\frac{\partial f}{\partial q_i} \frac{\partial g}{\partial p_i} - \frac{\partial g}{\partial q_i} \frac{\partial f}{\partial p_i} \right).$$

and (q_i, p_i), $(i = 1, \ldots, N)$ are the canonically conjugated variables. The Poincaré and the Melnikov integrals (6.65) and (6.66) are related,

$$M(t) = \partial P(t)/\partial t. \tag{6.67}$$

These integrals are usually called the *Poincaré–Melnikov* integrals.

Similar to the Poincaré integral defined Eq. (6.53) one can introduce the generalized Melnikov integral as the integral over the product $\{H_0, H_1\}$ taken along the unperturbed orbit $Z_0(t)$ on the given tori (I, H),

$$M(t, \Theta; I, H) = \int_{-\pi/\omega(I)}^{\pi/\omega(I)} \{H_0, H_1\} (Z_0(\tau + t; I, H, \tau) \, d\tau. \tag{6.68}$$

It is not difficult to show the integrals (6.53) and (6.68) are also related,

$$M(\Theta; I, H) = \frac{\partial P(\Theta; I)}{\partial t}. \tag{6.69}$$

Furthermore, we will also call the integral (6.68) the *Melnikov integral.*

Using the Fourier presentation of the perturbation Hamiltonian H_1 (6.9) one can obtain the Poincaré integral $P(\Theta; I)$ in a Fourier series. Introducing the notation, $n = (m_2, \ldots, m_{N+1})$, and the function

$$V_n(\vartheta, \Theta; I) = \sum_{m_1} H_m(I) e^{i m_1 \vartheta + i n \cdot \Theta}, \tag{6.70}$$

the Poincaré integral (6.53) can be rewritten as

$$P(\Theta; I) = \sum_n \int_{-\pi/\omega(I)}^{\pi/\omega(I)} V_n(\vartheta(\tau), \Theta(\tau + t), I) \, d\tau. \tag{6.71}$$

On the unperturbed orbits (6.5) we have $\Theta_i(\tau + t) = \Theta_i(\tau) + \Theta_i(t) - \Theta_{i0}$, $(i = 2, \ldots, N, N + 1)$ and therefore, one can write

$$V_n(\vartheta(\tau), \Theta(\tau + t), I) = V_n(\vartheta(\tau), \Theta(\tau), I) \, e^{i n \cdot \Theta(t) - i \chi_n},$$

where $\chi_n = n \cdot \Theta_0$ is the constant phase. Using the last formula in the integral (6.71) one obtains the following Fourier series,

$$
\begin{aligned}
P(\Theta; I, H) &= \sum_n R_n(I) e^{in\cdot\Theta - i\chi_n} \\
&= \sum_n \left[K_n(I)\cos\left(n\cdot\Theta - \chi_n\right) - L_n(I)\sin\left(n\cdot\Theta - \chi_n\right)\right], \qquad (6.72)
\end{aligned}
$$

where the function $R_n(I) = K_n(I) + iL_n(I)$ is defined by the integral over the function $V_n\left(\vartheta(\tau), \Theta(\tau); I\right)$,

$$
R_n(I) = \int_{-\pi/\omega(I)}^{\pi/\omega(I)} V_n\left(\vartheta(\tau), \Theta(\tau), I\right) d\tau. \qquad (6.73)
$$

Below we write down the Poincaré integral for the guiding-center motion described by the perturbed Hamiltonian (5.37) and (5.38). Recalling that in this case $\Theta = (\vartheta_\varphi, \Omega t)$, $I = (I_\varphi, h)$, and $n = (n, s)$ we have

$$
\begin{aligned}
P(t, \vartheta_\varphi; I_\varphi, h) &= \sum_{n,s} R_{ns}(I_\varphi, h) e^{in\vartheta_\varphi - is\Omega t - i\chi_n} \\
&= \sum_{n,s} \left[K_{ns}(I_\varphi, h)\cos\left(n\vartheta_\varphi - s\Omega t - \chi_n\right)\right. \\
&\quad \left. - L_{ns}(I_\varphi, h)\sin\left(n\vartheta_\varphi - s\Omega t - \chi_n\right)\right], \qquad (6.74)
\end{aligned}
$$

where the function $R_{ns} = K_{ns} + iL_{ns}$ is given by

$$
\begin{aligned}
R_{ns}(I_\varphi, h) &= \int_{-\pi/\omega(I)}^{\pi/\omega(I)} \tilde{V}_{ns}(\tau) d\tau, \\
\tilde{V}_{ns}(t) &= \sum_m H_{mns}(I_\varphi, h) e^{im\vartheta(t) + in\vartheta_\varphi(t) - is\Omega t}. \qquad (6.75)
\end{aligned}
$$

6.5.2 The Case of Field Lines

For the Hamiltonian system (3.9) and (3.13) for field lines the Poincaré integral $P(\varphi, \psi)$ is given by Eq. (6.62) as the integral over the perturbation Hamiltonian $\psi_1(z, p_z, \varphi)$ taken along unperturbed field line $(z(\varphi; \psi), p_z(\varphi; \psi))$ on the given magnetic surface $\psi = \text{const}$ for one poloidal turn. The Melnikov integral (6.68) for the field lines is given by

$$
M(\varphi; \psi) = \int_{-\pi q(\psi)}^{\pi q(\psi)} \{\psi_0, \psi_1\}\left(z(\varphi' + \varphi; \psi), p_z(\varphi' + \varphi; \psi), \varphi'\right) d\varphi'. \qquad (6.76)
$$

The relation (6.69) between $P(\varphi, \psi)$ and $M(\varphi, \psi)$ in this case reads as

$$M(\varphi; \psi) = \frac{\partial P(\varphi, \psi)}{\partial \varphi}. \tag{6.77}$$

For the perturbation Hamiltonian of type

$$\psi_1 (z, p_z, \varphi) = \sum_n \psi_n(z, p_z) \cos (n\varphi + \chi_n), \tag{6.78}$$

where χ_n is the phase of the toroidal mode n, the integral $P(\varphi; \psi)$ can be presented as

$$\begin{aligned} P(\varphi; \psi) &= \sum_n R_n(\psi) e^{in\varphi + i\chi_n} \\ &= \sum_n [K_n(\psi) \cos(n\varphi + \chi_n) - L_n(\psi) \sin(n\varphi + \chi_n)], \end{aligned} \tag{6.79}$$

with the Fourier coefficients,

$$\begin{aligned} R_n(\psi) &= K_n(\psi) + i L_n(\psi) \\ &= \int_{-\pi q(\psi)}^{\pi q(\psi)} \psi_n(z(\varphi, \psi), p_z(\varphi, \psi)) e^{in\varphi} d\varphi. \end{aligned} \tag{6.80}$$

The change of the poloidal flux through the magnetic surface $\psi =$ constant can be expressed via the integrals $R_n(\psi)$, i.e.,

$$\Delta\psi = \oint_{\psi\, =\, \text{const}} \frac{d\psi}{d\varphi} d\varphi = M(0; \psi) = \sum_n |R_n(\psi)| \sin (\chi_n). \tag{6.81}$$

6.5.3 Properties

Below we give a brief description of some important properties of the integrals $R_n(\psi)$ and their relations to the spectrum of perturbations $H_{mn}(\psi)$. More details can be found in Abdullaev (2005, 2006). Suppose the functions $\psi_n(z, p_z)$ in the perturbation poloidal flux (6.78) has the following Fourier expansion in the poloidal angle ϑ,

$$\psi_n(z, p_z) = \sum_m |H_{mn}| \cos(m\vartheta + \chi_n). \tag{6.82}$$

Then by integrating (6.80) one can obtain

$$R_n(\psi) = \pi q(\psi) \sum_m |H_{mn}(\psi)| \left[e^{i\chi_n} \frac{e^{ix_{mn}^{(+)}} - 1}{ix_{mn}^{(+)}} + e^{-i\chi_n} \frac{e^{ix_{mn}^{(-)}} - 1}{ix_{mn}^{(-)}} \right], \tag{6.83}$$

where we have introduced the notations $x_{mn}^{(\pm)} = 2\pi[m \pm nq(\psi)]$. As one can see that at the resonant surfaces ψ_{mn}, $q(\psi_{mn}) = m/n$, the integral $R_n(\psi)$ is determined by the spectra of perturbation $H_{mn}(\psi) = |H_{mn}(\psi)|\, e^{i\chi_n}$,

$$R_n(\psi_{mn}) = \pi q(\psi)\, |H_{mn}(\psi)|\, e^{-i\chi_n} = \pi q(\psi) H_{m,n}^*(\psi_{mn}). \tag{6.84}$$

Using this property the $R_n(\psi)$ can be presented as a sum of regular, $R_n^{(reg)}(\psi)$, and oscillatory, $R_n^{(osc)}(\psi)$, parts, i.e.,

$$R_n(\psi) = R_n^{(reg)}(\psi) + R_n^{(osc)}(\psi), \tag{6.85}$$

where $R_n^{(reg)}(\psi)$ is the regular part defined by extending the relation (6.84) for arbitrary values of ψ by replacing the integer number m to the continuous one $nq(\psi)$,

$$R_n^{(reg)}(\psi) = \pi q(\psi) H_{m=nq,n}^*(\psi). \tag{6.86}$$

The function $R_n^{(osc)}(\psi)$ describes the rapidly oscillating part of the $R_n(\psi)$. According to the definition the latter vanishes at the resonant surfaces ψ_{mn}, $q(\psi_{mn}) = m/n$, $R_n^{(reg)}(\psi_{mn}) = 0$.

The function $R_n^{(osc)}(\psi)$ can be determined by the asymptotical expansion of the integrals (6.80). Generic asymptotical formulas for $R_n^{(osc)}(\psi)$ near the separatrices are derived in Abdullaev (2005) (see, also Abdullaev 2006 and Sect. 9.3 and Appendix C). Particularly, near the separatrix $\psi_S = 0$ the leading term of the asymptotical expansion of $R_n^{(osc)}(\psi)$ in series of powers $|\psi|$ has the following asymptotics,

$$R_n^{(osc)}(\psi) \approx A|\psi|^{1/2} \sin(\pi nq(\psi)) + O(\psi), \tag{6.87}$$

where the constant parameter A does not dependent on ψ. The amplitude of oscillations $R_n^{(osc)}(\psi)$ also ceases when approaching the separatrix $\psi \to 0$ as $R_n^{(osc)}(\psi) \propto |\psi|^{1/2}$.

One should also note that near the hyperbolic X-point the function $R_n^{(osc)}(\psi)$ has a rescaling invariance property,

$$R_n^{(osc)}(\lambda\psi) = \sqrt{\lambda} R_n^{(osc)}(\psi), \tag{6.88}$$

where

$$\lambda = \exp(2\pi\gamma/n) \tag{6.89}$$

is the rescaling parameter. It is related with the logarithmic asymptotics of the safety factor $q(\psi)$ (2.51).

Near the separatrix and at the separatrix $\psi_S = 0$ the integrals $R_n^{(reg)}(\psi)$ decay exponentially with the toroidal mode number n. The analytical and numerical calculations show that

$$R_n(\psi_S) \propto \exp\left(-Cn/\gamma\right), \tag{6.90}$$

where C is constant parameter independent on γ.

Finally, one should note the important relation (6.84) between the Fourier coefficients $H_{mn}(\psi)$ in Eq. (3.12) taken at the corresponding resonant magnetic surfaces, ψ_{mn}, and the Poincaré integrals $R_n(\psi)$. It may be used to evaluate the spectrum of perturbation, H_{mn} at the different resonant surfaces (m, n) with the given toroidal mode n by the only one function $R_n^{(reg)}(\psi)$.

6.5.4 Example: Poincaré Integrals for the Model System

Below we calculate the Poincaré integrals of the model of magnetic field introduced in Sect. 1.3. The Poincaré integrals for the perturbation functions (3.34) can be easily found using the fact the $z(\varphi)$ is an even function of φ, while $p(\varphi)$ is an odd function of φ. Then according to the definition (6.80) the integral $R(\psi) = K(\psi) + iL(\psi)$ is equal to:

$$K(\psi) = \cos\alpha_0 \int_{-\pi q(\psi)}^{\pi q(\psi)} z(\varphi, \psi)\cos(n\varphi)d\varphi,$$
$$L(\psi) = \sin\alpha_0 \int_{-\pi q(\psi)}^{\pi q(\psi)} p(\varphi, \psi)\sin(n\varphi)d\varphi. \tag{6.91}$$

Using (1.49) one obtains the functions $K(\psi)$, $L(\psi)$ at the separatrix $\psi = 0$,

$$K(0) = \frac{\sqrt{2}z_a\cos\alpha_0}{\gamma}\int_{-\infty}^{\infty}\frac{\cos(n\tau/\gamma)d\tau}{\cosh\tau} = R(0)\cos\alpha_0,$$
$$L(0) = -\frac{\sqrt{2}z_a\sin\alpha_0}{\beta^2}\int_{-\infty}^{\infty}\frac{\sinh(\tau)\sin(n\tau/\gamma)d\tau}{\cosh^2\tau} = -\frac{n\sin\alpha_0}{\beta^2}R(0), \tag{6.92}$$

where

$$R(0) = \frac{\sqrt{2}\pi z_a}{\gamma\cosh(\pi n/2\gamma)}. \tag{6.93}$$

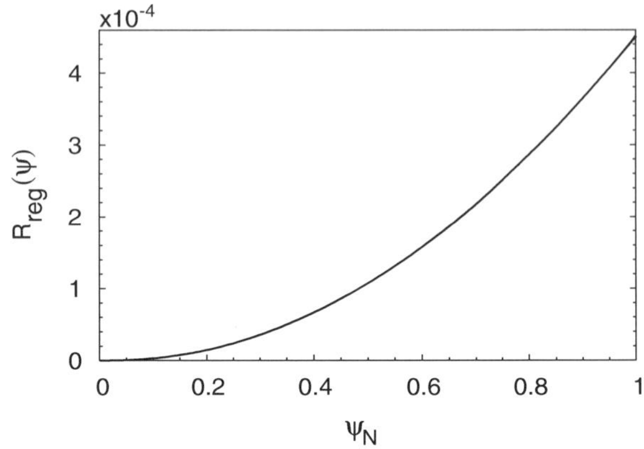

Fig. 6.4 Integral $R^{(reg)}(\psi)$ in (6.94) as a function of the normalized poloidal flux ψ_N. The values of parameters: the toroidal mode $n = 3$, $\gamma = 0.44$, $\alpha = 0.4553$, and $z_a = 1$

According to Eq. (6.84) the regular part $R^{(reg)}(\psi)$ of the integral $R(\psi) = R^{(reg)}(\psi) + R^{(osc)}(\psi)$ can be expressed through the Fourier components $H_m(\psi)$ (3.35) of the perturbation Hamiltonian,

$$
\begin{aligned}
K^{(reg)}(\psi) &= \pi q(\psi) H^{(c)}_{nq(\psi)}(\psi) = R^{(reg)}(\psi)\cos\alpha_0, \\
L^{(reg)}(\psi) &= \pi q(\psi) H^{(s)}_{nq(\psi)}(\psi) = -\frac{n\sin\alpha_0}{\beta^2} R^{(reg)}(\psi), \\
R^{(reg)}(\psi) &= \frac{\sqrt{2} z_a \pi}{\gamma\cosh\left(nC(\psi)\right)}.
\end{aligned}
\tag{6.94}
$$

In the limit $|\psi| \to 0$ the quantity $R^{(reg)}(\psi)$ coincides with $R(0)$ (6.93) obtained by the direct integration and it has the following asymptotics

$$
R^{(reg)}(\psi) \approx \frac{\sqrt{2} z_a \pi}{\gamma} \exp\left(-nC(\psi)\right). \tag{6.95}
$$

The dependence of the integral $R^{(reg)}(\psi)$ on the normalized poloidal flux ψ_N is plotted in Fig. 6.4.

The oscillatory parts of the Poincaré integrals $R^{(osc)}(\psi)$ can be estimated asymptotically. Near the separatrix the asymptotical expansions of $R^{(osc)}(\psi)$ in a series of powers of $|\psi|^{1/2}$ are given in Abdullaev (2005, 2006). According to these formulas the leading terms of these expansions of $R^{(osc)}(\psi)$ for the Hamiltonian perturbation (3.34) are

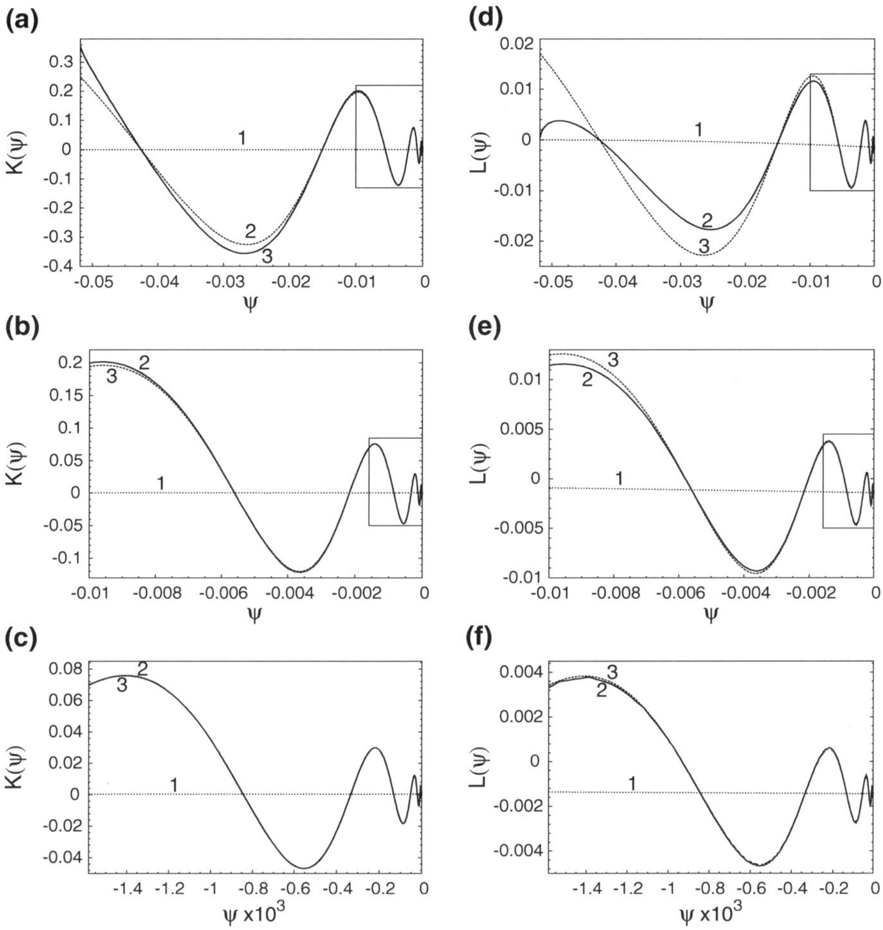

Fig. 6.5 Integrals $K(\psi)$ (**a**)–(**c**) and $L(\psi)$ (**e**)–(**f**) versus ψ: **a** *curves 1* correspond to the regular parts $K^{(reg)}(\psi)$ and $L^{(reg)}(\psi)$ given by Eq. (6.94), *solid curve 2* describes $K(\psi)$ obtained by the numerical integration of the integral (6.91), and *dashed curve 3* describes $K(\psi)$ obtained by the analytical formulae (6.94); **b** Expanded view of $K(\psi)$ in the rectangular box region shown in (**a**); **c** Expanded view of the rectangular box region shown in (**b**). The perturbation mode number $n = 3$. The angle $\alpha_0 = 0$ in (**a**)–(**c**), $\alpha_0 = \pi/2$ in (**e**)–(**f**). The rescaling parameter $\lambda = \exp(2\pi\gamma/n) = 2.51314$ and the other parameters are the same as in Fig. 1.6

$$
\begin{aligned}
K^{(osc)}(\psi) &= \frac{n}{\alpha} R^{(osc)}(\psi) \cos \alpha_0, \\
L^{(osc)}(\psi) &= \frac{\gamma}{\beta} R^{(osc)}(\psi) \sin \alpha_0, \\
R^{(osc)}(\psi) &\approx \frac{2\sqrt{2|\psi|} z_a}{n^2 + \gamma^2} \sin[\pi n q(\psi)].
\end{aligned}
\tag{6.96}
$$

Figure 6.5 show the dependencies of the integrals $K(\psi)$ and $L(\psi)$ on ψ obtained by the direct numerical integration of Eq. (6.91), as well as by the analytical formulae (6.94) and (6.96): the dotted curves 1 correspond to the regular part $K^{(reg)}(\psi)$ given by Eq. (6.94) solid curve 2 corresponds to the numerical calculations, dashed curve 3—to the analytical result. The perturbation mode number n is taken equal to $n = 3$. The corresponding rescaling parameter λ defined by Eq. (6.89) is equal to 2.51314.

As seen from Fig. 6.5 the analytical formulas (6.94) for the regular part $K^{(reg)}(\psi)$ and $L^{(reg)}(\psi)$ the asymptotical formula (6.96) for the oscillatory part, $K^{(osc)}(\psi)$ and $L^{(osc)}(\psi)$, well describe the behavior of the integral $K(\psi)$ near the separatrix with increasing the accuracy of approximation there. On the other hand the numerical calculations confirm also the rescaling property (6.88).

Figure 6.5 show also the rescaling invariance property (6.88) of the function $K^{(osc)}(\psi)$, $L^{(osc)}(\psi)$

$$R^{(osc)}(\lambda^2\psi) \approx \lambda R^{(osc)}(\psi), \tag{6.97}$$

i.e., a self-similarity behavior at the different scales near the separatrix.

6.6 Application of the Full-turn Transfer Mapping Near the Separatrix

We illustrate the application of the full–turn transfer (or the separatrix) mapping to obtain the Poincaré section of field lines near the X-point and the magnetic separatrix and the magnetic footprints at the divertor plates. We consider the model of magnetic field considered in Sect. 1.3.

The geometry of the mapping is shown in Fig. 6.6. The cross sections Σ_c consist of the segments of the z axis located near the farthest crossing points of the unperturbed separatrix with the x axis. The section Σ_s is located near the saddle point (z_s, p_s) and consists of two perpendicular to each other segments of the z and p axes with the center at $(z_s = 0, p_s = 0)$.

According to Eq. (6.79) the Poincaré integral for the Hamiltonian system (1.3) is given by

$$P(\varphi, \psi) = K(\psi)\cos(n\varphi + \chi_n).$$

The mapping (6.61) in this case reads as

$$\begin{aligned} \psi_{k+1} &= \psi_k \mp \epsilon n K(\psi_{k+1}) \sin\left(n\varphi_k \pm \pi n q(\psi_k) + \chi_n\right), \\ \varphi_{k+1} &= \varphi_k \pm \pi\left[q(\psi_{k+1}) + q(\psi_k)\right] \\ &\pm \epsilon \frac{dK(\psi_{k+1})}{d\psi_{k+1}} \cos\left(n\varphi_k \pm \pi n q(\psi_k) + \chi_n\right), \end{aligned} \tag{6.98}$$

where the function $K(\psi)$ is given by Eqs. (6.94) and (6.96). The upper (lower) signs in (6.98) correspond to the forward (backward) mapping.

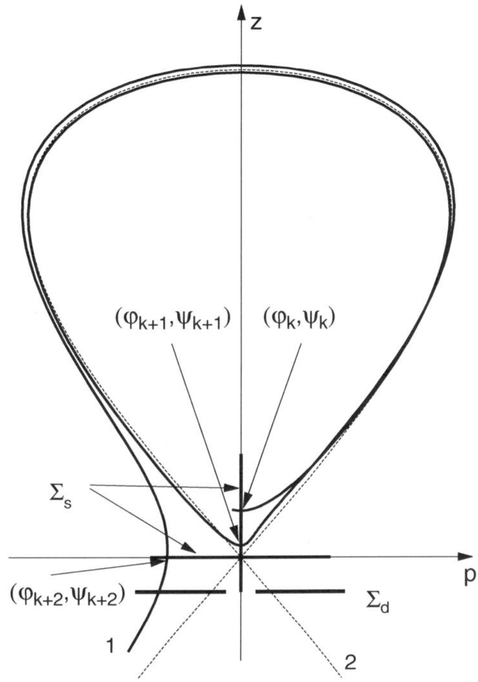

Fig. 6.6 Geometry of the full-turn transfer (separatrix) map. *Solid curve 1* describes the perturbed orbit, and *dotted curve 2* is the unperturbed separatrix

In the region close to the separatrix the mapping (6.98) encounters a difficulty related to solving the implicit equation with respect to ψ_{k+1}. This is because of the rapid oscillations of the oscillatory part $K^{(osc)}(\psi)$ of the integral $K(\psi)$ with approaching the separatrix. To overcome this problem one can neglect the oscillatory part $K^{(osc)}(\psi)$ retaining only the regular part $K^{(reg)}(\psi)$. This approach called a *primary resonant approximation* is justified due to the following arguments (see Abdullaev (2005, 2006)). The behavior of the system near the resonant tori $\psi = \psi_{mn}$ is mostly determined by the primary resonant terms $H_{mn}\cos(m\vartheta - \Omega t)$ of the perturbation Hamiltonian, i.e., by the amplitude H_{mn} near the resonant tori $\psi = \psi_{mn}$. According to the definition (6.84), (6.85) and (6.86), the oscillation part $R^{(osc)}(\psi)$ of the integral $R(\psi)$ vanishes at the resonant tori $\psi = \psi_{mn}$ and therefore it does not contribute to the resonant interaction of the system with the perturbation.

Further simplification of the mapping can be done near the small neighborhood of the separatrix. Using the asymptotics of $q(\psi)$ (2.51) at $|\psi| \to 0$ and replacing $K^{(reg)}(\psi)$ by its value at the separatrix $K(0)$ we obtain

$$\psi_{k+1} = \psi_k \mp \epsilon n K(0) \sin\left(n\varphi_k + \frac{n}{2\gamma}\ln\frac{Q}{|\psi_k|} + \chi\right),$$

$$\varphi_{k+1} = \varphi_k \pm \frac{1}{2\gamma}\left(\ln\frac{Q}{|\psi_k|} + \ln\frac{Q}{|\psi_{k+1}|}\right). \tag{6.99}$$

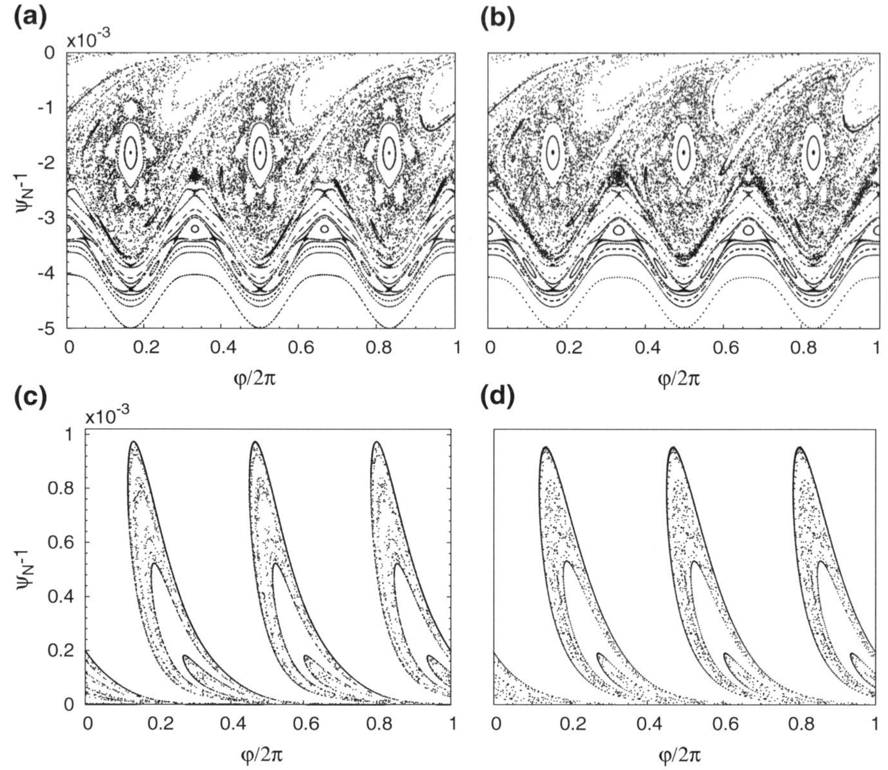

Fig. 6.7 Poincaré sections of the system in the (φ, ψ) plane of the section $p = 0$ (**a**), (**b**) and of the section $z = 0$ (**c**), (**d**): (**a**) and (**c**) are obtained by the small-step Runge–Kutta integrator of 5-th order; (**b**) and (**d**) are obtained by the mapping Eq. (6.98). The toroidal mode $n = 3$, the perturbation amplitude $\epsilon = 10^{-3}$, other parameters: $\alpha_0 = 0$, $\alpha = \beta = 1$, $z_a = 1$

Figure 6.7 show the Poincaré sections of system in the (φ, ψ_N) plane of the section $p = 0$ (a), (b) and of the section $z = 0$ (c), (d). Here $\psi_N = \psi/\psi_a - 1$ is the normalized poloidal flux. They are obtained by the small–step integration of the Hamiltonian system (1.37) using the 5-th order Runge–Kutta integrator [(a) and (c)] and the mapping (6.98) [(b) and (d)]. Field lines are terminated after crossing the section Σ_d. As seen from figure the mapping (6.98) well reproduces the results obtained by the direct small–step integration of the system (1.37).

6.7 Perturbed Field Lines Near the X-point

Since the mapping (6.61) has large steps it cannot be directly applied to map field lines to sections other than Σ_S, for instance to the divertor plates Σ_d. This problem can be resolved if the section is located close to the X-point where one can obtain the generic analytical solutions of the perturbed field lines.

Consider a magnetic system with one hyperbolic fixed point or X-point (R_s, Z_s) [or (z_s, p_s)] connected itself with a homoclinic orbit (separatrix) (see Fig. 2.6). Consider the effect of magnetic perturbations on field lines in a small neighborhood of the X-point. Expanding the full poloidal flux $\psi(z, p_z, \varphi) = \psi_0(z, p_z, \varphi) + \epsilon\psi_1(z, p_z, \varphi)$ near the X-point (z_s, p_s) in a series of powers of $(z - z_s)$, $(p_z - p_s)$ and using a linear coordinate transformation (2.39) it can be reduced to

$$\psi(\xi, \eta, \varphi) \approx \psi_0(z_s, p_s) - \frac{\alpha^2}{2}\xi^2 + \frac{\beta^2}{2}\eta^2 + \epsilon\left(a_\xi(\varphi)\xi + a_\eta(\varphi)\eta\right), \tag{6.100}$$

where $\alpha = \sqrt{|\lambda_1|}$, $\beta = \sqrt{\lambda_2}$ (see Sect. 2.3.1 and Eq. (2.40)). In (6.100) we have retained only the first nonzero terms and put $\psi_1(z_s, p_s, \varphi) = 0$, which does not change the equations of motion. The periodic coefficients $a_\xi(\varphi)$, $a_\eta(\varphi)$ are defined as

$$\begin{aligned} a_\xi(\varphi) &= a_z(\varphi)\cos\alpha + a_p(\varphi)\sin\alpha, \\ a_\eta(\varphi) &= -a_z(\varphi)\sin\alpha + a_p(\varphi)\cos\alpha, \end{aligned} \tag{6.101}$$

where

$$a_z(\varphi) = \frac{\partial\psi_1(z_s, p_s, \varphi)}{\partial z}, \qquad a_p(\varphi) = \frac{\partial\psi_1(z_s, p_s, \varphi)}{\partial p_z}. \tag{6.102}$$

The field line equations near the X-point (z_s, p_s):

$$\frac{d\xi}{d\varphi} = \frac{\partial\psi}{\partial\eta}, \quad \frac{d\eta}{d\varphi} = -\frac{\partial\psi}{\partial\xi}, \tag{6.103}$$

can be transformed to

$$\begin{aligned} \frac{dx}{d\varphi} &= \frac{\partial\psi}{\partial y} = \gamma x + \epsilon a_x(\varphi), \\ \frac{dy}{d\varphi} &= -\frac{\partial\psi}{\partial x} = -\gamma y + \epsilon a_y(\varphi), \end{aligned} \tag{6.104}$$

with the Hamiltonian

$$\psi = \gamma xy + \epsilon\left[a_x(\varphi)y - a_y(\varphi)x\right], \tag{6.105}$$

where $\gamma = \alpha\beta$, and

$$x = \frac{\alpha\xi + \beta\eta}{\sqrt{2\gamma}}, \quad y = \frac{-\alpha\xi + \beta\eta}{\sqrt{2\gamma}},$$
$$a_x(\varphi) = \frac{\alpha a_\eta - \beta a_\xi}{\sqrt{2\gamma}}, \quad a_y(\varphi) = -\frac{\alpha a_\eta + \beta a_\xi}{\sqrt{2\gamma}}. \tag{6.106}$$

The solutions of Eq. (6.104) with the initial conditions $x(\varphi_0, \epsilon) = x_0$, $y(\varphi_0, \epsilon) = y_0$ are given by

$$x(\varphi, \epsilon) \equiv x(\varphi; x_0, y_0, \varphi_0, \epsilon) = e^{\gamma(\varphi-\varphi_0)} \left[x_0 + \epsilon \int_{\varphi_0}^{\varphi} e^{-\gamma(\varphi'-\varphi_0)} a_x(\varphi') d\varphi' \right],$$
$$y(\varphi, \epsilon) \equiv y(\varphi; x_0, y_0, \varphi_0, \epsilon) = e^{-\gamma(\varphi-\varphi_0)} \left[y_0 + \epsilon \int_{\varphi_0}^{\varphi} e^{\gamma(\varphi'-\varphi_0)} a_y(\varphi') d\varphi' \right]. \tag{6.107}$$

Particularly, assuming that

$$a_x(\varphi) = a_1\sqrt{\gamma^2 + n^2} \cos(n\varphi + \chi_n),$$
$$a_y(\varphi) = a_2\sqrt{\gamma^2 + n^2} \cos(n\varphi + \chi_n), \tag{6.108}$$

where a_1 and a_2 are the constants, we obtain the following solution,

$$x(\varphi, \epsilon) = e^{\gamma(\varphi-\varphi_0)} \Big[x_0 + \epsilon a_1 \cos(n\varphi_0 + \chi_n + \phi) \Big] - \epsilon a_1 \cos(n\varphi + \chi_n + \phi),$$
$$y(\varphi, \epsilon) = e^{-\gamma(\varphi-\varphi_0)} \Big[y_0 - \epsilon a_2 \cos(n\varphi_0 + \chi_n - \phi) \Big] + \epsilon a_2 \cos(n\varphi + \chi_n - \phi). \tag{6.109}$$

The phase ϕ is given by

$$\phi = \arctan\frac{n}{\gamma}. \tag{6.110}$$

One should note that according to (6.101), (6.102), (6.106) and (6.108) the coefficients a_1 and a_2 are proportional to the magnitude $B_X^{(pert)}$ of the perturbation magnetic field at the X-point.

Field lines near the X-point are illustrated in Fig. 6.8. We have defined the area C enclosed by curve 4 near the X-point is defined by $|\psi - \psi_s| \leq \psi_c \sim \epsilon^2$, $|x| \leq x_c > 0$, $|y| \leq y_c > 0$. The field lines approaching very close to the unperturbed X-point have a complicated behavior. Field lines (curve 1) approaching the X-point but do not entering into the area C cross the ξ-axis only once at the point P_k and leave away. On the other hand field lines (curve 2) entering into the area C may be strongly affected by the unstable periodic orbit (curve 3) (see Sect. 7.5.1). Such field lines may leave the area C after several rotations around the periodic orbits (see Sects. 7.2 and 7.5.1).

The mapping (6.61) is not applicable for small values of $|\psi - \psi_s|$ of order of $\epsilon^2 \ll 1$. However, the area C for the typical magnetic perturbations in tokamaks is too small in comparison to the width of the stochastic layer, and can be neglected.

6.8 Mapping to the Arbitrary Section Near the X-point

The separatrix mapping determines field lines at the section Σ_s, consisting of two planes (ξ, φ) and (η, φ) crossing each other at the X-line (see Fig. 6.3b). To determine field lines at the arbitrary section Σ one needs to map field lines from the section Σ_s to the sections Σ. When the section Σ is located close to the X-point one can use the solutions (6.107) for perturbed field lines to map the crossing point P_k at the section Σ_s to the section Σ. The schemes of such mappings are shown in Fig. 6.9a, b for the forward (6.61) and the backward (6.63) mappings, respectively. The sections $\Sigma^{(+)}$ and $\Sigma^{(-)}$ are located on the right and the left sides of the axis ξ, respectively. The section Σ_d stands for the divertor plates.

6.8.1 General Relations

Let (ξ, η, φ) be coordinates of a field line at the crossing point P of the section Σ. Assume that the geometrical form of the section Σ is given by the function $F_\Sigma(R, Z) = f_\Sigma(x, y) = \text{const}$, where (x, y) are coordinates defined by Eq. (6.106). Since the section Σ is located close to the X-point the relative poloidal flux ψ at the

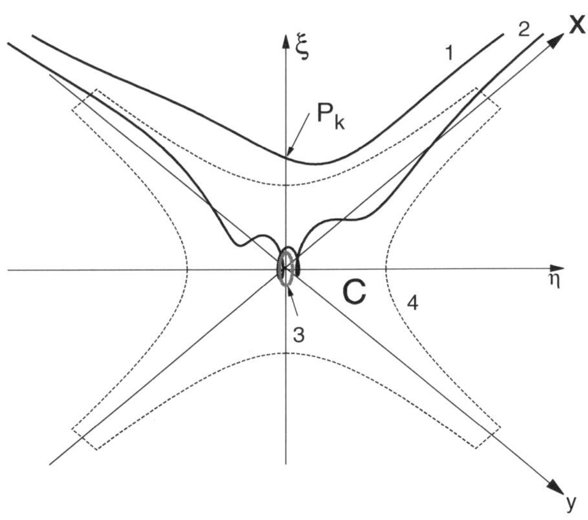

Fig. 6.8 Field lines near the X-point: *curve 1* corresponds to field lines crossing the section Σ_s once in one poloidal turn; *curve 2* corresponds field lines making several poloidal turns around the unstable periodic orbit shown by *curve 3*. Area C near the X-point is defined by $|\psi - \psi_s| \leq \psi_c \sim \epsilon^2$, $|x| \leq x_c > 0$, $|y| \leq y_c > 0$, *curve 4* is a boundary of the area C

section Σ can be presented as (6.105). The perturbed field lines in the coordinates (x, y) given by (6.107) we present in the form,

$$\begin{aligned} x(\varphi) &= x(\varphi; x_k, y_k, \varphi_k, \epsilon), \\ y(\varphi) &= y(\varphi; x_k, y_k, \varphi_k, \epsilon), \end{aligned} \tag{6.111}$$

with the coordinates (x_k, y_k, φ_k) of the point P_k of a field line at the section Σ_s as initial conditions (see Fig. 6.9a, b). We recall that at the axis $\xi = 0$ the poloidal flux $\psi_k > 0$, at the $\eta = 0$, $\psi_k < 0$, and the coordinates (x_k, y_k) are given by

$$\begin{cases} x_k = y_k = -\sqrt{\psi_k/\gamma}, & \text{at the axis } \xi = 0, \eta_k < 0 \psi_k > 0, \\ x_k = y_k = \sqrt{\psi_k/\gamma}, & \text{at the axis } \xi = 0, \eta_k > 0 \psi_k > 0, \\ x_k = -y_k = \sqrt{|\psi_k|/\gamma}, & \text{at the axis } \eta = 0, \psi_k < 0. \end{cases} \tag{6.112}$$

Imposing the boundary condition on the solution (6.111) at the section Σ,

$$\begin{aligned} f_\Sigma\left(x(\varphi), y(\varphi)\right) &= \text{const}, \\ \psi\left(x(\varphi), y(\varphi), \varphi\right) &= \psi, \end{aligned} \tag{6.113}$$

one finds the toroidal angle φ and the poloidal flux ψ at the section Σ as a function of ψ_k and φ_k.

The poloidal flux ψ and the toroidal angle φ on the section Σ is given by the mapping,

$$(\varphi, \psi) = \hat{M}\left(\varphi_k, \psi_k\right). \tag{6.114}$$

The latter can be presented as

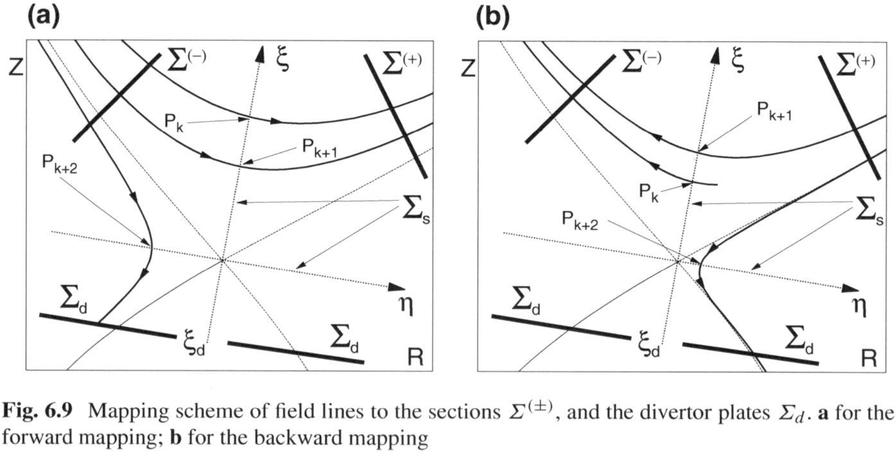

Fig. 6.9 Mapping scheme of field lines to the sections $\Sigma^{(\pm)}$, and the divertor plates Σ_d. **a** for the forward mapping; **b** for the backward mapping

$$\varphi = \varphi_k + \sigma \Delta\varphi(\psi_k, \epsilon),$$
$$\psi = \psi_k + \Delta\psi(\varphi, \psi_k, \varphi_k, \epsilon), \tag{6.115}$$

where $\Delta\varphi(\psi_k, \epsilon)$ and $\Delta\psi(\psi_k, \varphi, \varphi_k, \epsilon)$ are the increments of the toroidal angle φ and the poloidal flux ψ along the field line connecting the point (ψ_k, φ_k) at the section Σ_s to the point (ψ, φ) at the section Σ. The quantity σ determines the sign of the increment of the toroidal $\Delta\varphi$. It takes the following values

$$\sigma = \begin{cases} -1, \text{ for the mapping } P_k \to P, P \in \Sigma^{(-)}, \\ +1, \text{ for the mapping } P_k \to P, P \in \Sigma^{(+)}, \Sigma_d, \end{cases} \tag{6.116}$$

for the forward mapping (see Fig. 6.9a), and

$$\sigma = \begin{cases} +1, \text{ for the mapping } P_k \to P, P \in \Sigma^{(+)}, \\ -1, \text{ for the mapping } P_k \to P, P \in \Sigma^{(-)}, \Sigma_d, \end{cases} \tag{6.117}$$

for the backward mapping (see Fig. 6.9b).

6.8.2 Simplified Relations

In general the Eq. (6.113) for the toroidal angle φ and the poloidal flux ψ are transcendental which makes the determination of φ and ψ difficult. However, in the case of small perturbation amplitude ϵ the toroidal angle φ in Eq. (6.111) can be replaced by the one for unperturbed field lines, i.e, by putting $\epsilon = 0$ in Eq. (6.113).

We demonstrate this for the simple geometry of the section $\Sigma, \xi = \text{const}$ as shown in Fig. 6.9 by the divertor plate Σ_d. Putting $\epsilon = 0$ in (6.109) the Eq. (6.113) is then reduced to

$$x_k e^{\Delta\varphi} - y_k e^{-\Delta\varphi} = \frac{2\alpha}{\sqrt{2\gamma}}\xi,$$
$$\psi = \gamma x y = \gamma x_k y_k = \psi_k. \tag{6.118}$$

Using the relations (6.112) one obtains the following formulas for the increment $\Delta\varphi$

$$\Delta\varphi(\psi_k) = \frac{1}{\gamma} \begin{cases} -\sinh^{-1}(c), \text{ at the axis } \xi_k = 0, \eta_k < 0 \psi_k > 0, \\ \sinh^{-1}(c), \quad \text{at the axis } \xi_k = 0, \eta_k > 0 \psi_k > 0, \\ \cosh^{-1}(c), \quad \text{at the axis } \eta_k = 0, \psi_k < 0, \end{cases} \tag{6.119}$$

where $\sinh^{-1}(x)$, $\cosh^{-1} x$ are the inverse functions to $\sinh(x)$, $\cosh(x)$, and

$$c = \frac{\xi}{d_{cr}}, \qquad d_{cr} = \frac{\sqrt{2|\psi_k|}}{\alpha}.$$

Using (6.111) and (6.107) one can find the increment $\Delta\psi(\varphi, \psi_k, \varphi_k, \epsilon)$,

$$\Delta\psi = \epsilon\Bigg[y_k e^{\gamma(\varphi-\varphi_k)} \int_{\varphi_k}^{\varphi} e^{-\gamma(\varphi'-\varphi_k)} a_y(\varphi')d\varphi' + x_k e^{-\gamma(\varphi-\varphi_k)} \int_{\varphi_k}^{\varphi} e^{\gamma(\varphi'-\varphi_k)} a_x(\varphi')d\varphi' \Bigg], \quad (6.120)$$

where we have neglected the terms of order of $\epsilon^2 \ll 1$.

In particularly, when the functions $a_x(\varphi)$, $a_y(\varphi)$ are given by (6.108), the increment $\Delta\psi(\psi, \varphi, \varphi_0, \epsilon)$ is reduced to

$$\Delta\psi = \epsilon\Bigg\{ x_k a_2 \Big[e^{\gamma(\varphi-\varphi_k)} \cos(n\varphi + \chi_n - \phi) - \cos(n\varphi_k + \chi_n - \phi) \Big] - y_k a_1 \Big[e^{-\gamma(\varphi-\varphi_k)} \cos(n\varphi + \chi_n + \phi) - \cos(n\varphi_k + \chi_n + \phi) \Big] \Bigg\}. \quad (6.121)$$

The increment $\Delta\psi(\varphi, \psi_k, \varphi_k, \epsilon)$ is the oscillating function of the toroidal angle φ with the period $2\pi/n$. At the distances ξ much larger than d_{cr}, $\xi \gg d_{cr}$ (6.119) it can be approximated by

$$\Delta\psi \approx \epsilon D \cos(n\varphi + \chi_n - \phi), \quad (6.122)$$

where

$$D = \frac{\alpha\xi}{\sqrt{2|\psi|}} \begin{cases} x_k a_2, & \text{at } \sigma = +1 \\ -y_k a_1, & \text{at } \sigma = -1. \end{cases} \quad (6.123)$$

6.8.3 Determination of Magnetic Footprints

The heat and particle deposition patterns on the divertor plate are closely related to the magnetic footprints, i.e., the hitting points of open field lines with the divertor plates. To determine the magnetic footprints on divertor plates Σ_d one needs to map (6.115) field lines from the sections Σ_s to the sections Σ_d. That is field lines should be traced along perturbed field lines starting from the crossing point P_k of the segment on the $\eta-$axis, $\xi = 0$, of Σ_s until they reach the divertor plate Σ_d (see Fig. 6.9a, b).

Using the separatrix mapping (6.61) and the divertor mapping (6.114) one can obtain the analytical estimations for the radial extention of magnetic footprints on the divertor plates. A field line with the initial coordinates (φ_0, ψ_0), $(\psi_0 < 0)$ in the plasma region leaves the plasma when it crosses the axis η, i.e., $\psi_{k-1} < 0$ and $\psi_k > 0$ (see Fig. 6.3). According to (6.114) the maximal value of the poloidal flux

ψ_d carried by these field lines can reach the value

$$\Delta\psi_d = \max\ |\psi_k + \Delta\psi|\,. \tag{6.124}$$

Since $|\psi_k| \gg |\Delta\psi|$ and the increment $\Delta\psi$ is oscillating along the toroidal angle φ, the quantity $\Delta\psi_d$ oscillates around the maximal value of ψ_k with the amplitude of oscillations of $\Delta\psi$. The maximum of ψ_k according to the first equation in Eq. (6.61) is determined the maximum value of the derivative of the Poincaré integral $P\ (\varphi; \psi)$ (6.79) near the separatrix $\psi = 0$,

$$\begin{aligned} \max\ |\psi_k| &= \epsilon \max \left| \frac{\partial}{\partial \varphi_k} P\ (\varphi_k \pm \pi q(\psi_k); \psi_{k+1}) \right| \\ &= \epsilon \max \left| \sum_n n|R_n(0)| \sin\left(n[\varphi + \pi q(\psi_k) + \chi_n]\right) \right|. \end{aligned} \tag{6.125}$$

Particularly, if the magnetic perturbation has only one toroidal mode n, we have

$$\Delta\psi_d \approx \max\ |\psi_k| = \epsilon n |R_n(0)|. \tag{6.126}$$

Using (6.92) we obtain the following estimation of $\Delta\psi_d$ for the model of magnetic field (1.37),

$$\Delta\psi_d = \frac{\epsilon n \sqrt{2}\pi}{\cosh\ (\pi n/2)}. \tag{6.127}$$

The examples of magnetic footprints in the (φ, ψ) plane on the divertor plate obtained by the small-step integration of the Hamiltonian Eq. (1.37) using the small-step symplectic integrator and the mapping (6.98) are shown in Fig. 6.10a, b, respectively. The position of the divertor plate $\xi_d = 0.5$. The poloidal flux ψ is normalized to $\Delta\psi_d$ (6.127). The values of the perturbation parameter ϵ and the toroidal mode n are the same as in Fig. 6.7.

The magnetic footprints consists of spiral-like curves. The number of spirals is equal to the toroidal mode number $n = 3$. The spatial forms of the spirals are similar to the Poincaré section of field lines to the section $z = 0$ shown in Fig. 6.7b, but they are modulated along the toroidal angle φ. The amplitude of spatial modulation increases with the distance ξ_d of the divertor plates from the X-point. According to Eqs. (6.102), (6.108), (6.120)–(6.122) it also follows that these modulations are linearly proportional the magnitude $B_X^{(pert)}$ of the perturbation field at the X-point through the coefficients a_1 and a_2.

The separatrix mapping allows one not only very fast to calculate magnetic footprints on divertor plates poloidal divertor tokamaks, but also to find the analytical form of its spatial structure. As we will see in Sect. 7.5.4 the latter is determined by the stable and unstable manifolds of the magnetic field.

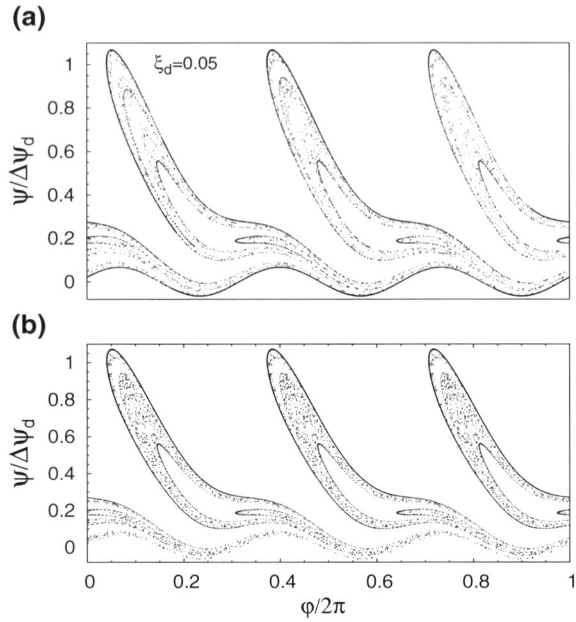

Fig. 6.10 Magnetic footprints on the divertor plate Σ_d (see Figs. 6.9 and 6.6). The width $\Delta\psi_d$ (6.127) for the perturbation parameter $\epsilon = 10^{-3}$ and the toroidal mode $n = 3$ is $\Delta\psi_d = 2.3945 \times 10^{-4}$. The position of the divertor plate $\xi_d = 0.05$: **a** the footprint obtained by the integration of the Hamiltonian Eq. (1.37) using the small-step symplectic integrator; **b** the one obtained by the mapping (6.98)

6.9 Mapping Models for Magnetic Field Lines

Mappings have been widely used to model and to study magnetic field lines in plasmas for a long time. They allow to avoid the time consuming integration of field line equations and at the same time to reveal the generic properties of field line behavior. Since in most cases the maps were not rigorously derived from the equations of field lines they are constructed in the symplectic form (6.43) in order to retain the flux-preserving property of magnetic field. Another requirement was that the mapping should have a simple form as possible.

Roughly one can distinguish two kind of mappings. The first kind of mappings, like the standard mapping or the tokamap are used to study the generic properties of field lines in plasmas, while the second kind of mappings, like the Martin–Taylor mapping [Martin and Taylor (1984)] or the specific mapping to study chaotic field lines in tokamaks with the magnetic separatrix [Punjabi et al. (1992); Kroetz et al. (2010)]. In Chap. 9 we describe some other models of magnetic fields created by the external magnetic perturbations.

Below we shortly describe some models for the Poincaré mapping of field lines,

$$(\vartheta_{k+1}, \psi_{k+1}) = \hat{M}(\vartheta_k, \psi_k), \tag{6.128}$$

where the variable (ϑ, ψ) stand for the poloidal angle and the toroidal flux.[7] The mappings have the following symplectic form

[7] In this section, for the sake of simplicity, we use the notation ψ for the toroidal flux.

$$\psi_{k+1} = \psi_k - \frac{\partial S(\vartheta_k, \psi_{k+1})}{\partial \vartheta_k},$$
$$\vartheta_{k+1} = \vartheta_k + \frac{2\pi}{q(\psi_{k+1})} + \frac{\partial S(\vartheta_k, \psi_{k+1})}{\partial \psi_{k+1}}, \tag{6.129}$$

with the generating function $S(\vartheta_k, \psi_{k+1})$.

6.9.1 The Class of Standard Mappings

This class of mappings corresponding to the symplectic mapping (6.129) with the generating function S depending only on the poloidal angle ϑ (Mendonča 1991). The general form of corresponding generating function is given by

$$\epsilon S(\vartheta) = \frac{K}{2\pi} \sum_m g_m \cos(m\vartheta + \chi_m), \tag{6.130}$$

where g_m and χ_m are the constant amplitudes and phases of perturbation modes. The parameter K describes the relative strength of magnetic perturbation. The corresponding mapping has the following form

$$\psi_{k+1} = \psi_k - \frac{K}{2\pi} \sum_m g_m \sin(m\vartheta_k + \chi_m),$$
$$\vartheta_{k+1} = \vartheta_k + \frac{2\pi}{q(\psi_{k+1})}. \tag{6.131}$$

Keeping in Eq. (6.131) the only one mode $m = 1$ with $g_m = 1$, $\chi_m = 0$ and choosing $q(\psi) = 1/\psi$ one obtains the *standard mapping* also known as the *Chirikov–Taylor mapping* (see Chirikov (1979)),

$$\psi_{k+1} = \psi_k - \frac{K}{2\pi} \sin \vartheta_k,$$
$$\vartheta_{k+1} = \vartheta_k + 2\pi\psi_{k+1}. \tag{6.132}$$

This kind of mappings can used to study a local behavior of field lines when one can neglect with the radial dependence of a perturbation field. The standard mapping has been widely used to model a field line stochasticity in tokamaks since late 1970s (see, e.g., Rechester et al. 1979; Rechester and White 1980; Rechester et al. 1981; Ichikawa et al. 1987; Rax and White 1992). Its simple form it allowed one to efficiently calculate the diffusion coefficients of field lines, the transport of test particles in a magnetic field with destroyed magnetic surfaces. The standard map has been also used to model the test particle transport in a tokamak caused by the drift-wave turbulence [Horton et al. (1998); Kwon et al. (2000)]. The more general mappings (6.131) proposed by Mendonča (1991) have been used by Tabet et al. (1998, 2000), Miskane et al. (2001) to study stochastic magnetic field lines in tokamaks.

6.9.2 The Wobig–Mendonça Mapping

The simplest form of a mapping which takes into account the radial dependence of the perturbation field has been proposed by Wobig (1987) and later generalized by Mendonc̃a (1991). This mapping called Wobig—Mendonc̃a mapping has the symplectic form (6.129) with $n = 1$ and the generating function is given by

$$\epsilon S(\vartheta, \psi) = -\frac{K}{2\pi} f(\psi) \sum_m g_m \cos m\vartheta. \tag{6.133}$$

The radial dependence of the perturbation field is described by the function $f(\psi)$. It is chosen in the form $f = 1 - \exp(-\psi)$. The Wobig—Mendonça mapping is reduced to the Wobig mapping at the limiting case $\psi \ll 1$. At regions far from the magnetic axis, when $\psi \gg 1$, the mapping approaches the mapping (6.131).

Balescu et al. (1998) has noted that the toroidal flux ψ in the standard mapping (6.132) may take negative values after some iterations of the mappings for arbitrary value of K, as well as the Wobig-Mendonça mapping for $K > 2\pi$. This fact is not compatible with the magnetic field line behavior in a toroidal system where the toroidal flux $\psi \sim r^2$ is always positive.

6.9.3 The Tokamap

Balescu et al. has proposed the specific form of the mapping called a *tokamap*, which is compatible with the toroidal geometry (Balescu et al. 1998; Balescu 1998). The tokamap has been constructed to satisfy constraints that (i) the mapping should be Hamiltonian (or symplectic); (ii) it should be *compatible with toroidal geometry* that the canonical momentum ψ (toroidal flux) be always positive number and the magnetic axis $\psi_0 = 0$ is mapped to itself. The tokamap has the symplectic form (6.129), where $n = 1$, i.e.,

$$\begin{aligned}
\psi_{k+1} &= \psi_k - \frac{K}{2\pi}\frac{\psi_k}{1+\psi_k}\sin\vartheta_{k+1}, \\
\vartheta_{k+1} &= \vartheta_k + \frac{2\pi}{q(\psi_k)} - \frac{K}{2\pi}\frac{1}{(1+\psi_k)^2}\cos\vartheta_{k+1},
\end{aligned} \tag{6.134}$$

with the generating function

$$\epsilon S(\vartheta, \psi) = -\frac{K}{2\pi} \frac{\psi}{1+\psi} \cos \vartheta. \tag{6.135}$$

The safety factor $q(\psi)$ can be chosen arbitrary. The toroidal flux ψ is normalized to its value at the plasma edge, i.e., $0 \leq \psi < 1$. The parameter K stands for the perturbation amplitude.

If the profile of $q(\psi)$ is a non–monotonic function of ψ then the tokamap becomes the *non-twist mapping* (see Sect. 8.4). The corresponding tokamap called *revtokamap* has been studied by Balescu (1998) for the following profile of the safety factor: $q(\psi) = q_m / \left[1 - a(\psi - \psi_m)^2\right]$, where q_m is a minimum value of q at $\psi = \psi_m$, and a is a parameter.

The tokamap and the revtokamap have been widely to study regular and chaotic behavior of field lines in tokamaks. Particularly, in works by Misguich (2001), Misguich et al. (2003), Constantinescu et al. (2005, 2007), Constantinescu (2005), Constantinescu and Constantinescu (2005), Petrisor (2007) this model is employed to investigate the internal transport barriers and radial transport of field lines, in Pavlenko et al. (2003) to study the magnetic reconnection during the sawtooth instability.

6.9.4 The Symmetric Tokamap

As was noted in Sect. 6.3.1 the symplectic mapping of the form (6.129) for Hamiltonian systems with delta-kick perturbations (6.42) has a generating function (6.48) with the complicated dependence on the perturbation field. It means that the tokamap cannot be derived from the field line equations with the simple dependence of perturbation $H_m(\psi)$ in Eq. (6.40) on the flux variable ψ, i.e., like $H_m(\psi) = \psi/(1+\psi)$. However, the latter perturbation gives rise the mapping in the symmetric form (6.47) with the generating function (6.135),

$$\begin{aligned}
\Psi_k &= \psi_k - \frac{K}{4\pi} \frac{\Psi_k}{1+\Psi_k} \sin \vartheta_k, \\
\Theta_k &= \vartheta_k - \frac{K}{4\pi} \frac{1}{(1+\Psi_k)^2} \cos \vartheta_k, \\
\bar{\Theta}_k &= \Theta_k + \frac{2\pi}{q(\Psi_k)}, \\
\psi_{k+1} &= \Psi_k - \frac{K}{4\pi} \frac{\Psi_k}{1+\Psi_k} \sin \vartheta_{k+1}, \\
\vartheta_{k+1} &= \bar{\Theta}_k - \frac{K}{4\pi} \frac{1}{(1+\Psi_k)^2} \cos \vartheta_{k+1}.
\end{aligned} \tag{6.136}$$

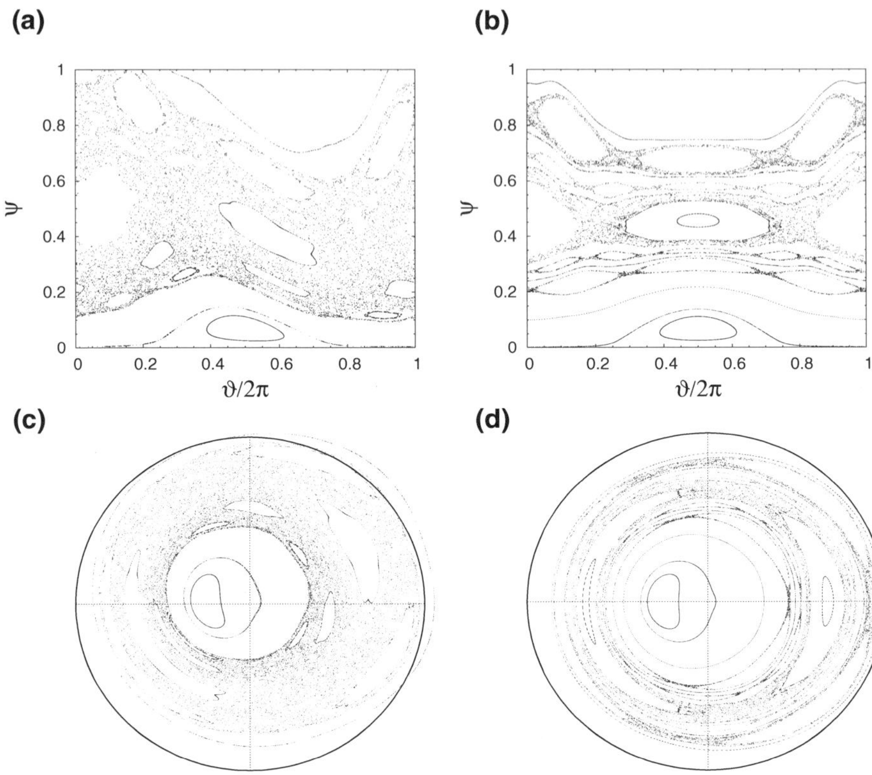

Fig. 6.11 Phase portraits of the tokamap (6.134) [(**a**), (**c**)] and the symmetric tokamap (6.136) [(**b**), (**d**)]: in the (ϑ, ψ)−plane (**a**), (**b**), and $(\sqrt{\psi}\cos\vartheta, \sqrt{\psi}\sin\vartheta)$-plane (**c**), (**d**). The q-profile is given by Eq. (6.137). The parameters $K = 4.5$, $q(0) = 1$

The symmetric tokamap is invariant with respect to the translation $\vartheta \leftrightarrow \pi - \vartheta$. It reflects the invariance of the symmetric mapping with respect to transformation $k \leftrightarrow k + 1$ with the simultaneous change $K \rightarrow -K$ and $\vartheta \rightarrow -\vartheta$.

Such a mapping has been proposed by Abdullaev (2004a) and called a *symmetric tokamap*. Its properties have been studied also by Wingen et al. (2005), Nasi and Firpo (2009), Schelin et al. (2011), Constantinescu and Firpo (2012).

The comparison of the phase space portraits of the tokamap and the symmetric tokamap is shown in Fig. 6.11 for the perturbation amplitude $K = 4.5$. The safety factor profile $q(\psi)$ is chosen in the following form (Balescu et al. 1998)

$$q(\psi) = \frac{4q(0)}{(2-\psi)(2-2\psi+\psi^2)}, \tag{6.137}$$

where $q(0)$ is the value of the safety factor on the magnetic axis $\psi = 0$.

It is shown that the properties of the symmetric tokamap are different from the tokamap. Particularly, it predicts higher values of the perturbation threshold for the destruction of the magnetic surface with "golden KAM curve", the most robust KAM barrier, than the tokamap (see Sect. 7.1 and Greene 1979). In the case of the plasma with the reversed magnetic shear the differences become even qualitatively different (see Abdullaev 2004a; Wingen et al. 2005; Nasi and Firpo 2009; Schelin et al. 2011; Constantinescu and Firpo 2012 for details).

6.9.5 Other Analytical Mapping Models of Field Lines

Below we give a short bibliographic guide to other analytical and numerical mapping models of a stochastic magnetic field. Since 1980s specific mapping models have been used to study regular and stochastic behavior of field lines in fusion devices.

The first mapping model to study the formation of ergodic zone at the plasma edge in the so-called ergodic limiter (or ergodic divertor) has been proposed by Martin and Taylor (1984) for the rectangular model of a tokamak. This model has been used by Regianni and Sakanaka (1994) to calculate the Lyapunov exponents and diffusion coefficients. The generalization of the Martin and Taylor mapping model to the toroidal plasma has been proposed by Viana and Caldas (1992), Caldas et al. (1996), Viana and Vasconcelos (1997), Ullmann and Caldas (2000), Portela et al. (2003) to study different statistical properties of magnetic field lines. More realistic and generic mapping models for magnetic field lines in the ergodic divertor, namely, the perturbed twist mappings have been proposed by Abdullaev et al. (1998), Fischer and Cooper (1998), Da Silva et al. (2001a, b), Da Silva et al. (2002a, b). The numerical study and analyses of magnetic field lines have been also studied in Viana (2000).

The mapping models have been also used by Bazzani et al. (1989, 1998) to study a field line diffusivity and transport in the Reversed Field Pinch (RFP). The RFP is another fusion device of a magnetically confined plasma. It is known as the equilibrium state of plasma with the minimal magnetic energy. The magnetic field of this state has intrinsic turbulent components due to a turbulent transient to the minimal energy state. The transport and diffusion processes in the RFP are mostly due to the magnetic field line stochasticity (D'Angelo and Paccagnella (1996); D'Angelo and Paccagnella (1999)).

The mapping approach to study chaotic field lines near the magnetic separatrix of poloidal divertor tokamaks, perhaps, has been first taken by Punjabi et al. (1992). A simple algebraic mapping called *tokamak divertor maps* was proposed to describe the field lines near the magnetic separatrix and on the divertor plates. Later these authors generalized this mapping to the symmetric form and the one which includes additional non-axisymmetric perturbations [see Punjabi et al. (1997, 2003, 2007) and references therein]. These mappings have been used to study generic features of the structure of field lines near X-points and the magnetic footprints affected by non-axisymmetric magnetic perturbations. Recently, the more sophisticated mapping to describe field lines in tokamaks with the magnetic separatrix has been proposed by Kroetz et al. (2010).

6.10 Bibliographic Comments

Since the introduction of the mapping concept by H. Poincarẽ (1892/1893/1899) it became an essential tool in the theory of dynamical systems, particularly, to study the regular and chaotic dynamics in Hamiltonian systems. It has been widely used in many areas of physics ranging from atomic and accelerator physics to dynamical astronomy (see references in Lichtenberg and Lieberman (1992) and Abdullaev (2006)). Especially the standard mapping (or the Chirikov–Taylor standard mapping) proposed by Chirikov (1969) and Taylor for the kicked Hamiltonian systems played a crucial tool to study the onset of chaotic motion in Hamiltonian systems.

The various methods to construct such mappings for the particle accelerators have been proposed by Dragt (1979, 1996), Dragt and Abel (1996), Dragt (2000), Berg et al. (1994), Warnock and Berg (1996) (see also references therein and Edwards and Syphers (1993).

Another powerful mapping, known *a separatrix mapping* has been proposed by Chirikov (1979) to study the dynamics of chaotic motion near separatrices. It has been extensively used not only to estimate the width of a stochastic layer near the separatrix (see Chirikov 1979; Zaslavsky et al. 1991; Vecheslavov 1996; Treshev 1998; Vecheslavov 1999; Vecheslavov and Chirikov 1998; Soskin and Mannella 2009; Shevchenko 2012), but also to study the mixing and transport processes in fluid dynamics [Weiss and Knobloch (1989); Ahn and Kim (1994); Latka and West (1995); Weeks et al. (1996) del Castillo-Negrete (1998); Kuznetsov and Zaslavsky (1998), plasma physics [Escande (1988); Lichtenberg and Wood (1989); Zaslavsky et al. (1991); Yamagishi (1995)] and in dynamical astronomy to study the dynamics of small planetary bodies, asteroids, in the Solar system, and rotational motion of satellites [Shevchenko and Scholl (1997); Shevchenko (1998, 1999)].

A number of works by Rom-Kedar (1994, 1995); Ahn et al. (1996); Treschev (2002, 2004) were devoted to the derivation of the separatrix mapping and to analyzes its properties. The different forms of the separatrix map or its generalizations have been discussed by Vecheslavov (1996, 1999), Luo and Han (2000), Shevchenko (2000), Luo and Han (2001), Luo (2002), Vecheslavov (2002). Among the mathematical studies on the separatrix mappings one should mention the works by Treschev (2002, 2004), Piftankin and Treshchev (2007), in which, particularly, a multidimensional separatrix mapping has been derived to study chaotic diffusion in the vicinity of asymptotic (separatrix) surfaces of a hyperbolic torus.

In spite of wide applications these mappings have one serious shortcoming: *the mapping variables do not coincide with the corresponding variables of original continuous system.* It means that these *mappings are not canonical*. Particularly, in the case of the separatrix mapping it has been clearly shown by its geometrical interpretation given by Escande (1988). It turns out that the main reason of this discrepancy is in the traditional derivation procedures of mappings from the corresponding Hamiltonian equations (see Abdullaev (2006) for the detailed discussions).

Perhaps, a *canonical separatrix mapping* has been first proposed by Abdullaev and Zaslavsky (1995) and used by Abdullaev and Zaslavsky (1996), Abdullaev and Finken (1998) to study the magnetic field lines near the magnetic separatrix of divertor tokamaks. The rigorous method to construct mappings based on the classical perturbation theory in finite–time intervals for the Hamilton–Jacobi equations of classical mechanics has been developed only a decade ago [see Abdullaev (1999, 2002, 2006)]. Using this method the canonical separatrix mappings has been also derived [Abdullaev (2004a, 2005)].

Beside analytical models the *numerical mapping* methods have been also developed to simulate heat and particle transport in realistic toroidal plasmas with a stochastic field lines. The method known as *interpolated cell mapping* has been originally developed by Tongue (1987), Montvai and Düchs (1993), de Rover et al. (1996a). This numerical mapping have been further developed by Kasilov et al. (1997), Runov et al. (2001), Feng et al. (2002), Kasilov et al. (2002), Kobayashi et al. (2004), Runov et al. (2004), Feng et al. (2005), Kobayashi et al. (2008), Feng et al. (2008), Frerichs et al. (2010), Frerichs, Clever, Feng, Lehnen, Reiter and Schmitz (2012), Frerichs, Reiter, Schmitz, Harting, Evans and Feng (2012) for Monte–Carlo simulations of transport processes in tokamak and stellarator plasmas with a stochastic magnetic field.

Chapter 7
Onset of Dynamical Chaos: Mathematical Aspects

The theory of chaos originated in the works by H. Poincaré is related to the problem of integrability of Hamiltonian systems under small periodic perturbations, i.e., to the fundamental problem of dynamics (see, Sect. 6.2.1). Chaos of magnetic field lines in magnetic fusion devices is an excellent example of dynamical chaos in Hamiltonian systems with one-and-half-degrees of freedom. In this chapter we discuss the some mathematical aspects of the onset of chaos in Hamiltonian systems.

A Hamiltonian system (6.5) is referred as an integrable if there exits a sufficient number of the integrals of motion I_i, $(i = 1, 2, \ldots, N)$. Orbits in such a system lie on the N—dimensional tori determined the integrals $I_i = \text{const}$. Under small perturbations (6.6) some of the integrals I_i (or all integrals except H) may not exist. It means the orbits do not stay on the tori but may fill the entire phase space $H = \text{const}$. The orbits become sensitive to the initial conditions. Particularly, two orbits with tiny differences in the initial conditions may greatly deviate from each other in the course of time. It means that it is practically impossible to predict orbit's behavior in such systems which leads the chaotic motion (Poincaré 2009). As was shown by H. Poincaré the main obstacles to the integrability are complex behavior of solutions of Hamiltonian systems, such as *creation of unstable periodic orbits, splitting of orbits asymptotically approaching the unstable periodic orbits*.

A half century later A. N. Kolmogorov gave a sketch the theorem that at certain conditions the Hamiltonian system (6.5) may be integrable, i.e., the conditionally periodic motion may conserve in the small neighborhood of noncritical points of phase space (Kolmogorov 1954). For the magnetic field lines described by Eqs. (3.9) and (3.12) it means survival of magnetic surfaces in the presence of small non-axisymmetric perturbations. However, in typical situations Hamiltonian systems may not possess the integrals of motion in whole phase space.

In this chapter we recall the some mathematical aspects of the onset of chaos in Hamiltonian systems emphasizing on its implications on the magnetic field lines and charged particle dynamics in magnetic fusion devices. First we formulate the Kolmogorov's theorem on the conservation of conditionally–periodic motion and its application to a magnetic field system and charged particle motion in a magnetic field.

S. Abdullaev, *Magnetic Stochasticity in Magnetically Confined Fusion Plasmas*, Springer Series on Atomic, Optical, and Plasma Physics 78,
DOI: 10.1007/978-3-319-01890-4_7, © Springer International Publishing Switzerland 2014

In Sect. 7.2 we describe methods to find the isolated periodic orbits, the Poincaré method and a more general method based on the mapping method described in Sect. 6.2. The features of the unstable periodic orbits, the stable and the unstable manifolds, their splitting will be discussed in Sect. 7.5. Finally, in Sect. 7.6 the onset of chaotic motion due to the creation of infinite number of unstable periodic points and the homoclinic structure of phase space is discussed.

7.1 Conservation of a Conditionally–Periodic Motion: The KAM Theory

In this section we give the main ideas of the Kolmogorov-Arnold-Moser (KAM) theory emphasizing on conditions of its applicability. This theory concerns the conservation of conditionally–periodic motion of Hamiltonian systems subjected to small perturbations in *infinite time nterval*. Kolmogorov (1954) formulated the corresponding theorem and gave the sketch of its proof. The theorem has been later proven by Arnold (1963a, b) and Moser (1962) (see, e.g., Arnold et al. 2006; Arnold 1989). It is now known the Kolmogorov-Arnold-Moser (KAM) theory (see, e.g., Arnold et al. 2006).

Consider an integrable Hamiltonian system described by Hamiltonian $H_0(I_1, \ldots, I_N)$. The orbits in phase space lie on the surface of the invariant torus determined by the actions (I_i = const) ($i = 1, \ldots, N$). The invariant torus is called *irrational* (or *non-resonant*), if the frequencies of motion, $\omega_i = \partial H_0(I)/\partial I_i$, ($i = 1, \ldots, N$) are rationally independent,

$$m \cdot \omega = m_1\omega_1 + m_2\omega_2 + \cdots + m_N\omega_N \neq 0, \tag{7.1}$$

for all integer numbers m_i not all of which are zero. If the frequencies ω_i ($i = 1, \ldots, N$) are rationally dependent then the tori are rational (or resonant). The unperturbed system is *nondegenerate* if the frequencies are functionally independent:

$$\det\left(\frac{\partial^2 H_0}{\partial I^2}\right) \neq 0. \tag{7.2}$$

Suppose that the integrable system is subjected to a small perturbation. The perturbed system is described by the system (6.5) with the Hamiltonian (6.4) and (6.6). The Kolmogorov's theorem (Kolmogorov 1954) states for the unperturbed system with H_0 satisfying the nondegenerate condition (7.2) the sufficiently small perturbations ϵ do not destroy the majority of *non-resonant tori*, but slightly deform them. That is the motion on these tori is still conditionally–periodic with the perturbed frequencies slightly different from the unperturbed ones $\omega(J)$. Orbits densely fill these invariant tori.

The condition of irrationality of nonresonant tori (7.1) is given by the Diophantine condition

$$|m \cdot \omega(I)| > c|m|^{-\nu}, \tag{7.3}$$

where ν and c are constants.

Formally, the existence of conditionally–periodic motion means that there exists such a canonical transformation of variables $(\vartheta, I) \to (\Theta, J)$,

$$\Theta = \vartheta + \frac{\partial S(J, \vartheta, \epsilon)}{\partial J}, \quad J = I - \frac{\partial S(J, \vartheta, \epsilon)}{\partial \vartheta}, \tag{7.4}$$

which transforms the Hamiltonian system (6.5) to the integrable Hamiltonian system

$$\frac{d\Theta}{d\tau} = \frac{\partial \mathcal{H}(J, \epsilon)}{\partial J} = w(J, \epsilon), \quad \frac{dJ}{d\tau} = -\frac{\partial \mathcal{H}(J, \epsilon)}{\partial \Theta} = 0, \tag{7.5}$$

with the new Hamiltonian $\mathcal{H}(J, \epsilon)$ which does not depend on the new angle variable Θ. The generating function $S(J, \vartheta, \epsilon)$ satisfies the Hamilton-Jacobi Eq. (6.12) in infinite time intervals. From (7.5) it follows that the new action variable J is a constant, and the angle variable Θ is a linear function of time

$$J = \text{const}, \quad \Theta = w(J; \epsilon)\tau + \Theta_0, \tag{7.6}$$

where $w(J; \epsilon)$ is a perturbed frequency of the invariant tori. Then the invariant tori is determined by new action variable $J_i = \text{const}$ $(i = 1, \ldots, N)$, and the motion on tori is conditionally−periodic with the frequencies $w(J, \epsilon)$. From (7.4) one obtains the form of the invariant tori in the old variable (ϑ, I),

$$I = J + \epsilon f(J, \Theta; \epsilon), \quad \vartheta = \Theta + \epsilon g(J, \Theta; \epsilon), \tag{7.7}$$

where f, g are continuously differentiable functions of period 2π in ψ: $f(J, \Theta; \epsilon) = f(J, \Theta + 2\pi; \epsilon)$, $g(J, \Theta; \epsilon) = g(J, \Theta + 2\pi; \epsilon)$, and of order of $f, g \sim 1$.

7.1.1 The Methods of KAM Theory

Formally the Kolmogorov's theorem formulates the conditions at which the canonical transformation (7.4) exists. The main difficulty in this proof arises from the following fact. The generating function $S(\vartheta, J, \epsilon)$ as a solution of the Hamilton-Jacobi Eq. (6.12) in the infinity time interval, unlike it's solution (6.15), (6.22 and 6.24) in a *finite time interval*, contains small denominators $m \cdot \omega(J)$ instead of the functions $a(x)$, $b(x)$. Particularly, the first term S_1 in the perturbation series (6.15) has the following form

$$S_1(\vartheta, J) = -\sum_m \frac{H_m(J)}{m \cdot \omega(J)}. \tag{7.8}$$

The latter become infinite or extremely large near the resonant tori, $J = J_m$, $m \cdot \omega(J) = 0$. This may lead to divergence of the perturbation series for the generating function S and causes the problem in proving its convergence. This problem has been resolved by Kolmogorov using the successive canonical changes of variables which eliminate phases in higher orders of ϵ.[1]

Below we briefly discuss this method of successive canonical transformations in the perturbation theory called *Kolmogorov's technique* (see Arnold et al. (2006), Lichtenberg and Lieberman (1992) for more details). It is based on successive canonical changes of variables in a perturbation Hamiltonian system and successively eliminates terms in the Hamiltonian of higher orders in the perturbation parameter ϵ. This method is an extremely rapidly convergent with increasing numbers of transformations of variables. Below we describe the *time–dependent version of superconvergent method.*

Consider again the perturbation Hamiltonian system (6.5) with the Hamiltonian (6.4), (6.6) and (6.25). Perform a change of variables $(\vartheta, I) \to (\Theta_1, J_1)$ via the generating function $F_1(\vartheta, J_1, \epsilon) = \vartheta J_1 + \epsilon S_1(\vartheta, J_1)$, i.e.,

$$I = J_1 + \epsilon \frac{\partial S_1}{\partial \vartheta}, \quad \Theta_1 = \vartheta + \epsilon \frac{\partial S_1}{\partial J_1}, \tag{7.9}$$

which transforms the Hamiltonian H to $\mathcal{H}^{(1)}$:

$$\mathcal{H}^{(1)}(\psi_1, J_1, t, t_0, \epsilon) = \mathcal{H}_0^{(1)}(J_1, \epsilon) + \epsilon^2 \mathcal{H}_1^{(1)}(\Theta_1, J_1, \epsilon), \tag{7.10}$$

where

$$\begin{aligned} \mathcal{H}_0^{(1)}(J_1, \epsilon) &= H_0(J) + \epsilon \mathcal{H}_1(J), \\ \mathcal{H}_1^{(1)}(\Theta_1, J_1, \epsilon) &= \frac{1}{2} \frac{\partial S_1}{\partial \vartheta} \cdot \frac{\partial^2 H_0}{\partial J \partial J} \cdot \frac{\partial S_1}{\partial \vartheta} + \frac{\partial H_1}{\partial J} \cdot \frac{\partial S_1}{\partial \vartheta} + O(\epsilon). \end{aligned} \tag{7.11}$$

Note that the new Hamiltonian (7.10) has a similar structure as (6.4) but the first term in the new Hamiltonian (7.10), $\mathcal{H}_0^{(1)}(J_1, \epsilon)$ does not depend on the new angle variable Θ_1, and the term depending on the phase Θ_1 has the second order in perturbation parameter ϵ^2.

Applying a second change of variables $(\Theta_1, J_1) \to (\Theta_2, J_2)$ to the new system with Hamiltonian (7.10) via the generating function

$$F_1(\Theta_1, J_2, \epsilon^2) = \Theta_1 J_2 + \epsilon^2 S_2(\Theta_1, J_2, \epsilon),$$

[1] Arnold (1963a) gave the proof of this theorem under assumption that the Hamiltonian $H(I, \vartheta)$ is an analytical function of its variables I, ϑ, while Moser had proved it for smooth Hamiltonian functions of class $C^r, r > 2N$, i.e., for the functions with a finite number r of continuous derivatives. Initially, for the case $N = 1$ Moser proved that $r \geq 333$, but later Herman showed that it is sufficient $r \geq 3$.

one transforms $\mathcal{H}^{(1)}$ into $\mathcal{H}^{(2)} = \mathcal{H}_0^{(2)} + \epsilon^4 \mathcal{H}_1^{(2)}$ pushing the phase-dependent perturbation terms to 4-th order in ϵ. After the mth change of variables: $(\psi_{m-1}, J_{m-1}) \rightarrow (\psi_m, J_m)$ the Hamiltonian becomes $\mathcal{H}^{(m)} = \mathcal{H}_0^{(m)} + \epsilon^{2^m} \mathcal{H}_1^{(m)}$ with a phase-dependent term of order ϵ^{2^m}.

As was shown by Kolmogorov (1954) and Arnold (1963a) this procedure converges quadratically, similar to Newton's root-finding method (see, e.g., Lichtenberg and Lieberman 1992). Such a quadratic convergence of perturbation series is known as *superconvergence*.

7.1.2 Conservation of Non-Resonant Magnetic Surfaces

Magnetic field system corresponds to the particularly case of Hamiltonian system with the $N = 2-$degrees of freedom. In this case the problem concerns the conservation of magnetic surfaces in a magnetic system in the presence of small non-axisymmetric magnetic perturbations.

Consider again the magnetic field line equations in the presence of small–amplitude magnetic perturbations breaking axisymmetry along the toroidal angle φ, i.e.,

$$\frac{d\vartheta}{d\varphi} = \frac{\partial \psi}{\partial \psi_t}, \quad \frac{d\psi}{d\varphi} = -\frac{\partial \psi}{\partial \vartheta},$$
$$\psi = \psi_0(\psi) + \epsilon \psi_1(\vartheta, \psi_t, \varphi), \tag{7.12}$$

where the perturbation parameter ϵ is assumed to be small, $\epsilon \ll 1$, the perturbation Hamiltonian $\psi_1(\vartheta, \psi_t, \varphi)$ is the periodic functions of the poloidal and toroidal angles (ϑ, φ),

$$\psi_1(\vartheta, \psi_t, \varphi) = \sum_{mn} H_{mn}(\psi_t) \cos(m\vartheta + n\varphi + \chi_{mn}). \tag{7.13}$$

In the absence of the perturbation ($\epsilon = 0$), the field lines densely covers the non-resonant magnetic surfaces with the irrational safety factor $q(\psi_{mn}) \neq m/n$ for any integer numbers m, n, but on the resonant magnetic surfaces, $\psi_{m,n}, q(\psi_{m,n}) = m : n$ field lines form closed orbits. Poincaré sections of field lines in the poloidal section $\varphi = \text{const}$ shown in Fig. 7.1a illustrates this situation.

From the Kolmogorov's theorem it follows that for sufficiently small perturbations, $\epsilon \ll 1$, the majority of irrational magnetic surfaces are survived with a slight deformation from the unperturbed surfaces as it illustrated in Fig. 7.1b. The irrational magnetic surfaces are called *Kolmogorov–Arnold–Moser* (KAM) surface. The applicability conditions of the KAM theory on the conservation of magnetic surfaces are:

(i) the perturbation amplitude ϵ should be sufficiently small;

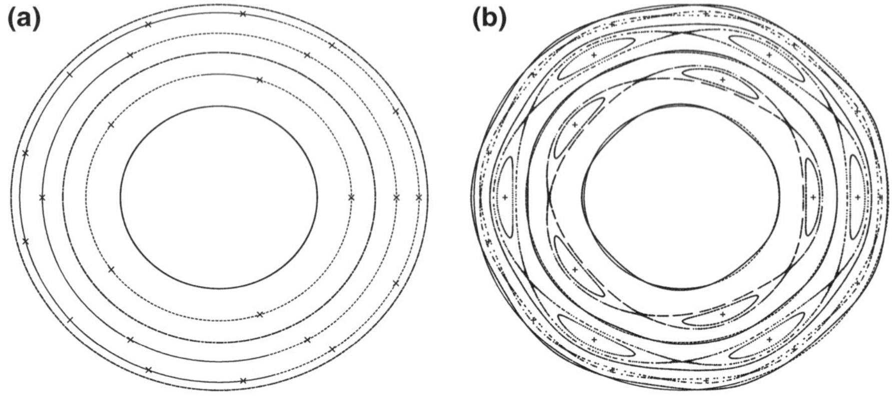

Fig. 7.1 Poincaré sections of field lines: **a** non-resonant and resonant orbits for $m : n = 5{:}1, 6{:}1$ and 13:2 in the absence of perturbation ($\epsilon = 0$); **b** Magnetic surfaces in the presence of small perturbation ($\epsilon \ll 1$). The irrational magnetic surfaces (*solid curves*) are slightly deformed near the unperturbed magnetic surfaces (*dashed curves*). The resonant orbits are destroyed

(ii) the unperturbed Hamiltonian $\psi_0(\psi_t)$ should be an analytical function of ψ_t. The magnetic perturbations $\psi_1(\psi_t, \vartheta, \varphi)$ is analytical or smooth function of variables $(\psi_t, \vartheta, \varphi)$ [2];

(iii) From the condition (7.2) it follows the safety factor $q(\psi)$ should satisfy the twist condition $dq^{-1}/d\psi = d^2\psi_0(\psi_t)/d\psi_t^2 \neq 0$; It means that there are no the KAM surface near the shearless magnetic surface where $dq/d\psi = 0$.

(iv) The irrationality condition of tori (7.3) means the safety factor $q(\psi)$ should be sufficiently irrational, i.e., $q(\psi_{mn}) \neq m/n$ for large integer numbers m, n that are without common dividers and satisfying the Diophantine condition

$$\left|q^{-1} - m/n\right| \geq C\sqrt{\epsilon}m^{-\nu}, \quad C = \text{const}, \nu = 2.5 \tag{7.14}$$

The most unbreakable magnetic surface or torus corresponds to the most irrational number $g^* = (\sqrt{5} - 1)/2$ known the golden KAM tori (see, Greene 1979).

The violation of one of these conditions near a given magnetic surface may break it and lead to complex behavior of field lines. Below we consider these cases.

One should that that perturbations ($\epsilon \neq 0$) may destroy the resonant magnetic surfaces and form a stochastic zone of field lines there (see Sect. 7.6). However, these field lines are confined in a finite area between invariant magnetic surfaces. This phenomenon has a crucial importance to achieve the magnetic confinement of fusion plasmas.

[2] The perturbation function $\psi_1(\psi_t, \vartheta, \varphi)$ should belong to the class C^k, $k \geq 2$. This condition on the smoothness of perturbations, perhaps, is satisfied for any magnetic perturbations.

7.1.3 Action Variable as an Adiabatic Invariant

Using the methods of KAM theory discussed above one can show that the action variable I_R in the Hamiltonian system (4.27) for the guiding–center motion is an adiabatic invariant. It allows one to eliminate a gyrophase ϑ_R from the Hamiltonian equations (4.27) and reduce its dimensions.

We assume that the ratio ρ_R/L of the gyroradius ρ_R to the characteristic length L of magnetic field variations plays role of small perturbation parameter $\varepsilon = \rho_R/L \ll 1$. Consider the Hamiltonian (4.27) for the guiding-center motion presented as a sum of two terms,

$$\mathcal{H} = \mathcal{H}_0(I_R, \mathbf{Q}, \mathbf{P}) + \varepsilon\mathcal{H}_1(\vartheta_R, I_R, \mathbf{Q}, \mathbf{P}), \tag{7.15}$$

[for simplicity we omitted subscripts "c"]. The first term which does not depend on the gyrophase ϑ_R can be considered as an unperturbed Hamiltonian, and the second one depending on ϑ_R is considered as the small perturbation Hamiltonian. This term can be eliminated using a successive canonical change of variables as discussed in Sect. 7.1.1. Such a change of variables

$$(\vartheta_R, I_R, \mathbf{Q}, \mathbf{P}) \to \left(\bar{\vartheta}_R, \bar{I}_R, \bar{\mathbf{Q}}, \bar{\mathbf{P}}\right)$$

transforms the Hamiltonian (7.15) into

$$\mathcal{H}_0\left(\bar{I}_R, \bar{\mathbf{Q}}, \bar{\mathbf{P}}; \varepsilon\right) + \varepsilon^2 H_2\left(\bar{\vartheta}_R, J_R, \bar{\mathbf{Q}}, \bar{\mathbf{P}}\right), \tag{7.16}$$

with the new radial action variable $\bar{I}_R$,

$$\bar{I}_R = I_R - \varepsilon\frac{\partial S_1}{\partial \vartheta_R}, \tag{7.17}$$

determined by the generating function $S_1 = S_1\left(\vartheta_R, \bar{I}_R, \mathbf{Q}, \bar{\mathbf{P}}\right)$.

The new Hamiltonian (7.16) contains the gyro–phase ϑ_R in the terms of order of ε^2. Further successive canonical transformations of variables shift the terms with the gyro–phase ϑ_R to the higher orders ε^4, ε^8 and so on. If such transformations converge then consecutive action variables I_R, $\bar{I}_R, \ldots$ converge to the invariant one J_R which will be an integral of motion. The latter is related to the action variable I_R,

$$I_R = J_R + \varepsilon g(\vartheta_R, J_R; \varepsilon), \tag{7.18}$$

where $g(\vartheta_R, J_R; \varepsilon)$ is a periodic function of ϑ_R with $\max|g| < c \lesssim 1$. The variation of the action I_R about the integral J_R is limited and of order of ε. Thus I_R is a *perpetual adiabatic invariant*.

The existence of such successive canonical transformations and thus of the invariant J_R is guaranteed by the KAM theory at certain conditions. As we have seen above it is necessary the condition (7.3) for frequencies should be fulfilled. Indeed, in systems with the strong magnetic field the frequency ω_R of radial gyro-oscillations is much larger than both slow drift frequencies ω_z and ω_φ along the poloidal and toroidal axes. Therefore, in typical tokamak plasmas the irrationality conditions of type

$$s\omega_R + m\omega_z + n\omega_\varphi \neq 0,$$

is satisfied for even moderately large mode numbers (m, n). Another condition of existence of the invariant J_R is smallness of the ratio $\varepsilon = \rho_R/L$.

However, in some cases these conditions may not be satisfied. For example, as was shown by Carlsson (2001), Yavorskij et al. (2002) in spherical tokamaks with a weak, but strongly spatially varying magnetic field the magnetic moment $\mu = m_0 v_\perp^2/2B$ of fast ions may be broken down. In these cases the invariant J_R also does not exist. This non–adiabaticity of the magnetic moment μ is due to relatively large value of the ratio $\varepsilon = \rho_R/L$ and occurrence of resonances

$$s\omega_R + m\omega_z + n\omega_\varphi = 0,$$

between gyro-oscillations and transit oscillations. Similar situation occurs in the presence of an electric field with strong gradients (Dirickx and Weyssow 1998). In this cases one cannot reduce the 6D system of guiding-center motion equations (4.24) to the 4D system (4.31). In such cases the drift surfaces may not exist which may lead to the deterioration of a particle confinement.

7.1.4 Stability of Drift Surfaces: Arnold Diffusion

Now we apply the KAM theory to the stability problem of drift surfaces in the presence of magnetic and electric field perturbations. As has been shown in Sect. 5.1 in axisymmetric toroidal systems guiding–center orbits lie on the drift surfaces. For the given particle energy E and the radial action I_x these surface similar to magnetic surfaces form a set of nested toroidal surfaces which do not cross other. However, *the drift surfaces corresponding to particles of different energies may cross each other.* There is an important question, "whether the irrational drift surfaces survive under perturbations breaking the toroidal symmetry". If so then do these invariant surfaces can confine particles in a finite area, like magnetic field lines? The answer on the question may come from the KAM theory.

The dynamics of guiding–center motion in the presence of time–dependent perturbations is described by the 4D non-autonomous Hamiltonian system (5.36) and (5.38). Introducing a pair of canonical variables $(\vartheta_t = \Omega t,\ p_t = -E/\Omega)$ instead of time t and the energy E can be presented as the 6D autonomous Hamiltonian system,

$$\begin{aligned}
\frac{d\vartheta}{d\tau} &= \frac{\partial \mathcal{H}}{\partial J} = \omega(J, I_\varphi) + \epsilon \frac{\partial H_1}{\partial J}, & \frac{dJ}{d\tau} &= -\frac{\partial \mathcal{H}}{\partial \vartheta} = -\epsilon \frac{\partial H_1}{\partial \vartheta}, \\
\frac{d\vartheta_\varphi}{d\tau} &= \frac{\partial \mathcal{H}}{\partial I_\varphi} = \omega_\varphi(J, I_\varphi) + \epsilon \frac{\partial H_1}{\partial I_\varphi}, & \frac{dI_\varphi}{d\tau} &= -\frac{\partial \mathcal{H}}{\partial \vartheta_\varphi} = -\epsilon \frac{\partial H_1}{\partial \vartheta_\varphi}, \\
\frac{d\vartheta_t}{d\tau} &= \frac{\partial \mathcal{H}}{\partial p_t} = \Omega + \epsilon \frac{\partial H_1}{\partial p_t}, & \frac{dp_t}{d\tau} &= -\frac{\partial \mathcal{H}}{\partial \vartheta_t} = -\epsilon \frac{\partial H_1}{\partial \vartheta_t},
\end{aligned} \tag{7.19}$$

with the Hamiltonian $\mathcal{H}$

$$\begin{aligned}
&\mathcal{H} = h_c(J, I_\varphi) + \Omega p_t + \epsilon H_1\left(\vartheta, \vartheta_\varphi, t, J, I_\varphi, p_t\right), \\
&H_1\left(\vartheta, \vartheta_\varphi, t, J, I_\varphi\right) = \sum_{m,n,s} H_{mns}(J, I_\varphi) e^{im\vartheta + in\vartheta_\varphi - is\vartheta_t}.
\end{aligned} \tag{7.20}$$

(For simplicity we have omitted subscripts "c", and introduced the notation H_1 instead of $h_c^{(1)}$). We recall that ω, ω_φ, are the transit frequencies of poloidal and toroidal direction on drift surfaces, and Ω is the frequency of oscillations in time.

First consider the time-independent perturbation when $\Omega = 0$. In this case the particle energy E is a constant of motion. From the Kolmogorov's theorem it follows that irrational drift surfaces (J, I_φ) satisfying the condition

$$m\omega + n\omega_\varphi \neq 0, \tag{7.21}$$

may survive at sufficiently small perturbations ϵ. The resonant drift surfaces (J_{mn}, $I_\varphi^{(mn)}$) for which $m\omega + n\omega_\varphi = 0$ are destroyed. Since the different resonant drift surfaces of particles with the same energy do not cross each other, the invariant drift surfaces confine the transport of chaotic particles formed near the destroyed resonant surfaces.

In the presence of time–dependent perturbations, $\Omega \neq 0$, the particle energy E are not conserved. In this case, according to KAM theory, the majority of irrational drift surfaces (J, I_φ) with $m\omega + n\omega_\varphi + s\Omega \neq 0$ may survive. However, these invariant drift surfaces may not confine chaotic particles formed near the destroyed resonant drift surfaces. The resonant drift surface defined

$$m\omega + n\omega_\varphi + s\Omega = 0, \tag{7.22}$$

and corresponding to particles with the different energies may cross each other forming the net. The motion along this net may transport particles far from initial confined area.

This kind of chaotic motion in many-dimensional Hamiltonian systems with the degrees of freedom $N > 2$ has been predicted by Arnold (1964) and known as the *Arnold diffusion*. The possibility of such diffusion of particles in toroidal systems has been studied by Lichtenberg (1992) and Pfirsch (2001). However, the rate of the Arnold diffusion is extremely small (see Lichtenberg andLieberman 1992,

Arnold et al. 2006 for more details). One expects that the Arnold diffusion of particles in a toroidal system can be neglected because it's insignificance in comparison with other mechanisms of transport.

7.2 Creation of Periodic Orbits

As was shown by Poincaré (1892/1893/1899) the main obstacles to the integrability of Hamiltonian systems are the creations of isolated periodic orbits and the splitting of separatrices in the presence of perturbations.[3] The breaking of magnetic surfaces by any small nonaxisymmetric perturbations and the creation of magnetic islands at the resonant magnetic surfaces $\psi_{m,n}$ is an important examples of this phenomenon. The behavior of system in this case is described by the Poincaré's last geometrical theorem proved by Birkhoff which is known as the *Poincaré–Birkhoff theorem* (Birkhoff 1927 (reprinted 1966)) (see, also Arnold 1989; Arnold et al. 2006).

7.2.1 Poincaré Method to Determine Isolated Periodic Orbits

There are many analytical and numerical methods to determine the periodic orbits of Hamiltonian systems. Below we describe the method of Poincaré (1892/1893/1899) developed for the problem of celestial mechanics (see, also Arnold et al. 2006; Kozlov 1996). This method determines only periodic orbits corresponding to the non-zero terms $H_m(I)$ in the perturbation Hamiltonian $H_1(I, \vartheta)$ (6.6). Furthermore, we call the corresponding (m, n) modes with $H_{mn}(I) \neq 0$ as *primary resonant tori*.

Consider the Hamiltonian system (6.5) and (6.6) subjected to the small–amplitude $2\pi-$periodic perturbation in ϑ. First we give a definition of periodic orbits. Let $Z(t) = (I(t), \vartheta(t))$ is the solution the Hamiltonian system (6.5) with the initial coordinates $Z_0 = (I_0, \vartheta_0)$ at $t = t_0$. Then the orbit is called a periodic with a period T if

$$Z(t_0 + T) = Z(t_0) = Z_0. \tag{7.23}$$

From the solution (6.2) of the unperturbed system (6.5), $(\epsilon = 0)$, it follows that if the products $\omega_i(I)T/2\pi$, $i = (1, \ldots, N)$, at a certain value $I = I_{m0}$ are equal to some integers numbers p_i then this solution is periodic with the period T. The corresponding action I_{m0} defines a resonant tori, and it is determined by the condition

$$m_0 \cdot \omega(I_{m0}) = \sum_{i=1}^{N} m_i^{(0)} \omega_i(I_{m0}) = 0, \tag{7.24}$$

---

[3] There are also topological obstacles to the integrability of Hamiltonian systems related with its complex phase space structure (Kozlov 1983; see also Kozlov 1996; Arnold et al. 2006).

where the integer number $m_0 = (m_1, m_2, \ldots, m_N)$. Suppose that

$$\det\left(\frac{\partial^2 H_0}{\partial I^2}\right) \neq 0, \quad \text{at } I = I_{m_0}. \tag{7.25}$$

For the field lines the latter condition means that the magnetic shear is finite, $dq^{-1}/d\psi \neq 0$ near the resonant magnetic surface ψ_0, $q(\psi_0) = m_0/n_0$.

The perturbation Hamiltonian (7.13) on the resonant tori ψ_0 is

$$H_1(I_{m_0}, \vartheta(t)) = H_1(I_0, \omega(I_{m_0})(t - t_0) + \vartheta_0), \tag{7.26}$$

where t_0 is the initial time t_0, and ϑ_0 are initial angles. It is a periodic function of t with a period T. The averaged function,

$$\langle H(t_0)\rangle = \frac{1}{T}\int_0^T H_1\left(I_{m_0}, \omega(I_0)(t - t_0) + \vartheta_0\right) dt, \tag{7.27}$$

will be a $2\pi-$periodic function of the initial time t_0. One can note that $\langle\psi_1(\varphi_0)\rangle$ (7.27) up to the constant factor is equal to the generating function $S_1(\vartheta, I, t, t - T)$ defined by Eq. (6.21), i.e.,

$$\langle H_1(t_0)\rangle = \frac{1}{T} S_1(\vartheta_0, I_{m_0}, T, 0), \quad \vartheta_0 = \omega(I_{m_0})t_0. \tag{7.28}$$

Poincaré had shown that when $d^2\langle H_1(t_0)\rangle/dt_0^2 \neq 0$ at $I = I_{m_0}$ the extremal points t_i of $\langle H_1(t)\rangle$, i.e., the minimal and maximal points, correspond the periodic orbits of the Hamiltonian system (6.5) and (6.6). Therefore, the destruction of a resonant tori leads to the creation at least two periodic orbits since the local minimum of the periodic function always its local maximum or other way around.

Consider as an example the Hamiltonian system,

$$\begin{aligned} H &= H_0(I) + \epsilon H_1(I, \vartheta, t), \\ H_1(I, \vartheta, t) &= \sum_m H_{mn}(I)\cos(m\vartheta + \chi_{mn})\sin(nt). \end{aligned} \tag{7.29}$$

Let $I = I_0$ be the resonant tori determined by the resonant condition $m_0\omega(I_0) = n_0$ with the integer numbers (m_0, n_0). Using Eq. (6.21) one can obtain the averaged function (7.27),

$$\begin{aligned} \langle H_1(t_0)\rangle &= -(1/2)H_{m_0n_0}(I_0)\sin\left(n_0t_0 + \chi_{m_0,n_0}\right) \\ &= -(1/2)H_{m_0n_0}(I_0)\sin\left(m_0\vartheta_0 + \chi_{m_0,n_0}\right). \end{aligned} \tag{7.30}$$

This function has n_0 local maxima at $t_0 = [(3\pi/2)s - \chi_{m_0n_0}]/n_0$ and n local minima at $t_0 = [(\pi/2)s - \chi_{m_0n_0}]/n_0$, $(s = 1, 2, \ldots, n_0)$ or m_0 local maxima at $\vartheta_0 = [(3\pi/2)s - \chi_{m_0n_0}]/m_0$ and m_0 local minima at $\vartheta_0 = [(\pi/2)s - \chi_{m_0n_0}]/m_0$, $(s = 1, 2, \ldots, m_0)$ in the angle variable ϑ.

7.2.2 Mapping Method to Determine Periodic Orbits

As was noted above the Poincaré method guarantees the existence of a large (may be also infinite) number of periodic orbits corresponding to the primary resonance tori.[4] Below we give another method to find periodic orbits of perturbed Hamiltonian system which allows to determine infinitely many periodic solutions but not only primary ones. The method is based on the construction of mapping described in Sect. 6.2. At the small perturbations, $\epsilon \ll 1$, it coincides with the Poincaré method.

Since the Poincaré mapping reduce a dimension of the system by one, then the determination of periodic orbits of continuous system is reduced to the determination of fixed points P_0 of the mapping (6.1), i.e.,

$$P_0 = (\hat{M})^p P_0, \tag{7.31}$$

where p is the smallest integer number for which Eq. (7.31) is satisfied.

Consider the mapping $P_k \to P_{k+1}$, $P_k = (\vartheta_k, I_k)$, after one full period $t_{k+1} = t_k + T$. Setting $\tau = t$ we choose the sections t_k as $t_k = t_0 - T/2$ and $t_{k+1} = t_0 + T/2$ the mapping (6.9)–(6.11) can be reduced to

$$\begin{aligned} I_{k+1} &= I_k - \epsilon \left[\frac{\partial S_k}{\partial \vartheta_k} - \frac{\partial S_{k+1}}{\partial \vartheta_{k+1}} \right], \\ \vartheta_{k+1} &= \vartheta_k + w(J_{k+1}, \epsilon)T + \epsilon \left[\frac{\partial S_k}{\partial J_k} - \frac{\partial S_{k+1}}{\partial J_k} \right], \end{aligned} \tag{7.32}$$

where $S_k \equiv S(\vartheta_k, J_k, t_k, t_0; \epsilon)$ is the solution of the Hamilton–Jacobi equation (6.12) with the initial condition (6.13) and $w(J, \epsilon)$ is the perturbed frequency given Eqs. (6.14) and (6.26). The intermediate variable J_k is given by

$$J_k = J_k - \epsilon \frac{\partial S_k}{\partial \vartheta_k}. \tag{7.33}$$

From Eq. (7.32) it follows that at the fixed points, $P_0 = (\vartheta_0, I_0)$ we have

---

[4] As asserted in Arnold et al. (2006) "the existence of infinitely many periodic solutions can be derived from Kolmogorov's theorem on the conservation conditionally periodic motions and the Poincaré–Birkhoff geometric theorem" (see also Moser 1973).

$$
\begin{aligned}
I_0 &\equiv I_{k+1} = I_k, \\
\vartheta_0 &\equiv \vartheta_{k+1} = \vartheta_k + 2\pi \mathcal{Q} = \vartheta_k, \,(\bmod 2\pi),
\end{aligned}
\tag{7.34}
$$

where $\mathcal{Q} = (\mathcal{Q}_1, \mathcal{Q}_2, \ldots, \mathcal{Q}_N)$ is the integer vector. Let $m_0 = (m_1^{(0)}, m_2^{(0)}, \ldots, m_N^{(0)})$ be the mode numbers corresponding to the resonant tori, $m_0 \cdot \omega\left(I_{m_0}\right) = 0$ on the resonant tori I_{m_0}. Then, the Eq. (7.32) for the fixed points (ϑ_0, I_0) can be reduced to the system

$$
\begin{aligned}
F_1(\vartheta, \Psi) &= \frac{\partial G(\vartheta, J)}{\partial \vartheta} = 0, \\
F_2(\vartheta, J) &= T w(J, \epsilon) - 2\pi \mathcal{Q} + \epsilon \frac{\partial G(\vartheta, J)}{\partial J} = 0,
\end{aligned}
\tag{7.35}
$$

where

$$
G(\vartheta, J) = S(\vartheta, J, t_0 - T/2, t_0; \epsilon) - S(\vartheta, J, t_0 + T/2, t_0; \epsilon). \tag{7.36}
$$

From the second set of equations in (7.35) one can obtain the following relation

$$
T m_0 \cdot w(J, \epsilon) + \epsilon m_0 \cdot \frac{\partial G(\vartheta, J)}{\partial J} = 0. \tag{7.37}
$$

Therefore, the determination of periodic orbits is reduced to the system of $2N$ algebraic equations (7.35) in the variables (ϑ, J). The relation between the variables I and J is given by (7.33).

From the first system of equations (7.35) it follows that the fixed points are determined by the extremal points of the generating function $G(\vartheta, J)$ along the angle variable ϑ. The latter according to Eqs. (6.15), (6.21) and (6.24), can be presented by a series of powers of the perturbation parameter ϵ,

$$
G(\vartheta, J) = G_1(\vartheta, J) + \epsilon G_2(\vartheta, J) + \cdots, \tag{7.38}
$$

where

$$
\begin{aligned}
G_1(\vartheta, J) &= \int_{t-T/2}^{t+T/2} H_1\left(\vartheta(t'), J, t'\right) dt', \\
G_j(\vartheta, J) &= \int_{t-T/2}^{t+T/2} F_j\left(\vartheta(t'), J, t'\right) dt', \qquad j \geq 2.
\end{aligned}
\tag{7.39}
$$

Here the integrals are taken along the unperturbed orbit $\vartheta(t') = \vartheta + (t' - t)\omega(J)$, $\vartheta = \vartheta(t)$.

In general the system (7.35) has the infinitely many solutions corresponding to the infinitely many periodic orbits. The primary periodic orbits correspond to the solutions of (7.35) with the generating function $G = G_1$, i.e., when the higher order terms G_j, $(j \geq 2)$ in ϵ are neglected. The function G_1 up to the constant factor coincides with the averaged Hamiltonian $\langle H_1 \rangle$ (7.27) and therefore these solutions correspond to the ones of the Poincaré method to find the primary periodic orbits. The secondary and higher order fixed points (periodic orbits) corresponding $(m, n)-$modes with $H_{mn} = 0$ in the perturbation Hamiltonian (7.13) can be found by taking into account the higher order terms G_j, $j \geq 2$ in the expansion (7.38) of the generating function G in ϵ.

Let us consider the system (7.35) for the case $N = 1$ and the resonant tori I_{m_0,n_0} determined by $m_0\omega(I_{m_0,n_0}) = n_0$. Let $\bar{\vartheta}_0$ is the root of the equation $F_1(\vartheta, J_{m_0n_0}) = 0$, at the resonant tori $I = I_{m_0n_0}$. Expanding the Eq. (7.35) in a series of powers of $(J - I_{m_0n_0})$, $(\vartheta - \bar{\vartheta}_0)$, one obtains

$$\begin{aligned} F_{1,\vartheta}^{(0)} (\vartheta - \bar{\vartheta}_0) + F_{1,J}^{(0)} (J - I_{m_0n_0}) &= 0, \\ T w_0' \left(J - I_{m_0n_0}\right) - \epsilon G_J^{(0)} &= 0. \end{aligned} \tag{7.40}$$

where $w_0' = w'(I_{m_0n_0}, \epsilon)$, where $w' = dw/dJ$. We have used the following notations:

$$F_{1,\vartheta}^{(0)} = \left.\frac{\partial F_1(\vartheta, J)}{\partial \vartheta}\right|_{\vartheta=\vartheta_0, J=I_{m_0n_0}}, \qquad F_{1,J}^{(0)} = \left.\frac{\partial F_1(\vartheta, J)}{\partial J}\right|_{\vartheta=\vartheta_0, J=I_{m_0n_0}},$$

$$G_J^{(0)} = \left.\frac{\partial G(\vartheta, J)}{\partial J}\right|_{\vartheta=\vartheta_0, J=I_{m_0n_0}}.$$

From Eq. (7.40) we get the fixed points (ϑ_0, J_0)

$$\begin{aligned} \vartheta_0 &= \bar{\vartheta}_0 - \frac{F_{1,J}^{(0)}}{F_{1,\vartheta}^{(0)}} \left(J_0 - I_{m_0n_0}\right), \\ J_0 &= I_{m_0n_0} - \frac{\epsilon G_J^{(0)}}{T w_0'}. \end{aligned} \tag{7.41}$$

The perturbed frequency $w(J, \epsilon)$ can be approximated by the unperturbed frequency $\omega(J)$ in (6.26). As seen from (7.41) the fixed points (ϑ_0, J_0) are shifted with respect to the ones $(\bar{\vartheta}_0, I_{m_0,n_0})$ given by the Poincaré method. The differences between them are of order of the perturbation amplitude ϵ.

Figure 7.2 shows the typical dependence of the generating function $G_1(\vartheta, J)$ (6.34) and (6.36) in the first order in ϵ (solid curve 1) and its derivative $\partial G_1(\vartheta, J)/\partial\vartheta$ (dashed curve 2) on ϑ at the resonant surface J_0. The extremal points ϑ_0 of $G_1(\vartheta, J)$ corresponding to the zeros of $\partial G_1(\vartheta, J)/\partial\vartheta$ are given by symbols $\circ$ for the elliptic fixed points and by the symbol $\times$ for the hyperbolic fixed points.

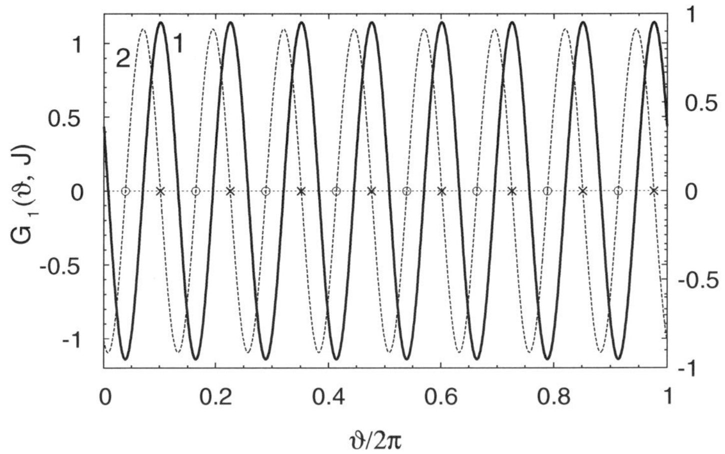

Fig. 7.2 Generating function $G_I(\vartheta, J)$ (*solid curve 1*) [left-hand axis] and its derivative $F_1(\vartheta, J) = \partial G_1(\vartheta, J)/\partial\vartheta$ (*dashed curve 2*) [right-hand axis] on the resonant surface $q(J_{mn}) = m/n$, $m = 8, n = 2$. Positions of fixed points are denoted by the symbols $\circ$ (*elliptic*) and $\times$ (*hyperbolic*)

7.3 Lyapunov Exponent

One of important characteristics of dynamical systems is the *Lyapunov exponent*. It gives the quantitative measure of the stability of dynamical systems. In this section we introduce the Lyapunov exponent and the method of its calculation via the symplectic mapping (6.9)–(6.11).

Let $\delta\vartheta_k = \vartheta(t_k) - \bar{\vartheta}(t_k)$ and $\delta I_k = I(t_k) - \bar{I}(t_k)$ be the infinitesimal separations of neighboring orbits $(\vartheta(t), I(t)$ and $(\bar{\vartheta}(t), \bar{I}(t)$ at the time instance t_k. We present them as a vector

$$\delta\mathbf{V}_k = \begin{pmatrix} \delta I_k \\ \delta\vartheta_k \end{pmatrix}. \tag{7.42}$$

The evolution of the vector $\delta\mathbf{V}_k$ in time can given by the following mapping

$$\delta\mathbf{V}_{k+1} = \mathbf{J}_k \delta\mathbf{V}_k, \tag{7.43}$$

where $\mathbf{J}_k$ is the $2N \times 2N$ Jacobian matrix of the mapping (6.7):

$$\mathbf{J}_k = \begin{pmatrix} \dfrac{\partial I_{k+1}}{\partial\psi_k} & \dfrac{\partial I_{k+1}}{\partial\vartheta_k} \\ \dfrac{\partial\vartheta_{k+1}}{\partial I_k} & \dfrac{\partial\vartheta_{k+1}}{\partial\vartheta_k} \end{pmatrix}. \tag{7.44}$$

We recall the property $\det \mathbf{J}_k = 1$. The calculation of the Jacobi matrix of the mapping (6.9)–(6.11)) in the two-dimensional phase space (ϑ, I) is given in Appendix E.

Consider the evolution of the distance $\delta s_k = \sqrt{(\delta I_k)^2 + (\delta\vartheta_k)^2}$ between orbits in time. According to (7.43) its square, $(\delta s_k)^2$, can be written as

$$(\delta s_k)^2 = \delta \mathbf{V}_k^T \delta \mathbf{V}_k = \delta \mathbf{V}_{k-1}^T \mathbf{J}_k^T \mathbf{J}_k d\mathbf{V}_{k-1}$$
$$= \delta \mathbf{V}_0^T \mathbf{J}_1^T \cdots \mathbf{J}_k^T \mathbf{J}_k \cdots \mathbf{J}_1 \delta \mathbf{V}_0. \quad (7.45)$$

where $\mathbf{J}_k^T$ is the transpose of the matric $\mathbf{J}_k$. Let $\lambda_1^{(k)}, \lambda_2^{(k)}$ be the eigenvalues of the matrix $\mathbf{J}_k$ determined by the eigenvalue problem,

$$\mathbf{J}_k \mathbf{A} = \lambda \mathbf{A}, \quad (7.46)$$

where $\mathbf{A}$ is the corresponding eigenvector. The eigenvalues are found as the solution of the equation,

$$\det \begin{pmatrix} \dfrac{\partial I_{k+1}}{\partial I_k} - \lambda & \dfrac{\partial I_{k+1}}{\partial \vartheta_k} \\ \dfrac{\partial \vartheta_{k+1}}{\partial I_k} & \dfrac{\partial \vartheta_{k+1}}{\partial \vartheta_k} - \lambda \end{pmatrix} = 0. \quad (7.47)$$

Furthermore, to find the eigenvalues λ we specifically consider the two– dimensional system in the phase space (ϑ, I). Then the solutions of 2×2 system (7.47) are

$$\lambda_{1,2}^{(k)} = D \pm \sqrt{D-1}, \quad (7.48)$$

where

$$D = \frac{1}{2}\left(\frac{\partial I_{k+1}}{\partial I_k} + \frac{\partial \vartheta_{k+1}}{\partial \vartheta_k}\right).$$

In general, the eigenvalues $\lambda_1^{(k)}, \lambda_2^{(k)}$ are functions of local coordinates (ϑ_k, I_k). In the case $D > 1$ the eigenvalues $\lambda_{1k}, \lambda_{2k}$ are real, and satisfy the condition

$$\lambda_1^{(k)} > 1, \quad \lambda_2^{(k)} < 1, \quad \lambda_1^{(k)}\lambda_2^{(k)} = 1.$$

In this case the orbits is locally unstable. If $D < 1$ the eigenvalues λ_{1k} and λ_{2k} are complex numbers, with the unity modules, $|\lambda_{1k}| = |\lambda_{2k}| = 1$. The orbits in this case are locally stable.

The *Lyapunov exponent*, σ, is defined as

$$\sigma = \lim_{k\to\infty} \frac{1}{k} \ln \frac{\delta s_k}{\delta s_0}. \quad (7.49)$$

It characterizes the degree exponential divergency of neighboring orbits in time,

$$\delta s(t_k) \sim e^{k\sigma}\, \delta s_0 = e^{t_k/t_D}\, \delta s_0, \;\; t_D = \frac{\Delta t}{\sigma},$$

with the characteristic time of divergency t_D. Here $t_k = k\Delta t$, and Δt is the time–step of the mapping.

From Eq. (7.45) it follows that it is determined by the largest eigenvalue, $\lambda^{(k)} = \max\,(\lambda_1^{(k)}, \lambda_2^{(k)})$, of the Jacobian matrix $\mathbf{J}_k$:

$$\sigma = \lim_{N\to\infty} \frac{1}{N} \ln \prod_{k=1}^{N} \lambda^{(k)}. \tag{7.50}$$

When the eigenvalue $\lambda^{(k)}$ does not depend on the local coordinates (ϑ_k, ψ_k), i.e., $\lambda^{(k)} = \lambda_1$, then $\sigma = \ln \lambda_1$. In general the Lyapunov exponent, σ, is positive for the unstable orbits since $\lambda^{(k)} > 1$, and it vanishes, $\sigma = 0$, for the stable orbit because $\lambda^{(k)} = 1$.

In Sect. 10.3 we calculate the Lyapunov exponents for the chaotic magnetic fields in a stochastic magnetic field created by the resonant magnetic perturbations.

7.4 Stability of Periodic Orbits (Fixed Points)

In order to find whether the periodic orbit (or the fixed point) is stable (elliptic) or unstable (hyperbolic) one should calculate the Lyapunov exponents at the fixed points by linearizing the mapping (7.32) near the fixed point (ϑ_0, I_0). For simplicity we consider the two-dimensional system in the phase space (ϑ, I). The linearization gives the equation for the deviations, $\delta I_k = I_k - I_0$, $\delta\vartheta_k = \vartheta_k - \vartheta_0$:

$$\begin{pmatrix} \delta I_{k+1} \\ \delta\vartheta_{k+1} \end{pmatrix} = \Lambda \begin{pmatrix} \delta I_k \\ \delta\vartheta_k \end{pmatrix}, \tag{7.51}$$

where Λ is equal to the Jacobi matrix $\mathbf{J}_k$ (7.44), taken at the fixed point (I_0, ϑ_0). The matrix Λ is known as a *monodromy matrix*. The eigenvalues λ of the matrix Λ known as *multipliers* are determined by the solution of the quadratic equation

$$\lambda^2 - 2A\lambda + 1 = 0, \quad A = \frac{1}{2}\mathrm{Tr}\, J = \frac{1}{2}\left(\frac{\partial I_{k+1}}{\partial I_k} + \frac{\partial\vartheta_{k+1}}{\partial\vartheta_k}\right). \tag{7.52}$$

In the first order of ϵ, when the terms of order of ϵ^k, $(k \geq 2)$ are neglected, using that $w(J) \approx \omega(J) = 1/q(J)$ we have

$$A \approx 1 + \epsilon C, \quad C = 2\pi m_0 \omega'(J_0) \frac{\partial^2 G}{\partial\vartheta^2} = -\frac{2\pi m_0 q'(\psi_0)}{q^2(\psi_0)} \frac{\partial^2 G}{\partial\vartheta^2}. \tag{7.53}$$

Suppose that $\omega'(J_0) < 0$ (for the safety factor $q'(J_0) > 0$) at the fixed point. Then from (7.52) it follows that at $G_{\vartheta\vartheta} > 0$, $(A < 1, C < 0)$, the eigenvalues $\lambda_{1,2}$ are complex numbers $\lambda_{1,2} = A \pm i\sqrt{1 - A^2}$ with $\lambda_1\lambda_2 = 1$. Therefore the minima of the generating functions $G(\vartheta, \psi_0)$ correspond to the elliptic fixed points , i.e., to the *stable periodic orbits*. Orbits P_k of the mapping (7.32) near these points lie on the closed elliptic curves as illustrated in Fig. 7.3a.

On the other hand at $G_{\vartheta\vartheta} < 0$ $(A > 1, C > 0)$ the eigenvalues $\lambda_{1,2}$ are real numbers $\lambda_{1,2} = A \pm \sqrt{A^2 - 1}$ with $\lambda_1\lambda_2 = 1$, and the maxima of the generating functions $G(\vartheta, \psi_0)$ correspond to the hyperbolic fixed points , i.e., to the *unstable periodic orbits*. The eigenvalues $\lambda_{1,2}$ can be presented as $\lambda_{1,2} = \exp\left(\sigma_{1,2}T\right)$, where the numbers $\sigma_{1,2}$ are characteristic exponents, $\sigma_1 = -\sigma_2 < 0$. Poincaré had shown that it can be expanded in the series of powers of $\sqrt{\epsilon}$,

$$\sigma_1 = \mu\sqrt{\epsilon} + \mu_1\epsilon + \cdots , \tag{7.54}$$

where $\mu, \mu_1, \ldots$ are constants. From (7.52) it is easy to find these constants, particularly, we have $\mu = \sqrt{2C}/T$, $\mu_1 = C/T$.

The orbits P_k near the fixed points lie on the hyperbolic curves as shown Fig. 7.3b. The asymptotes of hyperbolic orbits are the axes (x, y) obtained by rotating the axes (I, ϑ) around the fixed points (I_0, ϑ_0),

$$\begin{aligned} x &= \delta\vartheta\ \cos\phi + \delta\psi\ \sin\phi, \\ y &= -\delta\vartheta\ \sin\phi + \delta\psi\ \cos\phi, \end{aligned} \tag{7.55}$$

where the angle ϕ is determined by

$$\tan\phi = \frac{\lambda_1 - \Lambda_{11}}{\Lambda_{12}} = \frac{\Lambda_{21}}{\Lambda_{22} - \lambda_2}. \tag{7.56}$$

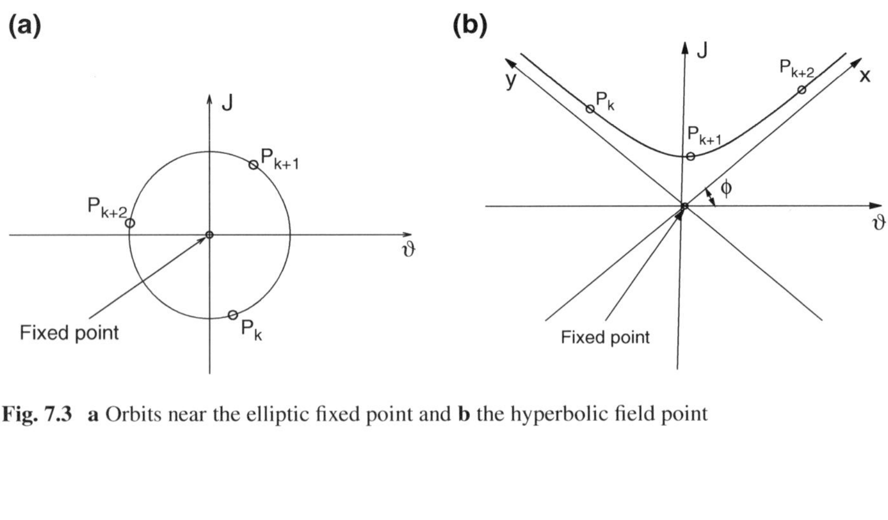

Fig. 7.3 **a** Orbits near the elliptic fixed point and **b** the hyperbolic field point

In the variables the dynamics of P_k is described by

$$
\begin{aligned}
x_{k+1} &= \lambda_1 x_k = \lambda_1^{k+1} x_0 = \exp\left((k+1)\sigma_1 T\right) x_0, \\
y_{k+1} &= \lambda_2 y_k = \lambda_1^{k+1} y_0 = \exp\left(-(k+1)\sigma_1 T\right) y_0,
\end{aligned} \tag{7.57}
$$

As seen from (7.57) the points P_k asymptotically approaches to the fixed point P_0 along the along the axis y, $(x_0 = 0)$ at $k \to \infty$. On the other hand they asymptotically approaches to P_0 along the along the axis x, $(y_0 = 0)$ at $k \to -\infty$. More detailed description of unstable periodic orbits will be given in Sect. 7.5.

The finite phase–space area $\delta\mathcal{A}(t_0) = \delta x_0 \delta y_0$ near the unstable periodic orbits (or hyperbolic fixed point) is strongly deformed due to the squeeze/stretch phenomenon. According to (7.57) it is stretched exponentially along the x-direction, and is squeezed along the y-direction, while the area $\delta\mathcal{A}(t) = \delta x(t)\delta y(t)$ is preserved, $\delta\mathcal{A}(t) = \delta\mathcal{A}(t_0)$. (Figure 10.1 in Chap. 10 illustrates this phenomenon).

For the reversed magnetic shear, $q'(J_0) < 0$ $(\omega'(J_0) > 0)$, the minima of the generating functions $G(\vartheta, J_0)$ correspond to the hyperbolic fixed points and the maxima correspond to the elliptic fixed points.

7.4.1 Fixed Points of the Separatrix Mapping

Fixed points can be also found using the full–turn transfer (separatrix) mapping (6.55) and (6.56). Below we illustrate it for the field lines. Assume that the Fourier presentation of the Poincaré integral $P(\varphi, \psi)$ (6.76) contains only one toroidal mode n with the constant coefficients $K_n(\psi) = K_0$. Then the mapping (6.95) can be reduced to

$$
\begin{aligned}
\psi_{k+1} &= \psi_k - \epsilon n R_0 \sin\left(n\varphi_k + \pi n q(\psi_k) + \chi_n\right), \\
\varphi_{k+1} &= \varphi_k + \pi\left[q(\psi_{k+1}) + q(\psi_k)\right],
\end{aligned} \tag{7.58}
$$

where we considered only the forward mapping, i.e., the upper signs. Introducing the variable $\phi = n\left(\varphi + \pi q(\psi)\right) + \chi_n$, we have

$$
\begin{aligned}
\psi_{k+1} &= \psi_k - \epsilon n R_0 \sin\phi_k, \\
\phi_{k+1} &= \phi_k + 2\pi n q(\psi_{k+1}).
\end{aligned} \tag{7.59}
$$

The primary fixed points of the mapping (7.59), i.e.,

$$
\psi_{k+1} = \psi_k, \quad \phi_{k+1} = \phi_k + 2\pi, \tag{7.60}
$$

are determined by $\sin\phi = 0$, $q(\psi) = m/n$, i.e., there is a pair of fixed points $\phi_o = \pi$ and $\phi_h = 2\pi$ on the resonant magnetic surfaces $\psi = \psi_{mn}$. The corresponding point at the toroidal axis φ are

$$\varphi_h^{(s)} = \pi(2s - m)/n - \chi_n \mod 2\pi, \quad s = 0, 1, \dots n - 1, \tag{7.61}$$

$$\varphi_o^{(s)} = \pi(2s + 1 - m)/n - \chi_n \mod 2\pi. \tag{7.62}$$

In order to find whether the fixed points are elliptic or hyperbolic one should calculate the eigenvalues of the Jacobi matrix (7.44) of the mapping (7.59) at these fixed points. The Jacobi matrix is given by

$$J = \begin{pmatrix} 1 & -\alpha \\ \beta & 1 - \alpha\beta \end{pmatrix}, \tag{7.63}$$

where $\alpha = \epsilon n R_0 \cos \phi_k$, $\beta = 2\pi n dq(\psi_{k+1})/d\psi$. The eigenvalues $\lambda_{1,2}$ of the matrix (7.44) are

$$\lambda_{1,2} = A \pm \sqrt{A^2 - 1},$$
$$A = 1 - \pi n^2 \epsilon R_0 \frac{dq(\psi_{k+1})}{d\psi} \cos \phi_k. \tag{7.64}$$

Suppose that $dq(\psi)/d\psi > 0$. Then at the fixed point $\phi_o = \pi$, the eigenvalues $\lambda_{1,2}$ are real with $\lambda_1 > 1$, $\lambda_2 < 1$, $(\lambda_1\lambda_2 = 1)$, and at $\phi_o = 2\pi$ we have complex eigenvalues $\lambda_{1,2}$ with $\lambda_1 = \lambda_2^*$, $(\lambda_1\lambda_2^* = 1)$. Therefore, the fixed points (7.61) are hyperbolic fixed points and correspond to the unstable periodic orbits. On the other hand the fixed points (7.62) are elliptic fixed points and they correspond to the stable periodic orbits.

In the case the reverse magnetic shear, $dq(\psi)/d\psi < 0$ the fixed points (7.61) become the elliptic ones, and the fixed points (7.62) are hyperbolic fixed points.

7.4.2 Numerical Methods

The above methods to determine to periodic orbits or fixed points of mappings are approximate and they are valid for small perturbations. At present the numerous numerical methods for finding periodic orbits have been developed (see, e.g., Nusse and Yorke 1998; Misguich et al. 2003). The methods of determination of fixed points of mappings is reduced to finding the roots of the system of nonlinear equations. Let (x_1, x_2) be the angular $(x_1 \equiv \vartheta)$ and the flux $x_2 \equiv \psi$ variables of the system. Then the fixed points of the mappings (7.31) are the roots of nonlinear system,

$$f_1(x_1, x_2) = x_1 - \hat{M}^p(x_1, x_2) - 2\pi n = 0,$$
$$f_2(x_1, x_2) = x_2 - \hat{M}^p(x_1, x_2) = 0. \tag{7.65}$$

There are many numerical methods solving these systems (Press et al. 1992; see, also Misguich et al. (2003) for detailed description). Particularly, the approximate fixed points found by the Poincaré method can be used as initial points in the iterative numerical procedure to solve the system (7.65).

7.4.3 Distribution of Fixed Points

The number of primary periodic orbits depends on the possible solutions of the equation $q(\psi) = m/n$ with respect to ψ for the all mode numbers (m, n) in the perturbation Hamiltonian (7.13). The number such resonant surfaces is finite for the plasmas with a limiter where the safety factor takes a finite value at the plasma edge. In poloidal divertor tokamaks in which the safety factor grows near the magnetic separatrix, like in Eqs. (2.51) or (2.74), the number of primary resonant surfaces are infinite. For instance, according to Eq. (2.51) in the standard divertor tokamak the resonant magnetic surfaces ψ_{mn} near the separatrix have universal asymptotics,

$$|\psi_{mn}| = Q \exp\left(-2\pi\gamma m/n\right), \tag{7.66}$$

for large mode numbers m. It is assumed that the poloidal flux $\psi = 0$ is at the separatrix.

Beside the primary periodic orbits there are a countable number of the secondary and higher–order periodic orbits (or fixed points). The distribution of these fixed points in phase space depends on the safety factor profile $q(\psi)$.

7.5 Unstable Periodic Orbits and Splitting of Separatrices

Poincaré had shown the splitting the asymptotical surfaces approaching to unstable periodic orbits also leads to the non-integrability of Hamiltonian systems. In this section we illustrate this phenomenon on the simple example of field lines near hyperbolic fixed points (X-point) and the magnetic separatrix. A rigorous treatment of this problem can be found in Arnold et al. (2006), Kozlov (1996) and Wiggins (2003).

7.5.1 Unstable Periodic Orbits Near the X-point

Near the X-point the unstable periodic orbits can be found using the generic formulas (6.104) and (6.106) for field lines given in Sect. 6.7. Consider, for simplicity, the solution (6.106). In the absence of perturbation, $\epsilon = 0$, the orbit on the y axis asymptotically approaches the X-point ($x = 0, y = 0$) at $\varphi \to \infty$, i.e., $\left\{x = 0,\, y = y_0 e^{-\gamma(\varphi-\varphi_0)}\right\}$ determines the stable manifold W_s. Similarly, the orbit on the x axis $\{x = x_0 e^{\gamma(\varphi-\varphi_0)}, y = 0\}$ asymptotically approaching the X-point $(x = 0, y = 0)$ at $\varphi \to -\infty$ determines unstable manifold W_u.

As seen from Eq. (6.106) any small nonzero perturbation, $\epsilon \neq 0$, creates the isolated *unstable periodic orbit* described by

$$x_p(\varphi; \epsilon) = -\epsilon a_1 \cos(n\varphi + \phi),$$
$$y_p(\varphi; \epsilon) = \epsilon a_2 \cos(n\varphi - \phi). \tag{7.67}$$

It is shown in Fig. 7.4 by curve 3.

Putting $x_0 = x_p(\varphi_0; \epsilon)$ in Eq. (6.106) one obtains the *stable manifold* W_s asymptotically approaching to the unstable periodic orbit (7.67) at $\varphi \to \infty$ (curve 1 in Fig. 7.4),

$$x(\varphi; \epsilon) = -\epsilon a_1 \cos(n\varphi + \phi),$$
$$y(\varphi; \epsilon) = e^{-\gamma(\varphi-\varphi_0)}\Big[y_0 - \epsilon a_2 \cos(n\varphi_0 - \phi)\Big] + \epsilon a_2 \cos(n\varphi - \phi). \tag{7.68}$$

The *unstable manifold* W_u asymptotically approaching to the unstable periodic orbit (7.67) at $\varphi \to -\infty$ is obtained by setting $y_0 = y_p(\varphi_0; \epsilon)$ in Eq. (6.106) (curve 2 in Fig. 7.4),

$$x(\varphi; \epsilon) = e^{\gamma(\varphi-\varphi_0)}\Big[x_0 + \epsilon a_1 \cos(n\varphi_0 + \phi)\Big] - \epsilon a_1 \cos(n\varphi + \phi),$$
$$y(\varphi; \epsilon) = \epsilon a_2 \cos(n\varphi - \phi). \tag{7.69}$$

7.5.2 Splitting of Separatrices

Consider the system possessing one (or two) magnetic X-points (or X-lines along the toroidal angle φ). Suppose that in the absence of nonaxisymmetric magnetic pertur-

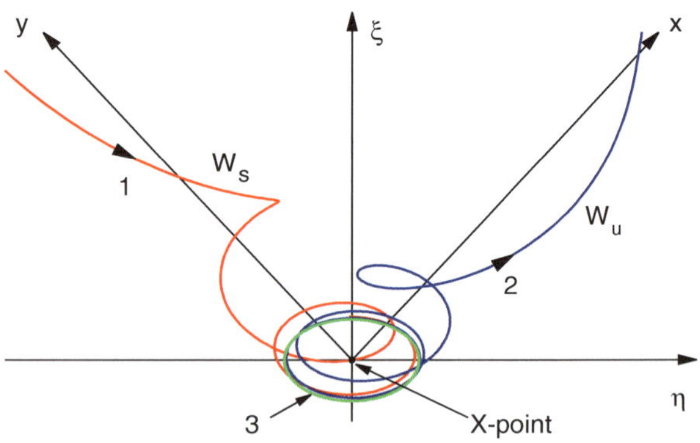

Fig. 7.4 Field lines near the hyperbolic fixed (X-) point. *Curve 1* (*red*) describes the stable manifold W_s asymptotically approaching the unstable periodic orbit at $\varphi \to \infty$, and *curve 2* (*blue*) describes the unstable manifold W_u asymptotically approaching the unstable periodic orbit at $\varphi \to -\infty$, *curve 3* describes the unstable periodic orbit

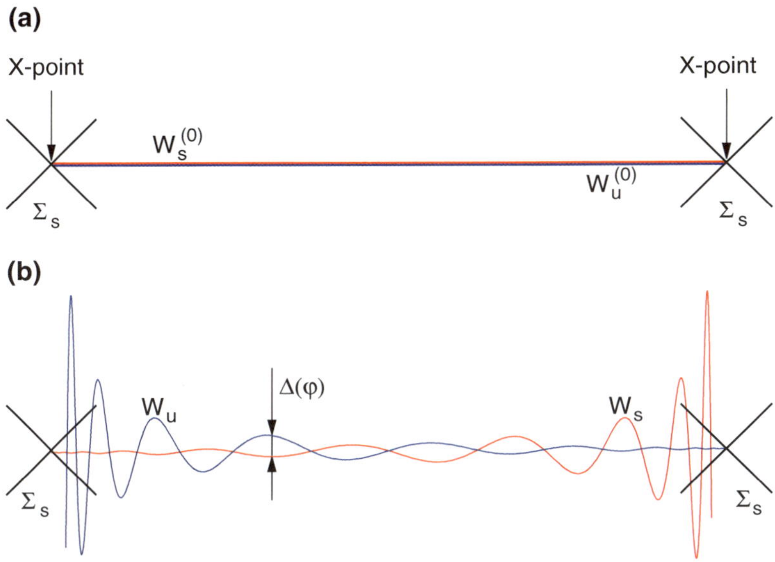

Fig. 7.5 **a** Stable $W_s^{(0)}$ and unstable $W_u^{(0)}$ manifolds in the absence of perturbations; **b** Splitting of stable W_u and unstable W_s manifolds in the presence of perturbations. The sections Σ_s are defined in Sect. 2.3.1 and in Fig. 2.4 (see also Sect. 6.4 and in Fig. 6.3)

bations the X-lines lie on the single magnetic separatrix surface. The poloidal section of the magnetic separatrix is schematically shown in Fig. 7.5a. The X-points coincide in the case of single–null divertor plasmas. The field lines $(z_s(\varphi, \psi_S), p_s(\varphi, \psi_S))$ on the magnetic separatrix asymptotically approach the X-points at the limit $\varphi \to \pm\infty$. We call the magnetic separatrix surface as a *stable manifold* W_s when the field lines on which asymptotically approach the X-point at the limit $\varphi \to +\infty$, and an *unstable manifold* W_u when it happens at the limit $\varphi \to -\infty$.

In the unperturbed system the stable and unstable manifolds coincide. As was found by Poincaré any small non-axisymmetric (periodic) perturbations split the stable and unstable manifolds. The magnetic separatrices W_s and W_u do not coincide each other. The splitting of the stable and the unstable manifolds is schematically illustrated in Fig. 7.5b.

Let $\psi^s(\varphi)$ and $\psi^u(\varphi)$ be the poloidal fluxes corresponding to the stable and unstable magnetic separatrix surfaces, respectively.

The measure of separatrix splitting can be described by the distance $\Delta(\varphi) = |\psi^s(\varphi) - \psi^u(\varphi)|$ between split separatrices at the toroidal section φ. Below we will show (see Sect. 7.5.3) that for the small magnetic perturbations $\Delta(\varphi)$ is given by the Poincaré integral $P(\varphi; \psi)$, and the Melnikov integral, $M(\varphi; \psi)$, introduced by Eqs. (6.62) and (6.63),[5]

[5] This formula has been first derived by Melnikov (1963b) (see also Guckenheimer and Holmes 1983; Arnold et al. 2006; Treschev 2006).

$$\Delta(\varphi) = \frac{dP(\varphi;\psi_S)}{d\varphi} = M(\varphi;\psi_S). \tag{7.70}$$

Let φ_k and φ_{k+1} are two successive zeros of the integral $M(\varphi)$, i.e., $M(\varphi_k;\psi_S) = [dP/d\varphi](\varphi_k) = 0$. The stable and the unstable manifolds cross each other at these points as it is illustrated in Fig. 7.6. The region enclosed by the manifolds W_u and W_s located between φ_k and φ_{k+1} is called a *lobe*. The area of the lobe is given by

$$\mathcal{A} = \int_{\varphi_k}^{\varphi_{k+1}} \Delta(\varphi)d\varphi = |P(\varphi_{k+1};\psi_S) - P(\varphi_k;\psi_S)|\,. \tag{7.71}$$

It is equal to the normalized poloidal flux through the lobe region, $\Delta\psi_{lobe}$. From Eq. (6.77) for the single toroidal mode n we have $P(\varphi) = K_n(0)\cos(n\varphi + \chi_n)$, and the magnetic flux is determined by the integral $K_n(0)$, i.e.,

$$\Delta\Psi_{lobe} = 2B_0R_0^2\epsilon|K_n(0)|. \tag{7.72}$$

7.5.3 Derivation of the Formula for the Separatrix Splitting

We recall that the magnetic field lines are equivalent to the Hamiltonian system (6.4) and (6.5) with the $N = 2$. The action–angle variables $(\vartheta_1, \vartheta_2, I_1, I_2)$ corresponds to the magnetic coordinates $(\vartheta, \varphi, \psi_t, \psi)$ (see Sect. 1.2.1 and Eq. 1.20). The behavior of field lines in the presence of magnetic perturbation $H_1(I,\vartheta) = \psi_1(\psi,\vartheta,\varphi)$ can be described by the mapping (6.9)–(6.11). We explicitly write down the mapping for the pair of canonical variables, the toroidal angle and the poloidal flux, (φ, ψ),

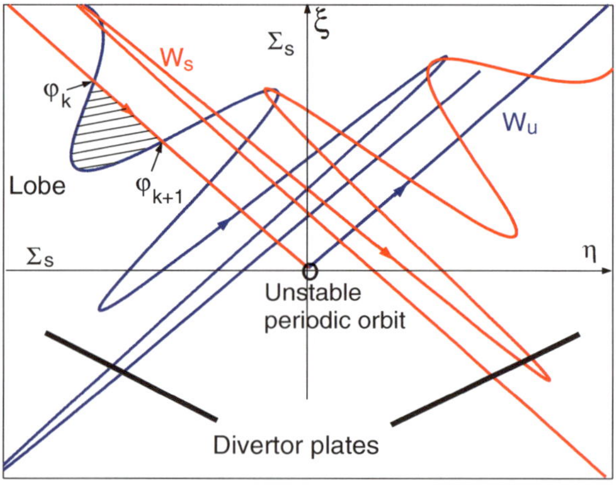

Fig. 7.6 Crossing of stable W_s (*red curve*) and unstable W_u (*blue curve*) manifolds and lobes near the X-point

$$\Psi_k = \psi_k - \epsilon \frac{\partial S_k}{\partial \varphi_k}, \quad \phi_k = \varphi_k + \epsilon \frac{\partial S_k}{\partial \Psi_k}, \tag{7.73}$$

$$\bar{\phi}_k = \phi_k + w_\varphi(\Psi_k, \epsilon)(\tau_{k+1} - \tau_k), \tag{7.74}$$

$$\psi_{k+1} = \Psi_k + \epsilon \frac{\partial S_{k+1}}{\partial \varphi_{k+1}}, \quad \varphi_{k+1} = \bar{\phi}_k - \epsilon \frac{\partial S_{k+1}}{\partial \Psi_k}, \tag{7.75}$$

where the generating function $S_k = S(\varphi(\tau_k), \Psi_k, \tau_k, \tau_0)$, according to (6.21) and (6.24), is determined by the integral,

$$S(\varphi, \vartheta, \Psi, \tau, \tau_0; \epsilon) = -\int_{\tau_0}^{\tau} F(\varphi(\tau'), \vartheta(\tau'), \Psi; \epsilon) d\tau', \tag{7.76}$$

$$F(\varphi, \vartheta, \psi; \epsilon) = \psi_1(\varphi, \vartheta, \psi) + \sum_{i \geq 2} \epsilon^{i-1} F_i(\varphi, \vartheta, \psi). \tag{7.77}$$

Here $\varphi(\tau') = \varphi + \omega_\varphi(\tau' - \tau)$, $\omega_\varphi = \partial H_0/\partial\psi = 1$. We have also used the fact that the function $F(\varphi, \vartheta, \Psi; \epsilon)$ does not explicitly depends on τ. Furthermore we set $\mathcal{H}_i = 0$, $i \geq 1$ in Eqs. (6.21) and (6.24), and therefore $w_\varphi = \omega_\varphi = 1$. Excluding the intermediate variable Ψ the evolution of the poloidal flux $\psi(\tau)$ can be presented as

$$\begin{aligned} \psi(\tau + t) &= \psi(\tau) + \epsilon \frac{\partial S(\tau + t)}{\partial \varphi(\tau + t)} - \epsilon \frac{\partial S(\tau)}{\partial \varphi(\tau)}, \\ \varphi(\tau + t) &= \varphi(\tau) + t - \epsilon \frac{\partial S(\tau + t)}{\partial \psi(\tau + t)} + \epsilon \frac{\partial S(\tau)}{\partial \psi(\tau)}, \end{aligned} \tag{7.78}$$

where we have set $t = \tau_{k+1} - \tau_k$, $\tau = \tau_k$ and $S(\tau_k) \equiv S_k$.

Let $\psi_s(\tau)$ and $\psi_u(\tau)$ be the stable and the unstable manifolds, respectively. Then for the arbitrary τ the mapping (7.78) determines the evolution of the stable and the unstable manifolds with the following asymptotics

$$\begin{cases} \psi(\tau) \to \psi_\epsilon(\tau), & \text{at } \tau \to \infty \text{ for the stable manifold,} \\ \psi(\tau) \to \psi_\epsilon(\tau), & \text{at } \tau \to -\infty \text{ for the unstable manifold,} \end{cases} \tag{7.79}$$

where $\psi_\epsilon(\tau)$ is the unstable periodic orbit.

Taking the limit $t \to \infty$ we obtain the solution for the stable manifold $\psi_s(\tau)$,

$$\begin{aligned} \psi_s(\tau) - \psi_\epsilon &= \epsilon \frac{\partial}{\partial \varphi} \left[\int_{\tau_0}^{\infty} F(\varphi + \tau' - \tau, \Psi; \epsilon) d\tau' - \int_{\tau_0}^{\tau} F(\varphi + \tau' - \tau, \Psi; \epsilon) d\tau' \right] \\ &= \epsilon \frac{\partial}{\partial \varphi} \int_0^{\infty} F(\varphi + \tau', \Psi; \epsilon) d\tau', \end{aligned} \tag{7.80}$$

where $\varphi \equiv \varphi(\tau)$. Similarly, taking the limit $t \to -\infty$ we obtain the equation for the unstable manifold $\psi_u(\tau)$,

$$\psi_u(\tau) - \psi_\epsilon = \epsilon \frac{\partial}{\partial \varphi} \int_0^{-\infty} F(\varphi + \tau', \Psi; \epsilon) d\tau'. \tag{7.81}$$

Subtracting Eq. (7.80) from (7.81) we have

$$\psi_u(\tau) - \psi_s(\tau) = \epsilon \frac{\partial}{\partial \varphi} \int_{-\infty}^{\infty} F(\varphi + \tau, \Psi; \epsilon) d\tau. \tag{7.82}$$

Putting $\tau = \varphi$ and using the definition of the Poincaré integral (6.62) and the expansion (7.77) we obtain the formula (7.70) in the first order of ϵ.

7.5.4 *Presentation of Stable and Unstable Manifolds by the Separatrix Mapping*

Using the separatrix mapping (6.58) and (6.60) one can found the analytical forms the stable and unstable manifolds of the section Σ_s near the separatrix. Actually, as seen from Figs. 7.5b and 7.6 it determines the crossing points of manifolds with the section Σ_s. Below, for simplicity we set $\psi_s = 0$ at the separatrix.

According to the definition the *stable manifold* is defined as a set of points in (φ, ψ) plane of the section Σ_s which are mapped by the forward mapping to $\psi_{k+1} = \psi_s = 0$ at $\varphi_{k+1} \to \infty$. From the first equation of Eq. (6.58) (with upper signs) it follows the equation for the stable manifold in the section Σ_s,

$$F^{(s)}(\varphi, \psi) \equiv \psi - \epsilon \frac{\partial}{\partial \varphi} P(\varphi + \pi q(\psi); 0) = 0. \tag{7.83}$$

The *unstable manifold* is a set of points (φ, ψ) mapped by the backward mapping to the point $\psi_{k+1} = 0$ at $\varphi_{k+1} \to -\infty$. From the second equation of Eq. (6.58) (with lower signs) it follows that the equation for the *unstable manifold*,

$$F^{(u)}(\varphi, \psi) \equiv \psi + \epsilon \frac{\partial}{\partial \varphi} P(\varphi - \pi q(\psi); 0) = 0. \tag{7.84}$$

Using the presentation (6.76) of the Poincaré integral $P(\varphi; \psi)$ the equations for the stable and the unstable manifolds take the form

$$F^{(s,u)}(\varphi, \psi) = \psi \mp \epsilon \sum_n n\Big[K_n(0)\sin(n\varphi \pm \pi n q(\psi) + \chi_n) - L_n(0)\cos(n\varphi \pm \pi n q(\psi) + \chi_n)\Big] = 0, \tag{7.85}$$

where the upper signs corresponds to the stable manifold, while the lower sign corresponds to the unstable manifolds, respectively.

Consider the specific case when the perturbation contains only one toroidal mode n. Since the manifolds are located close to the separatrix one can use the logarithmic asymptotics of the safety factor $q(\psi)$ (2.51). Then introducing the dimensionless flux variable h and the notations

$$h = \psi/\epsilon n R_n, \quad R_n = \sqrt{K_n^2(0) + L_n^2(0)}$$
$$A = \frac{n}{2\gamma}\ln\frac{Q}{\epsilon n R_n}, \quad \bar{\chi}_n = \chi_n - \arctan\frac{L_n(0)}{K_n(0)}, \tag{7.86}$$

Equation. (7.85) are reduced to the following implicit forms,

$$F^{(s,u)}(\varphi, h) = h \mp \sin\left(n\varphi \pm \frac{n}{2\gamma}\ln|h| + \bar{\chi}_n \pm A\right) = 0, \tag{7.87}$$

for the stable and the unstable manifolds, respectively.

One can also find the explicit dependence of the toroidal angle φ on h. There are several functions $\varphi(h)$ satisfying the Eq. (7.87). The number of these functions is equal to the toroidal mode n. The each of them has the two branches denoted by the superscripts "(b)" and "(t)", respectively. They are given by

$$\varphi^{(b)}(h) = \mp A \pm \frac{1}{2\gamma}\ln|h| + \frac{1}{n}(2\pi s + \arcsin h) + \bar{\chi}_n,$$
$$\varphi^{(t)}(h) = \mp A \pm \frac{1}{2\gamma}\ln|h| + \frac{1}{n}\left[(2s+1)\pi - \arcsin h\right] + \bar{\chi}_n,$$
$$s = 0, 1, 2, \ldots, n-1, \tag{7.88}$$

where the upper signs correspond to the stable manifold while the lower signs corresponds to the unstable manifold. These functions can be also presented in the parametric form

$$\varphi^{(b)} = \mp A \mp \frac{t}{2\gamma} + \frac{2\pi s}{n} + \frac{\sigma}{n}\arcsin e^{-t} + \bar{\chi}_n,$$
$$\varphi^{(t)} = \mp A \mp \frac{t}{2\gamma} + \frac{2\pi s}{n} + \frac{1}{n}\left[\pi - \sigma \arcsin e^{-t}\right] + \bar{\chi}_n,$$
$$h = \sigma e^{-t}, \quad 0 < t < \infty, \tag{7.89}$$

where $\sigma = +1$ for the manifold at the plane (η, φ) and $\sigma = -1$ for the manifold at the plane (ξ, φ) of the section Σ_s. Figure 7.7a, b show the unstable manifold in the (φ, h) plane and the coordinate system $(\sqrt{|h|}\sin\varphi, \sqrt{|h|}\cos\varphi)$, respectively, for the perturbation mode $n = 1$. The upper curve corresponds to the manifold at the plane (η, φ) and the lower curve for the manifold at the plane (ξ, φ) of the section Σ_s. The 3D view of the stable and the unstable manifolds in this section are plotted in Fig. 7.8 for the toroidal mode number $n = 3$: red curves correspond to the unstable manifolds, and blue curves to the correspond to the stable manifolds. The different manifolds are crossed each in the (ξ, φ) plane.

The area $\mathcal{A}$ of the region $\mathcal{L}$ enclosed by the lower and upper branches of the each spiral-like manifolds can be found by the integral

$$\mathcal{A} = \iint_{\mathcal{L}} d\varphi dh = \int_0^1 \left(\varphi^{(t)}(h) - \varphi^{(b)}(h)\right) dh$$
$$= \frac{2}{n}\int_0^1 \left(\frac{\pi}{2} - \arcsin h\right) dh = \frac{2}{n}. \tag{7.90}$$

The poloidal magnetic flux through this area is equal to

$$\Delta\Psi_{pol} = B_0 R_0^2 n \iint_{\mathcal{L}} d\varphi d\psi = B_0 R_0^2 \epsilon R_n(0)\mathcal{A}$$
$$= 2\epsilon R_n(0) B_0 R_0^2, \tag{7.91}$$

where B_0 is the magnitude of the toroidal magnetic field at the center of torus, R_0 is a major radius.

The width of the lobe along the toroidal direction Δl is given by (see Fig. 7.7a)

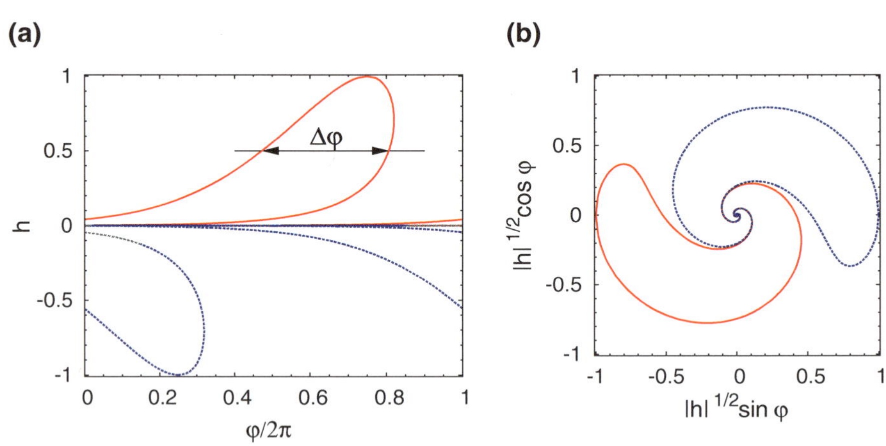

Fig. 7.7 **a** Unstable manifold of the mapping (7.87) in the (φ, h) plane; **b** Unstable manifold in the $(\sqrt{|h|}\sin\varphi, \sqrt{|h|}\cos\varphi)$ plane. Parameters are $A = 0$, $\gamma = 1/2$, $n = 1$

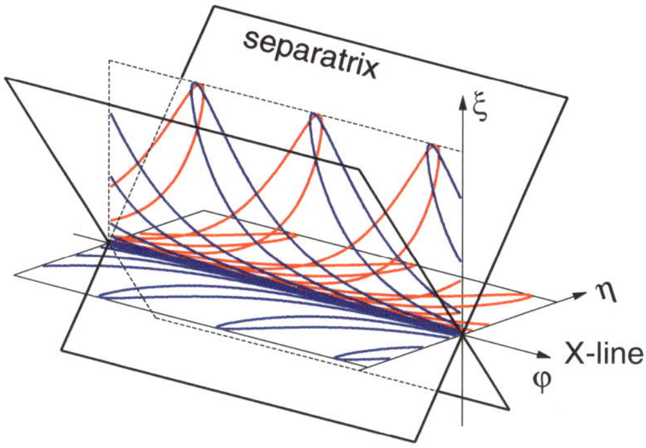

Fig. 7.8 Cross sections of stable and unstable manifold surfaces at the planes (φ, ξ) and (φ, η) of the (ξ, η, φ)-space. The toroidal mode $n = 3$

$$\Delta l = R_0 \Delta\varphi = |\varphi^{(t)}(h) - \varphi^{(b)}(h)| = \frac{2}{n}\left|\frac{\pi}{2} - \arcsin h\right|.$$

At the level of half of the radial width of lobe $h = 1/2$ one has $\Delta l = R_0 2\pi/3n$. For $R_0 = 3$ and $n = 3$ we have $\Delta l = 2\pi/3 = 2.094$ m.

7.5.5 Stable and Unstable Manifolds in 3D Space

From the analytical presentations (7.83) and (7.84) of the stable and unstable manifolds at the section Σ_s one can establish their analytical form in 3D space (R, Z, φ). For this it is necessary to map these forms along the perturbed field lines (6.108). Then, the analytical form of the manifolds in 3D space can be written as

$$F^{(s,u)}\left(\hat{M}^{-1}(\varphi, \psi)\right) = 0, \tag{7.92}$$

where $\hat{M}^{-1}$ is the mapping inverse to the mapping $\hat{M}$ (6.111). Near the X-point we can use the analytical solutions (6.104).

Below we consider the simple case when the magnetic perturbations at the X-point are small and can be neglected. Then in (7.92) one can use unperturbed field lines (6.104) neglecting the coefficients a_x, a_y. The inverse mapping $(\varphi_k, \psi_k) = \hat{M}^{-1}(\varphi, \psi)$ reads as

$$\begin{aligned}\varphi_k &= \varphi - \frac{1}{\gamma}\ln\frac{x}{x_k} = \varphi + \frac{1}{\gamma}\ln\frac{y}{y_k},\\ \psi_k &= \gamma x_k y_k = \psi = \gamma x y.\end{aligned} \tag{7.93}$$

Using the latter and (6.109) one can write the manifolds (7.92) in 3D space (x, y, φ) near the X-point as

$$F^{(s,u)}(\varphi, x, y) = \gamma xy \mp \epsilon \frac{\partial}{\partial \varphi} P\left(\varphi \pm \frac{1}{2\gamma} \ln \frac{Q}{\gamma |xy|} - \frac{1}{2\gamma} \ln \left|\frac{x}{y}\right|; 0\right) = 0, \quad (7.94)$$

where the asymptotics (2.51) of the safety factor $q(\psi)$ is used. Particularly, using (7.85) one can obtain the following formulas for the manifolds for the single mode perturbation,

$$F^{(s)}(\varphi, x, y) = \gamma xy - \epsilon n R_n \sin\left(n\varphi + \frac{n}{2\gamma} \ln \frac{Q}{\gamma x^2} + \bar{\chi}_n\right) = 0, \quad (7.95)$$

for the stable manifold. Similarly we have for the unstable manifold,

$$F^{(u)}(\varphi, x, y) = \gamma xy + \epsilon n R_n \sin\left(n\varphi - \frac{n}{2\gamma} \ln \frac{Q}{\gamma y^2} + \bar{\chi}_n\right) = 0. \quad (7.96)$$

Figure 7.9a, b illustrate the stable and unstable manifolds in the (ξ, η, φ)-space, obtained by the formulas (7.95) and (7.96).

7.5.6 Spatial Form of the Magnetic Footprints

As seen from Fig. 7.6 the set of crossing points of the manifolds with the divertor plates determines the spatial form of the magnetic footprints. The analytical form of magnetic footprints can be derived using the mapping (6.111) of the stable and the

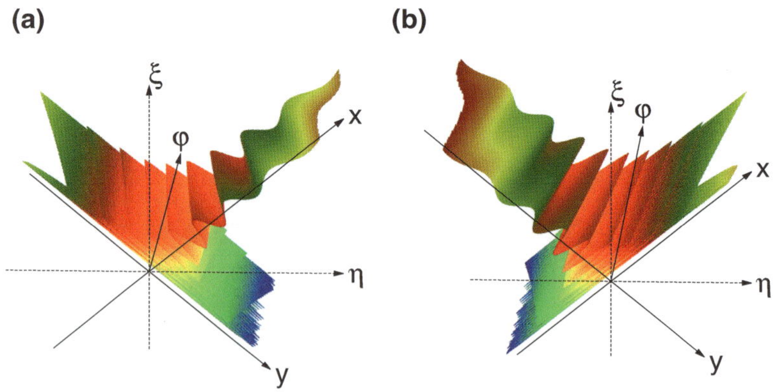

Fig. 7.9 3D- view of the stable (**a**) and unstable (**b**) manifolds near the X-line (or X-point) in the (ξ, η, φ)-space. The toroidal mode $n = 1$

unstable manifolds on the plane (η, φ) of the section Σ_s to the divertor plates Σ_d as schematically shown in Fig. 6.9. Then, the analytical form of the magnetic footprints on the divertor plates in given by (7.92) where the mapping $(\varphi_k, \psi_k) = \hat{M}^{-1}(\varphi_d, \psi_d)$ is given by

$$\begin{aligned} \psi_k(\varphi_d, \psi_d) &\approx h_d - D\cos(n\varphi_d + \chi_n - \phi), \\ \varphi_k(\varphi_d, \psi_d) &\approx \varphi_d - \Delta\varphi\left(\psi_d - D\cos(n\varphi_d + \chi_n - \phi)\right), \end{aligned} \tag{7.97}$$

where the function $\Delta\varphi(\psi)$ is defined by Eq. (6.116) and the constant D is given by (6.120) with $\xi = \xi_d$.

Therefore from (7.92) and (7.97) one obtains the following implicit analytical form of the magnetic footprints

$$F_{div}^{(s,u)}(\varphi, \psi) \equiv F^{(s,u)}\left(\varphi_k(\varphi, \psi), h_k(\varphi, \psi)\right) = 0. \tag{7.98}$$

In the case of the single toroidal mode perturbation n this equation, according (7.87), is reduced to

$$\begin{aligned} F_{div}^{(s,u)}(\varphi, h) &= \bar{h} \mp \sin\left(n\bar{\varphi} \pm \frac{n}{2\gamma}\ln|\bar{h}| + \tilde{\chi}_n^{(\pm)}\right) = 0, \\ \bar{h} &= h - \bar{D}\cos(n\varphi + \chi_n - \phi), \\ \bar{\varphi} &= \varphi - \bar{\Delta}\varphi\left(h - \bar{D}\cos(n\varphi + \chi_n - \phi)\right), \end{aligned} \tag{7.99}$$

where $\chi_n^{(\pm)} = \chi_n \pm A$,

$$\bar{D} = \frac{D}{\epsilon n R_n(0)}, \quad \bar{\Delta}\varphi(h) = \Delta\varphi\left(\epsilon n R_n(0) h\right).$$

The upper sign corresponds to the magnetic footprints on the inner divertor plate (left side), while the lower signs corresponds to the one on the outer divertor plate (right side) (see Fig. 6.9a, b).

Figure 7.10 shows the example of the unstable manifold $F_{div}^{(u)} = 0$ given by Eq. (7.99) at the section Σ_S ($\xi_d = 0$) and its mapping to the divertor plate Σ_d ($\xi_d \neq 0$) according to the scheme shown in Fig. 6.9. The spiral structure of the unstable manifold at the section Σ_S acquires a modulation along the toroidal angle φ similar to the magnetic footprints shown in Fig. 6.10a, b. Since $D \propto \xi_d a_2$ (see Eq. 6.119) the modulation amplitude in the flux coordinate h (or ψ) grows with the distance of the divertor plate from the X-point, as well as with the magnitude of the perturbation magnetic field at the X-point.

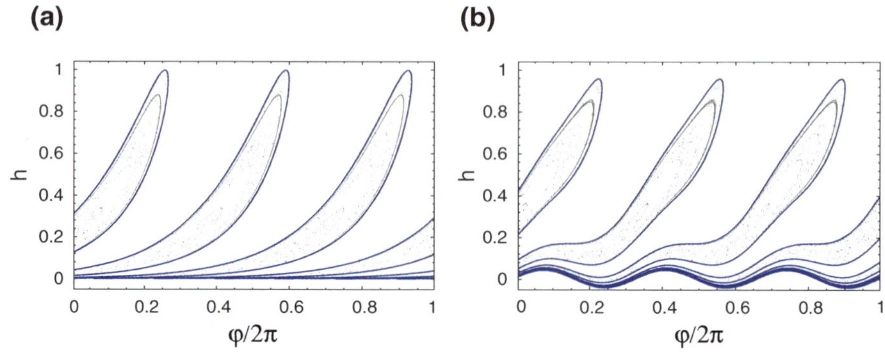

Fig. 7.10 Unstable manifolds (*blue curves*) at the (η, φ)-plane of the section Σ_s (**a**) and its mapping to the divertor plate Σ_d (**b**). The toroidal mode $n = 3$

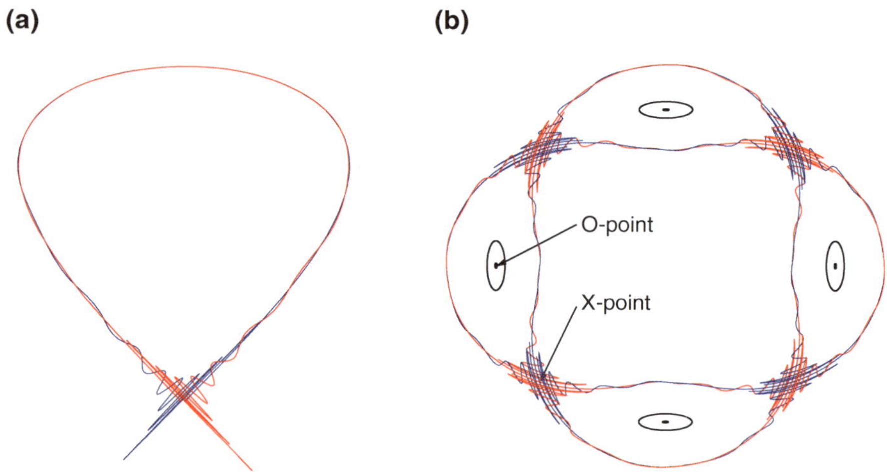

Fig. 7.11 **a** Homoclinic tangles near the separatrix of single-null divertor tokamak; **b** Poloidal cross section of periodic orbits and heteroclinic tangles near the separatrix of the resonant magnetic surface ψ_{mn}, $(m = 4, n = 1)$

7.5.7 Homoclinic (Heteroclinic) Tangles

As seen from Fig. 7.8b the stable (unstable) manifold does not cross itself, but the different manifolds cross each other infinite times. At each crossing point the unstable periodic points are created. The examples of such structures, called *homoclinic (or heteroclinic) tangles* are shown in Fig. 7.11a, b formed due to splitting of the separatrix of a single-null poloidal divertor tokamak, and the separatrices of the resonant magnetic island $m : n = 4 : 1$, respectively. A number of examples of homoclinic tangles and chaos of field lines in poloidal divertor tokamaks, can be found in many publications [see; e.g., Roeder et al. (2003), Evans et al. (2004, 2005)].

7.6 Onset of Chaotic Motion

As was first shown by Poincaré the creations of unstable periodic orbits and the splitting of separatrices in dynamical systems under periodic perturbation are main obstacle to the integrability, i.e., to the existence of integrals of motion. The motion of such a system becomes very sensitive to initial conditions: small changes in the initial conditions may give rise the dramatic changes in the final conditions and the prediction becomes practically impossible. Figure 7.12a, b illustrate such a motion near the unstable periodic orbit: (a) shows the orbits in the phase space (η, ξ); (b) shows the time evolution of the coordinate ξ. The two orbits with very close initial conditions (curves 1 and 2) approach the unstable periodic orbit (curve 3). After making several rotations near the unstable periodic orbits the orbits 1 and 2 exit this area in different directions. The exit direction of the orbit is very sensitive on the initial conditions and practically impossible to predict them.

Such a behavior of orbits near the separatrices is generic because of creation of homoclinic tangles (see Sect. 7.5.7) with the infinite number unstable periodic orbits. In general, any small error in initial conditions will be growing exponentially, that the prediction of results will be practically impossible. In other words the distance, $d(t)$, between two orbits with close initial conditions grows exponentially,

$$d(t) = d(0)e^{\sigma t}, \tag{7.100}$$

where $d(0)$ is an initial distance, and the exponent σ, $(\sigma > 0)$, is the measure of divergency of orbits. Figure 7.12 shows the example of such exponential divergence of two orbits. This phenomenon is known as a *dynamical chaos* or simply *chaos*.

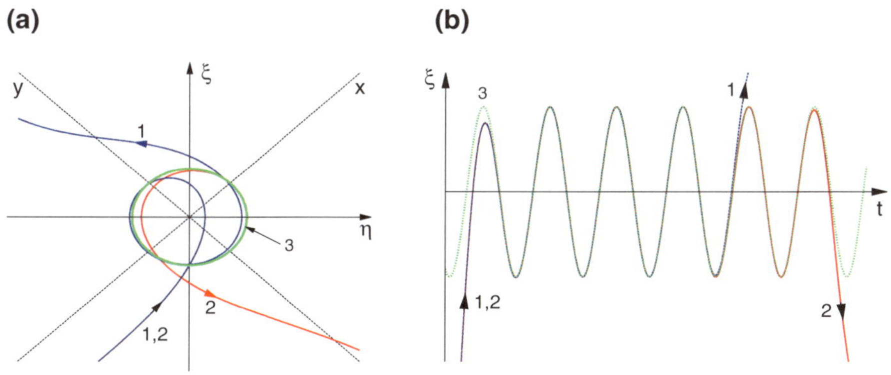

Fig. 7.12 Sensitivity of orbits near the unstable periodic and unperturbed separatrix with very small differences in initial conditions (*curves 1* and *2*). *Curve 3* corresponds to the unstable periodic orbit. **a** Orbits in the (η, ξ)-plane; **b** Time evolution of the ξ-coordinate

The rigorous mathematical proof of chaotic motion in dynamical systems is based on a *horseshoe map* of Smale (1998). The latter is a chaotic map of the square region into itself. It describes a chaotic behavior of dynamical systems with clear geometrical and analytical presentations. The horseshoe map possesses the following properties:

(i) there are an infinite number of periodic orbits;
(ii) periodic orbits of arbitrarily long period exist;
(iii) the number of periodic orbits grows exponentially with the period;
(iv) close to any point of the fractal invariant set there is a point of a periodic orbit.

(see, for example, Guckenheimer and Holmes (1983), Wiggins (2003) for the detailed description of the horseshoe map).

One should note the main feature of the horseshoe map is a *hyperbolicity*. The latter means that all periodic orbits (fixed points) of a dynamical system are hyperbolic. The example of such uniformly hyperbolic dynamical systems is the geodesic flow on manifolds of negative curvature studied by Anosov (1967).

One can assume the homoclinic (or heteroclinic) tangle structure formed due to transversely crossing of split stable and unstable manifolds possesses the properties of the horseshoe map (see Fig. 7.11). This may lead to the chaotic motion near the unperturbed separatrices. However, this assumption has not been rigorously proven for typical Hamiltonian systems. The latter, particularly, magnetic field systems, are not uniformly hyperbolic. They possess an equal number of elliptic and hyperbolic periodic orbits (or stable and unstable periodic orbits).[6] However, the numerical simulations show that typical Hamiltonian systems subjected to sufficiently large perturbations may be considered as chaotic. Majority of orbits are exponentially divergent with the property (7.100) and a single chaotic orbit may fill the large area of phase space.

7.7 Bibliographic Comments

At present there are many excellent textbooks, monographs, and popular books on the chaos theory, particularly, in Hamiltonian systems. For readers who interested in mathematical aspects of chaos theory in Hamiltonian systems one would recommend a book by Arnold et al. (2006). The general theory of dynamical systems has been given in books by Guckenheimer and Holmes (1983), Ott (2002) and Wiggins (2003).

Melnikov (1962, 1963a, b), perhaps, first had shown that the measure of separatrix splitting in Hamiltonian systems subjected to small perturbations, particularly, in a magnetic field system is determined by the certain integral over perturbation functions which bears his name.

---

[6] There are examples of Hamiltonian systems in which stable periodic orbits and the KAM stability islands appear even inside the lobes, the regions enclosed by the stable and unstable manifolds (Abdullaev 2000; Simó and Treschev 2008; Simó and Vieiro 2011). Particularly, these islands are responsible for the regular motions (even flights) of orbits in areas located deep inside the stochastic layer (Abdullaev 2000).

The history of development of chaos theory in the last half of century has been given by Sinai (2010), particularly, the history of the discovery of the horseshoe map has been described by Smale (1998).

Chapter 8
Onset of Dynamical Chaos: Physical Aspects

The physical aspects of dynamical chaos in Hamiltonian systems was developed by Chirikov (1959). It is based on the phenomena nonlinear resonance and the interaction of resonances (see Zaslavsky and Chirikov 1971, 1979). These physical principles of dynamical chaos were instrumental in the first studies of the problems of stability and destruction of magnetic surfaces in tokamaks and stellarators caused by the magnetic perturbations (Kerst 1962; Rosenbluth et al. 1966, Filonenko et al. 1967; Freis et al. 1973; Hamzeh 1974; Finn 1975; Matsuda and Yoshikawa 1975; Boozer and Rechester 1978). Chaos of magnetic field lines in a generic example of chaotic dynamics in Hamiltonian systems with one-and-half degrees of freedom. The nonlinear resonance corresponds to the formation of magnetic islands due to destruction of resonant magnetic surfaces in the presence of non-axisymmetric magnetic perturbations. The interaction of magnetic islands with the growing of a perturbations leads global chaotic behavior of field lines in a certain region of plasma.

This chapter is devoted to the detailed description of resonance phenomena occurring near the resonant tori and the onset of chaotic dynamics in the Hamiltonian systems subjected to the time–periodic perturbations. In Sect. 8.1 we analyze the behavior of field lines near the resonant magnetic surfaces. Beside the standard nonlinear resonance phenomenon when the behavior of field lines near the resonant surface can be reduced to the dynamics of the pendulum, we consider the cases of the magnetic field with a local zero shear and the strong radial dependence of a perturbation field, as well as the nonlinear resonance in a many–dimensional Hamiltonian system. The Chirikov's criterion of the onset of global chaos is presented in Sect. 8.2. The formation of the stochastic layer near the separatrix and their properties are discussed in Sect. 8.3. Particularly, we describe the rescaling invariance (or self–similarity) properties of the stochastic layer, estimate its width using the Chirikov's criterion and the separatrix mapping. in the final Sect. 8.4 the creation of chaos of field lines in a tokamak with a reversed magnetic shear is discussed.

S. Abdullaev, *Magnetic Stochasticity in Magnetically Confined Fusion Plasmas*, Springer Series on Atomic, Optical, and Plasma Physics 78,
DOI: 10.1007/978-3-319-01890-4_8,

8.1 Orbits Near Resonant Tori. Nonlinear Resonance

The creation of periodic orbits under non-axisymmetric magnetic perturbations drastically changes the field line geometry near the resonant magnetic surfaces. Such a phenomenon has been studied in classical mechanics as a resonant interaction of a nonlinear oscillatory system with a time—periodic force. In the physical literature this phenomenon is known as a *nonlinear resonance* (see Zaslavsky and Chirikov 1971; Chirikov 1979; Lichtenberg and Lieberman 1992). In this section we study this phenomenon. Beside the standard nonlinear resonance phenomenon we consider the cases when the validity conditions of the latter are violated. Particularly, the behavior of orbits near the resonant tori (or magnetic surfaces in the systems with the zero magnetic shear and the strong radial dependence of perturbations will be studied. Below we consider the structure of field lines near the isolated periodic orbits, i.e., the structure of magnetic islands.

Consider again the Hamiltonian system (3.9) with the Hamiltonian $\psi = \psi(\vartheta, \psi_t, \varphi)$,

$$\psi = \psi_0(\psi) + \epsilon \sum_{m,n} |H_{mn}(\psi_t)| \cos(m\vartheta - n\varphi + \chi_{mn}) \,. \tag{8.1}$$

The field lines in a small neighborhood of a resonant surface $\psi_{m,n}$, $q(\psi_{mn}) = m/n$ are mostly affected by the corresponding resonant (m, n)—term in (8.1). At the certain level of perturbation ϵ one can retain only this term and neglecting all other nonresonant terms in the perturbation Hamiltonian.

Rigorously, it can be done by the canonical change of variables $(\psi, \vartheta) \to (J, \phi)$ in the Hamiltonian system (3.9 and 8.1) via the generating function of mixed variables,

$$\begin{aligned} \phi &= \frac{\partial F}{\partial J} = m\vartheta - n\varphi + \chi_{mn}, \quad \psi_t = \frac{\partial F}{\partial \vartheta} = mJ, \\ F &\equiv F(\vartheta, J) = (m\vartheta - n\varphi + \chi_{mn})\, J. \end{aligned} \tag{8.2}$$

Then new Hamiltonian,

$$\mathcal{H}(\phi, J, \varphi) = \psi(\vartheta(\phi), J(\psi_t), \varphi) + \frac{\partial F}{\partial \varphi},$$

can be presented as

$$\mathcal{H}(\phi, J, \varphi) = H_{res}(\phi, J) + \epsilon \mathcal{H}_1'(\phi, J, \varphi, \epsilon), \tag{8.3}$$

$$H_{res}(\phi, J) = \psi_0(mJ) - nJ + \epsilon\, |H_{mn}|\,(mJ) \cos\phi, \tag{8.4}$$

where the term

$$\mathcal{H}'(\phi, J, \varphi) \equiv H_1'(\vartheta(\phi), \psi(J), \varphi), \tag{8.5}$$

is the perturbation terms in (8.1) from which the resonant (m, n) term is excluded. When these terms are neglected the system with the Hamiltonian $H_{res}(\phi, J)$ is integrable.

First we consider the dynamics of system near the resonant value ψ_{mn} neglecting the non-resonant term, $\mathcal{H}'_1(\Phi, \Psi, \varphi, \epsilon)$. Introducing the notations

$$\nu(J) = \frac{m}{q(mJ)} - n, \quad V(J) = H_{mn}(mJ),$$

the new Hamiltonian $\mathcal{H} = H_{res}(\phi, J)$ can be presented in the form

$$\mathcal{H}(J, \phi) = \int \nu(J) dJ + \epsilon V(J) \cos \phi. \tag{8.6}$$

Note that $\nu(J) = 0$ at the resonant value of $J = J_{mn} = \psi_{mn}/m$.

The frequency $\nu(J)$ of Hamiltonian (8.6) takes the positive values $\nu(J) > 0$ at $J > J_{mn} = \psi_{mn}/m$ and the negative values $\nu(J) < 0$ at $J < J_{mn}$. Then according to the Poincaré–Birkhoff geometrical theorem in certain neighborhood of the resonant action ψ_{mn} there exist at least two fixed points (Birkhoff 1927 (reprinted 1966); see also Arnold 1989). The fixed points of the system are determined by equations:

$$\begin{aligned} \dot{J} &= -\frac{\partial \mathcal{H}}{\partial J} = \epsilon V(J) \sin \phi = 0, \\ \dot{\phi} &= \frac{\partial \mathcal{H}}{\partial y} = \nu(J) + \epsilon \frac{dV(J)}{dJ} \cos \phi = 0, \end{aligned} \tag{8.7}$$

or

$$\phi = 0, \pi, \quad \nu(J) \pm \epsilon \frac{dV(J)}{dJ} = 0, \tag{8.8}$$

where the upper sign (+) in the second Eq. (8.8) corresponds to $\phi = 0$, the lower sign (−) corresponds to $\phi = \pi$.

8.1.1 Standard Nonlinear Resonance

If the perturbation ϵ and the derivative $dV(J)/dJ$ are small that one can neglect the product $\epsilon dV(J)/dJ$, one has the two fixed points $(\phi = 0,\ J = J_{mn})$ and $\phi = \pi,\ J = J_{mn}$, corresponding to the elliptic fixed point ($O-$point) and the hyperbolic fixed point ($X-$point), respectively. Then for the monotonic $q-$profile with $dq/d\psi_t \neq 0$ at $\psi_t = \psi_{mn}$ expanding the Hamiltonian (8.6) near the resonant value J_{mn} in a series of powers $J - J_{mn}$ the Hamiltonian system with $H_{res}(\phi, J)$ can be reduced to the pendulum problem:

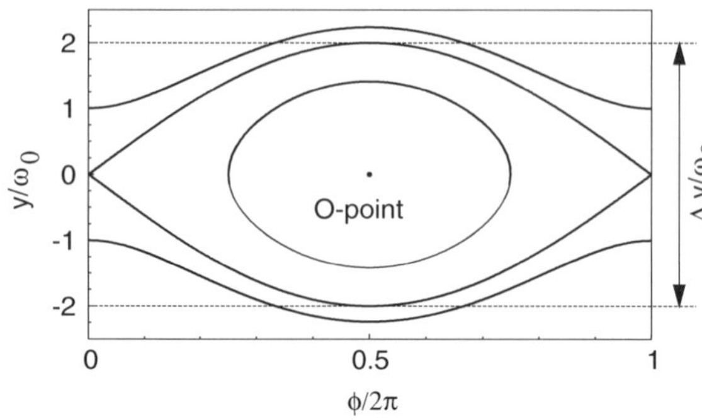

Fig. 8.1 Phase space of pendulum (8.9) and the definition of its width

$$H_{pend}(y,\phi) = \frac{y^2}{2} + \omega_0^2 \cos\phi, \tag{8.9}$$

where

$$y = m^2 (J - J_{mn}) \left.\frac{dq^{-1}}{d\psi_t}\right|_{J=J_{mn}}, \quad \omega_0 = m \left|\epsilon V(J_{mn}) \frac{dq^{-1}}{d\psi_t}\right|^{1/2}_{J=J_{mn}}.$$

We will call this case as a *standard nonlinear resonance*. The phase space of the pendulum was shown in Fig. 8.1. Since $\phi = m\vartheta - n\varphi$ there exist m elliptic points and m hyperbolic points in the original (ϑ, ψ_t) plane shown in Fig. 7.1b. The hyperbolic points are connected by heteroclinic orbits. They sharply divide the trapped field lines from the un-trapped field lines. The maximum variation of the trapped field lines in the toroidal flux variable ψ defines a *width of the nonlinear resonance* (or the *magnetic island*) $\Delta\psi_{mn}$. Since the width of the nonlinear resonance in y is $\Delta y = 4\omega_0$ one can easily obtain

$$\Delta\psi_{m,n} = 4 \left|\frac{\epsilon H_{mn}(\psi)}{dq^{-1}/d\psi_t}\right|^{1/2}_{\psi=\psi_{m,n}}. \tag{8.10}$$

The relation (6.81) allows to write the width of magnetic island through only one function, $R_n^{(reg)}(\psi)$, for a given toroidal mode number n:

$$\Delta\psi_{m,n} = 4 \left|\frac{\epsilon q(\psi_{mn}) R_n^{(reg)}(\psi_{mn})}{\pi dq/d\psi_t}\right|^{1/2}. \tag{8.11}$$

The width W_{mn} of the magnetic island in the poloidal flux ψ is given by

$$W_{mn} = \Delta\psi_{m,n}\frac{d\psi_p}{d\psi_t} = 4\left|\frac{\epsilon H_{mn}(\psi)}{dq/d\psi_t}\right|^{1/2}_{\psi=\psi_{m,n}} = 4\left|\frac{\epsilon R_n(\psi)}{\pi dq/d\psi_p}\right|^{1/2}_{\psi=\psi_{m,n}}, \tag{8.12}$$

where $dq/d\psi$ is a derivative with respect to the poloidal flux.

We should note that the above approximation is valid for the moderately nonlinear system satisfying the condition,

$$\epsilon \ll \alpha \ll \frac{1}{\epsilon}, \tag{8.13}$$

where

$$\alpha = \frac{I}{\omega(I)}\left|\frac{d\omega(I)}{dI}\right| = \frac{\psi_t}{q(\psi_t)}\left|\frac{dq(\psi_t)}{d\psi_t}\right|, \tag{8.14}$$

is the dimensionless parameter of the degree of nonlinearity (or the measure of the magnetic shear).

8.1.2 Field Lines Near the Zero Shear Magnetic Surface

From the condition (8.13) it follows that the standard nonlinear resonance approximation is not valid near the magnetic surface with the small or zero magnetic shear, where $dq(\psi_t)/d\psi_t = 0$ called a *shearless magnetic surface.* Below we study the field line structure in the latter case.

Consider again the system with the Hamiltonian (8.6). Let J_0 be the shearless action corresponding the magnetic surface with the zero shear, i.e., $dq/dJ = 0$ at $J = J_0$. Expanding the Hamiltonian (8.6) in a series of powers of $y = J - J_0$, it can be reduced to

$$H(\phi, y) = ay + \sigma\frac{y^3}{3} + b\cos\phi, \tag{8.15}$$

where

$$H = \frac{\mathcal{H} - \nu_0 J_0}{|\nu''(J_0)|}, \qquad a = m/q(mJ_0) - n,$$

$$b = \epsilon V(J_0)/\left|\nu''(J_0)\right|, \qquad \sigma = \operatorname{sgn}\left(\nu''(J_0)\right),$$

$$\nu''(J) = -\frac{m}{q^2(mJ)}\frac{d^2q(mJ)}{dJ^2}.$$

The quantity a describes mismatch between the safety factor $q(\psi)$ at the shearless magnetic surface $\psi_0 = mJ_0$ and the one $q(\psi_{mn}) = m/n$ at the resonant magnetic surface ψ_{mn}. The number $\sigma = 1$ (or $d^2q(mJ)/dJ^2 < 0$) when the safety factor $q(\psi)$

has a maximum at J_0, and therefore $a < 0$. The value $\sigma = -1$ (or $d^2q(mJ)/dJ^2 > 0$) when the safety factor $q(\psi)$ has a minimum at J_0 and $a > 0$. In (8.15) we have put $V(J) = V(J_0)$ by neglecting the derivative $dV(J)/dJ$.

The fixed points of the Hamiltonian system with (8.15) are determined by

$$\dot{\phi} = \frac{\partial H}{\partial y} = a + \sigma y^2 = 0,$$

$$\dot{y} = -\frac{\partial H}{\partial \phi} = b \sin \phi = 0.$$

For $\sigma = 1$ ($a < 0$) there are two kind of fixed points: elliptic fixed points at

$$\left(\phi_e^{(+)} = \pi,\, y_e^{(+)} = \sqrt{|a|}\right) \quad \text{and} \quad \left(\phi_e^{(-)} = 0,\, y_e^{(-)} = -\sqrt{|a|}\right).$$

The *up* and *down* hyperbolic fixed points are

$$\left(\phi_h^{(+)} = 0,\, y_h^{(+)} = \sqrt{|a|}\right) \quad \text{and} \quad \left(\phi_h^{(-)} = \pi,\, y_h^{(-)} = -\sqrt{|a|}\right),$$

respectively. Similarly, one can find fixed points for the case $\sigma = -1$. The phase space of Hamiltonian (8.15) is plotted in Fig. 8.2 for the different values of the perturbation parameter b.

For small values of b the up and down hyperbolic points are connected only with themselves by heteroclinic orbits similar to the case of the standard nonlinear resonance. At the certain level of perturbation b_c the up and the down hyperbolic points are reconnected. It occurs when

$$H\left(\phi = \pi,\, y = y_h^{(-)}\right) = H\left(\phi = 0,\, y = y_h^{(+)}\right).$$

According to the Hamiltonian (8.15) we obtain the reconnection threshold

$$b_c = 2|a|^{3/2}/3. \tag{8.16}$$

When $b > b_c$ the lower hyperbolic point starts to connect itself homoclinically.

The maximum variation of the action variable J near resonant orbits is proportional to $\epsilon^{1/3}$, which is larger than for the conventional nonlinear resonance where it is proportional to $\epsilon^{1/2}$ (see Eq. (8.10)).

8.1.3 Non-standard Nonlinear Resonance

If the perturbation function $V(J)$ sharply changes near the resonance surface ψ_{mn} the number fixed points may be doubled. We illustrate this case for the function $V(J)$ in the following form,

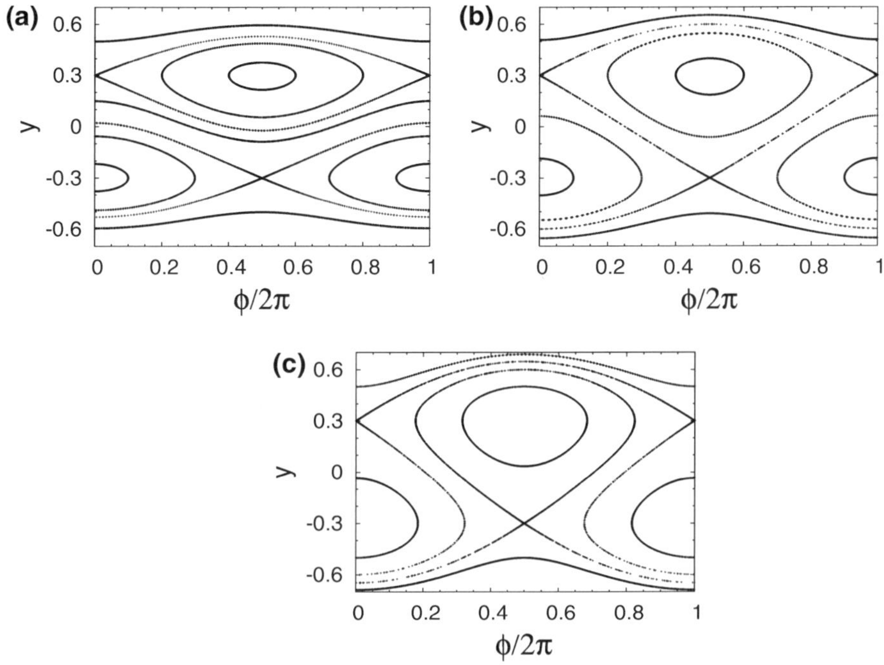

Fig. 8.2 Phase space of Hamiltonian (8.15) for the three different perturbation parameter b: **a** $b = 0.01$; **b** $b = 0.018$; **c** $b = 0.025$. The parameter $a = -0.09$

$$V(J) = \left[\cosh\left(\frac{1}{\Delta}\ln\frac{J}{J_0}\right)\right]^{-m\Delta/2}, \tag{8.17}$$

where the parameter Δ characterizes the sharpness of the peak of $V(J)$ at the extremal point at $J = J_0$. The parameter Δ can be called a sharpness parameter. At $|\ln(J/J_0)| \gg \Delta$ the function $V(J)$ has the following power-law asymptotics,

$$V(J) \approx \begin{cases} (J/J_0)^{m/2}, & \text{for } J/J_0 < e^{-\Delta}, \\ (J/J_0)^{-m/2}, & \text{for } J/J_0 > e^{\Delta}, \end{cases} \tag{8.18}$$

The dependence of $V(J)$ and its derivative,

$$f(J) = \frac{dV}{dJ} = \frac{m}{2J}V(J)\tanh\left(\frac{1}{\Delta}\ln\frac{J}{J_0}\right),$$

on the action J are shown in Fig. 8.3a, b (curves 1–3) for the different values of Δ. We assume that the position of extremal point J_0 is close to the resonant value J_{mn}.

The solution of Eq. (8.8) for the fixed points can be found graphically as crossing points of the function $f(J)$ with $\pm\nu(J)/\epsilon$. It is shown in Fig. 8.3b for the safety

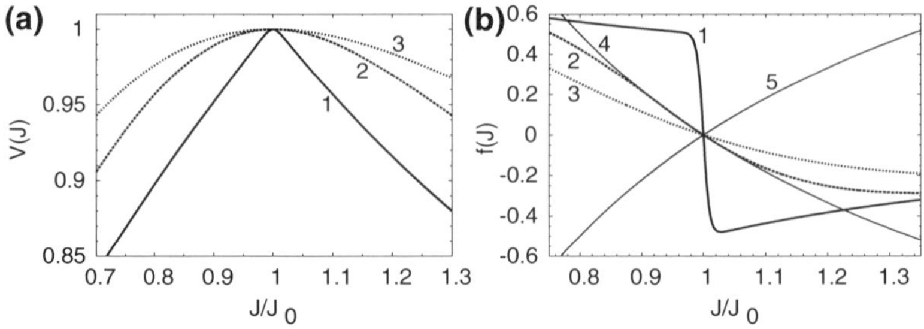

Fig. 8.3 **a** Non-monotonic profile of the perturbation function $V(J)$ (8.17). *Curve 1* corresponds to $\Delta = 0.01$; *curve 2* — $\Delta = 0.25$; *curve 3* — $\Delta = 0.5$; **b** Determination of fixed points. *Curves 1–3* describe the function $f(J)$ for the same values of Δ. *Curve 4* describes the function $\nu(J)/\epsilon$, and *curve 5*—the function $-\nu(J))/\epsilon$. The parameters are $m = 1$, $\epsilon = 0.5$, $J_{mn}/J_0 = 1$

factor $q(\psi_t) = (m/n)\psi_t/\psi_{mn}$. As seen from Fig. 8.3 at $\phi = 0$ the value of the action J of the fixed point is determined by the crossing of the curve 5 with curves 1–3. In this case there is only one solution of the Eq. (8.8) for the different values of Δ. However, at $\phi = \pi$ corresponding values of J are found by the crossing points of curve 4 with curves 1–3. As seen in this case it may exist one or three roots of the Eq. (8.8) depending on Δ and J_{mn}.

One can show that in the above example of the perturbation function $V(J)$ and the frequency $\nu(J)$ there exists two fixed points $(J_1, \phi = 0)$, $(J_2, \phi = \pi)$ for the value of $\Delta > \Delta_c$, where the critical value $\Delta_c = \epsilon J_{mn}/2J_0$. At $\Delta < \Delta_c$ there exist four fixed points $(J_1, \phi = 0)$, $(J_k, \phi = \pi)$, $(k = 2, 3, 4)$.

The phase space of Hamiltonian system (8.4) in the (ϑ, J) plane is shown in Fig. 8.4a, b for three values of Δ: red dots correspond to $\Delta = 0.01 < \Delta_c = 0.05$, green dots correspond to $\Delta = \Delta_c$, and blue dots correspond to $\Delta = 0.1 > \Delta_c$. Figure 8.4b shows the expanded view of the rectangular area in (a). The resonant action J_{mn} is taken equal to the extremal point $J_0 = 1$, and the perturbation parameter $\epsilon = 0.1$.

Consider first the case $\psi_{mn} = J_0$. As seen from Fig. 8.4a, b at the values of the sharpness parameter Δ exceeding the critical value Δ_c there are two fixed points, one elliptic fixed point P_1 located at $(\phi = 0, J = J_0)$ and one hyperbolic fixed point P_2 at $(\phi = \pi, J = J_0)$. The system bifurcates at the value $\Delta = \Delta_c$ where there are degenerate roots $J_k = J_0$ $(k = 2, 3, 4)$ of Eq. (8.8) at $\phi = \pi$. For the small values of $\Delta < \Delta_c$ the degeneracy disappears at this point and it gives rise three distinct fixed points P_k, $(k = 2, 3, 4)$: one elliptic fixed point P_2 at $(\phi = \pi, J = J_0)$ and two hyperbolic fixed points P_3, P_4 at $(\phi = \pi, J = J_3)$ and $(\phi = \pi, J = J_4)$, where $J_3 < J_0$, and $J_4 < J_0$ (red dots).[1]

[1] One should note that the bifurcation of a hyperbolic fixed point into an elliptic fixed point occurs also in the problem of the stability of the upper position of a pendulum with a vibrating suspension point. However, in this phenomenon discovered by Stephenson (1908), Bogolyubov (1950), Kapitsa (1951) (see a book by AKN (2006) for more details and additional references), such a bifurcation

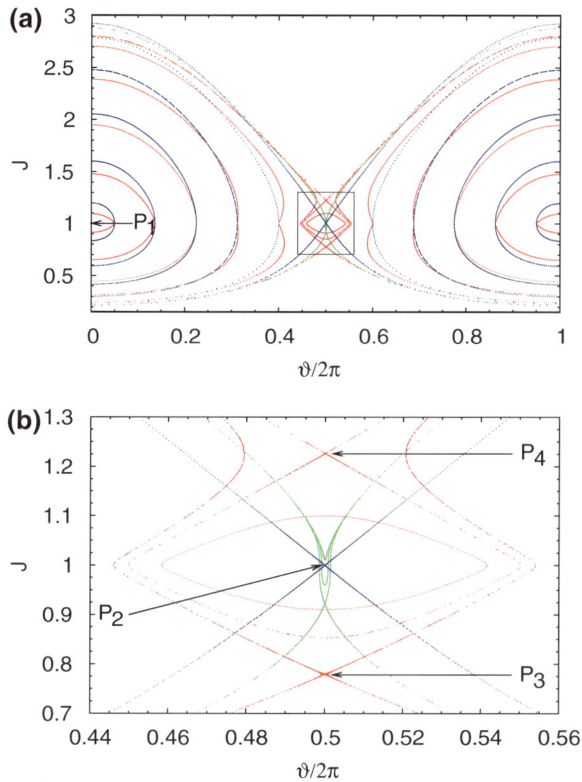

Fig. 8.4 Poincaré sections of Hamiltonian system (8.4). The safety factor is $q(\psi_t) = (m/n)\psi_t/\psi_{mn}$, the resonant numbers $m = n = 1$, the perturbation parameter $\epsilon = 0.5$. **a** Whole phase space in the (ϑ, J) plane; **b** Expanded view of the rectangular area near the fixed point ($\phi = \pi$, $J_2 = 1$). *Red curve* corresponds to $\Delta = 0.01$, *green curve* $-\Delta = 0.25$, and *blue curve* $-\Delta = 0.5$

In general, the hyperbolic fixed points P_3 and P_4 do not have a mirror symmetry with respect to line $J = J_0$. Each of them connects itself by two homoclinic orbits. In the latter case the trapped motion occurs in regions: the large one with the center at the elliptic fixed point P_1 and the small one with center at the elliptic fixed point P_2.

The described phenomenon, perhaps, occurs in the reconnection processes of magnetic field lines in plasmas. Particularly, it may happen when the amplitude of a magnetic field created by a thin current layer at the resonant magnetic surface has an extremal point with a sufficiently sharp peak. The magnetic island in this case may have a structure similar to one shown in Fig. 8.4. Such a magnetic structure of the

(Footnote 1 continued)
appears when the system is subjected to the external time–periodic perturbation, while in our case it occurs in a fully integrable autonomous system.

internal magneto–hydrodynamic (MHD) modes in tokamaks has been also studied by Chudnovskiy (2004).

8.1.4 Nonlinear Resonance in Many-Dimensional Systems

Up to now we considered the nonlinear resonance phenomenon describing the magnetic field structure near the resonant magnetic surfaces under nonaxisymmetric magnetic perturbation. Similar we can study the structure of drift surfaces for the guiding center motion near the resonant drift surfaces. As was shown in Sect. 5.2 the drift motion is described by Hamiltonian in axisymmetric tokamak is described by the two-degrees-of-freedom Hamiltonian system (5.17). Now we study the nonlinear resonance phenomenon in this two-dimensional system. In the presence of time–dependent nonaxisymmetric perturbations the Hamiltonian of the drift motion can be presented as[2]

$$H = H_0(I_z, I_\varphi) + \epsilon H_1(\vartheta_z, \vartheta_\varphi, I_z, I_\varphi, t), \tag{8.19}$$

where the perturbed Hamiltonian ϵH_1 can be presented as a Fourier series in $\vartheta_z, \vartheta_\varphi$:

$$\epsilon H_1 = \epsilon \sum_{m,n} H_{mn}(I_z, I_\varphi) \cos(m\vartheta_z - n\vartheta_\varphi + \Omega t). \tag{8.20}$$

Here Ω is the frequency of perturbations, $H_{mn}(I_z, I_\varphi)$ is the spectral components of the perturbation Hamiltonian. The perturbed drift motion is governed by

$$\begin{aligned} \frac{d\vartheta_\alpha}{dt} &= \frac{\partial H}{\partial I_\alpha} = \omega_\alpha(I_z, I_\varphi) + \epsilon \frac{\partial H_1}{\partial I_\alpha}, \\ \frac{dI_\alpha}{dt} &= -\epsilon \frac{\partial H_1}{\partial \vartheta_\alpha}, \quad \alpha = z, \varphi\ . \end{aligned} \tag{8.21}$$

According to the KAM (Kolmogorov-Arnold-Moser) theory most drift surfaces may survive under the perturbation (8.20) if the perturbation is sufficiently small and far from the resonant drift surfaces. The latter are determined by the condition

$$m\omega_z(I_z, I_\varphi) - n\omega_\varphi(I_z, I_\varphi) + \Omega = 0, \tag{8.22}$$

with certain integer numbers m, n. The resonant drift surfaces are significantly affected by the perturbation. Consider the drift orbits near the (m, n)-resonance surface. We introduce the new canonical phases Θ, Θ_φ and the actions J, J_φ, with the help of the generating function $F = (m\vartheta_z - n\vartheta_\varphi + \Omega t)J + \vartheta_\varphi J_\varphi$. Then, taking into account only the dominating single resonance perturbation $H_1 =$

[2] Here the notations I_z, ϑ_z, and ω_z stand for J, ϑ, and ω, respectively, used in (5.17).

$\epsilon H_{mn}\cos(m\vartheta_z - n\vartheta_\varphi + \Omega t)$ the canonical equations can be transformed to new ones with the new Hamiltonian function

$$H_J(\psi, J, J_\varphi) = H_0(mJ, -nJ + J_\varphi) + \Omega J + \epsilon H_{mn}(mJ, -nJ + J_\varphi)\cos\Theta. \tag{8.23}$$

The relations between the old action variables I_z, I_φ and the new one J are

$$\Theta = \frac{\partial F}{\partial J} = m\vartheta_z - n\vartheta_\varphi + \Omega t, \qquad \Theta_\varphi = \frac{\partial F}{\partial J_\varphi} = \vartheta_\varphi, \tag{8.24}$$

$$I_z = \frac{\partial F}{\partial \vartheta_z} = mJ, \quad I_\varphi = \frac{\partial F}{\partial \vartheta_\varphi} = -nJ + J_\varphi. \tag{8.25}$$

From (8.23) it follows that $nI_z + mI_\varphi = J_\varphi$ is an integral of motion. Suppose that the resonance condition (8.22) is satisfied for $J = J_{mn}$. The canonical equations for the new variables $(\Theta, \Theta_\varphi, J, J_\varphi)$ can be simplified by expanding the Hamiltonian (8.23) in series of powers of $\Delta J = J - J_{mn}$, and neglecting the terms proportional to $\epsilon\Delta J$. Then we obtain

$$\frac{d\psi}{dt} = \frac{\partial H_J}{\partial J} = \kappa_{mn}\Delta J, \qquad \frac{d\Delta J}{dt} = -\frac{\partial H_J}{\partial \Theta} = \epsilon V_{mn}\sin\Theta, \tag{8.26}$$
$$H_J = \frac{\kappa_{mn}}{2}(\Delta J)^2 + \epsilon V_{mn}\cos\Theta,$$

where $V_{mn} \equiv H_{mn}(mJ_{mn}, -nJ_{mn} + J_\varphi)$ and

$$\kappa_{mn} = \left(m\frac{\partial}{\partial I_z} - n\frac{\partial}{\partial I_\varphi}\right)\tilde{\Omega}. \tag{8.27}$$

Here, $\tilde{\Omega} = m\omega_z - n\omega_\varphi$. Equation (8.26) correspond to a pendulum with a frequency (for small amplitude oscillations) $\nu_{mn} = \sqrt{|\epsilon V_{mn}|/\kappa_{mn}}$. For $-\kappa_{mn}\nu_{mn}^2 < H_J < \kappa_{mn}\nu_{mn}^2$ the particle orbits are trapped by the chain of drift islands. The widths of the islands, $W_{mn} = 2\max|\Delta J|$, are

$$W_{mn} = 4\nu_{mn} = 4\sqrt{|\epsilon V_{mn}|/\kappa_{mn}}\ . \tag{8.28}$$

In the old action variables I_z, I_φ we have $\delta I_z = mW_{mn}$ and $\delta I_\varphi = nW_{mn}$. Taking into account that near the resonance the variables are not independent, one can easily show that the quantity κ can be written as $\kappa_{mn} = \partial\tilde{\Omega}/\partial\bar{J}$, where $\bar{J} = I_z/m - I_\varphi/n = 2I_z/m - J_\varphi$.

8.1.5 Stochastic Layer

In the above approximation of the nonlinear resonance we have taken into account only the resonant term $H_{res}(\Phi, \Psi)$ in the Hamiltonian (8.3) neglecting the perturbation term $\mathcal{H}_1'(\Phi, \Psi, \varphi, \epsilon)$. The latter affects on the structure of the nonlinear resonance shown in Fig. 8.1. It splits the separatrices at any small magnitude of perturbation ϵ. The heteroclinic tangle structure like one shown in Fig. 7.11b creates the chaotic orbit (see Sects. 7.5.7 and 7.6).

The region near the separatrix created by one single chaotic orbit is called a *stochastic layer* . It constitutes a seed or as called by B. V. Chirikov an *embryo* of chaos (Chirikov 1979). For sufficiently small perturbation ϵ the width of the stochastic layer is exponentially small and the chaotic motion is confined in this very small region. The chaotic orbits cannot diffuse far from its initial positions because of existence of invariant curves. The properties of the stochastic layer will be studied in Sect. 8.3.

Figure 8.5 shows the Poincaré section in the phase space of the periodically-driven pendulum described by the Hamiltonian

$$H(y, \Phi, t) = H_{pend}(y, \Phi) + \epsilon\omega_0^2 \cos(\Phi - \Omega t + \chi), \tag{8.29}$$

where $H_{pend}(y, \Phi)$ is Hamiltonian of the pendulum given by Eq. (8.9), Ω is the perturbation frequency. As seen from Fig. 8.5 the stochastic layer is formed near the separatrix of the pendulum.

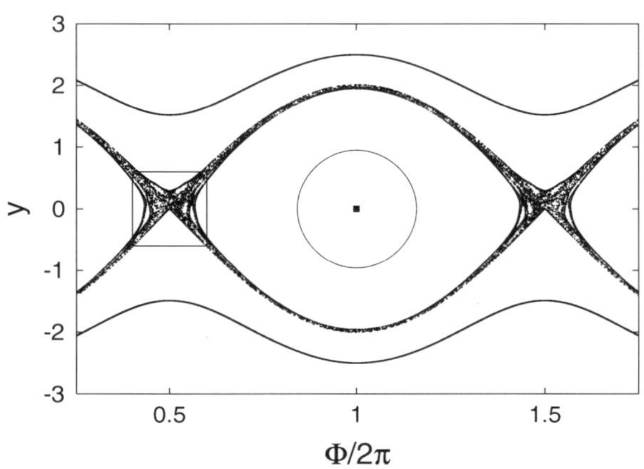

Fig. 8.5 Stochastic layer near the separatrix of the pendulum

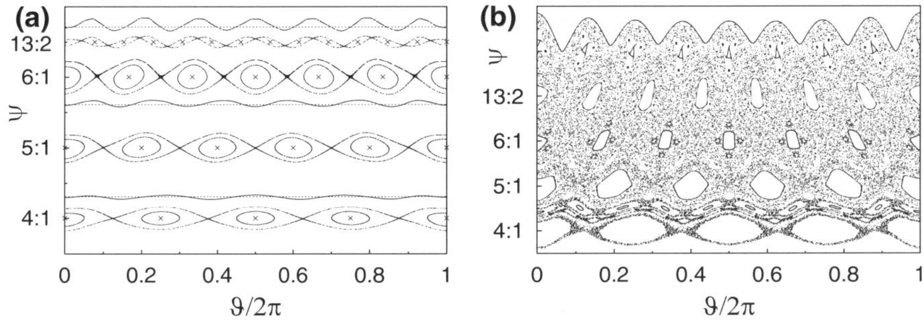

Fig. 8.6 Poincaré sections of regular and chaotic field lines: **a** a case of small perturbations; **b** a case of the sufficiently large perturbation. The numbers $(m : n) = 4 : 1 \ldots$ at the vertical axis ψ show the positions of the resonant surfaces ψ_{mn}

8.2 Onset of Chaotic Motion: Chirikov's Criterion

Chirikov (1959, 1979) proposed a qualitative idea of the onset of chaos in Hamiltonian systems and the method to estimate the critical perturbation at which the chaotic motion occurs. It is based on the nonlinear resonance phenomenon described in Sect. 8.1.1. Below we shortly describe the Chirikov method known as the *Chirikov's criterion.*[3,4]

Consider the Hamiltonian system (7.12) and (7.13). The any small perturbation $\epsilon \ll 1$ destroys the resonant surfaces ψ_{mn}, $q(\psi_{mn}) = m/n$, forming the chain of magnetic islands as shown in Figs. 7.1b and 8.6a. For the sufficiently small perturbations the neighboring islands are well isolated, and in the first order approximation they can be described by the nonlinear pendulum (see Sect. 8.1.1). However, the neglected perturbation terms of the Hamiltonian destroy the separatrix of the pendulum and thereby creating the chaotic motion in the confined layer near the unperturbed separatrix (see Sect. 8.1.5).

With increasing the magnitude of magnetic perturbations, ϵ, the width of magnetic islands W_{mn} (8.10) grow and the neighboring magnetic islands start to interact destroying invariant magnetic surfaces between them. It may lead to the global chaos and unrestricted behavior of field lines in a certain region of the plasma as shown in Fig. 8.6b. If the stochastic region is located at the plasma edge the field lines leave the plasma region.

---

[3] We should note that in some cases the Chirikov criterion cannot be applied to study onset of global chaos. Particularly, since it uses the pendulum approximation (see Sect. 8.1.1) it is not valid near the region with small or zero shears $dq/d\psi$ (see, the article "Chirikov criterion" by Shepelyansky 2009, and also Sect. 8.4).

[4] There are rigorous approaches to determine the onset of chaos: the Greene's residue criterion conjectured by Greene (1979) and the renormalization procedure developed by Escande (1982) and MacKay (1983) (see also Lichtenberg and Lieberman (1992), MacKay (1992), Elskens and Escande (2003)).

The criterion of the onset of global chaos can be qualitatively described by the overlapping of magnetic islands (Chirikov 1959, 1979),

$$\sigma_{Chir} = \frac{\Delta\psi_{m,n} + \Delta\psi_{m+1,n}}{2|\psi_{m+1,n} - \psi_{m,n}|} \geq \sigma_c, \tag{8.30}$$

i.e., the distance between neighboring magnetic islands $\delta\psi = |\psi_{m+1,n} - \psi_{m,n}|$ becomes smaller than their average width $\Delta\psi = \left(\Delta\psi_{m,n} + \Delta\psi_{m+1,n}\right)/2$. The overlapping parameter σ_{Chir}, which is often called the *Chirikov parameter*, should exceed a certain critical value $\sigma_c \approx 1$. However, as we will see in the next section the numerical calculations show that σ_c may take lower values.

For the standard mapping (6.129) the Chirikov parameter σ_{Chir} is equal to $2/\pi K^{1/2}$. The global chaos occurs at the values $K > K_c = 0.98$ (see MacKay 1983).

8.3 Properties of the Stochastic Layer

The stochastic layer is not completely chaotic, but it has regions with a regular motion called KAM stability islands. It's structure, i.e., the mutual positions of the KAM stability islands changes with the perturbation parameter ϵ. It has been found that the topological structure of the stochastic layer near the X-points has a non-trivial rescaling invariance property with respect to the change of perturbation parameter ϵ.

8.3.1 Rescaling Invariance of Phase-Space Near the Hyperbolic X-Point

Below we describe the universal rescaling invariance property of the stochastic layer near the hyperbolic saddle point (or X-point): the phase–space structure of the system is invariance under a certain transformation of the perturbation amplitude and phase. It is related with the universal logarithmic behavior of the safety factor (2.51) or the period of motion (5.58) near the separatrix.

Let $(x = \Phi - \pi, y)$ a phase space coordinates near the X-point $\Phi = \pi$, $y = 0$. Then the phase space of the stochastic layer near the X-point is invariant with respect to the following change of the perturbation parameter ϵ and its phase χ:

$$\begin{aligned} &\epsilon_a \to \epsilon_b = \lambda\epsilon_a, \quad \chi_a \to \chi_b = \chi_a + \pi, \\ &x_a \to x_b \approx \lambda^{1/2} x_a, \quad y_a \to y_b \approx \lambda^{1/2} x_a, \end{aligned} \tag{8.31}$$

where λ is a rescaling parameter determined the parameter γ (2.45) describing the behavior of unperturbed system near the X-point and the frequency of perturbation,

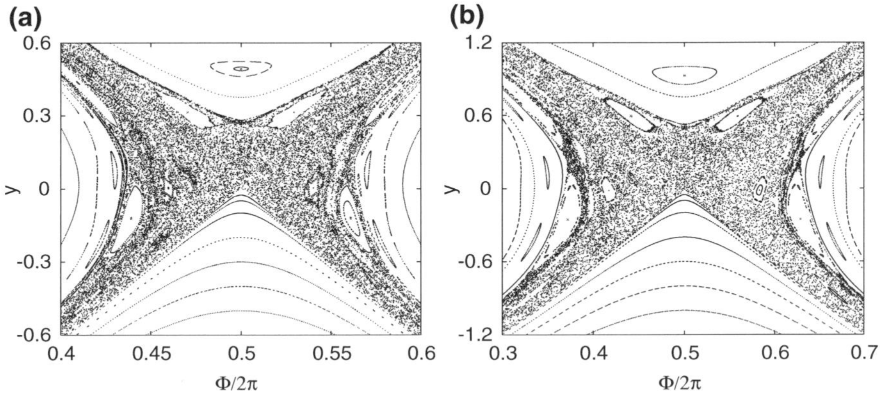

Fig. 8.7 Phase space near the X-point for the two different perturbation parameters: **a** $\epsilon_a = 0.02$, $\chi_a = 0$; **b** $\epsilon_b = 0.08$, $\chi_b = \pi$

Ω:

$$\lambda = \exp\left(\frac{2\pi\gamma}{\Omega}\right). \tag{8.32}$$

The similarity of the structures is seen from Fig. 8.7a, b where the expanded views of the rectangular area near the X-point in Fig. 8.5 are shown: (a) corresponds to the parameters ($\epsilon_a = 0.02$, $\chi = \chi_a = 0$); (b) – ($\epsilon_b = \lambda\epsilon_a = 0.08$, $\chi_b = \chi_a + \pi$). The rescaling parameter λ is chosen equal to 4. In Fig. 8.7a, b the coordinates x, y are rescaled by factor $\lambda^{1/2} = 2$.

Obviously, that the homoclinic (or heteroclinic) tangle structure near the X-point is also invariant to the rescaling transformation (8.31).

Consider an example of the rescaling invariance of field lines near the magnetic separatrix in the model of magnetic field (1.37) subjected to the perturbation (3.34) (see Sect. 1.3 and 3.3). Figure 8.8a, b present the Poincaré sections of field lines at the axis $z > 0$ shown in Fig. 6.6 for the two values of the perturbation amplitudes and phases: (a) $\epsilon_a = 10^{-3}$, $\chi_a = 0$; (b) $\epsilon_b = \lambda\epsilon_a = 2.513 \times 10^{-3}$, $\chi_b = \pi$. It is obtained using the mapping (6.95). The rescaling parameter of this system with the toroidal mode n is $\lambda = \exp(2\pi\gamma/n)$. The structures of the stochastic layer corresponding to these two different perturbation amplitudes are similar with the rescaled of the vertical coordinate $z_b = \lambda^{1/2} z_a$.

The rescaling invariance property of the system can be simply explained using the separatrix mapping (6.58). Using the asymptotics of the safety factor $q(\psi)$ given by Eq. (2.51) we have

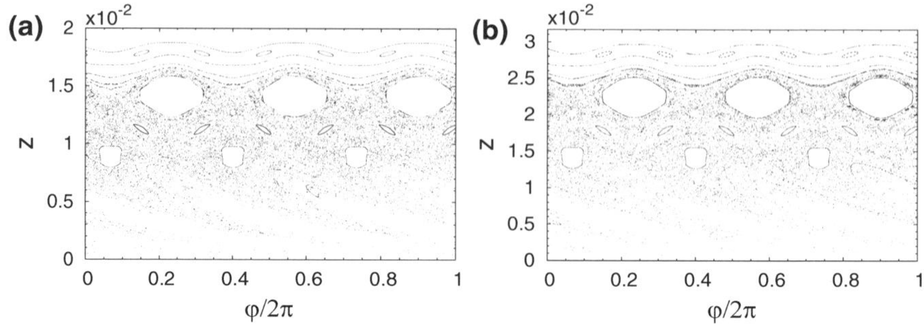

Fig. 8.8 Poincaré sections of field lines at the axis $z > 0$ in the model of magnetic field (1.37) subjected to the perturbation (3.34) (see Sects. 1.3 and 3.3) for the different perturbation amplitudes: **a** $\epsilon_a = 10^{-3}$, $\chi_a = 0$; **b** $\epsilon_b = \lambda\epsilon_a = 2.513 \times 10^{-3}$, $\chi_b = \pi$. The rescaling parameters $\lambda = \exp(2\pi\gamma/n) = 2.513$. The parameter $\gamma = 0.44$ and the toroidal mode $n = 3$

$$\begin{aligned} \psi_{k+1} &= \psi_k \mp \epsilon M \left(\varphi_k \pm \frac{1}{2\gamma} \ln \frac{Q}{|\psi_k|}; 0 \right), \\ \varphi_{k+1} &= \varphi_k \pm \pi \left[q(\psi_k) + q(\psi_{k+1}) \right]. \end{aligned} \tag{8.33}$$

The Melnikov integral $M(\varphi; h)$ is a 2π–periodic function of φ and according to Eqs. (6.74 and 6.76) is given by

$$M(\varphi; \psi) = -\sum_n n \left[K_n(\psi) \sin(n\varphi + \chi_n) + L_n(\psi) \cos(n\varphi + \chi_n) \right]. \tag{8.34}$$

It is easy to see that the system (8.33) with the Melnikov integral (8.34) and the safety factor

$$q(\psi) = \frac{1}{2\pi\gamma} \ln \frac{Q}{|\psi|} \tag{8.35}$$

is invariant with respect to the transformation

$$\begin{aligned} \epsilon \to \epsilon' &= \lambda\epsilon, \quad \chi_n \to \chi_n' = \chi_n + \pi, \\ \psi \to \psi' &= \lambda\psi, \quad \lambda = \exp(2\pi\gamma), \end{aligned} \tag{8.36}$$

which is equivalent to the rescaling transformations (8.31) with $\Omega = 1$.

If the perturbation contains only a single toroidal mode n then the phase space is invariant with respect to the transformation (8.36) with the rescaling parameter $\lambda = \exp(2\pi\gamma/n)$. This can be also shown by rewriting the mapping in the normalized variables

$$h = \frac{\psi}{\epsilon n K_0}, \quad A = \ln\left(\frac{Q}{\epsilon n K_0}\right)^{n/\gamma},$$
$$x = n\left(\varphi - \frac{1}{2\gamma}\ln|h|\right). \tag{8.37}$$

In these variables the mapping (8.33) reads as $(x_k, h_k) \to (x_{k+1}, h_{k+1})$,

$$h_{k+1} = h_k \mp \sin \bar{x}_k,$$
$$\bar{x}_{k+1} = \bar{x}_k \pm \left(A - \frac{n}{\gamma}\ln|h_{k+1}|\right), \tag{8.38}$$

where

$$\bar{x} = x + \frac{1}{2}A + \chi_n. \tag{8.39}$$

At the given ratio n/γ the mapping (8.38) is invariant with respect to transformation $A \to A + 2\pi$, $\chi_n \to \chi_n + \pi$. The latter is equivalent to the transformation (8.36). Therefore the structure of the stochastic layer near the X-point is a periodic function of A.

We should note that the rescaling invariance property is valid only in the small neighborhood of the X-point. It may be violated for the large values of the perturbation amplitude ϵ and the rescaling parameter λ. At the regions far from the X-point where the logarithmic asymptotics the safety factor (2.51) or the period of motion (2.58) is violated this property cannot be observed (see Abdullaev 2006 for details).

8.3.2 Periodicity of Transport Characteristics

According to the definition of A (8.37) it follows that at the given values of n/γ and K_0 the structure of stochastic layer changes periodically when the logarithm of the perturbation parameter, $\ln \epsilon$ (at the fixed value of Q) or the parameter $\ln Q$ (at the fixed value of ϵ) varies with the period $\ln \lambda$. It also leads to the periodic or quasiperiodic dependence of statistical characteristics of field lines, such as the average connection lengths, on the magnitude of perturbation field. Below we illustrate this property on the quasiperiodic dependence of the average length $\langle L \rangle$ of field lines on the perturbation amplitude ϵ and the parameter Q.

Consider again the model of the magnetic field (1.37) with the perturbation (3.34). Chaotic field lines formed near the separatrix connect the left and the right divertor plates Σ_d (see Fig. 6.6) making a number of poloidal turns inside the stochastic layer. The connection length L along an individual field line is a random function of the initial toroidal angle $\varphi^{(0)}$ and it depends on the structure of the stochastic layer. Some field lines may be trapped by the magnetic islands embedded in the stochastic

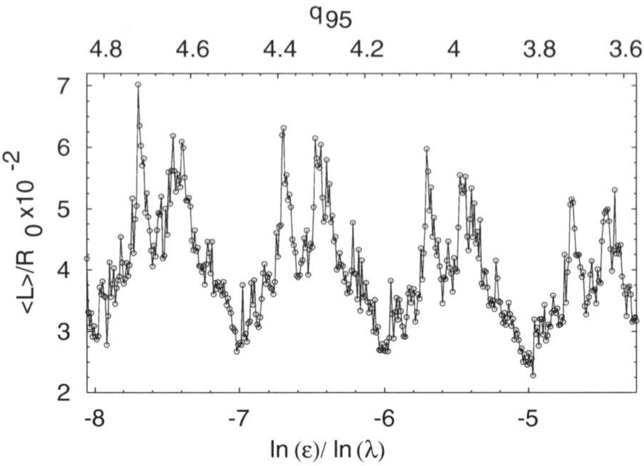

Fig. 8.9 Dependence of the average connection length $\langle L\rangle$ of field lines on the perturbation amplitude ϵ (*lower horizontal axis*) and on q_{95} (*upper horizontal axis*) for the model of a perturbation magnetic field (1.37 and 3.34) (see Sects. 1.3 and 3.3). The rescaling parameters $\lambda = \exp(2\pi\gamma/n) = 2.513$. The parameter $\gamma = 0.44$ and the toroidal mode $n = 3$

layer. The connection length $\langle L\rangle$ averaged over a large number N, ($N \gg 1$), of field lines can be found by

$$\langle L\rangle = R_0 \frac{R_0}{N} \sum_{i=1}^{N} \left(\varphi_i^{(f)} - \varphi_i^{(0)}\right), \tag{8.40}$$

where $\varphi_i^{(0)}$ and $\varphi_i^{(f)}$ are the toroidal angles of the i-th field line at the left and the right divertor plates, respectively.

Since the averaged connection length $\langle L\rangle$ has a quasiperiodic dependence on $\ln Q$ it should have a periodic dependence on the safety factor $q(\psi)$. According to (8.35) its period in q is equal to

$$\Delta q = q\,(\psi;\, Q + \ln\lambda) - q\,(\psi;\, Q) = \frac{1}{n}. \tag{8.41}$$

Figure 8.9 shows the dependence of the averaged connection length $\langle L\rangle$ of field lines on the perturbation amplitude ϵ (lower horizontal axis) and on the value of q_{95} (upper horizontal axis). It is obtained using the mapping (6.95). One clearly can the periodic dependence of $\langle L\rangle$ on $\ln\epsilon$ with the period $\ln\lambda$ and on q_{95} with the period $\Delta q = 1/n = 0.333$.

In the recent experiments in the DIII-D tokamak by Schmitz et al. (2009, 2012) and in the JET tokamak by Liang et al. (2010) the quasi-periodic dependence of plasma parameters, an electron pressure and the frequency of edge localized modes on the value of q_{95} in the presence of the RMPs have been observed. These effects are probably due to the just described quasiperiodical dependence of the structure of the stochastic layer on the safety factor. Particularly, the experimentally found period

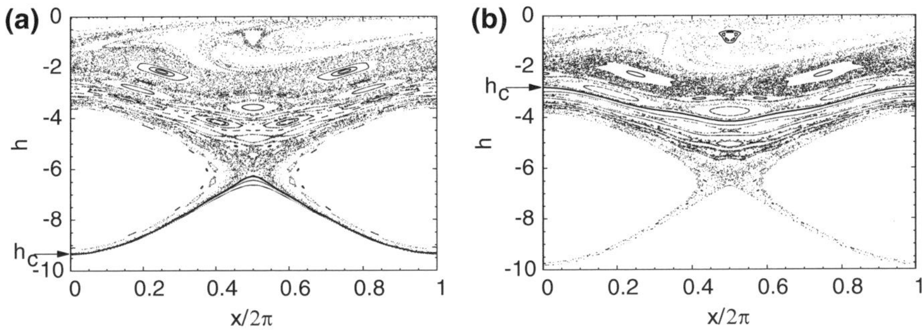

Fig. 8.10 Poincaré sections of the mapping (8.38) in the (x, h)−plane: **a** for $A/2\pi = 0.87$; **b** $A/2\pi = 0.9$. The toroidal mode $n = 3$, and the parameters $\gamma = 1$. The *solid curve* are invariant surface separating the stochastic layer from *inner field lines*

in q_{95} in the DIII-D tokamak with the toroidal mode of the RMPs $n = 3$ is in a close agreement with $\Delta q = 0.333$ predicted by (8.41).

The described property of the stochastic layer may have an important role in controlling of heat and particle transport at the plasma edge by the resonant magnetic perturbations. The detailed description of the statistical properties of motion near the separatrix of Hamiltonian systems can be found in Abdullaev (2000, 2006).

8.3.3 The Width of Stochastic Layer

The structure of the phase space near the unperturbed separatrix is complex. It consists of the stochastic layer with embedded in it large number of regions of regular orbits called KAM stability islands. Figure 8.10a, b illustrate the structures of the stochastic layer in the (x, y)−plane obtained by the mapping (8.38) for the two values of the parameter A: (a) for $A/2\pi = 0.87$; (b) $A/2\pi = 0.9$. The invariant curves plotted in Figures are barriers which prevent the chaotic orbit from the stochastic layer to penetrate to inner regions. The extend (or the width) of the stochastic layer along the poloidal flux ψ can be characterized by the maximal deviation, ψ_c, of the chaotic orbit with respect to the unperturbed separatrix $\psi = 0$. Below we estimate the width ψ_c using the qualitative Chirikov criterion and compare it with the numerical simulations based on the separatrix mapping.

To be specific we consider the perturbation with the single toroidal mode n in the the standard divertor configuration. The resonant magnetic surfaces, ψ_{mn}, defined by $q(\psi_{mn}) = m/n$, near the separatrix become dense, i.e., the distance between neighboring resonant magnetic surfaces, $\delta\psi = |\psi_{m+1,n} - \psi_{mn}|$, goes asymptotically to zero with increasing the poloidal mode number m, or equivalently with decreasing $|\psi|$. Using the logarithmic asymptotics of the safety factor $q(\psi)$ (2.51), we have for large m,

$$\psi_{mn} \approx -Q \exp\left(-\frac{2\pi\gamma m}{n}\right),$$
$$\delta\psi \approx \frac{2\pi\gamma}{n}|\psi|. \tag{8.42}$$

Using Eq. (8.12) for the width of the (m, n)—magnetic island in the poloidal flux ψ, the Chirikov criterion (8.30) for the onset of chaos can be reduced to,

$$\sigma^2(\psi) \approx \left(\frac{W_{mn}}{\delta\psi}\right)^2 = \frac{4n^2\epsilon|R_n(\psi)|}{\pi^2\gamma|\psi|} \gtrsim \sigma_c^2, \tag{8.43}$$

where σ_c is a critical parameter of overlapping islands.

The condition Eq. (8.43) is satisfied for the motion near the separatrix with $|\psi| < \psi_c$, where ψ_c is the boundary value of the stochastic layer determined by the equation, $\sigma(\psi_c) = \sigma_c$:

$$|\psi_c| \approx \frac{4n^2\epsilon|R_n(0)|}{\pi^2\gamma\sigma_c^2}, \tag{8.44}$$

where the Poincaré integral $R_n(\psi)$ is replaced by its value at the separatrix $R_n(0)$.

The boundary of the stochastic layer ψ_c can be calculated by the separatrix mapping (8.38). It allows to estimate the quantity σ_c. The dependencies of the quantities $h_c = \psi_c/\epsilon n K_0$, ($K_0 = R_n(0)$) and $\sigma_c = 2\sqrt{n/(h_c\gamma)}/\pi$ on the parameter A are presented in Fig. 8.11 a, b for the several values of γ. As seen that the critical value of the Chirikov parameter σ_c is confined in the interval (0.4, 0.6) for all possible values of the parameter A.

Numerical calculations performed by Shevchenko (2008) for the driven pendulum shows that the quantity σ_c takes the value ≈0.5 which is close to ones from our calculations. This value is also in agreement with the numerical calculations of the stochastic layer formed near the magnetic separatrix in poloidal divertor tokamaks (see, e.g., Abdullaev et al. 2008).

The width of the stochastic layer is linearly proportional to the perturbation parameter ϵ, i.e., $|\psi_c| \propto \epsilon$. Using the relation Eq. (2.40) one can obtain the width of the stochastic layer along the spatial coordinate $X = R_0\xi$: $X_c = \sqrt{2|\psi_c|}/\alpha \propto \epsilon^{1/2}$. As we will see below the condition $\sigma_c \gtrsim 1$ is satisfied only in the outermost region of the stochastic layer.

From Eqs. (8.44 and 6.87) it follows that the width of the stochastic layer has the following dependence on the toroidal mode number n

$$|\psi_c| \propto n^2 \exp(-nC), \tag{8.45}$$

with the constant C weakly dependent on n. The maximum width ψ_c is achieved at n close to the $n_m = 2/C$ and it decreases with the increasing the toroidal mode n. The value of n_m is $n_m \simeq 2$ since the constant C takes the values of order of $C \simeq 1$ in the divertor tokamaks (see Sect. 3.4 and Figs. 3.13 and 3.14). The value of ψ_c is inversely proportional to the parameter γ which according to (2.51) determines the safety factor profile, $q(\psi)$, near the separatrix.

In the case of the *snowflake divertor configuration* with the safety factor asymptotics (2.74) we have

$$\psi_{mn} \approx Q^3 \left(\frac{n}{m}\right)^3, \quad \delta\psi \approx \frac{3}{nQ}|\psi|^{4/3}. \tag{8.46}$$

The Chirikov criterion in this case reads as

$$\sigma^2(\psi) \approx \frac{8n^2 Q\epsilon |R_n(0)|}{3\pi|\psi|^{4/3}} \gtrsim \sigma_c^2, \tag{8.47}$$

from which one gets the estimation for the boundary ψ_c of the stochastic layer

$$\psi_c \approx \left(\frac{8n^2 Q\epsilon |R_n(0)|}{3\pi\sigma_c^2}\right)^{3/4}. \tag{8.48}$$

In contrast to the standard divertor the dependence of ψ_c on the perturbation amplitude ϵ has a power-law, $\psi_c \propto \epsilon^{3/4}$. To confirm such a dependence we calculate ψ_c using the mapping (8.33) with the Melnikov integral $M(\varphi) = \epsilon n K_0 \sin(n\varphi)$ and the safety factor $q(\psi)$ corresponding to the snowflake divertor considered in Sect. 2.6 and Fig. 2.11,

$$q(\psi) = \frac{Q}{|\psi|^{1/3}} + a_0 + a_1\psi. \tag{8.49}$$

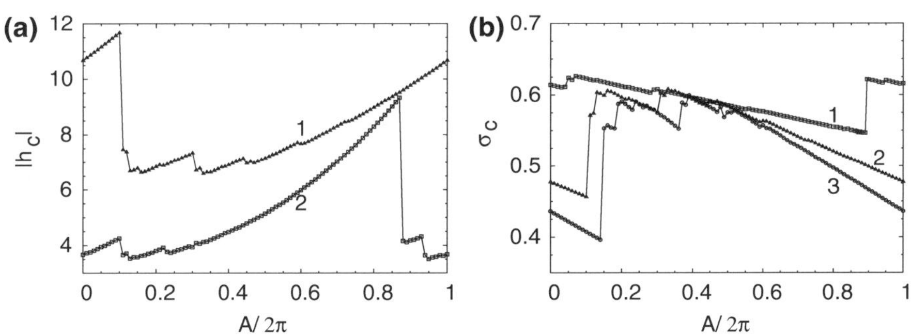

Fig. 8.11 **a** Maximal deviation h_c of chaotic orbits from the separatrix $h = 0$ as a function of A: *curve 1* corresponds to $\gamma = 0.5$, *curve 2*—to $\gamma = 1$; **b** Critical overlapping parameters σ_c as a function of A: *curve 1* corresponds to $\gamma = 0.2$, *curve 2*—to $\gamma = 0.5$; and *curve 3*—to $\gamma = 0.8$. The toroidal mode $n = 3$

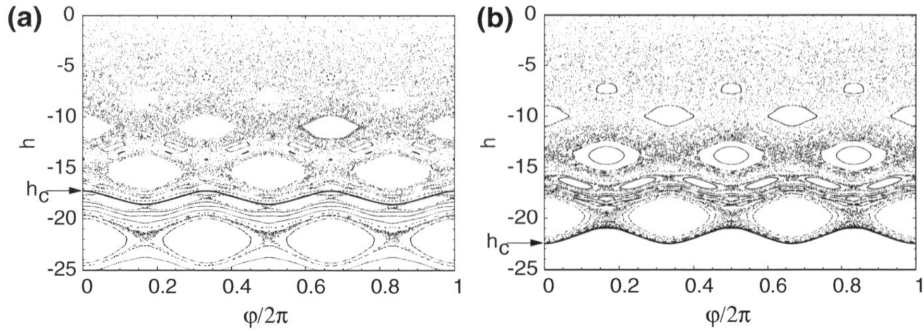

Fig. 8.12 Poincaré sections of the mapping (8.33) with the safety factor (8.49) corresponding to the snowflake divertor configuration. The vertical axis is $h = \psi/\epsilon n K_0$. **a** for $\epsilon = 0.9\times10^{-5}$; **b** $\epsilon = 10^{-5}$. The toroidal mode $n = 3$, and the parameters $Q = 0.19694$, $a_0 = 1.09577$, $a_1 = 174.675$. The *solid curve* are invariant surface separating the stochastic layer from *inner field lines*

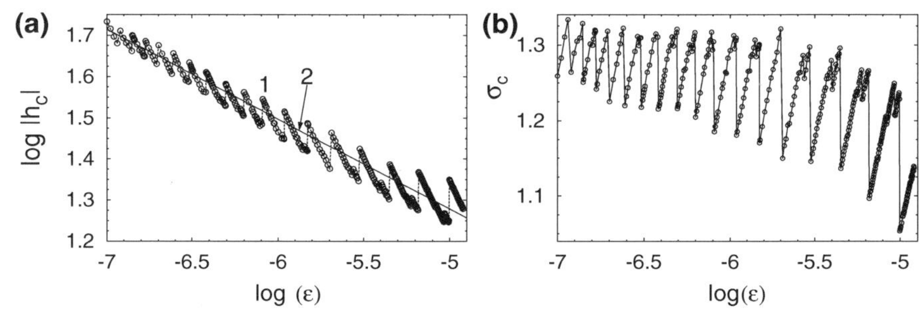

Fig. 8.13 **a** Maximal deviation $h_c = \psi_c/\epsilon n K_0$ of chaotic orbits from the separatrix $h = 0$ as a function of the perturbation amplitude *epsilon* (log − log dependence): *curve 1* corresponds to the numerical calculations, *curve 2* corresponds to its fitting via the function $h_c = a\epsilon^b$ with $a = 1.2786$, $b = -0.218$; **b** Critical overlapping parameters σ_c as a function of ϵ. The toroidal mode $n = 3$

Figure 8.12a, b present the Poincaré sections of field lines in the $(\varphi, h = \psi/\epsilon n K_0)-$ plane for the two different value of the perturbation parameter ϵ: (a) $\epsilon = 0.98\times10^{-5}$; (b) $\epsilon = 10^{-5}$.

As seen the small change of the perturbation amplitude ϵ leads to the drastic increase of the width h_c because attaching the large marginal island to the stochastic layer. This explains the dependence of h_c on the perturbation amplitude ϵ with discontinuities at certain its values as shown in Fig. 8.13a. The fitting of this dependence with the power law $h_c = a\epsilon^b$ gives the exponent $b = -0.218$. Therefore, we have $\psi_c \propto \epsilon h_c \propto \epsilon^{\tilde{b}}$ with the exponent $\tilde{b} = b + 1 = 0.788$ which is close to the one 3/4 predicted by Eq. (8.48). The Chirikov parameter σ_c calculated from these values of h_c (or ψ_c) using (8.48) are presented in Fig. 8.13b. For the perturbation amplitude ϵ ranging from 10^{-7} to 10^{-5} the parameter σ_c oscillates between 1.05 and 1.3, i.e., close to the qualitatively expected value of unity.

8.4 Magnetic Fields with a Reversed Shear

Consider the case of a tokamak magnetic configuration with a so–called *reversed magnetic shear* when the the safety factor $q(\psi)$ is non-monotonic function of ψ. The profile of the safety factor $q(\psi)$ has a minimum not at the magnetic axis $\psi = 0$, but at a normalized radius of 0.3–0.6 and regularly increases towards the center ($\psi = 0$) and the edge of the plasma $\psi = \psi_a$. Recent tokamak experiments (see reviews by Litaudon 1998, Wolf 2003) have shown the appearance of improved confinement regimes, *internal transport barriers* (ITB's), in the presence of a reversed magnetic shear. A non-monotonic profile of a rotational transform $\iota = 1/q(\psi)$ occurs also in fusion machines like a Heliac (Hayashi et al. 1994).

In Hamiltonian systems with the non-monotonic safety factor the KAM theory is not applicable in a small neighborhood of the minimum of the safety factor, since the twist condition $q'(\psi) = dq(\psi)/d\psi \neq 0$ is violated.

The structure of field lines near the shearless magnetic surface has been studied in Sect. 8.1.2 and was shown in Fig. 8.2a–c. Below we present a specific example of a magnetic field with the reversed shear and study the field lines near the magnetic surface ψ_m with the minimal value of the safety factor $q(\psi)$, i.e., $q'(\psi_m) = 0$. To be specific we consider the profile $q(\psi) = q_m + a(\psi - \psi_m)^2$ plotted in Fig. 8.14a for the parameters $q_m = 1.95$, $\psi_m = 0.416$, and $a = 11.86$. The corresponding value of the safety factor $q(\psi)$ at the magnetic axis $\psi = 0$ and at the plasma edge $\psi = 1$ are $q(0) = 4$ and $q(1) = 6$, respectively.

In order to emphasize the effect we have chosen a safety factor profile with the magnetic surfaces $q = 2$ located below and the above shearless magnetic surface ψ_m. For simplicity consider a magnetic perturbation containing a several poloidal $m : n$ modes including $m : n = 2 : 1$ mode resonant to the magnetic surfaces $q = 2$, i.e.,

$$\epsilon H_1(\psi, \vartheta, \varphi) = \epsilon \sum_{m=1}^{5} \cos(m\vartheta - \varphi)\,. \tag{8.50}$$

We will study the behavior of field lines in a neighborhood of the shearless magnetic surface ψ_m. The evolution of the field lines with the increase the perturbation amplitude ϵ is shown in Fig. 8.14b–d: (b) $\epsilon = 5 \times 10^{-4}$; (c) $\epsilon = 1.1028 \times 10^{-3}$; (d) $\epsilon = 3 \times 10^{-3}$. At the sufficiently small perturbations ϵ the majority nonresonant magnetic surfaces are survived. The pair of the resonant magnetic surfaces $q = 2$ are destroyed creating magnetic islands.

At the perturbation amplitudes ϵ less than the critical value $\epsilon_c = 1.1028 \times 10^{-3}$, these islands are separated by the invariant KAM surfaces (see Fig. 8.14b). Increasing the perturbation ϵ leads to the break of the invariant KAM surface and the reconnection of separatrices. At the critical value $\epsilon = \epsilon_c$ the X-points of the top and bottom magnetic islands are connected as shown in Fig. 8.14c. At $\epsilon > \epsilon_c$ the X-points of the bottom resonance reconnect itself with the separatrix loops which encircle the O-points of the top resonances and vice versa. The modified structure of islands is separated by new invariant surfaces. These invariant surfaces are "hard" to break. This case is shown in Fig. 8.14d when $\epsilon = 3 \times 10^{-3}$. The invariant surfaces prevent a penetration of field lines from the chaotic bottom region to the chaotic region above the shearless surface. These surfaces can be destroyed at larger perturbation amplitudes ϵ. The described picture of transition to global chaos in the system with the non-monotonic safety factor is completely different from the one with monotonic safety factor considered in the previous section.

Finally, we would like to note the following observation. A careful look of Fig. 8.14b–d shows the two islands near the shearless surface is so close that the Chirikov criterion (8.30) is formally well satisfied. However, no chaotic transport through the shearless surface appears. Thus, the Chirikov criterion is not applicable near the shearless surface where $dq/d\psi$ is small or zero.

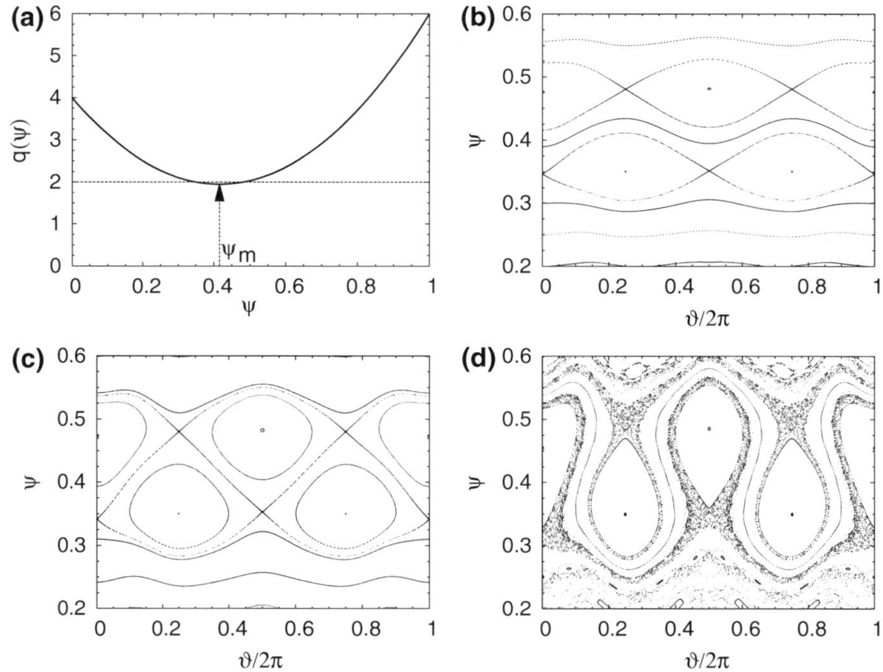

Fig. 8.14 **a** Non-monotonic profile $q(\psi) = q_m + a(\psi - \psi_m)^2$ at the values: $q(0) = 4, q_m \equiv q(\psi_m) = 1.5$, and $q(\psi_a) = 6$; Parameter a is expressed through these values; **b–d** Poincaré sections of field lines at the different perturbation amplitudes: **b** $\epsilon = 5 \times 10^{-4}$; **c** $\epsilon = 1.1028 \times 10^{-3}$; **d** $\epsilon = 3 \times 10^{-3}$

8.5 Bibliographic Comments

The rescaling invariance property of phase space near the hyperbolic fixed points (or X-points) of Hamiltonian systems has been first discovered in numerical simulations of Hamiltonian systems (Abdullaev and Zaslavsky 1994; Zaslavsky and Abdullaev 1995). It has been explained in Abdullaev and Zaslavsky (1995, 1996) by constructing the canonical separatrix mapping. Particularly, the homoclinic (or heteroclinic) tangle structure near the X-point has also such a property (Kuznetsov and Zaslavsky 2002; Kuznetsov 2003). The detailed description of the rescaling invariance property in Hamiltonian systems is given by Abdullaev (2000, 2006).

The dynamics of Hamiltonian systems with non-monotonic frequency of motion also known as *nontwist systems* or *nontwist mappings* has been extensively investigated during the last two decades (see the papers by Howard and Hohs 1984; Castillo-Negreto and Morrison 1993; del Castillo-Negreto et al. (1996, 1997); Soskin 1994; Abdullaev 1994) and a review paper Soskin et al. (2003). The behavior of field lines in tokamaks with the reversed magnetic shear has been studied by Balescu (1998), Morrison (2000), Misguich et al. (2003), Abdullaev (2004b), Wingen et al. (2005), Petrisor (2007), Roberto et al. (2004, 2005), El Mouden et al. (2007).

Chapter 9
Magnetic Field Structure Induced by Resonant Magnetic Perturbations in Tokamaks

In this chapter we consider essential features of the structure of magnetic field in tokamaks in the presence of the externally applied resonant magnetic perturbations (RMPs). In most tokamaks the two types of external coils are installed: a set of helical coils and a set of saddle coils. The first ones have been installed in limiter tokamaks with circular cross-sections. The aim of this is to control heat and particle transport by creating the stochastic zone of field lines at the plasma edge. This concept known as *ergodic divertor* will be discussed in Sect. 9.1.

In divertor tokamaks with the magnetic separatrix the RMPs are mainly created by sets of saddle coils. Initially the main goal of this setting has been to compensate the error magnetic fields caused by imperfection of the toroidal coils. Later they have been used to control various undesirable magneto-hydrodynamic (MHD) modes generated by plasma instabilities and to mitigate the so-called edge localized modes (ELMs) accompanying high-confinement regimes known as the H-modes in plasmas. We consider the effect of the RMPs generated by sets of saddle coils on the magnetic field structure in Sect. 9.2.

The generic models of the RMPs in the both types of tokamaks with the ergodic divertor and the magnetic separatrix will be presented in Sects. 9.1.3 and 9.3. These analytical models of RMPs are useful to analyze the generic features of the RMPs, the structure of the stochastic field lines and their statistical properties.

The fractal properties of the chaotic (open) field lines created by the RMPs will be studied in Sect. 9.4. The concepts of the chaotic scattering system will be used to analyze the fractal structures of magnetic field lines at the plasma edge and on the magnetic footprints on the divertor plates. We will also shortly discuss the experimental observations of the structures of a temperature field at the plasma edge and the heat deposition patterns on the divertor plates formed in the presence of the RMPs. They confirm that these structures predominantly follows the structure of stochastic field lines.

S. Abdullaev, *Magnetic Stochasticity in Magnetically Confined Fusion Plasmas*, Springer Series on Atomic, Optical, and Plasma Physics 78,
DOI: 10.1007/978-3-319-01890-4_9,

9.1 Ergodic Divertor Concept

The idea of the *ergodic limiter* or *ergodic divertor* concept is based on the creation of a perturbed magnetic field at the plasma edge by a special coil system installed outside the plasma vessel [Engelhardt and Feneberg (1978); Feneberg and Wolf (1981); Samain et al. (1982)]. The perturbation field creates the zone of chaotic field lines at the plasma edge which are open to the plasma wall. These field lines guide ionized exhaust particles to the special divertor plates. The schematic view of the poloidal section of ergodic divertor tokamak is shown in Fig. 9.1. In the literature on plasma physics this zone is more known as an *ergodic zone*. Below we describe some generic features of ergodic divertors installed in Tore Supra and TEXTOR tokamaks.

9.1.1 Ergodic Divertor of Tore Supra

Tore Supra is a large size tokamak with superconducting toroidal coils. Its major radius is $R_0 = 2.38$ m and the minor radius $a = 0.8$ m. The toroidal magnetic field reaches $B_t = 3.03$ T and the plasma current $I_p = 1.5$ MA. The ergodic divertor (ED) coil configuration consists of six identical modules located on outer board of the torus, i.e., on the low field side, and equidistantly spaced along the toroidal angle φ [Samain et al. (1984); Nguyen et al. (1995)] which is shown in Fig. 9.2. Each module has a poloidal extension $\Delta\theta \approx 2\pi/3$ and a toroidal extention $\Delta\varphi \approx 2\pi/14$. The current flows from the feeder to the inner side of the winding and returns through the outer side of the winding. It's scheme is shown in Fig. 9.2b. The average minor radius of the windings is $r_c = 85$ cm. The maximal ED current is $I_{ed} = 45$ kA.

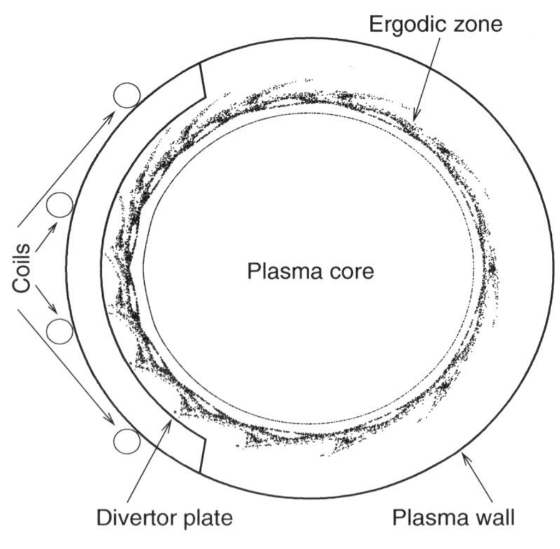

Fig. 9.1 Schematic view of the section of an ergodic divertor tokamak

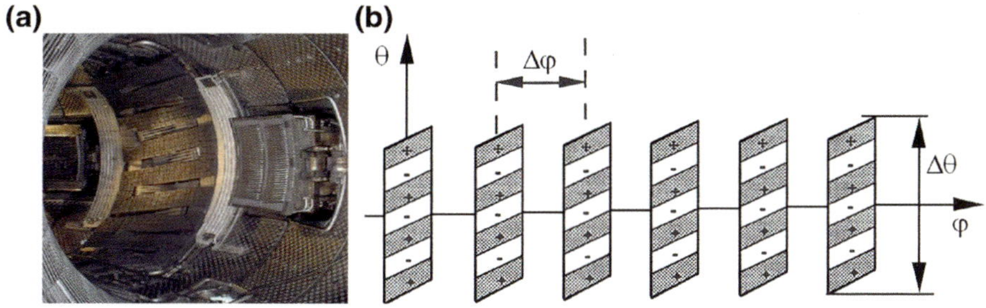

Fig. 9.2 **a** Picture of the ergodic divertor coils of the Tore Supra tokamak. **b** Scheme of coil modules (adopted from Ghendrih (1995))

Such a coil configuration creates the magnetic perturbations which can be described by Eq. (3.18) with the main toroidal mode $n = 6$ and its harmonics sn, ($s = 2, 3, \dots$). The perturbation harmonics $A_{mn}(r)$ is roughly proportional to (see, Ghendrih et al. (1996), also Appendix A.2 and Abdullaev et al. (2008))

$$A_{m,sn} \propto A_0 \left(\frac{r}{r_c}\right)^m \frac{\sin([m - sm_0]\Delta\theta/2)}{[m - sm_0]\Delta\theta/2} \frac{\sin(s[n - \alpha m_0]\Delta\varphi/2)}{s[n - \alpha m_0]\Delta\varphi/2}, \tag{9.1}$$

where A_0 is a constant, $A_0 \sim I_{ed}\Delta\theta\Delta\varphi$, the parameter $m_0 \approx 10$, and α is the slope of helical coils with respect to horizon, and $\alpha m_0 = n = 6$. From Eq. (9.1) it follows that the poloidal spectrum of perturbation in geometrical coordinates for a given toroidal mode sn is localized near the central mode $m_c = sm_0$ with the width $\Delta m = 2\pi/\Delta\theta \approx 3$. Using Eq. (3.18) one can show that the perturbation field $B_r = \partial A_\varphi/\partial\theta$ has the radial dependence $B_r \propto (r/r_c)^{sm_0-1}$, i.e., the perturbation field components corresponding to higher toroidal modes sn, $s > 1$, hardly penetrate into the plasma region.

Since the perturbation coils are located on the LHS of the torus, according to Eq. (3.29), the spectrum of magnetic perturbations $H_{mn}(\psi)$ in flux coordinates (ψ, ϑ) has the following asymptotical behavior

$$H_{m,sn} \propto A_0 \left(\frac{r}{r_c}\right)^{m/\gamma_1} \frac{\sin([m/\gamma_1 - sm_0]\Delta\theta/2)}{[m/\gamma_1 - sm_0]\Delta\theta/2}, \tag{9.2}$$

where γ_1 is the pitch of field lines on the LFS (see Eq. (3.24)). The typical values of γ_1 for the Tore Supra plasma lie approximately between 1.7 and 1.8. According results of Sect.3.2 for these values of γ_1 the perturbation spectrum of poloidal modes for the given toroidal mode $n = 6$ is localized near the central mode $m_0^* = sm_0\gamma_1 \approx s(17 \div 18)$ with the width $\Delta m^* \approx \Delta m\gamma_1 = \pi\gamma_1/\Delta\theta \approx 9 \div 10$. The contour plot of the perturbation spectrum is plotted in Fig. 9.3a for the following plasma parameters: $B_t = 3.03$ T, $I_p = 1.5$ MA.

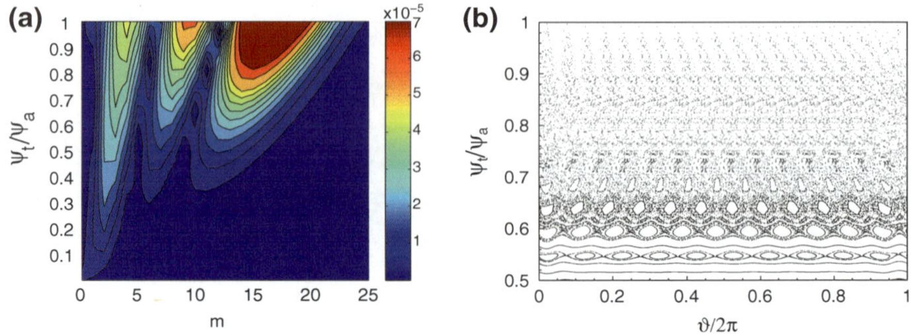

Fig. 9.3 **a** Contour plot of poloidal spectrum of perturbation field, $\epsilon|H_{mn}|$, in the (m, ψ_t)-plane for the toroidal mode $n = 6$. **b** Poincaré sections of field lines in the (ϑ, ψ_t)-plane at the perturbation current $I_p = 9\,\text{kA}$

Such a perturbation field with broad poloidal modes destroys the resonant magnetic surfaces $\psi_{mn}, q(\psi_{mn}) = m/n$ creating a chain of $m : 6$, $(m = 13 \div 20)$, islands. At the certain perturbation current I_{ed} the interaction of these magnetic islands creates a stochastic layer located near $q = 3$ surface at the plasma edge. The example of the stochastic layer is shown in Fig. 9.3b by the Poincaré section of field lines at the perturbation current $I_{ed} = 9\,\text{kA}$.

9.1.2 Dynamic Ergodic Divertor of the TEXTOR

The TEXTOR is a medium size tokamak with a limiter configuration. Its major radius $R_0 = 1.75$ m, a minor radius $a = 0.46$ cm, the toroidal field $B_t = 2.25$ T, and the plasma current $I_p = 200 - 400\,\text{kA}$.

In contrast to the ED of Tore Supra the perturbation coils in the dynamic ergodic divertor (DED) are installed inboard side of the TEXTOR vessel as shown in Fig. 9.4a, b. It consists of a quadruple set of 16 helical coils at the minor radius $r_c = 53.25$ cm aligned parallel to the magnetic field at the nearby $q = 3$ surface. The coils are bundled such that the outlets are at 4 toroidal locations, 4 on top of TEXTOR and 4 at the bottom. This grouping of the coils is technically favorable, but it requires the installation of a pair of compensation coils which are indicated in a green color (C). The coils, numbered by $j = 1, \ldots, 16$, cover a poloidal section $\Delta\theta = 2\theta_c \approx 70°$ (including compensation coils $\Delta\theta \approx 80°$) (see Fig. 9.4b).

The TEXTOR-DED has the two new features in comparison with the ED of Tore Supra. Firstly, the DED can operate in three different operational modes with the main toroidal mode $n = 1$, $n = 2$, and $n = 4$, called the $m : n = 3 : 1, 6 : 2$, and $12 : 4$ modes. This can be done by changing the distribution of currents in the DED-coils. Secondly, as was mentioned above, it also permits operation with a rotation (time-varying) perturbation magnetic field.

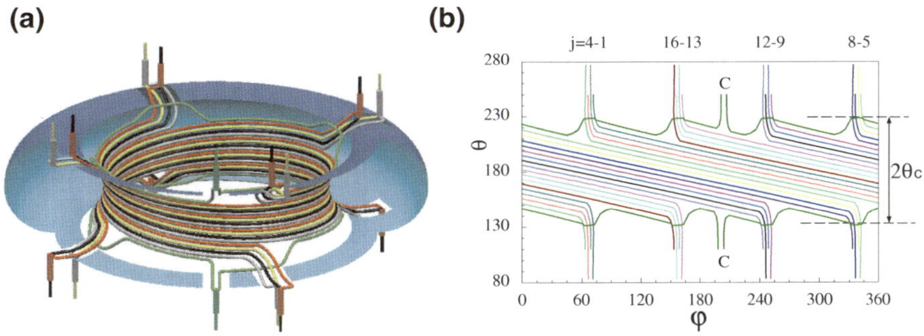

Fig. 9.4 **a** Sketch of the DED coils for the $n = 4$ mode configuration. **b** Coil configuration in the (φ, θ) plane

The perturbation harmonics $A_{mn}(r)$, created by the DED coils, can be roughly presented by (see, e.g., a report by Finken et al. (2005a) and Sect. A.1)

$$A_{m,n} \propto A_0(-1)^m \left(\frac{r}{r_c}\right)^m \frac{\sin([m + nm_0/4]\theta_c)}{[m + nm_0/4]\theta_c}, \tag{9.3}$$

where A_0 is a constant $\propto I_d\theta_c$, the parameter $m_0 \approx 20$. The poloidal spectrum of perturbation in geometrical coordinates for a given toroidal mode n is localized near the central mode $m_c = nm_0/4$ with the width $\Delta m = \pi/\theta_c \approx 10$. The perturbation field B_r has the radial dependence $B_r \propto (r/r_c)^{nm_0/4-1}$.

Since the perturbation coils are located on the high field side (HFS) of the torus, according to Eq. (3.29), the spectrum of magnetic perturbations $H_{mn}(\psi)$ in flux coordinates (ψ, ϑ) has the following asymptotical behavior

$$H_{m,n} \propto (-1)^m A_0 \left(\frac{r}{r_c}\right)^{nm/4\beta_1} \frac{\sin([m/\beta_1 + nm_0/4]\theta_c)}{[m/\beta_1 + nm_0/4]\theta_c}, \tag{9.4}$$

where β_1 is the pitch of field lines on the HFS with the typical values of $\beta_1 \approx 0.5 \div 0.6$ [see Eq. (3.24)]. The perturbation spectrum for the toroidal mode n is localized near the central mode $m_0^* = nm_0\beta_1/4 \approx n(11 \div 13)/4$ with the width $\Delta m^* \approx \Delta m\beta_1 \approx n(3 \div 4)/4$. Figure 9.5a shows the contour plot of the perturbation spectrum for the plasma parameters: $B_t = 1.9\,\text{T}$, $I_p = 382\,\text{kA}$.

Unlike the ED of Tore Supra the stochastic layer (or ergodic zone) in the TEXTOR-DED is formed at the plasma edge due to interaction of only a few magnetic islands. Figure 9.5b shows the Poincaré section of field lines in the case of the $n = 4$ operational mode. The ergodic zone in this case is created by the $m : 4$ magnetic islands $m = 11, 12, 13$.

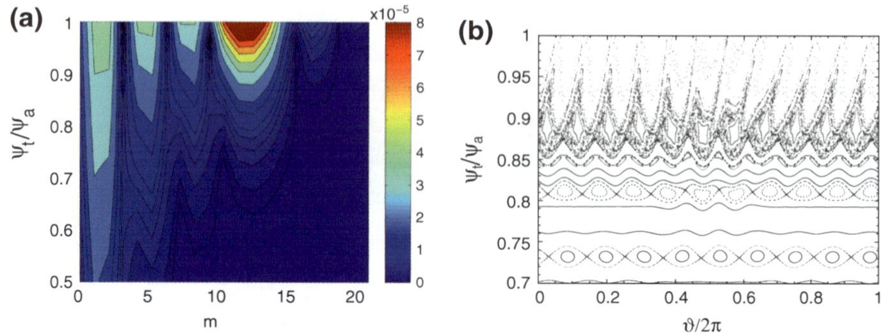

Fig. 9.5 **a** Contour plot of poloidal spectrum of perturbation field, $\epsilon|H_{mn}|$, in the (m, ψ_t)-plane in the TEXTOR–DED for the toroidal mode $n = 4$. **b** Poincaré sections of field lines in the (ϑ, ψ_t)-plane for the $n = 4$ operational mode. The parameters: $B_t = 1.9$ T, $I_p = 382$ kA, the DED current $I_d = 4.5$ kA

9.1.3 Model of Magnetic Perturbations in Ergodic Divertors

Considered above the ergodic divertors in Tore Supra and the TEXTOR tokamaks allows to propose an analytical model for a magnetic perturbation created by the set of helical coils. Such a model would be useful to study a generic features of magnetic field lines at the plasma edge, particularly, their fractal and statistical properties, transport of field lines and particles in the stochastic magnetic fields.

Based on the general features of magnetic perturbations created by a helical coils, studied in Sect.3.2, and the spectra of magnetic perturbations for the ED Tore Supra (9.2) and the TEXTOR-DED (9.4), one can write the following general formula for the poloidal spectrum H_{mn},

$$H_{mn}(\psi_t) = C_m \left(\frac{r}{r_c}\right)^{m/\gamma} \frac{\sin\left([m/\gamma - m_0]\theta_c\right)}{m[m/\gamma - m_0]\theta_c},$$
$$r = r(\psi_t) = R_0\sqrt{2\psi_t}, \tag{9.5}$$

where r_c, γ, θ_c, m_0 are parameters of the model. The physical meaning of these parameters are the similar to those ones in the previous sections. The parameter γ corresponds to γ_1 or β_1 defined by Eqs. (3.24) and (3.26), respectively. The quantity θ_c is a half of the poloidal extension of coils $\Delta\theta$. One can assume that γ and θ_c do not depend on the toroidal flux ψ. The parameter $C_m = 1$ if the perturbation is localized on the LFS as in the TORE SUPRA-ED. In this case $\gamma > 1$. If the perturbation is localized on the HFS like in the TEXTOR-DED one has $C_m = (-1)^m$ and $\gamma < 1$ and the poloidal spectrum $H_{mn}(\psi)$ is localized near the central mode $m_0^* = m_0/\gamma$. By the appropriate choice of γ one can localize the central resonant magnetic surface ψ_{mn}, $q(\psi_{mn}) \approx m_0^*/n = m_0/n\gamma$. The width of the spectrum Δm is determined by θ_c, i.e., $\Delta m \approx \pi\gamma/\theta_c$. One can model the safety factor profile as

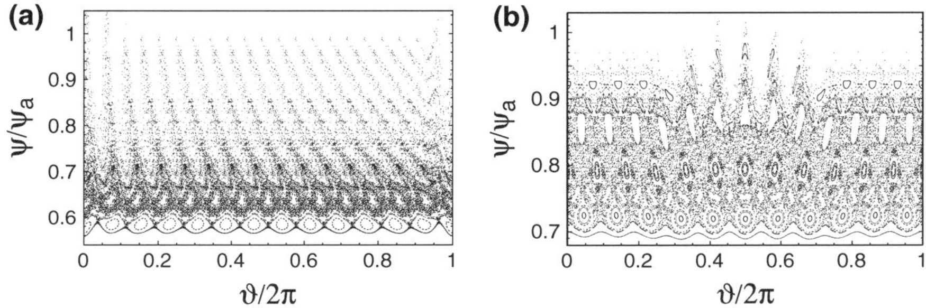

Fig. 9.6 Poincaré sections of field lines in the (ϑ, ψ)-plane: **a** for the model of Tore Supra ED with the toroidal mode $n = 6$, perturbation parameter $\epsilon = 0.002$, $R_0 = 2.38$ m, $a = 0.8$ m, $r_0 = 0.85$ m, $\gamma_1 = 1.81$, $\theta_c = \pi/7$, $m_0 = 10$, $q_0 = 0.9$, $q_a = 3.6$; **b** for the model of the TEXTOR-DED with $n = 4$, $R_0 = 1.75$ m, $a = 0.47$ m, $r_c = 0.5325$ m, $\gamma = 0.6$, $\theta_c = \pi/7$, $m_0 = 20$, $q_0 = 0.9$, $q_a = 3.5$, and $\epsilon = 0.008$

$$q(\psi_t) = q_a \frac{r^2/a^2}{1-(1-r^2/a^2)^\mu} = q_a \frac{x}{1-(1-x)^\mu}, \qquad x = \frac{\psi_t}{\psi_a}, \tag{9.6}$$

where q_a and $\psi_a = a^2/2R_0^2$ are a safety factor and the toroidal flux at the plasma boundary $r = a$, respectively, μ is a parameter, related to the value of $q(\psi)$ at the magnetic axis $r = 0$, i.e., $q(0) = q_a/\mu$.

Figure 9.6a, b show the Poincaré sections of magnetic field lines at the plasma edge for the ED models like to the Tore Supra ED and the TEXTOR-DED, respectively.

9.2 RMPs in Poloidal Divertor Tokamaks

The typical structure of the stochastic zone of field lines created by the external magnetic perturbations near the separatrix is shown in Fig. 9.7: (a) the entire cross section; (b) the expanded view of the area near the X-point. The red dots correspond to the field lines run in the positive direction of the toroidal angle φ, and the green dots correspond to the field lines in the opposite direction. As seen from Fig. 9.7b the structure of field lines near the X-point is determined by the stable and unstable manifolds as was illustrated in Fig.7.6.

9.2.1 RMP Coils in the DIII-D Like Plasmas

Below we discuss some features of poloidal spectrum and stochastic magnetic field lines near the separatrix in the DIII-D like plasma with the I-coils. Particularly, we will use the asymptotical formula found in Sect. 3.4 for the poloidal spectrum.

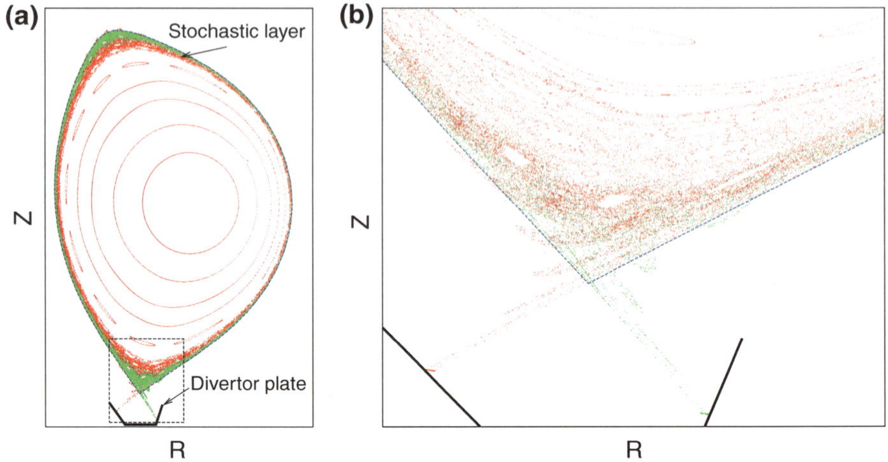

Fig. 9.7 **a** Typical Poincaré section of field lines near the separatrix; **b** Expanded view of of the stochastic layer near the X-point. Red dots correspond to field lines along positive direction of the toroidal angle φ and green dots correspond to the ones along the opposite direction of φ

We simulate the DIII-D like plasma with the analytical equilibrium (2.33). The plasma parameters and the coefficients c_i in the solution (2.33) are given in Table 9.1.

The internal (I-) coils consist of the two sets of six saddle coils located above (upper) and below (down) the equatorial plane, as shown in Fig. 9.8. The coordinates of coils are chosen as follows: $R_1 = R_4 = 2.207\,\mathrm{m}$, $Z_1 = -Z_4 = 1.0448\,\mathrm{m}$, $R_2 = R_3 = 2.42\,\mathrm{m}$, $Z_2 = -Z_3 = 0.49\,\mathrm{m}$. The primary toroidal mode number n is equal 3. The safety factor profile and the relation between the geometrical poloidal angle θ and intrinsic poloidal angle ϑ are plotted in Fig. 9.9.

Table 9.1 Parameters of the DIII-D like plasmas.

Parameters		Coefficients c_i
Plasma current I_p	1.5 MA	$c_1 = 7.38313049 \times 10^{-2}$
Toroidal field B_t	1.85 T	$c_2 = -7.63418022 \times 10^{-2}$
Major radius R_0	1.69 m	$c_3 = -1.48401237 \times 10^{-1}$
Minor radius a	0.6 m	$c_4 = -4.03431412 \times 10^{-2}$
Plasma β_{pol}	1.2	$c_5 = 1.10785700 \times 10^{-1}$
Elongation κ	1.8	$c_6 = -1.33643576 \times 10^{-1}$
Triangularity δ	0.36	$c_7 = -6.19373915 \times 10^{-3}$
X-point (R_s, Z_s)	$(1.27, -1.065)$ m	$c_8 = 4.02569166 \times 10^{-2}$
The value of q_{95}	3.714	$c_9 = 2.83700281 \times 10^{-1}$
The value of Ψ_0	5.3024	$c_{10} = -1.31678345 \times 10^{-1}$
The value of $\Psi_{p0} = \Psi_0 \psi_a$	−0.2322	$c_{11} = -2.39136437 \times 10^{-2}$
A	−0.072	$c_{12} = 5.97966667 \times 10^{-3}$

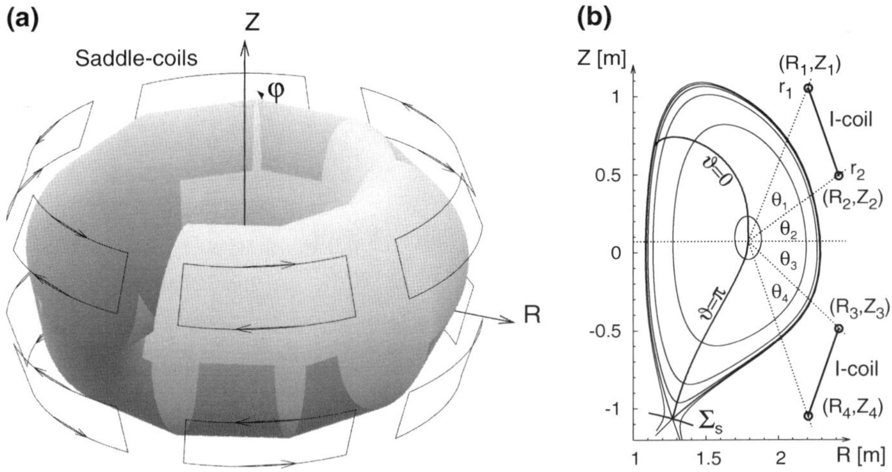

Fig. 9.8 Plasma configuration and saddle coils for the DIII-D-like plasma with the internal (I-)coils: **a** 3D-view in the cylindrical coordinate system (R, Z, φ); **b** in the (R, Z) plane

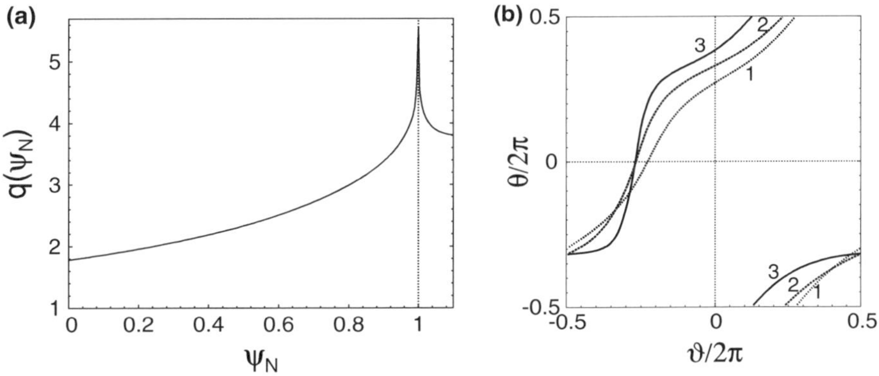

Fig. 9.9 **a** Safety factor profile. The value of q_{95} is 3.714. **b** Relation between the geometrical poloidal angle θ and the intrinsic poloidal angle ϑ for the plasma configuration shown in Fig. 9.8 at the three magnetic surfaces: $\psi_N = 0.1$ (*curve 1*), $\psi_N = 0.5$ (*curve 2*), and $\psi_N = 0.95$ (*curve 3*). The angles θ and ϑ are calibrated to have the values $\theta = 0$ at the outer equatorial plane and $\vartheta_s = -\pi$ at the section Σ_s, respectively

The perturbation magnetic field of saddle coils is given by the perturbation poloidal fluxes $\psi_n(R, Z)$ (3.44) corresponding to the horizontal segments $j = 1, 2, 3, 4$ of saddle coils and by the perturbation function $G(R, Z)$ (3.51) corresponding to the vertical segments of saddle coils. The dependencies of the functions $\psi_n(R, Z)$ and $G_n(R, Z)$ on the geometrical poloidal angle θ and the intrinsic poloidal angle ϑ at the magnetic surface $\psi_N = 0.95$ are plotted Figs. 9.10a, b and 9.11a, b for the primary toroidal mode $n = 3$ and the perturbation current $I_c = 1\,\text{kA}$. They are described by

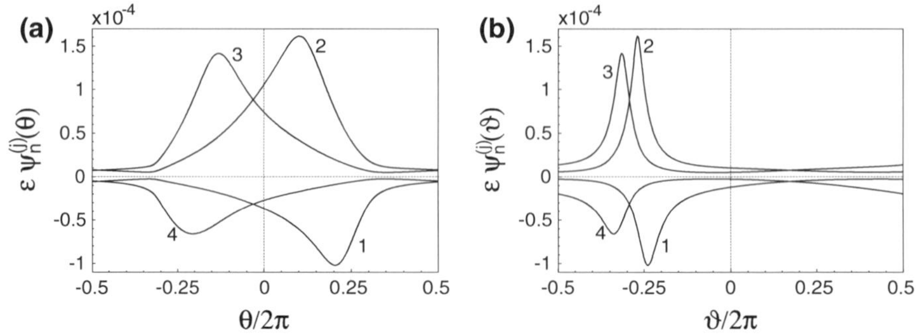

Fig. 9.10 Angular dependencies of the perturbation poloidal flux $\psi_n(R, Z)$ at the magnetic surface $\psi_N = 0.95$ created by the horizontal segments of saddle coils shown in Figs. 9.8: **a** versus the geometrical angle θ, **b** versus the intrinsic poloidal angle ϑ. The current of the coils is $I_c = 1\,\text{kA}$, the toroidal mode $n = 3$. The numbering of curves corresponding to the one in Fig. 9.8b

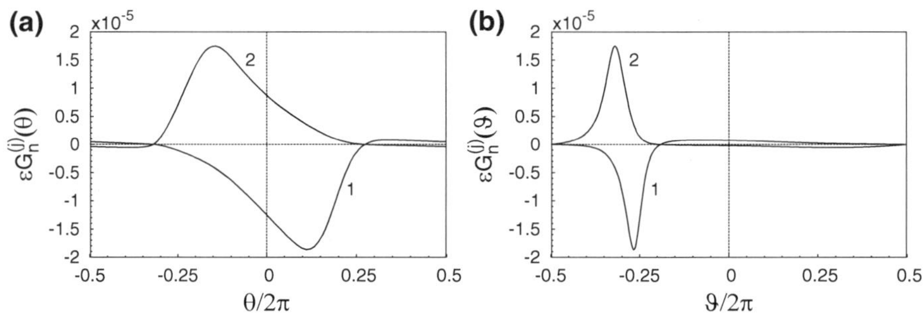

Fig. 9.11 The same as in Fig. 9.10 but for the perturbation function $G_n(R, Z)$ corresponding to the vertical segments of saddle coils: **a** versus the geometrical angle θ, **b** versus the intrinsic poloidal angle ϑ. *Curve 1* corresponds to the upper set of saddle coils, *curve 2*—to the lower set of saddle coils

the pulse-like functions of ϑ. The width of pulses decreases with approaching the separatrix $\psi_N \to 1$ Fig. 9.11.

9.2.2 Poloidal Spectra of the RMPs

As was shown in Sect.3.4 the poloidal spectra of magnetic perturbations created by a set of saddle coils has universal asymptotical behavior described by the asymptotical formulae (3.54) and (3.58). The poloidal spectra corresponding to each horizontal segments ($j = 1, 2, 3, 4$) are given by

$$H_{mn}^{(j)}(\psi) = \frac{A_j(\psi)}{2q(\psi)} \exp\left(-\frac{mC_j(\psi)}{q(\psi)} - im\vartheta_j(\psi) + \chi^{(j)}\right), \tag{9.7}$$

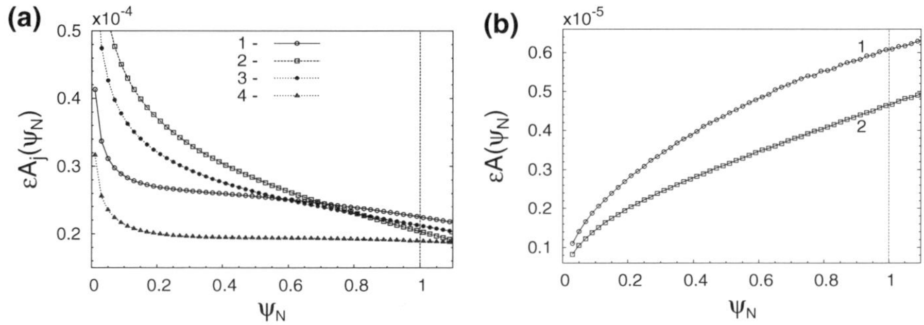

Fig. 9.12 Radial profiles of the amplitudes $\epsilon A_j(\psi_N)$ in the asymptotical formulae (9.7) and (9.8) corresponding to the horizontal segments (**a**) and the vertical segments (**b**) of saddle coils, respectively. In (**a**) the numbering of curves corresponding to the one in Fig. 9.8; **b** *curve 1* corresponds to the upper saddle coils, *curve 2*—to the lower one

where $\chi^{(1)} = \chi^{(3)} = 0, \chi^{(2)} = \chi^{(4)} = \pi$. The coefficients $A_j(\psi), C_j(\psi)$ are smooth functions of flux variable ψ with finite values at the separatrix $\psi_N = 1$. We recall that the angles ϑ_j and Θ_j are related by Eq. (3.63). The radial profiles of these functions in the interval normalized poloidal flux $0 < \psi_N < 1$ are shown in Figs. 9.12a, 9.13a and 9.14a.

The poloidal spectra corresponding to the vertical segments of saddle coils are described by the asymptotical formulae

$$H_{mn}^{(s)}(\psi) = \frac{nA_s(\psi)}{2m} \exp\left(-\frac{mC_s(\psi)}{q(\psi)} - im\vartheta_s(\psi) + \frac{\pi}{2}\right), \tag{9.8}$$

corresponding to the upper ($s = 1$) and lower ($s = 2$) sets of saddle coils. The functions $A_s(\psi)$, $C_s(\psi)$, and $\vartheta_s(\psi)$ are similar to those in (9.7). Their radial profiles are plotted in Figs. 9.12b, 9.13b,and 9.14b, respectively.

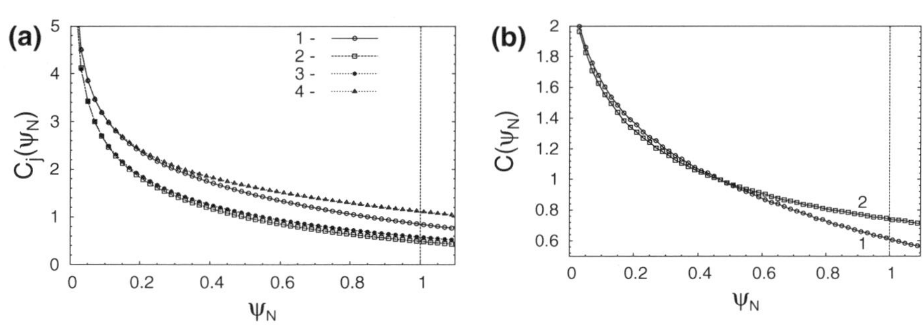

Fig. 9.13 The same as in Fig. 9.12 but for the functions $C_j(\psi_N)$ in the asymptotical formulae (9.7) and (9.8) corresponding to the horizontal segments (**a**) and the vertical segments (**b**), respectively

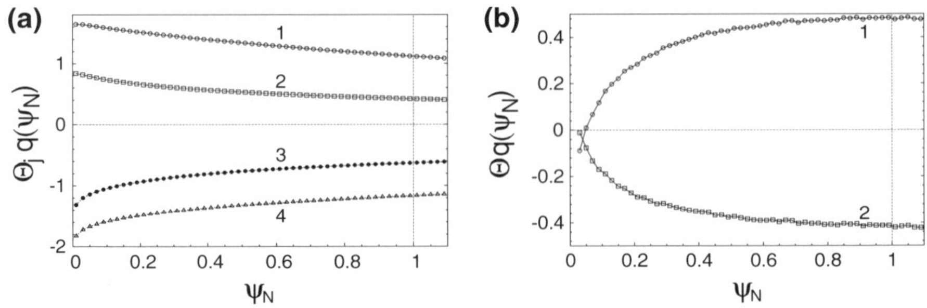

Fig. 9.14 The same as in Fig. 9.12 but for the angles $\Theta_j(\psi_N)$ defined (3.63): **a** correspond to the horizontal segments, **b** corresponds to the vertical segments

As seen from Fig. 9.12a, b the amplitudes A_j corresponding horizontal segments of sets of saddle coils are approximately one order of magnitude larger than the amplitudes of A_s, $(s = 1, 2)$ corresponding to the vertical segments. Because of up/down asymmetry of the equilibrium plasma with a single-null configuration the mode amplitudes A_j and the exponents C_j corresponding to the horizontal segments located symmetrically with respect to the equatorial plane $Z = 0$ are not equal. The amplitude A_j corresponding to the horizontal segments $j = 2$ and $j = 3$ located close to the equatorial plane $Z = 0$ take the largest values with close magnitudes. The corresponding exponents C_j, $(j = 2, 3)$, take also the smallest values with close magnitudes (see Fig. 9.13a). The amplitude A_4 and the exponent C_4 corresponding the horizontal segment $j = 4$ located the most close to the X-point has the smallest value A_4 and the largest C_4.

9.2.3 Even- and Odd-Parity Configurations

There are two possible setup of current directions in the upper and lower sets of coils called the up/down parity [Evans et al. (2005)]. The latter is referred to as "odd" ("even") parity when the upper and lower set of coils have opposite (same) polarity at each toroidal angle φ. The poloidal mode spectrum $H_{mn}^{(e,o)}$ of the even (e) and odd (o) parities can be presented as a linear combination of $H_{mn}^{(\pm)}$ corresponding to the set of coils located upper (+) and lower (−) equatorial plane, respectively,

$$H_{mn}^{(e,o)} = H_{mn}^{(+)} \pm H_{mn}^{(-)}. \tag{9.9}$$

Below we give a qualitative analysis of behavior of these coils configurations using the asymptotical formulae (9.7) and (9.8) for the each segments of sets of saddle coils. For simplicity we neglect the contributions from the vertical segments of saddle coils since their amplitudes, A_K, are much smaller than ones from horizontal ones (see Fig. 9.12a, b). Further, assume that $A_2 \approx A_3$, $C_2 \approx C_3$, for the odd-parity configuration and $A_2 \approx -A_3$ for even-parity configuration, and neglect the terms corresponding to the segments $j = 1$ and $j = 4$ which give a several times smaller contributions. Then, we have

$$H_{mn}^{(e,o)} \approx \frac{A_2}{2q} e^{-mC_2/q} \left(e^{-im\vartheta_2} \mp e^{-im\vartheta_3} \right)$$
$$= \frac{2A_2}{q} e^{-mC_2/q - im\bar{\vartheta}} \begin{cases} i \sin(m\Delta\vartheta), & \text{even parity,} \\ \cos(m\Delta\vartheta), & \text{odd parity,} \end{cases} \quad (9.10)$$

where $\bar{\vartheta} = (\vartheta_2 + \vartheta_3)/2$, $\Delta\vartheta = (\vartheta_3 - \vartheta_2)/2$.

Consider the primary toroidal mode $n = 3$ which gives the main contribution of the RMP field. Figure 9.15a shows the dependence of $m|H_{mn}|$ on the poloidal mode number m at the magnetic surface $\psi_N = 0.887$ calculated using the formulae (9.7): curve 1 corresponds to the odd-parity, curve 2—to the even-parity configurations, curves 3 and 4 correspond to the upper and lower sets of saddle coils. The quantity $m|H_{mn}|$ is proportional to the poloidal spectrum B_{mn} of the perturbation magnetic field strength. As seen from Fig. 9.15a for the even parity configuration $m|H_{mn}|$ has local maximum at $m = m_0 \approx 10$, while for the odd parity configuration it has the notch near this mode number. Such a behavior of poloidal spectrum of the RMPs is found by numerical calculations in the DIII-D tokamak [Evans et al. (2005)].

The analytical formula (9.10) explains this feature of the spectrum, and allows to express the mode m_0 through the angles ϑ_j. According to Eq. (9.10) one has

$$|H_{mn}^{(e)}| \propto \left| \sin\left(m \frac{\alpha_3(\psi) - \alpha_2(\psi)}{2q(\psi)} \right) \right|,$$

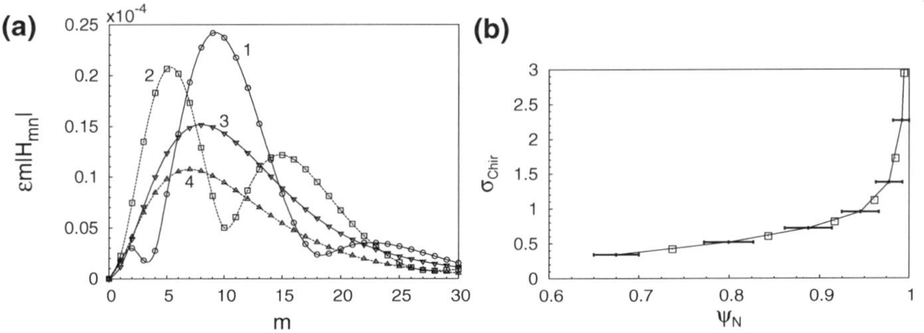

Fig. 9.15 **a** Poloidal spectrum of RMPs $m|H_{mn}|(\psi)$ at the resonant magnetic surface $q(\psi_{mn}) = 10/3$, ($\psi_{mn} = 0.887$) for the DIII-D like plasma: *curve 1* corresponds the odd parity coil configuration, *curve 2* corresponds the even parity coil configuration, *curves 3* and *4* correspond to the upper and the lower sets of saddle coils, respectively. The current amplitude $I_0 = 1$ kA. **b** The Chirikov parameter σ_{Chir} (□) and the magnetic island widths W_{mn} (*horizontal* segments) corresponding to the even parity coil configuration. The current amplitude $I_0 = 4.4$ kA

for the even parity configuration with the maximum at $m \approx m_0 = \pi/2\Delta\vartheta$, $[\alpha_j(\psi) = \Theta_j(\psi)q(\psi)$, $j = 2, 3$.] For the odd parity one has

$$|H_{mn}^{(o)}| \propto \left|\cos\left(m\frac{\alpha_3(\psi) - \alpha_2(\psi)}{2q(\psi)}\right)\right|,$$

with the minimum at $m = m_0$. Using the numerical values of α_2 and α_3 at $\psi_N = 0.887$ we have $m_0 = 9.875$ which is close to the numerically found values 10 and 11, respectively (see Fig. 9.15a).

9.2.4 Structure of Stochastic Layer and Magnetic Footprints

The RMPs create the magnetic islands on the resonant magnetic surfaces. The width, $\Delta\psi_{mn}$, of the magnetic island at the given resonant magnetic surface ψ_{mn}, $q(\psi_{mn}) = m/n$, is determined by the value of $H_{mn}(\psi)$, at ψ_{mn}, and given by Eq. (8.10) in the toroidal flux ψ_t and (8.12) in the poloidal flux ψ. The latter using the relation $R_n^{(reg)}(\psi) = \pi q(\psi) H_{mn}^*(\psi)$ between the spectrum $H_{mn}(\psi)$ and the regular part of the Poincaré integral $R_n^{(reg)}(\psi)$ can be reduced to

$$W_{mn} = \Delta\psi_{mn}\frac{d\psi}{d\psi_t} = F_n(\psi_{mn}), \tag{9.11}$$

where

$$F_n(\psi) = 4\left|\frac{\epsilon R_n^{(reg)}(\psi)}{\pi dq/d\psi}\right|^{1/2}. \tag{9.12}$$

The function $F_n(\psi)$ expressed via the Poincaré function $R_n^{(reg)}(\psi)$ does not explicitly contain the poloidal mode number m. As seen from Eq. (9.11) the values of the continuous function $F_n(\psi)$ at the resonant surfaces ψ_{mn} give the width W_{mn} of magnetic islands.

Using the asymptotical formulas (9.7) for $H_{mn}(\psi)$, one has

$$R_n^{(reg)}(\psi) = \frac{\pi}{2}\sum_{j=1}^{4}\delta_j A_j e^{-nC_j + in\alpha_j}, \tag{9.13}$$

where unit coefficients δ_j are defined by

$$\delta_j = \begin{cases} (+1, -1, +1, -1) & \text{for the even parity,} \\ (+1, -1, -1, +1) & \text{for the odd parity,} \end{cases}$$

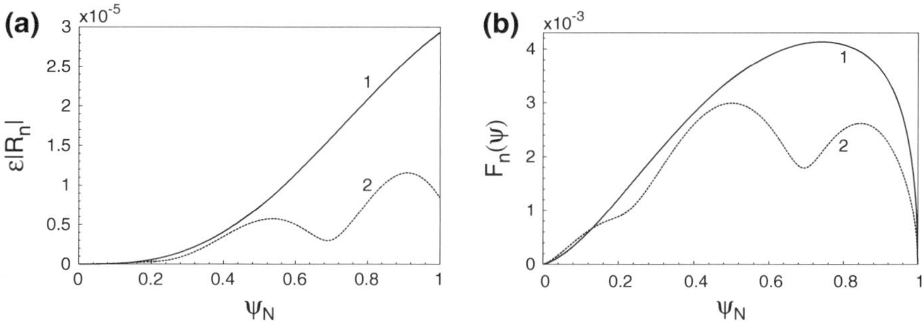

Fig. 9.16 Radial profiles of the Poincaré integrals $|R_n(\psi)|$ (9.13) and the function $F_n(\psi)$ (9.12) for the even (*curve 1*) and the odd (*curve 2*) parity configurations. The current amplitude $I_0 = 1.0\,\mathrm{kA}$

for the even and the odd parity configurations, respectively. Here $\alpha_j(\psi) = \vartheta_j q(\psi)$.

Figure 9.15b shows the width of magnetic islands W_{mn} (horizontal segments) as a function of ψ_{mn} and the Chirikov parameter (8.30), $\sigma_{Chir} = 0.5(W_{m+1,n} + W_{m,n})/|\psi_{m+1,n} - \psi_{m,n}|$ (the symbols □) as a function of $(\psi_{m+1,n} + \psi_{m,n})/2$ for the even parity configuration of coils. The width W_{mn} is normalized to the constant $\Psi_{pa}/B_0 R_0^2$ (see Eq. (2.36)). The current of the RMPs coils is chosen $I_0 = 4.4\,\mathrm{kA}$. The radial profiles of the Poincar integrals $|R_n(\psi)|$ (9.13) and the function $F_n(\psi)$ (9.12) are shown in Fig. 9.16a, b.

The analytical formulas for the perturbation fluxes $\psi_n(\psi, \vartheta)$ (3.12) are given by Eqs. (3.67) and (3.69). In our case according to (3.43) the perturbation poloidal flux created by the horizontal segments of the saddle coils can be presented as a sum of the perturbation fluxes corresponding to the each segments j, ($j = 1, 2, 3, 4$),

$$\psi^{(pert)}(\psi, \vartheta, \varphi) = \epsilon \sum_{j=1}^{4} \delta_j \psi_1^{(j)}(\psi, \vartheta, \varphi), \tag{9.14}$$

where

$$\psi_1^{(j)}(\psi, \vartheta, \varphi) = \sum_{n=N}^{\infty} \frac{A_j}{2q(\psi)\Lambda_j} \left(1 - e^{-2C_j/q(\psi)}\right) \sin(n\varphi). \tag{9.15}$$

where the function $\Lambda_j \equiv \Lambda_j(\psi, \vartheta)$ is defined in Eq. (3.66). The perturbation parameter ϵ is defined by Eq. (3.46).

These formulas are used for the integration of Hamiltonian field line equations (3.9) via the symplectic mappings (6.29). For the convenience we introduce the Hamiltonian function h by normalizing the poloidal flux ψ (3.10) to $\Psi_{pa}/B_0 R_0^2$ and adding a unit constant 1. Then the normalized Hamiltonian function $h = \psi B_0 R_0^2/\Psi_{pa} + 1$ coincides with ψ_N and it can be written as

$$h = h_0(J) + \epsilon_{eff} h_1(\vartheta, J, \varphi), \tag{9.16}$$

where the unperturbed Hamiltonian $h_0(J)$ can be written through the safety factor $q(\psi_t)$,

$$h_0(J) = \int_0^J \frac{dJ}{q(h(J))}. \tag{9.17}$$

The perturbation Hamiltonian $h_1\,(\vartheta, J, \varphi)$, and the effective perturbation parameter ϵ_{eff} are given by

$$h_1\,(\vartheta, J, \varphi) = \sum_{j=1}^{4} \delta_j \psi_1^{(j)}(\psi, \vartheta, \varphi),$$
$$\epsilon_{eff} = \left(\frac{B_0 R_0^2}{\Psi_{pa}}\right)\epsilon. \tag{9.18}$$

Here J is the normalized toroidal flux (or the action variable conjugated to the poloidal angle ϑ). The relation between J and h is given by

$$J = \int_0^h q(h)dh. \tag{9.19}$$

The unperturbed Hamiltonian h_0 takes the value $h_0 = 0$ at the magnetic axis and $h_0 = 1$ at the separatrix.

One should recall that the poloidal flux $\psi = \Psi/B_0 R_0^2$, where Ψ is the non-normalized poloidal flux satisfying the Grad-Shafranov equation (2.3). The normalized poloidal flux ψ_N is defined by Eq. (2.36), where Ψ_{pa} is the value of the poloidal flux at the magnetic axis (see Table (9.1)).

The examples of the Poincaré sections of field lines at the plasma edge in the (ϑ, ψ_N)-plane are shown in Fig. 9.17 for the even-parity (a) and for the odd-parity (b) coil configurations with the current amplitude $I_0 = 4.4$ kA. In (9.15) we have retained only the term $n = N$ corresponding the primary toroidal mode $n = N = 3$. The plasma parameters and the coil configuration are the same as in Fig. 9.15b.

The Poincaré sections in the (φ, ψ_N)-plane can be obtained by the the separatrix mapping (6.58). To apply this mapping one should notice that the phase χ_n in the presentation (6.75) is equal to $\chi_n = -\pi/2$ (compare Eqs. (6.75) and (9.15)). The Poincaré integral $P(\varphi; \psi)$ (6.76) is reduced to

$$P(\varphi; \psi) = \sum_{n=N}^{\infty} R_n(\psi) e^{in\varphi - i\pi/2}$$
$$= \sum_{n=N}^{\infty} \left[K_n(\psi)\sin(n\varphi) + L_n(\psi)\cos(n\varphi)\right], \tag{9.20}$$

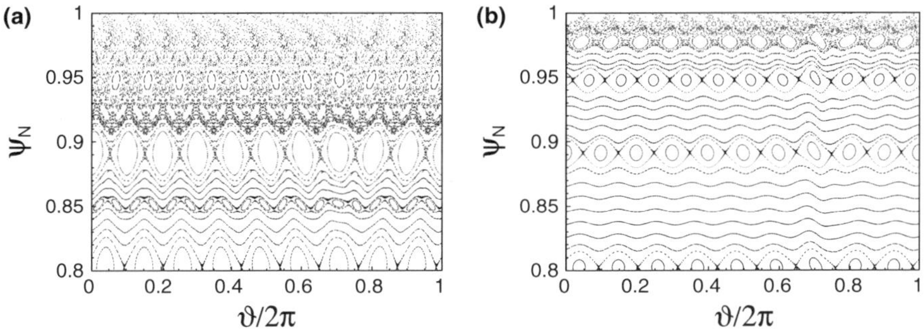

Fig. 9.17 Poincaré sections of field lines in the (ϑ, ψ_N) plane: for the the even parity (**a**) and the odd parity (**b**) coil configurations in the DIII-D like plasma. The perturbation current $I_0 = 4.4\,\mathrm{kA}$

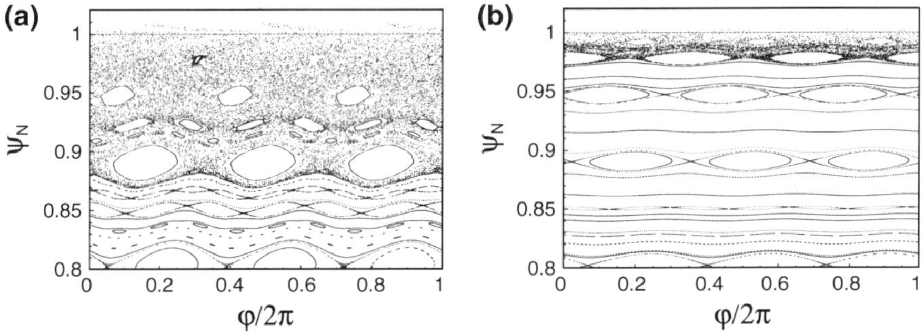

Fig. 9.18 The same as in Fig. 9.17 but in the (φ, ψ_N) plane

Furthermore, we neglect the oscillatory part $R_n^{(osc)}(\psi)$ of the integrals $R_n(\psi) = R_n^{(reg)}(\psi) + R_n^{(osc)}(\psi)$, retaining only the regular part $R_n^{(reg)}(\psi)$ given by (9.13).

The Poincaré sections in the (φ, ψ_N)-plane obtained by the separatrix mapping (6.58) are shown in Fig. 9.18 for the same coil configurations as in Fig. 9.17.

Using the separatrix mapping (6.58) one can also easily calculate the magnetic footprints on the divertor plates (see Sect.6.8.3 and Fig.6.9). According to (6.123) the width of the magnetic footprints are determined by the Poincaré integrals $R_n^{(e,o)}(0)$. In the normalized poloidal flux they are determined by

$$\Delta\psi_d^{(e,o)} = \epsilon_{eff} n |R_n^{(e,o)}(0)|. \tag{9.21}$$

The magnetic footprints corresponding to the even-parity and the odd-parity configurations are given in Fig. 9.19a, b, respectively, for the same values of parameters as in Figs. 9.17 and 9.18. The normalized distance $\xi_d = z_d/R_0$ of the divertor plate from the X-point is chosen $\xi_d = 0.1$. As seen from Fig. 9.19 the magnetic footprint consists of several helical stripes. The number of the latter is equal to the toroidal mode number $n = 3$ of the perturbation magnetic field.

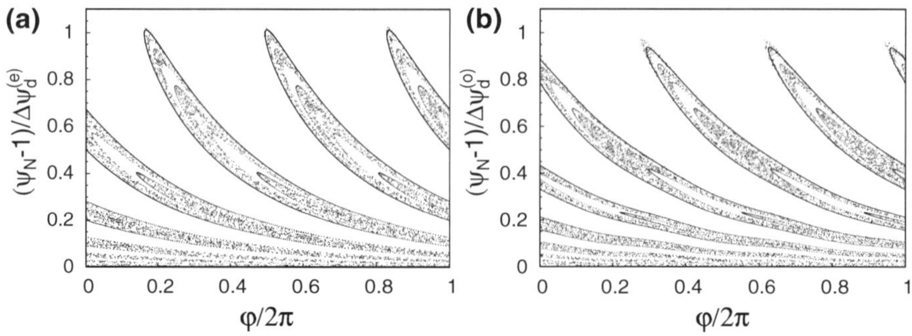

Fig. 9.19 Magnetic footprints in the (φ, ψ_N) plane for the the even parity (**a**) and the odd parity (**b**) coil configurations in the DIII-D like plasma. The parameters are the same as in Fig. 9.17a, b. The widths of the magnetic footprint $\Delta\psi_d^{(e)} = 8.6 \times 10^{-3}$ and $\Delta\psi_d^{(e)} = 2.69 \times 10^{-3}$, for the even and the odd parity configurations, respectively

One should note that although the Poincaré section of field lines give a quite clear structure of magnetic field lines in magnetic islands and the inner boundary of the stochastic layer, it cannot resolve the fine structure and properties of open field lines at the very plasma edge where they leave the plasma region. These open field lines have a fine fractal structure near the separatrix and on the divertor plates. We will study such field lines in Sect. 9.4.

9.3 Generic Models of the RMP Field in Poloidal Divertor Tokamaks

The most studies of the RMP fields are based on the numerical field line tracing codes for the realistic magnetic configurations of the equilibrium plasma and the external perturbation coils. Although these approaches allow one to consider specific fusion devices, like DIII-D, JET or ITER, it would be useful to have an analytical model of magnetic field near the magnetic separatrix of poloidal divertor tokamaks. Such a model would allow fast and rigorously to study the structure and statistical properties of stochastic field lines at the plasma edge in the presence of RMPs. This model should be simple as possible, and at the same time it should capture essential generic features of the magnetic field of the equilibrium plasma and RMPs near the magnetic separatrix.

As we have shown in Sect.3.4 the poloidal spectra of the RMPs, $H_{mn}(\psi)$, created by the saddle coils have an universal behavior and can be presented by the asymptotical formulas (3.54) and (3.58). These formulas can serve as a base for a generic model of the RMPs. Such models have been proposed in Abdullaev (2009, 2010).

Let n be the predominant toroidal mode number of the RMPs. We suppose that the variable ψ stands for the poloidal flux ψ and takes value $\psi = 0$ at the magnetic axis and $\psi = 1$ at the separatrix. The perturbation poloidal flux corresponding to this field, ψ_1, can be presented as

$$\epsilon\psi_1(\psi, \vartheta, \varphi) = \epsilon\psi_n(\vartheta, \psi)\sin(n\varphi), \tag{9.22}$$

where

$$\psi_n(\vartheta, \psi) = \sum_{\substack{m=-\infty \\ m\neq 0}}^{\infty} H_{mn}(\psi)e^{im\vartheta} = 2\sum_{m=1}^{\infty} |H_{mn}(\psi)|\cos(m\vartheta + \chi_{mn}). \tag{9.23}$$

Here ϵ stands for the dimensionless perturbation parameter, the complex amplitude $H_{mn}(\psi) = |H_{mn}(\psi)|e^{i\chi_{mn}}$ describes the magnitude $|H_{mn}(\psi)|$ and the phase χ_{mn} of the (m, n)-perturbation mode. There are two simplest forms of $H_{mn}(\psi)$ corresponding to the asymptotical formulas (3.54) and (3.58). We call them as models I and II and present them in the forms

$$|H_{mn}(\psi)| = \frac{1}{2}\exp\left(-m\frac{C(\psi)}{q(\psi)}\right)\begin{cases} 1/q(\psi), & \textbf{Model I}, \\ n/m, & \textbf{Model II}. \end{cases} \tag{9.24}$$

Model I corresponds to the perturbation field generated by the horizontal segments of saddle coils, while Model II corresponds to the one created by the vertical segments of saddle coils.

The function $C(\psi)$ has a finite value at the separatrix, $\psi = 1$, $C_0 = C(1) \neq 0$. Based on the numerical radial dependence of $C(\psi)$ given by Eq. (3.61) we choose the function $C(\psi)$ for our models in the following form,

$$C(\psi) = C_0\psi - (1/2)q(\psi)\ln\psi, \tag{9.25}$$

which gives a plausible asymptotics of $H_{mn}(\psi)$,

$$|H_{mn}(\psi)| \propto \begin{cases} \psi^{m/2}, & \text{at } \psi \ll 1, \\ \exp\left[-2\pi m C_0/\ln(Q/|h|)\right], & \text{at } |h| \ll 1, \end{cases} \tag{9.26}$$

near the magnetic axis $\psi \ll 1$ and the separatrix $|h| \ll 1$, respectively. In (9.26) h is the relative poloidal flux $h = \psi - 1$. At the separatrix $h = 0$, and $h < 0$ in the plasma region, and $h > 0$ outside the plasma region.

The phase χ_{mn} is presented as a linear function of m,

$$\chi_{mn} = -m\vartheta_0(\psi), \vartheta_0(\psi) = D_0/q(\psi). \tag{9.27}$$

We suppose that the coefficient D_0 is constant and $\vartheta_0(\psi)$ is inverse proportional to the safety factor $q(\psi)$. As we will see below the latter determines a value of the poloidal angle near which the perturbation field is localized.

9.3.1 Perturbation Fluxes

Putting $A_j = 1$ in Eqs. (3.67) and (3.69), and subtracting the term H_{mn} with $m = 0$ we can obtain the analytical formula for the perturbation function $\psi_n(\vartheta, \psi)$. For Model I one has

$$\psi_n(\vartheta, \psi) = \frac{e^{-b}}{q(\psi)\Lambda}\left[\cos(\vartheta - \vartheta_0) - e^{-b}\right]. \tag{9.28}$$

For Model II we have

$$\psi_n(\vartheta, \psi) = -n \arctan \frac{\sin(\vartheta - \vartheta_0)}{\sin\left(\vartheta - \vartheta_0 - e^b\right)}. \tag{9.29}$$

In Eq. (9.28) and (9.29) the following notations are introduced

$$\Lambda = 1 + e^{-2b} - 2e^{-b}\cos(\vartheta - \vartheta_0), b = \frac{C(\psi)}{q(\psi)}. \tag{9.30}$$

The parameter b weakly goes to zero when one approaches the separatrix $\psi \to 1$ (or $h \to 0$) according to the law, $b = 2\pi\gamma C_0 / \ln(Q/|h|)$.

9.3.2 The Safety Factor

Because of the generic asymptotics of the safety factor $q(\psi)$ (2.51) of equilibrium plasma in standard divertor tokamaks we can present it as

$$q(\psi) = \frac{1}{2\pi\gamma}\ln\frac{Q}{|h|} + a_1 h + a_2 h^2, \tag{9.31}$$

where γ, Q, a_1, and a_2 are constant parameters. Using (9.31) one can find the relation between ψ and the toroidal flux ψ_t,

$$\psi_t(\psi) = \int_0^{\psi} q(\psi)d\psi = \psi_a + \frac{h}{2\pi\gamma}\left(\ln\frac{Q}{|h|} + 1\right) + \frac{a_1}{2}h^2 + \frac{a_2}{3}h^3, \tag{9.32}$$

where $\psi_a = \psi_t(1)$ is the toroidal flux at the separatrix. The parameters γ, Q, a_1, and a_2 in (9.31) can be chosen by fitting the safety factor found by experimental measurements or equilibrium code calculations with the formula (9.31). On the other hand at the given values of γ, Q, the constants a_1 and a_2 can be found from the values of the safety factor at the magnetic axis, $q(0)$, and the value of ψ_a.

Figure 9.20 shows the dependencies of the safety factor q on ψ (curve 1 on the left-hand axis) and the toroidal flux ψ_t on ψ (curve 2 on the right-hand axis) for

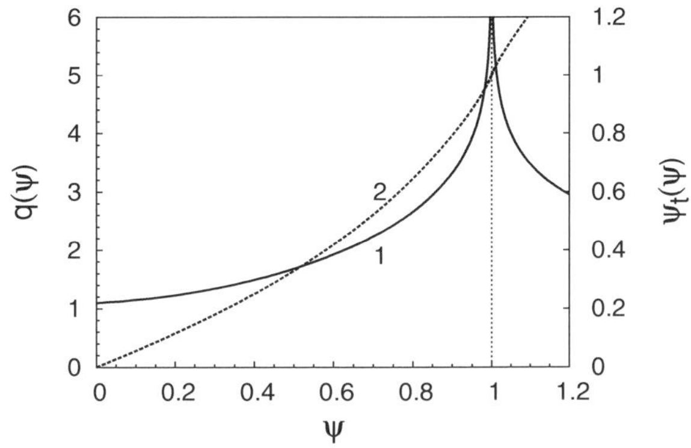

Fig. 9.20 Safety factor $q(\psi)$ given by Eq. (9.31). Parameters: $\gamma = 0.21$, $Q = 9.8$, $\psi_a = 2$, $q_0 = 1.1$, $q_{95} = 3.92$

the following parameters: $\gamma = 0.21$, $Q = 9.8$, $\psi_a = 2$, $q_0 = 1.1$. The value of $q_{95} = 3.92$. We put $a_1 = 0$, $a_2 = 0$ outside the separatrix.

9.3.3 *Poincaré Integrals*

The full-turn transfer mapping (6.58) can be effectively used to integrate the field line equations for our models. According to Eqs. (6.59) and (9.22) the Poincaré integral $P(\varphi, \psi)$ in the mapping has a single toroidal mode n,

$$P(\varphi, \psi) = K_n(\psi)\sin(n\varphi) + L_n(\psi)\cos(n\varphi). \tag{9.33}$$

The calculations of the functions $K_n(\psi)$ and $L_n(\psi)$ for our models are presented in Appendix C. Using (9.27) and (9.30) they can be presented as for the both models,

$$\begin{aligned} R_n(\psi) &= R_n^{(reg)}(\psi) + R_n^{(osc)}(\psi), \\ R_n^{(reg)}(\psi) &= \pi e^{-nC(\psi)+inD_0}, \\ R_n^{(osc)}(\psi) &= \sin(\pi n q(\psi))\, F(\psi, \vartheta_0), \end{aligned} \tag{9.34}$$

where $F(\psi, \vartheta_0)$ is the function of ψ, ϑ_0. This function for each models is found by the asymptotical expansions in the series of powers of $1/nq(\psi)$. The details of calculations are given in Appendix C.1. The asymptotical formulae for the functions $R_n^{(osc)} = K_n^{(osc)} + iL_n^{(osc)}$ in leading order of $1/nq(\psi)$ are given by Eq. (C.20) for Model I and by Eq. (C.24) for Model II.

Figures 9.21 and 9.22 shows the radial dependencies of $K_n(\psi)$ and $L_n(\psi)$ for Model I and Model II, respectively.

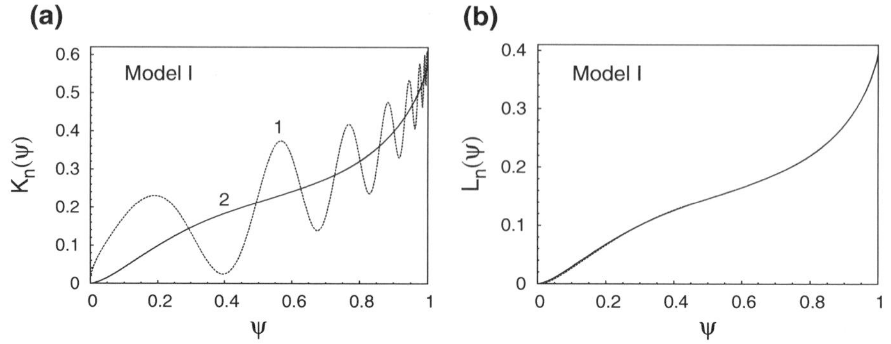

Fig. 9.21 Dependencies of $K_n(\psi)$ (**a**) and $L_n(\psi)$ (**b**) on ψ for Model I: *Curve 1* (*solid*) corresponds $K_n(\psi)$ (**a**) and $L_n(\psi)$ Eq. (9.34), *curve 2* (*dashed*) corresponds to the regular part $R_n^{(reg)}(\psi)$. The parameters of the model are the same as in Fig. 9.20 and $D_0 = 0.2$

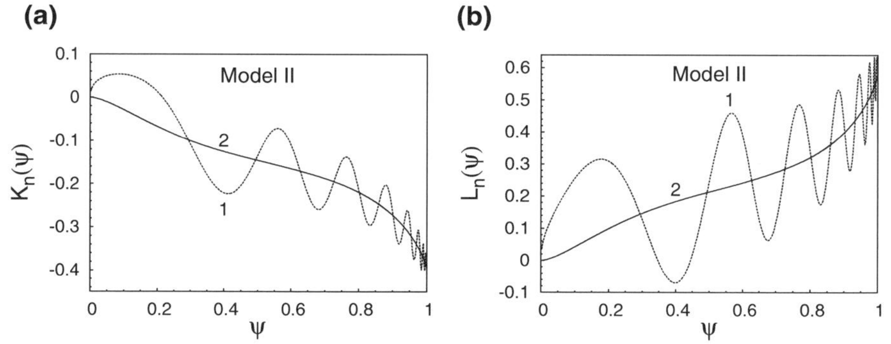

Fig. 9.22 The same as in Fig. 9.21 but for Model II

9.3.4 Perturbation Magnetic Field

According to Eqs. (1.28) and (9.22) the strength of the magnetic field perturbation normal to the magnetic surface, B_ψ, is given by

$$B_\psi = B_\varphi \epsilon F(\vartheta, \psi) \sin(n\varphi), \qquad F(\vartheta, \psi) = -\frac{\partial \psi_n(\vartheta, \psi)}{\partial \vartheta}. \tag{9.35}$$

From Eqs. (9.28) and (9.29), one has

$$F(\vartheta, \psi) = -\frac{e^{-\alpha}}{\Lambda} \begin{cases} \dfrac{\sin(\vartheta - \vartheta_0)\left(1 - e^{-2\alpha}\right)}{q(\psi)\Lambda}, & \text{Model I,} \\ n\left[\cos(\vartheta - \vartheta_0) - e^{-\alpha}\right], & \text{Model II.} \end{cases} \tag{9.36}$$

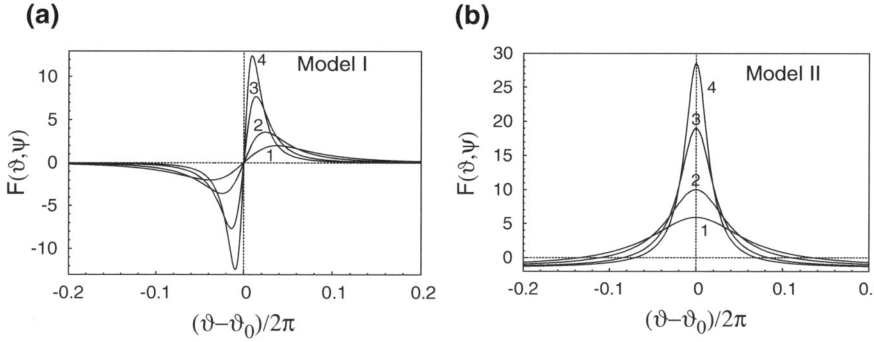

Fig. 9.23 Poloidal dependence of the perturbation function $F(\vartheta, \psi, \varphi)$ (9.39) at the several values of the poloidal flux ψ and at the fixed $\varphi = \pi/2$ for the model I (**a**) and the model II (**b**): *curve 1* corresponds to $\psi = 0.6$, *curve 2*—$\psi = 0.8$, *curve 3*—$\psi = 0.95$, *curve 4*—$\psi = 0.99$. The toroidal mode $n = 3$, $C_0 = 0.5$, and other parameters are the same in Fig. 9.20

The magnetic field perturbation is localized in a finite interval of poloidal angles ϑ with the central angle ϑ_0. The width of this interval can be estimated from the position of extremal points $\vartheta_m^{(\pm)}$ of the functions $F(\vartheta, \psi, \varphi)$, where $\partial F(\vartheta, \psi, \varphi)/\partial\vartheta = 0$. For Model I the extremal points $\vartheta_m^{(\pm)}$ are given by

$$\vartheta_m^{(\pm)} = \vartheta_0 \pm \begin{cases} \arccos\left(\frac{1}{2}\left[\sqrt{\cosh^2\alpha + 8} - \cosh\alpha\right]\right), & \text{for Model I,} \\ \arcsin\left(\tanh\alpha\right), & \text{for Model II.} \end{cases} \tag{9.37}$$

Particularly, the maximum values of F for Model II is

$$\max|F| = \frac{n}{2\sinh\alpha}. \tag{9.38}$$

The distance between peaks $\Delta\vartheta = \vartheta_m^{(+)} - \vartheta_m^{(-)}$ depends on the parameter α. At small values of α we have $\Delta\vartheta \approx \alpha/\sqrt{3}$ for Model I and $\Delta\vartheta \approx 2\alpha$ for Model II.

Using Eqs. (9.23) and (9.35) the perturbation field B_ψ can be presented as a Fourier series

$$B_\psi = \sum_{m=1}^{\infty} B_{mn}\sin\left(m\vartheta + \chi_{mn}\right)\sin(n\varphi),$$
$$B_{mn}(\psi) = B_\varphi \epsilon m|H_{mn}(\psi)|, \tag{9.39}$$

where $B_{mn}(\psi)$ is a poloidal spectrum of the perturbation field B_ψ.

Figure 9.23a, b show the angular dependencies of the perturbation function $F(\vartheta, \psi)$ for Model I (a) and Model II (b) at the several magnetic surfaces ψ. The parameters of the plasma are the same as in Fig. 9.20 and the toroidal mode $n = 3$.

As seen from Fig. 9.23b that at the plasma edge $\psi = 0.95$ the magnitude of the perturbation field (9.39) is of order of $B_\psi \approx B_0 10\epsilon$, where B_0 is the toroidal field magnitude. For $B_0 = 2\,\mathrm{T}$ and the perturbation parameters $\epsilon = 10^{-4}$ the perturbation field B_ψ takes values 20 G.

9.4 Fractal Properties of Chaotic Field Lines

In this section we describe generic properties of chaotic field lines at the plasma edge created by the RMPs. Behavior of these field lines has a fractal nature.[1] The main characteristic features of these field lines are (i) a finite radial extent of the stochastic layer; (ii) openness of chaotic field lines to the wall of device. Such chaotic systems are called *chaotic scattering systems*. These systems with a finite domain of motion should be studied differently than the chaotic systems in infinite domain of motion. Below we describe the main methods to study open chaotic field lines at the plasma edge.

9.4.1 Open Field Lines: Chaotic Scattering System

The magnetic field lines at the plasma edge with initial coordinates located in the stochastic layer eventually leave the plasma region hitting the divertor plate (expect for those field lines being trapped inside the magnetic islands). Therefore the stochastic zone of *open field lines at the plasma edge can be viewed as a chaotic scattering system*. The magnetic field lines enter into the plasma edge from the divertor plate and leave when hitting it again after a certain number of poloidal turns, N_p. It is illustrated in Fig. 9.24, where a several field lines entering into the plasma region and leaving it after null, one, and two poloidal turns.

In chaotic scattering systems, a trajectory may leave a system in one of several different ways. The space of initial coordinates corresponding to the various exit ways are separated by a boundary, which may be a fractal (Bleher et al. (1990)). The set of initial conditions for which trajectories leave the system in a particular way is called a *basin* of a particular mode. In the case of the ergodic divertor it is convenient to classify field lines by the number of poloidal turns N_p. Indeed, the perturbation field created by divertor coils are localized on the HFS (or the LFS), and field lines enter into the plasma and leave it on this side making almost full poloidal turns. Therefore, the set of initial conditions for which field lines cross the section $\theta = 0$ (mod 2π) with the same number of times may be referred as the basin of a particular number of poloidal turns N_p. Figure 9.24a, b show the schematic sketch of field line connections in the ergodic divertor tokamak and the tokamak with the magnetic

[1] The notion *fractal* has been introduced by B. Mandelbrot (1982) to describe complex, irregular behavior of geometrical objects in nature and physical phenomena.

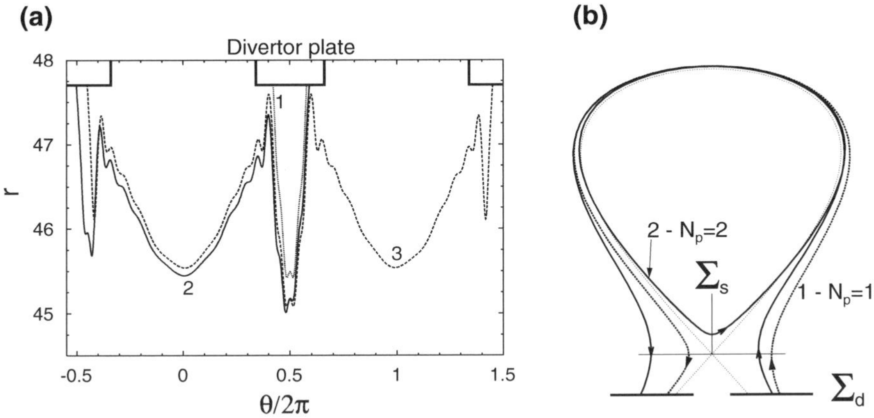

Fig. 9.24 Schematic sketch of field line connections: **a** in the ergodic divertor tokamak. *Curve 1* corresponds $N_p = 0$, *curve 2*—$N_p = 1$, and *curve 3*—$N_p = 2$; **b** in the tokamak with the magnetic separatrix. *Curve 1* corresponds to $N_p = 1$, *curve 2*—$N_p = 2$

separatrix, respectively. Field lines with $N_p = 0$ that do not cross the section $\theta = 0$ belong to a special set called a *private flux zone* (see Fig. 9.24a). In the tokamak with the magnetic separatrix the field lines with $N_p = 1$ do not penetrate into the plasma although they are making a full poloidal turn (see Fig. 9.24b).

The spatial structures of boundaries of basins belonging to different N_p reveal fine details of the structure of field lines which cannot be seen in Poincaré sections. The structure of these basins can be studied by plotting the contours of N_p within the plasma edge and on the divertor plate.

9.4.2 Basin Boundary Structure at the Plasma Edge: Laminar Plots

9.4.2.1 Laminar Zone in Limiter Tokamaks

First we consider the structure of basin boundaries referring to the basin as the set of points (ϑ, ψ) at the given poloidal section $\varphi =$ constant which are crossed by field lines with a particular number of poloidal turns N_p. The procedure to obtain these plots is the following.

At the poloidal plane $\varphi = 0$ the field line with a given initial coordinate (ϑ, ψ_t) is traced by iterating the map (6.29) along the positive and negative directions of the toroidal angle φ until a field line reaches the divertor plate. Then we determine a fractional number of poloidal turns N_{pol} as the ratio of the total change of the poloidal angle $\Delta\theta$ to the full circle 2π, i.e., $N_{pol} = \Delta\theta/2\pi$. The values of N_{pol} computed in this way are close to integer numbers although they are not exactly integer. Let

N_p be the integer number closest to N_{pol}. Areas in the (ϑ, ψ_t)-plane with different poloidal turns N_p are topologically different. The dependence of N_{pol} on the initial coordinates (ϑ, ψ_t) is displayed by a contour plot with contour lines separating the basins of different poloidal turns N_p. This plot is called as *laminar plot*.

The examples such plots are presented in Figs. 9.25 and 9.26 for the ED of Tore Supra (see Rompuy (2009); Abdullaev et al. (2008)) and the TEXTOR-DED (see Finken et al. (2005a)), respectively. The whole laminar plot has a lattice or quasiperiodic structure consisting of crossing each other elongated stripes. Such a structure is determined by the distribution of the primary unstable periodic fixed points in the (ϑ, ψ_t)-plane, plotted by symbols $\times$ in Fig. 9.25b. In limiter tokamaks the number of these fixed points is finite and determined by the safety factor $q(\psi)$ (see Sect. 7.4.3).

The private flux zones with $N_p < 0.5$ form the quasiperiodic chain of islands surrounded by the long dark elongated stripes. The basin with one poloidal turn has a non-fractal boundary with the private flux zone. But it may have fractal boundaries with the basins corresponding to two and more poloidal turns. For $N_p \geq 2$ there are several topologically different basins related to the same N_p. As seen from Figs. 9.25 and 9.26 the relatively large basins of a few poloidal turns $N_p \leq 3$ at the plasma edge are clearly separated by non-fractal boundaries. But they are alternating with the stripes containing the basins for a few poloidal turns up to very large N_p. At the HFS some of these stripes are radially extended toward the divertor plate.

Often these stripes are called "*fingers*" and the area of the stochastic layer with a small numbers of N_p is called a *laminar zone* [Eich et al. (2000)]. The transport of heat and particles along the field lines in this zone has more a convective or laminar nature rather than the diffusive nature. The quantitative measure of the laminar zone will be given in Sect. 10.2.3 based on the radial profiles of diffusion coefficients of field lines like one shown in Fig. 10.4b.

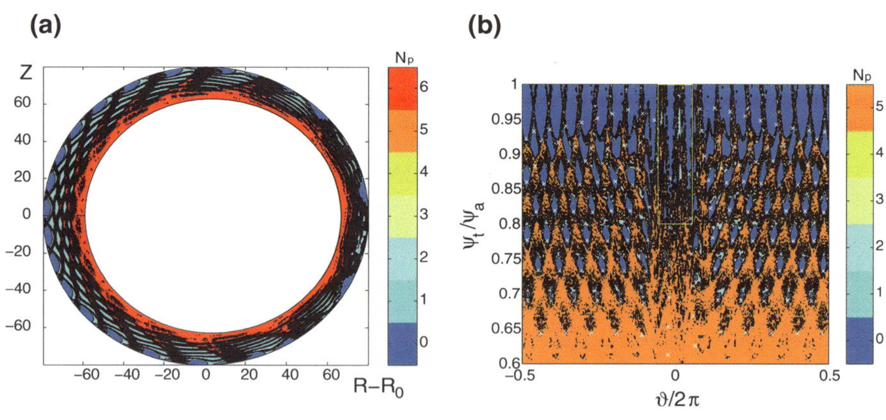

Fig. 9.25 Contour plots of connection lengths in poloidal turns N_p for the ergodic divertor of Tore Supra: **a** in the (R, Z)-plane; **b** in the (ϑ, ψ_t)-plane. Colorbar shows the number of poloidal turns N_p. The parameters are the same as in Fig. 9.3. The ED perturbation current $I_d = 22.5$ kA

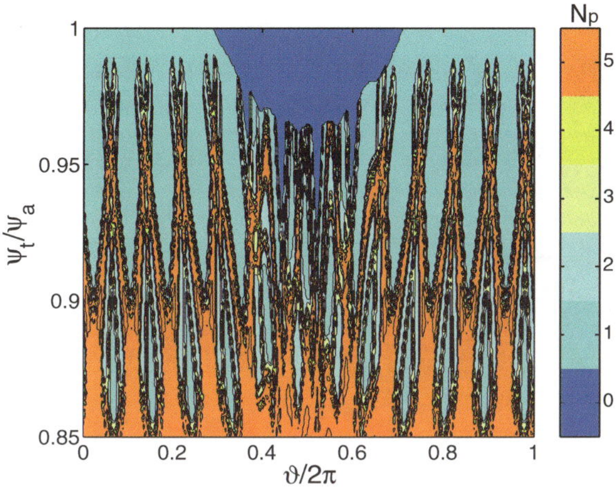

Fig. 9.26 The same as in Fig. 9.25, but for the TEXTOR DED. The parameters are the same as in Fig. 9.5. The ED perturbation current $I_d = 7\,\text{kA}$

To study the fine fractal structure of stripes we have magnified the area of the stripe with the fine resolution of basins of higher poloidal turns. Figure 9.27 illustrates the typical fractal structure of stripes: the blow up of the rectangular areas in (a) reveals a more fine details. It shows that the basins at the stripe are highly elongated and the boundaries between them have fractal structure, i.e., the stripes consists of layered basins of different poloidal turns with a self−similar behavior at different spatial scales. These fractal structures of the basin boundaries have the so-called *Wada property*: every boundary point has an arbitrarily small neighborhood containing points of all the basins [Kennedy and Yorke (1991); Nusse and Yorke (1996a, b); Viana et al. (2011)].[2] As seen from Fig. 9.27b the basins of field lines with a few poloidal turns N_p are "sandwiched" between basins for field lines with large numbers of poloidal turns $N_p \gg 1$.

9.4.2.2 Structure of Laminar Zone Near the Magnetic Separatrix

In poloidal divertor tokamaks with the magnetic separatrix the laminar plot reveals the structure of open field lines similar to the ones in the limiter tokamaks. However, in contrast to the latter it consists of very long helical stripes. The typical example of such a structure consisting of the two sets of helical stripes of opposite helicity is shown in Fig. 9.28 in the (φ, ψ)-plane of the section Σ_s shown in Fig. 9.24b. The corresponding the Poincaré section of field lines was shown in Fig. 9.18a. The number of helical stripes of a given helicity in the (φ, ψ)-plane is determined by the toroidal

[2] The description of fractal structures in nonlinear dynamical systems, particularly, their Wada property can be found, for example, in a review by Aguirre et al. (2009).

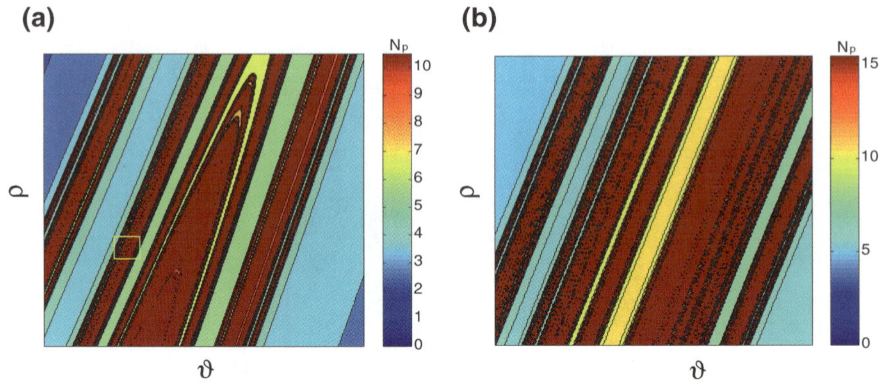

Fig. 9.27 Fractal structure of a typical "finger": **a** a whole finger; **b** the expanded view of the rectangular area in (**a**)

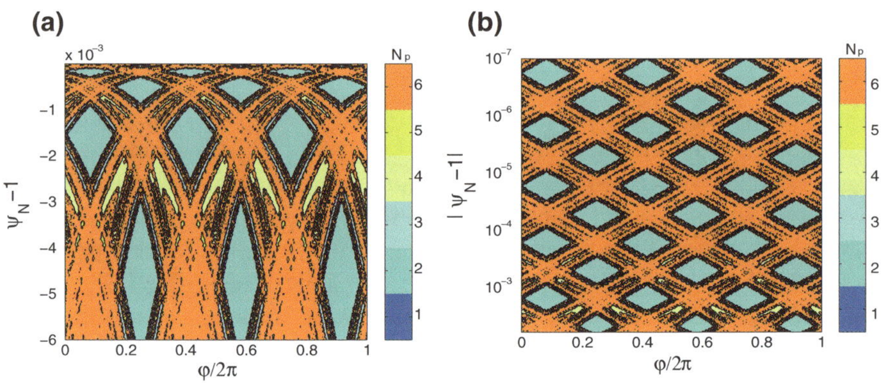

Fig. 9.28 Typical contour plot of connection lengths in the (φ, ψ)-plane (**a**) and $(\varphi, \ln|\psi|)$-plane (**b**) near the separatrix of poloidal divertor tokamaks. The calculations are made using the separatrix mapping (6.29)

mode n. One can see that the laminar plot of the stochastic layer near the magnetic separatrix has the periodic structure in logarithmic scale, $\ln|\psi|$, (see Fig. 9.28b) unlike the corresponding structures in limiter tokamaks shown in Figs. 9.25 and 9.26.

In divertor tokamaks unlike the limiter ones the number of the primary unstable fixed points is infinite. Near the separatrix they are approximately determined by Eq. (7.66). The radial coordinates ψ_{mn} of fixed points in the logarithmic scale are a linear function of the mode number m: $\ln|\psi_{mn}| = \ln Q - m(2\pi\gamma/n)$ which leads to the periodic structure shown in Fig. 9.28b.

The fractal structure of helical stripes can be also studied by plotting the dependence of N_p on the one of coordinates at the certain sections. The typical dependence of N_p on the toroidal angle φ on the given magnetic surface $\psi = \text{const}$ is plotted in Fig. 9.29. It shows the layered fractal structure of each stripes.

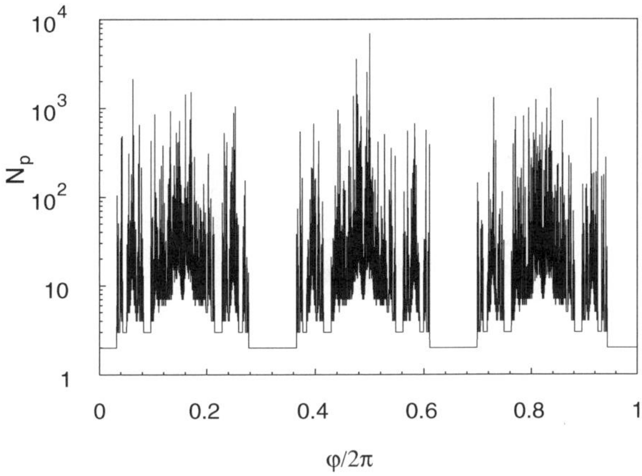

Fig. 9.29 Dependence of N_p on the toroidal angle φ at the given magnetic surface $\psi = -4 \times 10^{-4}$ shown in Fig. 9.28

9.4.3 *Fractal Structure of Magnetic Footprints*

In order to study the basin boundary structure on the divertor plate we will use the following procedure. We follow a field line which enters into the plasma starting from the divertor plate with a given initial coordinate (φ, θ) and returns back to the plate after a certain number of poloidal turns N_p. The set of initial conditions (φ, θ) with a particular number N_p determines a basin. The whole picture of basin boundaries with $N_p \geq 1$ on the plasma wall determines the *magnetic footprints*. Similar to the stripe at the plasma edge the magnetic footprints have a fractal structure as well.

9.4.3.1 Ergodic Divertor Tokamaks

The example of magnetic footprints are displayed in Fig. 9.30 for the Tore Supra (Rompuy (2009))and the TEXTOR-DED (Finken et al. (2005a)) tokamaks. The corresponding Poincaré sections and laminar plots were given in Fig. 9.30.

One can see from Fig. 9.30a, b that the field lines can enter into the plasma (or hit the divertor plate from the plasma side) only along the pairs of narrow helical stripes, equal to the toroidal mode number n, $n = 6$ for the Tore Supra and $n = 4$ for the TEXTOR-DED. (Dark blue areas in Figs. correspond to the field lines in a private flux zone). The distance between stripes of each pairs depends on the plasma current I_p (see Abdullaev et al. (2001)). Each helical stripe has a fractal structure and it consists of layered basins of different poloidal turns (see Fig. 9.30a, b). The width of layers is changing along the toroidal direction φ. The area of the basin with one poloidal turn $N_p = 1$ is the largest. For the higher $N_p > 1$ corresponding areas of basins are drastically decreased in size. The boundaries between these basins are fractal.

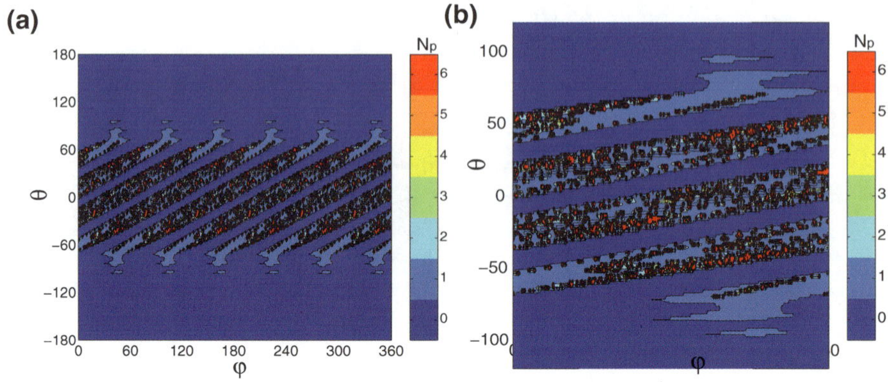

Fig. 9.30 Typical structure of magnetic footprints (basin boundary structure) on the (φ, θ)-plane at the divertor plates for the Tore Supra (**a**) and the TEXTOR-DED (**b**)

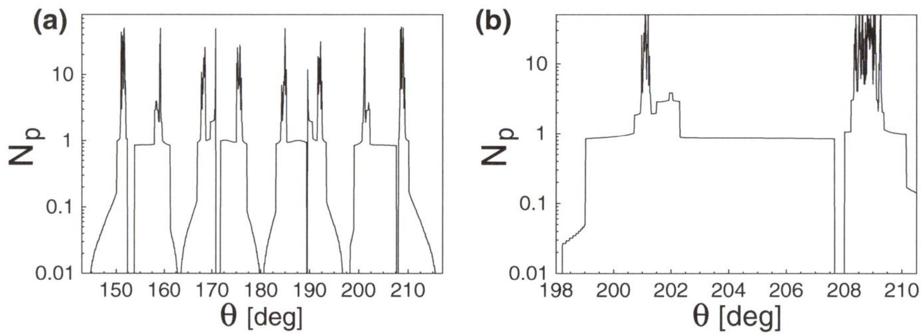

Fig. 9.31 **a** Fractal dependence of N_p on θ at fixed toroidal angle $\varphi = 180°$ corresponding the magnetic footprint in Fig. 9.30a; **b** a view of the one helical stripe

The fine structure of helical stripes can be revealed by studying the dependence of N_p on the poloidal angle θ for fixed toroidal angle φ. Such a dependence of N_p on θ is described by Cantor-like, fractal curves. It is presented in Fig. 9.31 at the fixed toroidal angle $\varphi = 180°$. The curve in Fig. 9.31a describes the poloidal section of the all four helical stripes shown in Fig. 9.30a, while the expanded view of the area along the one helical stripe is shown in Fig. 9.31b. They clearly show areas of field lines connecting plate to plate in one, two, three and more poloidal turns N_{pol}. These areas are described by almost horizontal steps in the fractal curve. The width of layers becomes smaller with increasing N_{pol}.

The structure of the helical stripes plays an important role for heat and particle deposition on the divertor plates. Indeed, the basins with $N_p \gg 1$ correspond to the field lines coming from deep within ergodic zone. These field lines may bring high energetic particles to the wall because the particles predominantly move along field lines. Therefore one can expect that the spatial distribution of power deposition within the helical stripes will depend on the spatial structure of basins with $N_p > 1$. The cross-field diffusion of particles broadens the spatial distribution of power deposition around the maxima located at the basins corresponding to large number of poloidal turns.

9.4.3.2 Divertor Tokamaks with Magnetic Separatrix

The typical fractal structure of magnetic footprints in the divertor tokamaks is shown in Fig. 9.32a, b by the contour plot of the connection poloidal turns N_p in the (ψ, φ)-plane of the section Σ_d shown in Fig. 9.24b. As in the ergodic divertor case it consists of several spirals, but with the very long length along the toroidal angle φ (see also Sect. 9.2.4 and Fig. 9.19). One can note that the structure of magnetic footprints shown in a logarithmic scale is similar to the periodic structure of chaotic field lines at the plasma edge shown in Fig. 9.28b.

The contour plots show the fractal boundary structure of basins of different N_p. More detailed structure of fractal boundaries can be revealed by plotting the dependence of N_p on the poloidal flux ψ at the section the fixed toroidal angle φ or on the toroidal angle φ at the section with $\psi =$ constant. Such plots are shown in Fig. 9.33a and b, respectively.

More detailed analysis of the fractal structure of the magnetic footprints can be found in studies by Wingen et al. (2009); Evans et al. (2010) (and in references therein).

9.5 Experimental Observation of the Magnetic Structure

Since particles predominantly follow magnetic field lines one expects that the plasma particle distributions at the plasma edge and the heat deposition patterns on the diver-

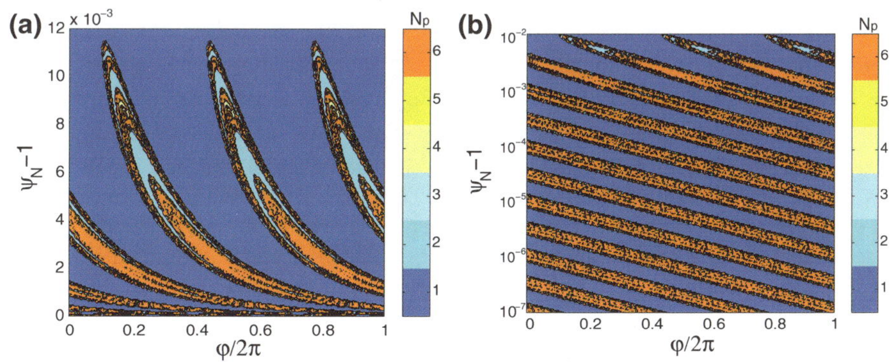

Fig. 9.32 Typical contour plot structure of magnetic footprints in poloidal divertor tokamaks (**a**); **b** shows the same as in (**a**) but in a logarithmic scale log ψ

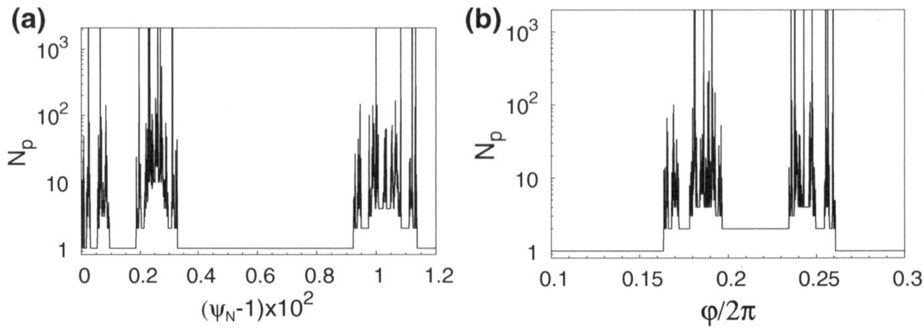

Fig. 9.33 Fractal dependencies of N_p on ψ_N at the fixed toroidal angle $\varphi = 0.24\pi$ (**a**) and on φ in the interval $0.1 \leq \varphi/2\pi \leq 0.3$ at the fixed $\psi_N = 1.007$ (**b**)

tor plates would resemble the magnetic field structures. This would allow one to experimentally observe the structures of magnetic field at the plasma edge imposed by the RMPs by measuring the temperature field patterns. Modern advanced measurement devices allows to measure those characteristics with high temporal and spatial resolutions. The comparison of observations with the computed magnetic field structures would also verify the validity of the so-called vacuum approximation in which the reaction of the plasma on the external magnetic perturbations is neglected.

Such experimental observations have been extensively performed in the TEXTOR-DED by Finken et al. (2005a); Lehnen et al. (2005); Jakubowski et al. (2006a); Jakubowski et al. (2006b); Wingen et al. (2007); Jakubowski et al. (2007) (see also Abdullaev et al. (2004)). Experimental observations of lobe structures near the magnetic separatrix in the MAST tokamak have been also recently reported by Kirk et al. (2012, 2013) and Chapman et al. (2013). Below we briefly present some of the results obtained in the TEXTOR-DED.

Figure 9.34a, b show an image of the ion CIII light emission, recorded by means of CCD camera equipped with an interference filter at $\lambda = 465$ nm. The figures clearly show the magnetic field structure formed by the RMPS created by the DED coils like ones shown in Fig. 9.26. Superimposed to the figure is a Poincaré plot visualizing the structure of the magnetic field lines. One observes that the distribution of the CIII spectral line intensity follows the magnetic structures: the finger-like structure of the magnetic field shown by the laminar plot in Fig. 9.34c is experimentally observed in the spatial distribution of the CIII carbon shown in (b).

The regular patterns of heat deposition on the divertor plate have been observed since very first experiments in the TEXTOR-DED. Typical example of the infrared images recorded by the infrared camera is shown in Fig. 9.35a. The images present the temperature distribution over the divertor target plates.

The generic features of the DED heat flux density pattern are the stripe-like structures parallel to the DED coils. One can see four pairs of strike zones formed by the incoming heat fluxes, where the half width of one of stripes is of order of 3 cm. The

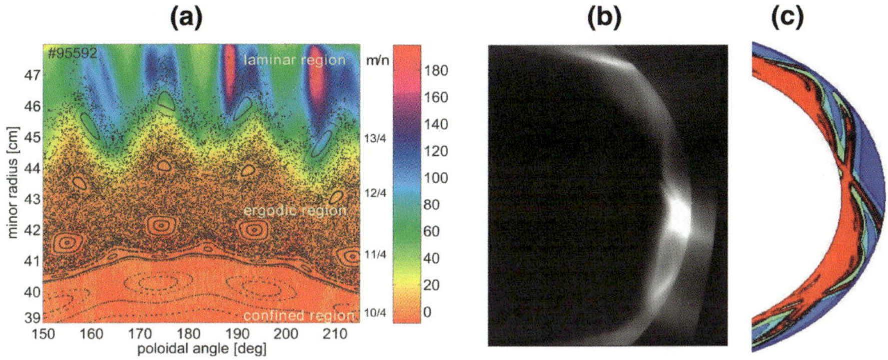

Fig. 9.34 **a** The two-dimensional distribution of the CIII emission in the front of the DED coils for the discharge #95592 with the Poincaré section visualizing the calculated structure of the magnetic field. The colorbar indicates the intensity of the CIII line at 465 nm in the arbitrary units. The m/n numbers on the right hand side of the figure indicate the island chains (Jakubowski et al. (2005)). **b** CIII carbon emission in front of the divertor plates for the discharge #94809 Lehnen et al. (2005); **c** Laminar plot of field lines calculated for the plasma parameters corresponding to the discharge #94809: $I_p = 300\,\text{kA}$, $B_t = 2.25$, $\beta_{pol} = 0.3$ and for the $n = 3$ mode TEXTOR–DED operation with the divertor current $I_d = 3.75\,\text{kA}$

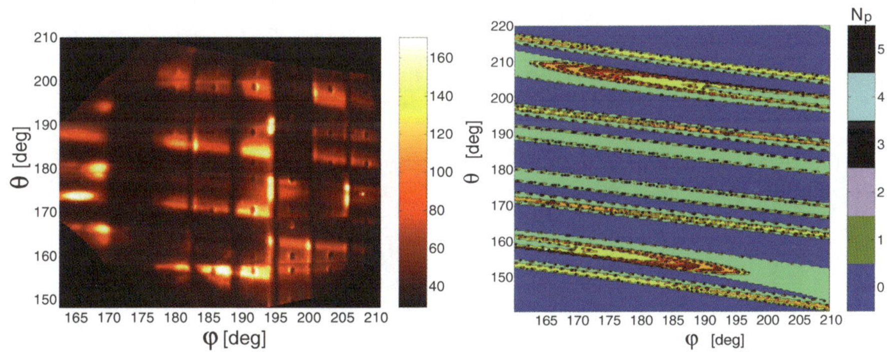

Fig. 9.35 **a** Temperature distribution over the divertor target plates in the pressure measured by the infrared camera; **b** The magnetic footprints calculated for the temperature distribution in (**a**)

temperature distribution qualitatively resembles the footprint structure of the field lines on the divertor wall, which is presented in Fig. 9.35b.

To perform a direct comparison of the heat deposition pattern with the structure of magnetic footprints a heat flux profile along the poloidal angle was evaluated (Jakubowski et al. (2004)). In Fig. 9.36 the power flux profile calculated from the temperature evolution along the tiles at the toroidal angle $\varphi = 187°$ is shown. For the same toroidal position the connection length profile along the DED tiles has been calculated. The blue curve visualize the connection lengths of the magnetic field lines (values are on the left ordinate); the green line presents the heat fluxes towards the

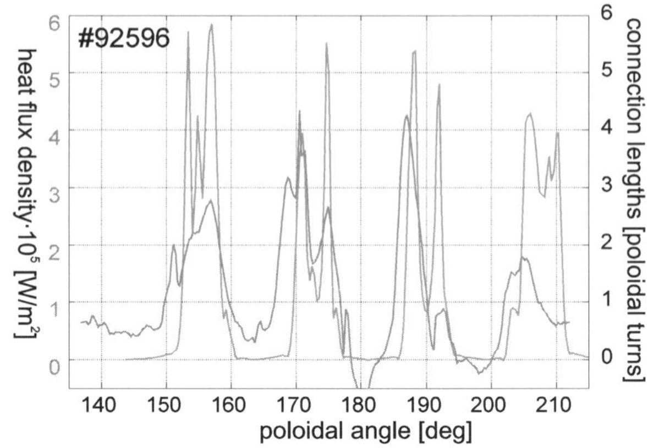

Fig. 9.36 Comparison of the magnetic structures predicted theoretically and measured heat fluxes: *green curve* presents the power flux profile evaluated from the measured temperature distribution (the left hand axis); *blue curve* (the right hand axis) represents the connection lengths of the field lines starting along the line, for which the power flux was evaluated

divertor target plates calculated from the temperature distribution (the right ordinate is the corresponding one). The power flux profile is in fair agreement with modeled connection length profile. Indeed the middle power flux stripes are composed of the two hotter "channels", where the heat is transported from the ergodic region to the divertor wall. The connection length profiles of the outer stripes are composed only from the long length field lines. Therefore the heat flux density distribution is more uniform. The heat flux density profiles are broader than the footprints, which is consistent with the assumption of a cross-field diffusion of the particles from the relatively thin fingers to the laminar zone. Most of the heat is brought from the plasma core to the wall by the field lines with relatively long connection lengths (more then 4 poloidal turns).

9.6 Bibliographic Comments

Ergodic divertors. During last two decades the ergodic divertor concepts were implemented in the Texas Experimental Tokamak (TEXT) in Austin, Texas, USA (more known as an ergodic limiter [Gentle (1981); McCool et al. (1990)], Tore-Supra in Cadarache, France [Samain et al. (1984, 1990); Ghendrih et al. (1992, 1996); Nguyen et al. (1995); de Michelis et al. (1995); Grosman (1999); Ghendrih and Grosman (2009)], the *Dynamic Ergodic Divertor* (DED) in the TEXTOR tokamak [see Finken (ed.) (1997); Finken et al. (1999, 2005a, b); Abdullaev et al. (2003)]. Ergodic divertors has been also installed in several small fusion devices, for example, JFT-2M [Shoji et al. (1992); Evans et al. (1989)], CSTN- II [Takamura et al. (1987, 1989)], HYBTOK-II [Shen et al. (1989)], TBR-1 Caldas et al. (2002), TCABR [Pires et al. (2005), Kawamura et al. (1982); Hattori et al. (1984)].

RMPs in divertor tokamaks. Many modern fusion devices, tokamaks are equipped with external coils to control magnetohydrodynamic (MHD) instabilities in plasmas. For example, internal (I-) coils in the DIII-D tokamak are used to stabilize the resistive wall mode (RWM) and to suppress the large edge-localized modes (ELMs) in high-confinement (H-mode) regimes (see, for example, the papers by Jackson et al. (2003); Okabayashi et al. (2005); Evans et al. (2004, 2006)). To the similar purpose serves the Error Field Coils (EFC) installed in the spherical tokamak, National Spherical Tokamak Experiment (NSTX) Gates et al. (2009). Control coils have been also installed in the ASDEX Upgrade tokamak [Suttrop et al. (2009)], and are foreseen to install in the International Thermonuclear Experimental Reactor (ITER) to control the RWM and to mitigate the ELMs (see, e.g., Refs. Schaffer et al. (2008)).

The external coils are typically installed inside or outside of a plasma vacuum vessel as a set of toroidally distributed saddle coils as shown in Figs. 3.6 and 9.8. The resonant magnetic perturbations (RMPs) created by these coils are non-axisymmetric along the poloidal and toroidal directions. The calculations of the toroidal and poloidal spectra of RMPs, $H_{mn}(\psi)$, in Eqs. (3.12) and (3.13) are the important first step in studies of the effect of magnetic perturbations on various properties of plasmas in tokamaks (see Refs. Evans et al. (2006); Yan and Evans (2007); Schaffer et al. (2008); Suttrop et al. (2009)).

One of the successful applications of externally created resonant magnetic perturbations is a suppression of the so-called edge localized modes (ELMs) in tokamak plasmas (see Evans et al. (2006, 2008); Liang et al. (2007a, b) and references therein). The ELMs are large, repetitive magnetohydrodynamics instabilities appeared due to steep edge pressure gradients in high-confinement regimes (or H-modes) of plasmas. Type-I ELMs produce a rapid loss energy and particles from the plasma edge region resulting in large impulsive loads on divertor target plates. Due to their potential to damage divertor target plates it is necessary to control this type of ELMs in a future fusion reactor, ITER.

Chaotic scattering systems. Pioneering studies on the chaotic scattering systems has been done by Eckhard (1988); Bleher et al. (1989). There is also a special issue of "Chaos" on this subject edited by Ott and Tel (1993). Scattering properties of magnetic field lines have been investigated by Lau and Finn (1991); Pomphrey and Reiman (1992); Reiman (1996); Abdullaev and Zaslavsky (1996); Abdullaev et al. (2001); Da Silva et al. (2002b) and others. Some latest references on the scattering properties of field lines can be found in reviews by Viana et al. (2011); Altmann et al. (2013).

Stochasticity due to internal MHD activities. The above mentioned magnetohydrodynamics instabilities developed at the edge plasmas act as a source of non-axisymmetric magnetic perturbations. Recent large scale magnetohydrodynamics numerical simulations (see Sugiyama (2012) and references therein) show that the toroidal numbers n of these modes is sufficient low to destroy magnetic separatrix and create a quite large stochastic layer of magnetic field lines. The latter may significantly affect plasma confinement and macroscopic stability at the plasma edge.

Chapter 10
Transport of Field Lines and Particles in a Stochastic Magnetic Field

In this chapter the transport of field lines and charged particles in a stochastic magnetic field will be investigated. To study this problem we will use a test particle model in which no reaction of a plasma medium to particle motion is taken into account. In this simplified approach we derive the analytical formulas of the diffusion coefficients describing the radial chaotic transport of particles and investigate their validity using the numerical simulations. The collisionless and collisional transport regimes will be considered. In the latter case we reexamine the existing models using the results of numerical simulations. We propose a simple empirical formula for the collisional radial diffusion coefficient of particles in a stochastic magnetic field.

In Sect. 10.1 we briefly present the basics of the kinetic description of the chaotic transport in Hamiltonian systems. Particularly, the quasilinear approximation will be considered. The chaotic transport of field lines and particles in a stochastic magnetic field in collisionless regimes will be studied in Sect. 10.2 using the numerical simulations. The diffusion coefficients of field lines are compared with the corresponding quasilinear formulas, and their limitations are found. The effect of collisions on the particle transport in a stochastic magnetic field is studied in Sect. 10.4.

10.1 Chaotic Transport in Hamiltonian Systems

In dynamically chaotic systems the prediction of trajectories becomes practically impossible due to exponential divergence of orbits: any small error in initial conditions leads to the enormous error in the final conditions. Therefore an individual orbit gives a little knowledge about a system. To describe such systems their statistical description becomes more appropriate. This approach is based on the concepts and methods of statistical mechanics, particularly, the kinetic description. To apply these methods to study the dynamically chaotic systems the latter should satisfy a certain properties, mainly, the ergodicity and mixing properties. In this section we recall these notions and the kinetic equations for the dynamically chaotic Hamiltonian systems.

S. Abdullaev, *Magnetic Stochasticity in Magnetically Confined Fusion Plasmas*, Springer Series on Atomic, Optical, and Plasma Physics 78,
DOI: 10.1007/978-3-319-01890-4_10,

10.1.1 Ergodicity and Mixing

Let $(q(t), p(t))$ be the position of system in phase space (q, p) at the time instant t. The time–evolution of system over time period T is given by the map

$$(q_{k+1}, p_{k+1}) = \hat{M}(q_k, p_k), \tag{10.1}$$

where $(q_k, p_k) \equiv (q(t_k), p(t_k))$, $(t_k = kT)$. Consider the evolution of the arbitrary integrable function $f(q, p)$ of (q, p). Suppose the system's orbits lie in a finite domain W of phase space. Then the motion is called *ergodic* if the time averaging of the function f defined as

$$\bar{f} = \lim_{N\to\infty} \frac{1}{N} \sum_{k=1}^{N} f(q_k, p_k), \tag{10.2}$$

is equal to its value, $\langle f \rangle$ obtained by averaging over phase space,

$$\langle f \rangle = \int_W f(q, p) dq dp, \tag{10.3}$$

i.e.,

$$\bar{f} = \langle f \rangle. \tag{10.4}$$

To introduce the mixing property of dynamical systems we define the *correlation function*. Let f and g be two arbitrary integrable functions of (q, p): $f = f(q, p)$ and $g = g(q, p)$. These functions may describe some physical properties of the system. The correlation function $G_{fg}(t)$ of the quantities f and g is defined as

$$G_{fg}(t) = \overline{M^s f \cdot g} - \bar{f}\bar{g}, \qquad t = sT, \tag{10.5}$$

where

$$\begin{aligned} \overline{M^s f \cdot g} &\equiv \overline{f(q_{k+s}, p_{k+s}) \cdot g(q_k, p_k)} \\ &= \lim_{N\to\infty} \frac{1}{N} \sum_{k=1}^{N} f(q_{k+s}, p_{k+s}) g(q_k, p_k). \end{aligned} \tag{10.6}$$

The correlation function $G_{fg}(t)$ gives a quantitative measure of the correlation of the functions f and g separated by time delay t. The dynamical system is called a *mixing* if the correlation function (10.5) vanishes at the limit $t \to \infty$:

$$G_{fg}(t) = 0, \qquad \text{for } t \to \infty. \tag{10.7}$$

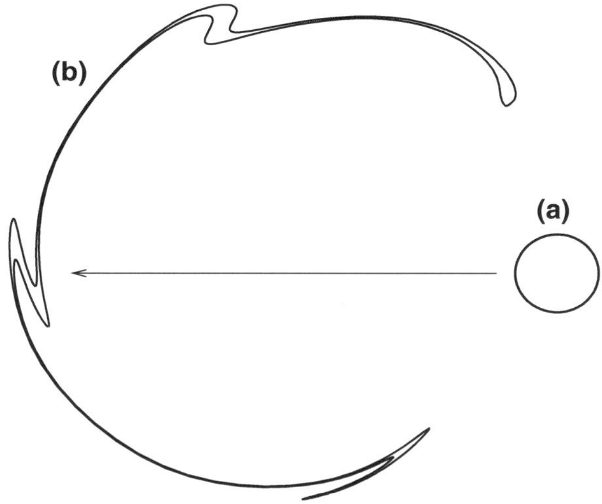

Fig. 10.1 Example of mixing property of system: initial phase space area (**a**) of a circular shape turns into the stretched and squeezed area (**b**) with a complicated shape

The mixing represents a more subtle property of system than the ergodicity and it is related with the exponential divergence of neighboring orbits (7.100). In systems with the mixing property the initial small finite domain of phase space acquires the complicated shape with the same area (due to Liouville theorem) but with the stretched and bended regions as illustrated in Fig. 10.1. The ergodicity of system follows from the mixing property. However, the opposite statement that the mixing property is the consequence of the ergodicity is not generally true.

The correlation function $G_{fg}(t)$ (10.5) is defined as time–averaged quantity. It can be also defined as a quantity averaged over phase space. Let $(q(t; q_0, p_0), p(t; q_0, p_0))$ be an orbit of the system with the initial condition $(q_0, p_0) = (q(t_0), p(t_0))$ at $t = t_0$. The correlation function of the functions f and g separated by time delay t can be defined as the integral

$$G_{fg}(t) = \langle f(t)g(t_0)\rangle - \langle f(t)\rangle\langle g(t_0)\rangle, \tag{10.8}$$

where

$$\langle f(t)g(t_0)\rangle = \int\limits_W \left[f\left(q(t; q_0, p_0), p(t; q_0, p_0)\right) g(q_0, p_0)\right] dq_0 dp_0.$$

For ergodic systems the both definitions of the correlation function (10.5) and (10.8) should coincide. The examples of the correlation function will be given in the forthcoming sections.

The law according to the correlation function tends to zero significantly depends on the structure of phase space. In the case of systems with a highly developed chaotic zone with no KAM stability islands the correlation function decays exponentially with t,

$$G_{fg}(t) \sim \exp(-t/\tau_c), \tag{10.9}$$

where τ_c is the *correlation time*. Usually, it is inverse proportional to the Lyapunov exponent σ (7.50) averaged over the phase space area W:

$$\tau_c \sim \frac{1}{\langle\sigma\rangle}.$$

However, in typical partially chaotic Hamiltonian systems with the embedded KAM stability islands in the chaotic zone the correlation function decays slowly according to power-like law,

$$C_{fg}(t) \sim t^{-p}, \tag{10.10}$$

with the constant exponent p which significantly depends on the structure of chaotic zone.

In order to quantify the measure chaos in dynamical chaotic systems Kolmogorov (1958, 1959), and Sinai (1959) introduced the metric entropy, h_K, which has an analogy to the Shennon entropy in the information theory. It gives a measure of the time rate of creation of information as a chaotic system evolves and it is called the *Kolmogorov entropy* [also the *Kolmogorov-Sinai entropy*]. It was found that the metric entropy, h_K, is related to the averaged Lyapunov exponent $\langle\sigma\rangle$,

$$h_K \sim \langle\sigma\rangle \sim \tau_c^{-1}. \tag{10.11}$$

For dynamically chaotic systems the Kolmogorov entropy is positive, $h_K > 0$. The quantity

$$\tau_K = 1/h_K, \tag{10.12}$$

(with an appropriate factor having a time scale) is also called the Kolmogorov time, which characterizes the correlation time (see, e.g., Lichtenberg and Lieberman (1992); Ott (2002)).

In Sects. 10.2–10.4 we will calculate the correlation time τ_c, the Lyapunov exponent σ, and the Kolmogorov time τ_K for the chaotic field lines and particles in a stochastic magnetic field created by the resonant magnetic perturbation in tokamaks.

10.1.2 Kinetic Equation

Consider the Hamiltonian system (6.4)–(6.6) and suppose that in the certain domain D of the action variables I_i, $(i = 1, \ldots, N)$, the orbits are chaotic. Therefore, there is no sense to follow individual orbits and it is more appropriate the statistical description of an ensemble of orbits. Let $f(I, \vartheta, t)$ be a probability distribution function (PDF) $f(I, \vartheta, t)$ in the phase–space of $(I, \vartheta) = (I_1, \vartheta_1, \cdots, I_N, \vartheta_N)$ at the time instant t. The quantity $dP = f(I, \vartheta, t)dId\vartheta$ is a number of particles in the volume element $dV = dId\vartheta$. The time evolution of the PDF $f(I, \vartheta, t)$ is governed by the Liouville's equation,

$$\frac{\partial f}{\partial t} + \sum_{i=1}^{N} \frac{\partial f}{\partial \vartheta_i} = \epsilon \sum_{i=1}^{N} \left[\frac{\partial f}{\partial I_i} \frac{\partial H_1}{\partial \vartheta_i} - \frac{\partial f}{\partial \vartheta_i} \frac{\partial H_1}{\partial I_i} \right], \tag{10.13}$$

which follows from the Hamiltonian equations (6.5). Similar to the latter the Liouville's equation is reversible in time.

We introduce the distribution function $F(I, t)$, obtained by averaging over phases ϑ_i,

$$F(I, t) = \frac{1}{(2\pi)^N} \int_0^{2\pi} \cdots \int_0^{2\pi} f(I, \vartheta, t) d\vartheta_1 \cdots d\vartheta_N. \tag{10.14}$$

The kinetic description of the chaotic Hamiltonian system is based on the kinetic equation describing the irreversible time-evolution of the PDF $F(I, t)$ in the space of slowly–varying action variables I_i,

$$\frac{\partial F}{\partial t} = -\sum_{i=1}^{N} \frac{\partial}{\partial I_i} [u_i(I)F] + \sum_{i,j=1}^{N} \frac{\partial^2}{\partial I_i \partial I_j} \left[D_{ij}(I)F \right], \tag{10.15}$$

known as the Fokker–Planck equation (or also the Fokker–Planck–Kolmogorov equation). The quantities $u_i(I)$ and $D_{i,j}(I)$ are the convection and the diffusion $D_{i,j}(I)$ coefficients defined through the first and the second moments of displacements of orbits in the space of the action variables I, i.e.,

$$\begin{aligned} u_i(I) &= \frac{\langle \Delta I_i \rangle}{\Delta t}, \\ D_{ij}(I) &= \frac{\sigma_{ij}^2(t)}{2\Delta t} \equiv \frac{\langle \Delta I_i \Delta I_j \rangle}{2\Delta t}, \qquad \Delta t > \tau_c, \end{aligned} \tag{10.16}$$

where $\langle (\cdots) \rangle$ means averaging over initial angles, $\vartheta_0 = \vartheta(0)$, on a given level of the tori, $I \equiv I(0)$, and $\Delta I_i = I_i(t + \Delta t) - I_i(t)$ is the increment of the action variable I_i during time Δt. In the derivation of the kinetic equation it assumes that the phases $\vartheta_i(t + \Delta t)$ at the time instant $t + \Delta t$ become statistically independent on it's values at the previous time t.

For the dynamical chaotic Hamiltonian system (6.4)–(6.6) one can obtain the kinetic equation (10.15). The convection $u_i(I)$ and the diffusion $D_{ij}(I)$ coefficients expressed via the Hamiltonian functions (6.4) and (6.6) are given by [1]

[1] One should notice that in the works by Filonenko et al. (1967); Zaslavsky and Chirikov (1971); Sagdeev et al. (1988); Chirikov (1979) the diffusion coefficient D is defined as $D = \langle (\Delta I)^2 \rangle / \Delta t$, i.e., the factor $1/2$ is omitted. However this factor is put into the right hand side of the Fokker-Planck equation (10.15).

$$u_i^{(Q)} = \sum_j \frac{dD_{ij}(I)}{dI_j}, \tag{10.17}$$

$$D_{ij}^{(Q)}(I) = \frac{\pi\epsilon^2}{8} \sum_m m_i m_j |H_m|^2 \delta\left(m \cdot \omega(I)\right). \tag{10.18}$$

Using the relation between u_i and D_{ij} the kinetic equation (10.15) can be presented in the flux form

$$\frac{\partial F}{\partial t} = \sum_{i=1}^{N} \frac{\partial \Gamma_i}{\partial I_i}, \tag{10.19}$$

where

$$\Gamma_i = \sum_{j=1}^{N} D_{ij}(I) \frac{\partial F}{\partial I_j} \tag{10.20}$$

is the flux along the action variable I_i. Thus the kinetic equation describe the chaotic transport of "particles" across the tori.

Furthermore we will call $D_{ij}^{(Q)}(I)$ given by (10.18) as the *quasilinear diffusion coefficient* since it coincides with the one in the Fokker–Planck equation in the *quasilinear theory* of plasmas. The latter is the simplest kinetic description of particle-wave interaction in a turbulent plasma in a self–consistent way. However, the kinetic equation (10.15) corresponds to the non–self–consistent dynamics and describes a slow evolution of a distribution function $F(I,t)$ on a time scale longer than transit times $T \sim 1/\omega(I)$. It assumes that the particle motion does not affect back on the perturbation field (6.4) and (6.6).

One should note that the important step in the derivation of the kinetic equation for the chaotic Hamiltonian system is its mixing property. Particularly, for the 1D Hamiltonian system with the action angle variables (I, ϑ) it means that the phase correlation functions defined by

$$\begin{aligned} G_m(t) &= \langle \exp[im(\vartheta(t,\vartheta_0) - \vartheta_0)] \rangle \\ &= \frac{1}{2\pi} \int_0^{2\pi} \exp[im(\vartheta(t,\vartheta_0) - \vartheta_0)] d\vartheta_0, \end{aligned} \tag{10.21}$$

vanish at $t \to \infty$. Here $\vartheta(t,\vartheta_0)$ is the phase at the time instant t as a function of the initial phase $\vartheta_0 = \vartheta(0,\vartheta_0)$ at $t = 0$. For the strongly developed chaotic systems the correlation function $G_m(t)$ decays exponentially with t,

$$G_m(t) \sim \exp(-t/\tau_c), \tag{10.22}$$

where τ_c is the correlation time of phases.

10.1.3 Derivation of the Quasilinear Diffusion Coefficients

Below we present the derivation of the diffusion and the convection coefficients (10.18) and (10.17) for the one-dimensional Hamiltonian system in the presence of time–periodic perturbation. We rewrite the Hamiltonian system (6.4)–(6.6) in the form

$$\frac{d\vartheta}{dt} = \omega(I) + \epsilon V'(I, \vartheta, t), \qquad \frac{dI}{dt} = -\epsilon V(I, \vartheta, t), \tag{10.23}$$

where

$$V(I, \vartheta, t) = \sum_{m,n} m H_{mn}(I) \sin(m\vartheta - n\Omega t),$$
$$V'(I, \vartheta, t) = \sum_{m,n} \frac{dH_{mn}(I)}{dI} \cos(m\vartheta - n\Omega t). \tag{10.24}$$

where Ω is a frequency of perturbation. Using the unperturbed solutions $I = I_0 =$ const, $\vartheta^{(0)}(t) = \vartheta_0 + \omega(I)t$, one find the solutions $\vartheta^{(1)}(t)$, $I^{(1)}(t)$ up to the first order of ϵ,

$$I^{(1)}(t) = I_0 + \epsilon \int_0^t V(I_0, \vartheta^{(0)}(t'), t')dt',$$
$$\vartheta^{(1)}(t) = \vartheta^{(0)}(t) + \epsilon \int_0^t V'(I_0, \vartheta^{(0)}(t'), t')dt'. \tag{10.25}$$

Then the second Equation in (10.23) for the solution of $I(t)$ up to ϵ^2 is reduced to

$$\frac{dI^{(2)}}{dt} = \epsilon V(I^{(1)}(t), \vartheta^{(1)}(t), t) \approx \epsilon V(I^{(0)}(t), \vartheta^{(0)}(t), t)$$
$$+ \epsilon^2 \left[\frac{\partial V}{\partial I} \int_0^t V(I_0, \vartheta^{(0)}(t'), t')dt' \right.$$
$$\left. + \frac{\partial V}{\partial \vartheta}(I_0, \vartheta^{(0)}(t), t) \int_0^t V'(I_0, \vartheta^{(0)}(t'), t')dt' \right]. \tag{10.26}$$

Using these solutions one can write the following formulas for the convection and the diffusion coefficients,

$$u = \left\langle \frac{d\Delta I}{dt} \right\rangle \approx \epsilon \left\langle V(I^{(0)}(t), \vartheta^{(0)}(t), t) \right\rangle$$
$$+ \epsilon^2 \left\langle \frac{\partial V}{\partial I}(I_0, \vartheta^{(0)}(t), t) \int_0^t V(I_0, \vartheta^{(0)}(t'), t')dt' \right.$$
$$\left. + \frac{\partial V}{\partial \vartheta}(I_0, \vartheta^{(0)}(t), t) \int_0^t \frac{\partial V}{\partial I}(I_0, \vartheta^{(0)}(t'), t')dt' \right\rangle, \qquad (10.27)$$

$$D = \frac{1}{2} \left\langle \frac{d(\Delta I)^2}{dt} \right\rangle \approx \epsilon^2 \left\langle V(I_0, \vartheta^{(0)}(t), t) \int_0^t V(I_0, \vartheta^{(0)}(t'), t')dt' \right\rangle, \qquad (10.28)$$

where $\Delta I = I(t) - I_0$. Furthermore using the definition of the function $V(I, \vartheta, t)$ (10.24) in Eqs. (10.27) and (10.28), performing some lengthly calculations, one obtains for th convection coefficient,

$$\left\langle \frac{dI}{dt} \right\rangle = \epsilon^2 \sum_m \sum_{n,n'} m^2 H_{mn'} \frac{dH_{mn}}{dI} \frac{1}{\mu_{mn'}} \left\{ \sin\left[(n - n')\Omega t\right] + \sin(\mu_{mn} t) \right\},$$

where $\mu_{mn} = m\omega - n\Omega$. For only the single n mode we have

$$\left\langle \frac{dI}{dt} \right\rangle = \epsilon^2 \sum_m m^2 H_{mn} \frac{dH_{mn}}{dI} \frac{\sin(\mu_{mn} t)}{\mu_{mn}}.$$

At the times $t \gg |\mu_{mn}|^{-1}$ using the relation,

$$\frac{\sin(\mu_{mn} t)}{\mu_{mn}} \approx \pi \delta(\mu_{mn}), \text{ at } t \gg |\mu_{mn}|^{-1},$$

one obtains the following formula

$$\langle I - I_0 \rangle = u(I)t,$$
$$u(I) = \pi \epsilon^2 \sum_m m^2 H_{mn}(I) \frac{dH_{mn}}{dI}(I) \delta(m\omega - n\Omega). \qquad (10.29)$$

Performing the similar calculations one can obtain the following formula for the diffusion coefficient,

$$D(I, t) = \epsilon^2 \frac{1}{2} \sum_m \sum_{n,n'} m^2 H_{mn'} H_{mn} \frac{1}{\mu_{mn'}} \left\{ \sin\left[(n - n')\Omega t\right] + \sin(\mu_{mn} t) \right\}. \qquad (10.30)$$

For the single mode n one has

$$\left\langle (I - I_0)^2 \right\rangle = \epsilon^2 \sum_m m^2 |H_{mn}|^2 \frac{[1 - \cos((m\omega - n\Omega)t)]}{(m\omega - n\Omega)^2} \approx 2D(I)t, \qquad (10.31)$$

$$D(I) = \frac{\pi\epsilon^2}{2} \sum_m m^2 |H_{mn}|^2 \delta(m\omega - n\Omega). \qquad (10.32)$$

From Eqs. (10.29) and (10.32) one can readily follows the relation (10.17) between u and $D : u(I) = dD(I)/dI$. According to (10.29) it follows that the radial convection transport with the finite velocity $u \neq 0$ appears only for the radially inhomogeneous perturbations, i.e., $dH_m/dI \neq 0$.

One expects that the formulas (10.29) and (10.32) are valid for the sufficiently wide perturbation spectrum $H_{mn}(I)$ in m. However, there is no the rigorous proof of this statement. Nonetheless the numerical calculations of the transport coefficients performed in the next section show the accuracy of the quasilinear formulas is higher in the case of stochastic systems with the wider spectrum of perturbations (see Sect. 10.2.3, Figs. 10.4 and 10.6).

10.2 Diffusion of Particles and Field Lines in a Stochastic Magnetic Field

In this section we consider the simplified approach to the particle transport in a stochastic magnetic field created at the plasma edge by the applied RMPs. The ergodic divertors studied in Sects. 9.1 and 9.2 are examples of such a system. For simplicity, we study the transport of electrons in the chaotic magnetic field neglecting particle collisions and the effect of the small-scale magnetic and electric field fluctuations, i.e. the plasma turbulence. We also neglect the deviation of drift orbits from the magnetic surfaces. The effect of particle drifts on the transport will be briefly discussed in Sect. 10.2.4.

10.2.1 Particle Diffusion in Collisionless Plasmas

As we have seen in Sect. 5.1.3 for the thermal electrons of energy of several keV the deviation of the electron drift surfaces from the magnetic ones is of order of the electron gyroradius $\rho_c \sim 10^{-4}$ m. Therefore the electrons predominantly follow magnetic field lines and the radial transport of electrons across magnetic surfaces is determined by the radial deviation of field lines from the magnetic surfaces.[2]

Let X be the spatial coordinate normal to the magnetic surfaces. Then the radial convection and diffusion coefficients of particles are defined as $v_\perp = \langle \Delta X \rangle / \Delta t$ and

[2] However, this assumption is valid only for the passing particles. It does not take into account the trapped particles whose orbits does not lie on the magnetic surfaces.

$D_\perp = \langle(\Delta X)^2\rangle/2\Delta t$, where ΔX is a random radial advance of particle during time period Δt, and $\langle(\cdots)\rangle$ stands for averaging over magnetic surface. Suppose that $v_\parallel$ is the thermal velocity of electrons along the field lines. Particles make a full toroidal turn in time $\Delta t = \Delta l/v_T$ where $\Delta l = 2\pi R_0$ is the connection length of field line. Therefore we have

$$v_\perp = \frac{\langle\Delta X\rangle}{\Delta t} = \frac{\Delta l}{\Delta t}\frac{\langle\Delta X\rangle}{2\Delta l} = u_{FL}v_\parallel,$$
$$D_\perp = \frac{\langle(\Delta X)^2\rangle}{2\Delta t} = \frac{\Delta l}{\Delta t}\frac{\langle(\Delta X)^2\rangle}{2\Delta l} = D_{FL}v_\parallel, \tag{10.33}$$

where

$$u_{FL} = \frac{\langle\Delta X\rangle}{\Delta l}, \qquad D_{FL} = \frac{\langle(\Delta X)^2\rangle}{2\Delta l}, \tag{10.34}$$

are the convection and the diffusion coefficients of field lines. It means that in the collisionless plasma the convection and the diffusion of charged particles are mainly determined by the radial convection and diffusion of field lines.

The convection u_{FL} and the diffusion coefficient D_{FL} can be expressed through the corresponding convection $u(\psi_t)$ and diffusion coefficients $D(\psi_t)$ in the magnetic flux coordinate ψ_t,

$$u(\psi_t) = \frac{\langle\Delta\psi_t\rangle}{\Delta\varphi}, \qquad D(\psi_t) = \frac{\langle(\Delta\psi_t)^2\rangle}{2\Delta\varphi}. \tag{10.35}$$

To do this one should use the relation between the radial coordinate X and the toroidal flux ψ_t (or the poloidal flux ψ). Then

$$u_{FL} = \frac{\langle\Delta X\rangle}{\Delta l} = \frac{1}{R_0}\frac{dX}{d\psi_t}u(\psi_t),$$
$$D_{FL} = \frac{\langle(\Delta X)^2\rangle}{2\Delta l} = \frac{1}{R_0}\left(\frac{dX}{d\psi_t}\right)^2 D(\psi_t). \tag{10.36}$$

In the plasmas with the circular cross sections the radius of the magnetic surface r can be chosen as the radial coordinate X (see Sects. 2.2.1, 2.2.2, and Fig. 2.1). In the plasma with the elongated shape the coordinate X can be chosen along the section Σ_S connecting the X-point with the magnetic axis as shown Fig. 2.4a.

10.2.2 Transport of Chaotic Field Lines

The kinetic equation for chaotic field lines can be obtained from the Hamiltonian system (3.9) and (3.10). Replacing the time t by the toroidal angle φ and the action I by the toroidal flux ψ_t the kinetic equation (10.15) for the magnetic field lines can be presented as

$$\frac{\partial F}{\partial \varphi} = -\frac{\partial}{\partial \psi_t}\left[u(\psi_t)F\right] + \frac{\partial^2}{\partial \psi_t^2}\left[D(\psi_t)F\right]. \tag{10.37}$$

In the quasilinear approximation (10.18) the diffusion coefficients $D(\psi_t)$ is given by

$$u^{(Q)}(\psi_t) = \frac{dD(\psi_t)}{d\psi_t}, \tag{10.38}$$

$$D^{(Q)}(\psi_t) = \frac{\pi\epsilon^2}{2}\sum_{m,n} m^2 |H_{m,n}|^2 \delta\left(\frac{m}{q(\psi_t)} - n\right). \tag{10.39}$$

Using the relation (3.17) the diffusion coefficient $D^{(Q)}(\psi_t)$ is rewritten in terms of the spectra of magnetic perturbations b_{mn},

$$D^{(Q)}(\psi_t) = \frac{\pi\epsilon^2}{2}\sum_{m,n} |b_{mn}|^2 \delta\left(\frac{m}{q(\psi_t)} - n\right). \tag{10.40}$$

As seen from Eq. (10.39) the presence of the δ—functions makes the diffusion coefficient $D^{(Q)}(\psi_t)$ defined only at the resonant magnetic surfaces ψ_{mn}, $q(\psi_{mn}) = m/n$. Using the relation (6.83) between the Fourier coefficients H_{mn} of the perturbation Hamiltonian and the Poincaré integrals R_n, one can extend the definition of $D^{(Q)}(\psi_t)$ for the arbitrary values of the toroidal flux ψ_t,

$$D^{(Q)}(\psi_t) = \frac{\epsilon^2 q(\psi_t)}{8\pi}\sum_{n} n^2 |R_n^{(reg)}(\psi_t)|^2. \tag{10.41}$$

This formula allows one to estimate the diffusion coefficients of field lines in the chaotic systems using only one function, the Poincaré integral $R_n^{(reg)}(\psi)$, corresponding to the each toroidal mode n. Unlike Eq. (10.39) it is a continuous function of the flux variable ψ_t.

10.2.3 Numerical Diffusion Coefficients

Below we calculate the transport of coefficients $u(\psi_t)$ and $D(\psi_t)$ of chaotic field lines for the model of the ergodic divertor proposed in Sect. 9.1.3. The typical examples of the Poincaré sections of chaotic field lines at the plasma edge created by the RMPs were shown in Fig. 9.6a, b for the Tore Supra-like and the TEXTOR-DED plasmas. One should note two distinctive features of these stochastic zones. Firstly, the stochastic zone is formed in a radially limited domain with the upper boundary at the plasma wall. Secondly, such a stochastic zone is open and field lines should be terminated after hitting the wall. These features allows one introduce only the local transport coefficients defined above.[3]

The convection and the diffusion coefficients of chaotic field lines can be directly calculated by the numerical integration of the Hamiltonian equations (3.9) and (3.10). Before one should note that the transport coefficients are local functions of the magnetic flux coordinate ψ_i. Therefore, the first and the second moments of displacements of field lines in the definitions of $u(\psi_t)$ and $D(\psi_t)$ (10.35) should take this fact into account.

Let ψ_0 be the toroidal flux of the given magnetic surface. We will define the first and the second moments of displacements $\psi_t(\varphi)-\psi_0$ of field lines by averaging over the large number M, ($M \gg 1$), of field lines launched from the magnetic surface ψ_0 with uniformly distributed initial poloidal angles $\vartheta_{i0} = (2\pi/M)i$, ($i = 1, \dots, M$), at the poloidal section $\varphi = 0$,

$$M_1(\varphi) \equiv \langle \psi_t(\varphi) \rangle - \psi_0 = \frac{1}{M}\sum_{i=1}^{M} \psi_t(\varphi, \vartheta_{i0}) - \psi_0,$$

$$\sigma^2(\varphi) \equiv \langle (\psi_t(\varphi) - \langle \psi_t(\varphi) \rangle)^2 \rangle = \frac{1}{M}\sum_{i=1}^{M} [\psi_t(\varphi, \vartheta_{i0}) - \langle \psi_t(\varphi) \rangle]^2 . \qquad (10.42)$$

where $\psi_t(\varphi, \vartheta_{i0})$ is the toroidal flux of a field line at the section $\varphi = 0$ with the initial poloidal angle ϑ_{i0} at $\varphi = 0$.

10.2.3.1 Phase Correlation Function

The correlation function of the phase $G_m(l)$, $l = R_0\varphi$, (10.21) for the field lines is calculated by the formula

$$G_m(l) = \frac{1}{M}\sum_{i=1}^{M} \exp\left[im(\vartheta(l, \vartheta_{i0}) - \vartheta_{i0})\right] . \qquad (10.43)$$

The e-folding decay length l_{cr} of the correlation function,

$$G_m(l) \sim \exp\left(-l/l_{cr}\right) , \qquad (10.44)$$

defines the correlation length l_{cr} of the perturbation magnetic field. Figure 10.2 shows the typical dependencies of $G_m(l)$ on the length along the field line l for the model

[3] We should note that in the unlimited stochastic domain one can define the diffusion coefficient as $D = \sigma^2(t)/2t$, at $t \to \infty$. If such a diffusion coefficient has a finite value the process called a normal Gaussian process. For the partially chaotic motion with the KAM stability islands the second moment $\sigma^2(t)$ may have the following asymptotics $\sigma^2(t) \sim t^\gamma$, at $t \to \infty$ with the $\gamma \neq 1$. Such chaotic processes is called *subdiffusive* if $\gamma < 1$, and *superdiffusive* if $\gamma > 1$.

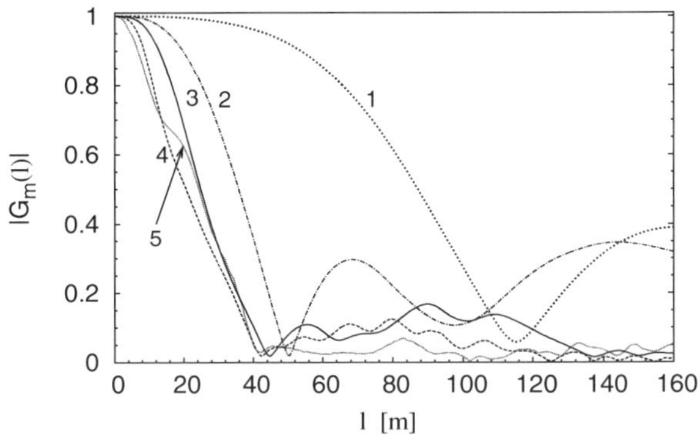

Fig. 10.2 Typical dependencies of the correlation function $G_m(l)$ (10.43) on the distance $l = R_0\varphi$ along the toroidal angle φ at the different values of the perturbation parameter ϵ: *curve 1* corresponds to $\epsilon = 10^{-4}$, *curve 2* corresponds to 5×10^{-4}, *curve 3* corresponds to 10^{-3}, *curve 4* corresponds to 2×10^{-3}, and *curve 5* corresponds to 4×10^{-3}. Field lines are launched from the magnetic surface $\psi = 0.79\psi_a$. The mode number is $m = 18$. Other parameters are the same as Fig. 9.6a

of the magnetic field proposed in Sect. 9.1.3. The plasma parameters correspond to the ED of Tore Supra and given in Fig. 9a. The mode number m is taken equal to the central mode $m_c = m_0\gamma_1 = 18$.

As one can see that the e-folding decay length l_{cr}, determined at the $1/e$ level of $G_m(l)$, decreases with increasing the perturbation parameter ϵ and for $\epsilon \gtrsim 10^{-3}$ reaches its saturated value l_{cr} of order of the connection length $l_c = \pi q(\psi)R_0 = 21$ m. One expects that in tokamaks with the localized RMPs along the poloidal angle θ (or ϑ) as shown, for example in Figs. 3.9, A.2, A.5 the correlation length l_{cr} should not take a value less than the connection length l_c. Field lines can be significantly affected by perturbation fields at least in a half of poloidal turn after which the phase information can be lost. Therefore the corresponding length is of order of the connection length l_c.

10.2.3.2 Diffusion Coefficient

According to the relation (6.83) and Eq. (9.5) for the spectrum of the perturbation H_{mn} we obtain the following formula for the Poincaré integrals $R_n(\psi)$,

$$R_n^{(reg)}(\psi) = \frac{2\pi C_m}{n}\left(\frac{r}{r_{ed}}\right)^{nq(\psi)/\gamma_1} a(y), \qquad r = R_0\sqrt{\psi_t},$$

$$a(y) = \frac{\sin y}{y}, \qquad y = [nq(\psi)/\gamma_1 - m_0]\theta_c, \tag{10.45}$$

where the safety factor $q(\psi)$ is defined by Eq. (9.6). Using Eq. (10.36) and the relation between the radial coordinate r and the flux coordinate we obtain the convection and the diffusion coefficients in the quasilinear approximation,

$$u_{FL}^{(Q)} = \frac{R_0}{r} u^{(Q)}(\psi_t), \qquad D_{FL}^{(Q)} = \frac{R_0^3}{r^2} D^{(Q)}(\psi_t), \tag{10.46}$$

where $u^{(Q)}(\psi_t)$ and $D^{(Q)}(\psi_t)$ are given by

$$u^{(Q)}(\psi_t) = \frac{dD^{(Q)}(\psi_t)}{d\psi_t}, \quad D^{(Q)}(\psi_t) = \frac{\epsilon^2 n^2 q(\psi_t)}{8\pi} |R_n^{(reg)}(\psi_t)|^2. \tag{10.47}$$

Figure 10.3a, b shows the typical dependencies of the first and the second moments of the radial displacements, $\langle \Delta r \rangle$, $\langle (\Delta r)^2 \rangle$ on the toroidal angle φ for the ED of Tore Supra. Calculations are made according the definitions (10.42) using the relation $\Delta r = (dr/d\psi_t)\Delta\psi_t$. At the same figure the dependence of the correlation functions of angles $G_m(l)$ on the toroidal angle φ for the central mode $m = m_0^* = 18$ is shown [right-hand axis].

As seen from Fig. 10.3a, b the first and second moments $\langle \Delta r \rangle$, $\langle (\Delta r)^2 \rangle$ grow linearly at the initial interval of φ, $0 < \varphi < \varphi_1$. The upper limit φ_1 slightly larger than the decay length $\varphi_c = l_{cr}/R_0$ of the correlation function $G_m(\varphi)$. There is the second interval of φ, $\varphi_1 < \varphi < \varphi_2$, with the linear dependencies of the moments on φ. For the values $\varphi > \varphi_2$ the second moment $\langle (\Delta r)^2 \rangle$ reaches its saturated value and does not grow anymore.

Using the definition (10.35) we determine the coefficients $u(\psi_t)$ and $D(\psi_t)$ by fitting the moments $M_1(\varphi)$ and $\sigma^2(\varphi)$, with the linear functions $u_0 + u_{FL}\varphi$ and $D_0 + 2D_{FL}\varphi$, respectively, at the initial linear stage of their growth, i.e., at the interval $0 < \varphi < \varphi_1$. Since the upper limit of the interval φ_1 is of order of the correlation length φ_c of angle, the field lines are not completely decorrelated, i.e., they are not fully chaotic. Therefore, following to Elskens and Escande (2003) we call the coefficient D_{FL} defined above as the *non-chaotic diffusion coefficients*. On the other hand the coefficients D_{FL} determined in the second interval $\varphi_1 < \varphi < \varphi_2$ of the linear growth of $\sigma^2(\varphi)$ we call the *chaotic diffusion coefficients* since the field lines are fully decorrelated there.

Figure 10.4a, b shows the typical dependencies of the convection u_{FL} and non-chaotic diffusion D_{FL} coefficients, respectively, on the normalized toroidal flux coordinate ψ/ψ_a for the stochastic zone of the ED of Tore Supra. The dotted curves 1–3 present the numerical calculations for the three different perturbation amplitudes $\epsilon = 5 \times 10^{-4}$, 10^{-3}, 2×10^{-3}, respectively. The solid curves 4–6 present the corresponding dependencies given by the quasilinear formula (10.46).

As seen from Fig. 10.4a, b the numerically calculated convection and the diffusion coefficients closely follow the quasilinear approximation (10.46) at the certain radial intervals $\psi_1 < \psi_t < \psi_{lam}$. For the large perturbation ϵ the numerical values of D_{FL} start to deviate from the quasilinear formula (see curves 3 and 6 in Fig. 10.4b). However, for $\psi_t > \psi_{lam}$ the numerical coefficients abruptly decay and completely deviate from the quasilinear formula. The area of the stochastic layer $\psi_t > \psi_{lam}$ can define the quantitative measure of the laminar zone introduced in Sect. 9.4.2.

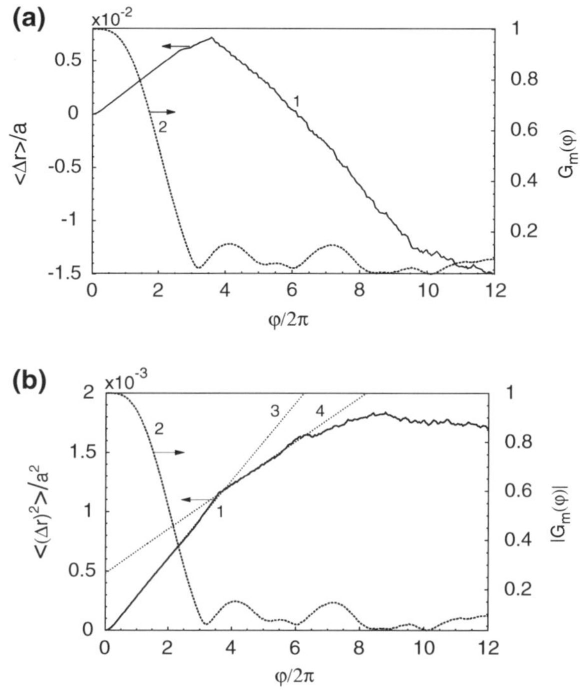

Fig. 10.3 Typical dependencies of the first moment $\langle \Delta r \rangle$ (**a**) and the second moment $\langle (\Delta r)^2 \rangle$ (**b**) of the radial displacement Δr and the correlation function $G_m(\varphi)$ (10.43) [*right-hand axis*] along the toroidal angle φ. Field lines are launched from the magnetic surface $\psi = 0.85\psi_a$. The toroidal mode $n = 6$, the perturbation parameter $\epsilon = 10^{-3}$, and other parameters are the same as Fig. 9.6a. *Curve 1* corresponds to the numerical calculations, *curve 2* describes the correlation function $G_m(\varphi)$, $\varphi = l/R_0$, for the central mode $m = 18$, and *curve 3* describes the linear function $f(\varphi) = d_0 + 2D\varphi$ with the coefficients d_0, D_r, obtained by fitting to the second moment $\langle (\Delta r)^2 \rangle$ at the initial linearly growing range of φ, and *curve 4* is a linear function $f_c(\varphi) = d_1 + 2D_c\varphi$ with the coefficients d_1, D_c obtained by fitting in the second linearly growing range of $\langle (\Delta r)^2 \rangle$

The lower boundary of this zone ψ_{lam} depends on the perturbation amplitude ϵ: it decreases with increasing ϵ.

As was noted even in the region $\psi_1 < \psi_t < \psi_{lam}$ the numerical diffusion coefficients deviate from the quasilinear formula (10.46) with increasing the perturbation amplitude ϵ. This is shown in Fig. 10.5 by plotting the diffusion coefficient D_{FL} versus ϵ at the fixed magnetic surface $\psi = 0.72\psi_a$. Beside of that we have plotted also the chaotic diffusion coefficient D_c versus ϵ (curve 3) found in the second interval of the linear dependences of σ^2 on φ (see, Fig. 10.3b). As seen from Fig. 10.5 the non-chaotic diffusion coefficient monotonically deviates from the quasilinear value with increasing ϵ. At the same time the chaotic diffusion coefficient changes irregular for larger ϵ although it coincides with the non-chaotic one at small ϵ.

Such a behavior of the transport coefficients is related with openness of the field lines to the wall. As was shown in Sect. 9.4.2 this creates the laminar zone of

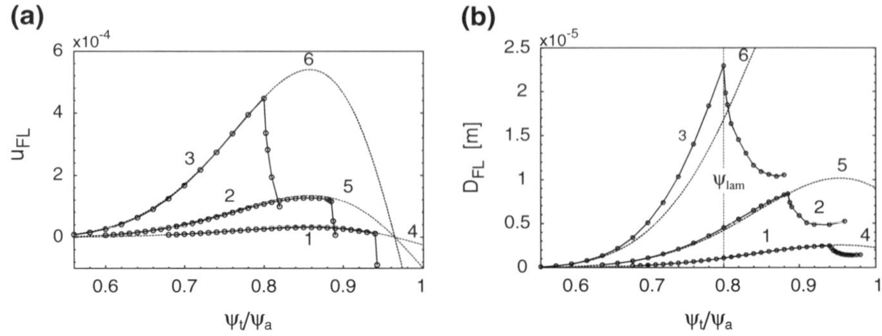

Fig. 10.4 Radial profiles of local field line convection u_{FL} (**a**) and diffusion D_{FL} (**b**) coefficients for the ED of Tore Supra. The toroidal mode $n = 6$, perturbation parameter $\epsilon = 5 \times 10^{-4}$ (*curves 1* and *4*), $\epsilon = 10^{-3}$ (*curves 2* and *5*), $\epsilon = 2 \times 10^{-3}$ (*curves 3* and *6*). Other parameters are the same as in Fig. 9.6a. *Curves 1–3* correspond to the numerical calculations, *curves 4–6* correspond to the quasilinear values calculated by Eq. (10.46)

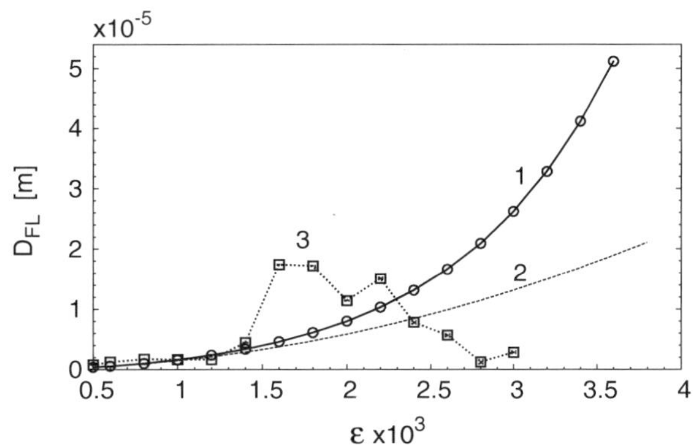

Fig. 10.5 Dependence of the nonchaotic (*curve 1*) and the chaotic (*curve 3*) diffusion coefficients D_{FL} on the perturbation parameter ϵ at the given $\psi_t = 0.72\psi_a$. A quasilinear dependence of D_{FL} on ϵ is given by *curve 2*. The parameters are the same as in Fig. 10.4

stochastic field lines possessing the fractal structures like ones presented in Fig. 9.25. The majority of field lines at the very edge of plasma, i.e., the laminar zone, have short connection lengths and they are not fully chaotic. The transport of particles along these field lines are not chaotic, therefore cannot be fully described by the convection and the diffusion coefficients. The existence of the boundary value ψ_{lam} which sharply separates the two different regimes of transport of field lines gives one to quantitatively define the laminar zone as $\psi_t > \psi_{lam}$.

The typical radial profiles of the convection and diffusion coefficients in the TEXTOR-DED are presented in Fig. 10.6a, b. In this case the width of the poloidal spectrum of the RMP and the radial width of stochastic zone in the TEXTOR-DED are smaller than of those in the ED Tore Supra (see Sects. 9.1, 9.1.1, 9.1.2 and Fig. 9.6a, b). In spite of that the quasilinear formulas well describes the transport coefficients u_{FL} and D_{FL} in the inner regions of the stochastic zone, although their accuracy is lower that in the case of the ED of the Tore Supra.

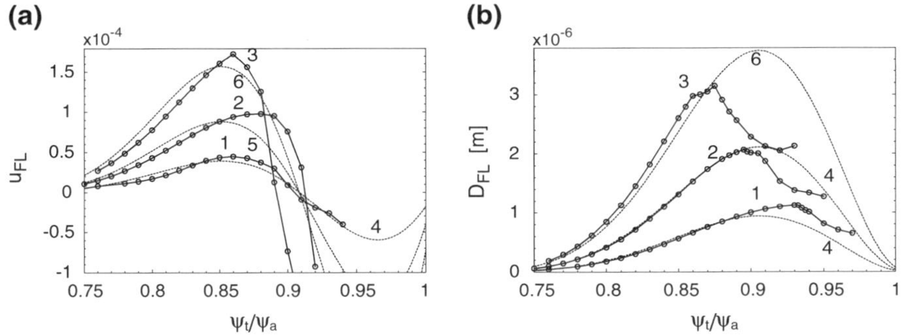

Fig. 10.6 The same as in Fig. 10.4a, b but for the TEXTOR-DED tokamak. The toroidal mode $n = 4$, perturbation parameter $\epsilon = 4 \times 10^{-3}$ (*curves 1* and *4*), $\epsilon = 6 \times 10^{-3}$ (*curves 2* and *5*), $\epsilon = 8 \times 10^{-3}$ (*curves 3* and *6*). Other parameters are the same as in Fig. 9.6a. *Curves 1–3* correspond to the numerical calculations, *curves 4–6* correspond to the quasilinear values calculated by Eq. (10.46)

10.2.4 Drift Effects of Particle Diffusion

The above consideration of collisionless particle diffusion in a stochastic magnetic field assumes that particle orbits closely follow magnetic field lines. This assumption is valid for the passing low energetic particles, especially for the thermal electron. With increasing the energy of particles their orbits start to deviate from magnetic field lines. Since the direction of this shift of guiding-center orbits is different for co–passing and counter–passing particles the magnetic field perturbation affect differently on these of types of particles. Therefore the diffusion of these passing particles in a stochastic magnetic field will deviate from the ones given by Eqs. (10.33) and (10.34), and it will depend on the direction of particle motion.

The effect of orbit drift on particle diffusion can be studied by the slight modifications of the above equations: the diffusion coefficient of field lines D_{FL} in (10.33) should be replaced by the diffusion coefficient of orbits. This leads to the dependence of diffusion coefficients on the direction of particle motion and on an electric field. These effects of particle drifts on the transport of ions in a stochastic magnetic field of the TEXTOR–DED have been studied by Abdullaev et al. (2006a) (see, also references in Sect. 10.5).

10.3 The Kolmogorov Length

As we have seen in Sect. 7.3 the *Lyapunov exponent* gives the quantitative measure of exponential divergence of nearby orbits or field lines in the ergodic zone. As we have recalled in Sect. 10.1 the Lyapunov exponent is related with another statistical characteristics of chaotic dynamical systems, namely, with the Kolmogorov time τ_K which characterizes the loss of information on initial state of system in time.

For the chaotic field lines one can introduce the *Kolmogorov length* L_K which is an analog of the Kolmogorov time τ_K defined by Eq. (10.12), since the length along the toroidal direction $l = R_0\varphi$ plays a role of time t. It gives a quantitative measure of the distance at which chaotic field lines are de-correlated.

In this section we estimate the Lyapunov exponent and the Kolmogorov length of magnetic field lines in the model of a stochastic magnetic field considered in the previous sections. Since field lines in the stochastic zone are open and have finite connection lengths we cannot determine the global Lyapunov exponent σ as defined by Eq. (7.50). Therefore we need to introduce a *finite–time Lyapunov exponent* and a *finite-length Kolmogorov length.*

We define the finite–time Lyapunov exponent, $\sigma(N;\vartheta_0,\psi_0)$, for finite steps N of the mapping,

$$\sigma(N,\vartheta_0,\psi_0) = \frac{1}{N}\ln\frac{\delta s_N}{\delta s_0} = \frac{1}{N}\ln\prod_{k=1}^{N}\lambda^{(k)}. \tag{10.48}$$

It characterizes the *local divergency of orbits* in a chaotic system and depends on the initial coordinates (ϑ_0,ψ_0).

Furthermore, we will consider a system of magnetic field lines described by the mapping (6.29)–(6.31). We should notice that since the toroidal angle φ plays a role of time–like variable t the Lyapunov exponent σ should have no dimension and the quantity l_s/σ characterizes the e-folding length of exponential separations of neighboring field lines along the toroidal axis $l = R_0\varphi$. Here l_s is the length of a field line per one mapping step.

Therefore the quantity defined as

$$L(N,\vartheta_0,\psi_0) = \frac{l_s}{\sigma(N,\vartheta_0,\psi_0)}, \tag{10.49}$$

gives a quantitative measure of the finite–time (or finite–length) e-folding length of exponential separation of neighboring orbits. The quantity $L(N;\vartheta,\psi)$ as well as $\sigma(N;\vartheta,\psi)$ sensitively depends on the initial condition (ϑ,ψ). Particularly, its dependence on the initial poloidal angle ϑ has a fractal nature. A typical dependence of $L(N;\vartheta,\psi)$ on ϑ is plotted in Fig. 10.7 for the stochastic layer of the magnetic field model in Sect. 9.1.3. The perturbation parameter is taken $\epsilon = 10^{-3}$, and other parameters are the same as in the Fig. 9.6a. Initial field lines are taken on the magnetic surface $\psi = 0.79\psi_a$, and they are run up to 240 toroidal turns or until they reach the divertor plate. As seen from Fig. 10.7a, b the dependence $L(N;\vartheta,\psi)$ on the initial poloidal angle has indeed a fractal nature.

As we have seen the length $L(N;\vartheta,\psi)$ for open chaotic field lines depends on initial conditions. Below we introduce the characteristic e-folding length of exponential divergence of neighboring field lines of given magnetic surface ψ by averaging over initial poloidal angles. By analogy to the Kolmogorov time τ_K (10.12) we define the Kolmogorov length, $L_K(N,\psi)$, as the inverse of the finite-time Lyapunov exponent, $\bar{\sigma}(N;\psi)$, averaged over the magnetic surface ψ:

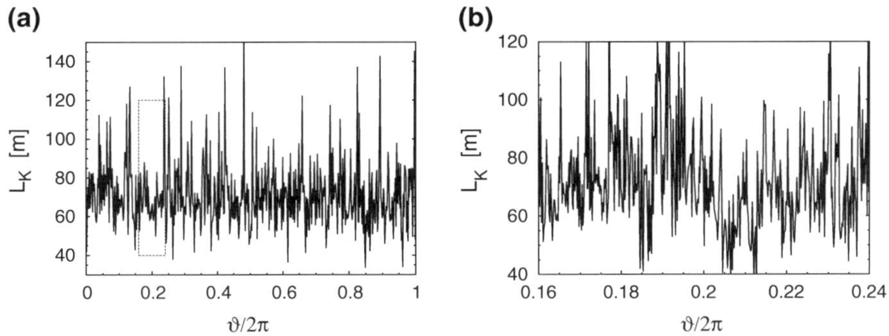

Fig. 10.7 E-folding length $L(N; \vartheta, \psi)$ versus the initial poloidal angle ϑ. Initial field lines taken on the magnetic surface $\psi_t = 0.79\psi_a$. The perturbation parameter $\epsilon = 10^{-3}$. **a** Shows the whole interval of ϑ. **b** Shows the expanded view in the interval $0.16 \leq \vartheta/2\pi \leq 0.24$ shown in (**a**) by a *rectangular area*

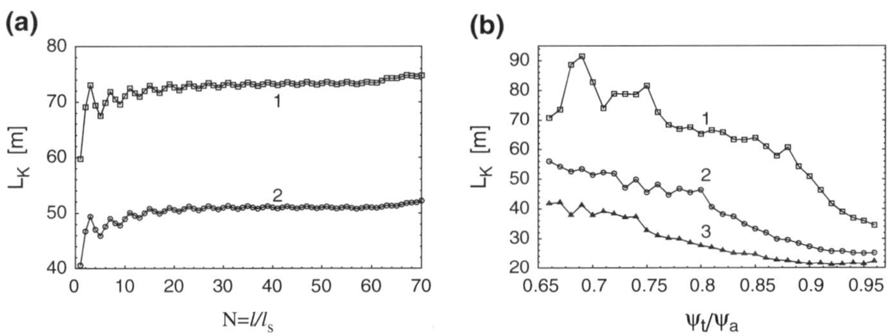

Fig. 10.8 **a** Finite-time Kolmogorov length $L_K(N)$ versus the mapping step $N = l/l_s$ at the given magnetic surface $\psi_t = 0.79\psi_a$: *curve 1* corresponds to the perturbation parameter $\epsilon = 10^{-3}$ and *curve 2*—$\epsilon = 2 \times 0^{-3}$; **b** The saturated values of the Kolmogorov length $L_K(N)$ for the *three* different values of the perturbation parameter ϵ: 10^{-3} (*curve 1*), 2×10^{-3} (*curve 2*), and 3×10^{-3} (*curve 3*). The mapping step $l_s = R_0(2\pi/24)$

$$L_K(N, \psi) = \frac{l}{\bar{\sigma}_N},$$

$$\bar{\sigma}_N = \frac{1}{M}\sum_{i=1}^{M} \sigma(N, \vartheta_i, \psi), \tag{10.50}$$

where $\sigma(N, \vartheta_i)$ (10.48) with the initial coordinate $(\vartheta_i = 2\pi i/M, \psi)$, and M is a number of field lines. According to the definition (10.50) is a function of poloidal flux ψ of a magnetic surface.

Figures 10.8a shows the dependence of the Kolmogorov length $L_K(N, \psi)$ on the mapping step N. The length of the mapping step is taken equal to $l_s = R_0(2\pi/24)$. One can see that $L_K(N, \psi)$ saturates at $N > 40$ which corresponds to a several toroidal turns. The saturated values of the Kolmogorov length $L_K(N, \psi)$ depend on the positions of magnetic surfaces ψ as shown in Fig. 10.8b. The value of $L_K(N, \psi)$ decreases with the increase of the perturbation ϵ. With increasing ψ the Kolmogorov length $L_K(N, \psi)$ overall decreases since the perturbation field grows outwardly (see curves 1–3 in Fig. 10.8b). However, $L_K(N, \psi)$ has some local maxima located near the remnants of magnetic islands which can be seen in the Poincaré sections of field lines shown Fig. 10.9. This is because of the regular field lines in magnetic islands which effectively decrease the Lyapunov exponents $\sigma(N, \vartheta, \psi)$ or (increase $L(N, \vartheta, \psi)$. The influence of the structure of the stochastic layer becomes weaker for the higher magnitudes of perturbations ϵ because of shrinking the sizes of magnetic islands.

10.4 Effect of Collisions on a Particle Diffusion in a Stochastic Field

Up to now we considered the collisionless particle transport in a stochastic field. As we have seen such a transport is determined by the transport properties of magnetic field lines. The particle collisions and the plasma turbulence, however, modifies these transport properties. In general the problem of particle and heat transport in a stochastic magnetic field is very complex. It has been a subject of many investigations starting from the middle of 1970s and it remains still one of the important issues of the plasma transport in magnetic fusion devices.

In this section we study the effect of collisions on the transport of passing particles in a stochastic magnetic field [Abdullaev (2013)]. We will not take into account the trapped particle transport. We will use the model a stochastic magnetic field employed in the previous section.

First we recall some important quantity describing collisional plasmas. Let $\chi_\parallel$ and $\chi_\perp$ be the parallel and the perpendicular diffusion coefficients of particles in a

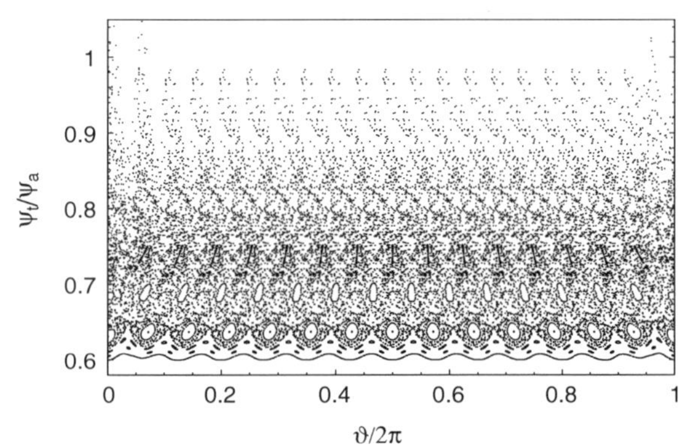

Fig. 10.9 Poincaré section of field lines for the model of the magnetic field in Sect. 9.1.3 with the perturbation parameter $\epsilon = 10^{-3}$. Other parameters are the same as in Fig. 9.6a

collisional plasmas in the absence of magnetic perturbations. The mean–free path λ_{mfp} of a particle is related to the parallel diffusion coefficient, $\chi_{\|} = v_{\|}\lambda_{mfp}/2$. For the electrons it is given by the following formula

$$\lambda_{mfp} = \lambda_e = v_{T_e}\tau_e = 8.5 \times 10^{21}(T_e^2/n) \text{ m}, \tag{10.51}$$

where τ_e is the electron collision time (with ions), T_e is the electron temperature in keV, n is a plasma density in the unit m^{-3}, and the Coulomb logarithm is taken equal to $\ln \Lambda = 17$ (see Wesson (2004); Huba (2000)). At a given temperature the ion mean free path $\lambda_i = v_{T_i}\tau_i$ is equal λ_e, where τ_i is the ion collision time.

For simplicity the parallel velocity $v_{\|}$ is taken equal to the thermal velocity $v_T = (kT_i/m_i)^{1/2} = 3.09 \times 10^5 T_i^{1/2}$ m/s, where k is the Boltzmann's constant.

10.4.1 Collisional Diffusion Regimes

The effect of particle collisions on the particle transport in a stochastic magnetic field has been considered in the numerous analytical and numerical works (see references in Sect. 10.5). Since it is difficult to perform the analytical calculations in dynamically chaotic systems the analytical theories are mostly based on the heuristic arguments rather than on the rigorous mathematical treatments. In such theories the stochastic magnetic field is introduced by its property of the exponential divergency of magnetic field lines with close initial coordinates and it is characterized by the Kolmogorov length L_K. The predictions of these theories for the radial diffusion coefficient D_r have only a qualitative nature. Particularly, the different regimes of the transport have been identified. Below we recall some well–known regimes.

As we have seen in Sect. 10.2.1 in the *collisionless regime* the diffusion coefficient D_r is determined only by the diffusion coefficient of magnetic field lines, D_{FL}, $D_r = v_T D_{FL}$. This regime is valid when the mean free path λ_{mfp} is much longer than the Kolmogorov length L_K.

For moderately collisional plasmas with $\lambda_{mfp} \lesssim L_K$ one expects the Rechester-Rosenbluth regime [Rechester and Rosenbluth (1978)] with the diffusion coefficient is determined by the mean free path λ_{mfp} and the Kolmogorov length L_K

$$D_r = \frac{v_T D_{FL}\lambda_{mfp}}{L_{K\delta}} = \frac{D_{FL}2\chi_{\|}}{L_{K\delta}}, \tag{10.52}$$

where the characteristic length $L_{k\delta}$ is determined by

$$L_{k\delta} = L_K \ln\left[\frac{l_{cr}}{L_K}\left(\frac{\chi_{\|}}{\chi_{\perp}}\right)^{1/2}\right]. \tag{10.53}$$

Here l_{cr} is the de-correlation length of the perturbation magnetic field across unperturbed field line (see Eq. (10.44)). According to Laval (1993) this regime is valid at the conditions $\lambda_{mfp} < L_K < l_{cr} \left(\chi_{\parallel}/\chi_{\perp}\right)^{1/2}$.

One expects that in extremely collisional plasmas the Kadomtsev-Pogutse regime is valid [Kadomtsev and Pogutse (1979)]. The corresponding diffusion coefficient is given by

$$D_r = D_{FL} 2 l_{cr}^{-1} \sqrt{\chi_{\parallel}\chi_{\perp}}. \tag{10.54}$$

The validity conditions of the this regime are $L_K > l_{cr} \left(\chi_{\parallel}/\chi_{\perp}\right)^{1/2}$ and $D_{FL} 2\sqrt{\chi_{\parallel}/\chi_{\perp}} > l_{cr}$. The detailed discussions of these regimes are reviewed by Krommes et al. (1983); Laval (1993); Isichenko (1991a).

Laval (1993) considered a so-called a long mean free path regime with the diffusion coefficient

$$D_r = \frac{v_{\parallel} D_{FL}}{1 + l_{c\delta}/\lambda_{mfp}}, \tag{10.55}$$

where

$$l_{c\delta} = L_K \ln \left[v_{\parallel} l_{cr}^{-1} \left(\chi_{\parallel}\chi_{\perp}\right)^{1/2} \right]. \tag{10.56}$$

It is valid at the conditions $L_K < \chi_{\parallel}/v_{\parallel} = \lambda_{mfp}/2$. The diffusion coefficients of all these regimes and their applicability conditions are listed in Table 10.1.

One should note the following feature on these regimes of collisional transport. At the limit of vanishing the perpendicular diffusion coefficient $\chi_{\perp}$, i.e., $\chi_{\perp} \to 0$, the diffusion coefficients D_r become extremely small, i.e., $D_r \sim 1/|\ln \chi_{\perp}|$ for the Rechester–Rosenbluth and Laval regimes, $D_r \sim \chi_{\perp}^{1/2}$ for the Kadomtsev–Pogutse regime. However, this contradicts to the fact that the radial transport of particles in

Table 10.1 Collisional diffusion regimes in a stochastic magnetic field

Regimes	Range of applicability	Diffusion coefficient D_r
Collisionless	$\lambda \gg L_K$	$D_r = v_{\parallel} D_{FL}$
Rechester–Rosenbluth	$\lambda_{mfp} < L_K$	$D_r = \dfrac{v_{\parallel} D_{FL}}{L_{K\delta}}$
	$L_K < l_{cr} \left(\dfrac{\chi_{\parallel}}{\chi_{\perp}}\right)^{1/2}$	$L_{k\delta} = L_K \ln \left[\dfrac{l_{cr}}{L_K} \left(\dfrac{\chi_{\parallel}}{\chi_{\perp}}\right)^{1/2} \right]$
Kadomtsev–Pogutse	$L_K > l_{cr} \left(\dfrac{\chi_{\parallel}}{\chi_{\perp}}\right)^{1/2}$	$D_r = D_{FL} \dfrac{2\sqrt{\chi_{\parallel}\chi_{\perp}}}{l_{cr}}$
	$D_{FL} 2\sqrt{\dfrac{\chi_{\parallel}}{\chi_{\perp}}} > l_{cr}$	
Laval	$\lambda_{mfp} > 2L_k$	$D_r = \dfrac{v_{\parallel} D_{FL}}{1 + l_{c\delta}/\lambda_{mfp}}$
		$l_{c\delta} = L_K \ln \left[v_{\parallel} l_{cr}^{-1} \left(\chi_{\parallel}\chi_{\perp}\right)^{1/2} \right]$

a stochastic magnetic field occurs independent on whether or not the perpendicular diffusion $\chi_\perp$ disappears.

The verification of these collisional diffusion regimes were made in a number of works. Among them one can note the work by Rax and White (1992) and its generalization by Rack et al. (2012). They have used the standard mapping (6.129) as a model for the stochastic magnetic field. The collisions are introduced by random kicks across magnetic surfaces.

However, this model does not quite well represent the stochastic magnetic field in real tokamak experiments. As we have seen in the previous chapters the stochastic magnetic field created by the externally applied magnetic perturbations covers a finite region at the plasma edge with open magnetic field lines. The magnitude of the magnetic field perturbations strongly depends on the spatial coordinates. These features of the perturbation magnetic field in real tokamak plasmas are not captured by the standard mapping model.

Below we present a numerical study of collisional diffusion in a realistic model of a stochastic magnetic field. Particularly, the model of the magnetic perturbation field for the ergodic divertor proposed in Sect. 9.1.3 will be employed.

10.4.2 Random Walk Model for the Particle Collisions

In order to introduce the particle collisions in the transport of particles we consider the following simple model of random walk of test particles. Let y be a coordinate along the magnetic field line, while x be a coordinate perpendicular to the field lines. We assume that a particle motion is a random walk process: it moves freely along the field line with a step l_s. After each step it collides with other particle with the probability p, $0 < p < 1$. After the collision the particle changes its direction to the opposite one simultaneously being displaced on the distance $\pm\rho$ along the x-axis, i.e., across the field lines.

Let $P(y, n)$ be a probability density function (PDF) to find a particle at the interval $[y, y+dy]$ along the coordinate y after n steps. Similarly we define the PDF $P(x, n)$ for the displacement along the x-axis. The PDFs $P(y, n)$ and $P(x, n)$ after a large number of steps n, $n \gg 1$, are given by the following asymptotical formulas (see, e.g., Rax and White (1992))

$$P(y,n) = \frac{1}{\sqrt{2\pi \bar{D}_\parallel n}} e^{-y^2/2\bar{D}_\parallel n}, \qquad \bar{D}_\parallel = l_s^2 \frac{1-p}{p}, \tag{10.57}$$

$$P(x,n) = \frac{1}{\sqrt{4\pi \bar{D}_\perp n}} e^{-x^2/4\bar{D}_\perp n}, \qquad \bar{D}_\perp = \frac{1}{2}\rho^2 p. \tag{10.58}$$

Let $v_\parallel$ is a parallel velocity of a particle. Replacing the step number n by the time t, $t = nl_s/v_\parallel$, the PDFs $P(y, n)$ and $P(x, n)$ can be presented as

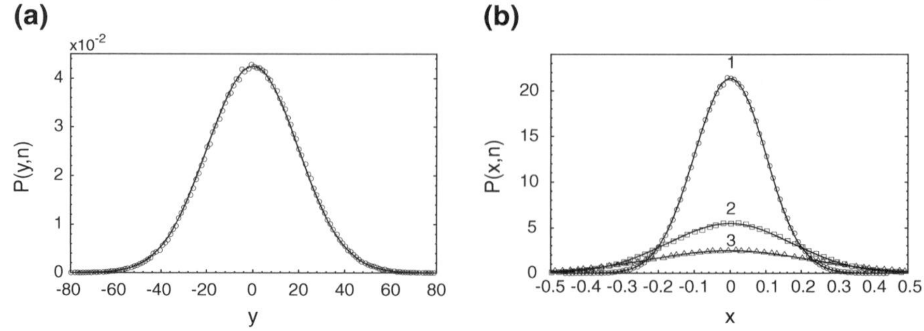

Fig. 10.10 **a** Distribution function $P(y, n)$ of a random walk along the y-axis with $p = 0.5$. The asymptotical PDF $P(y, n)$ (10.57) is plotted by *solid curve*, symbol ⊙ correspond to the numerical simulation, the coefficient $\bar{D}_{\|} = 0.3906$; **b** Distribution function $P(x, n)$ of the perpendicular diffusion along the x-axis: *curve 1* and symbol ∘ correspond to $p = 0.1$, $\bar{D}_{\perp} = 5 \times 10^{-6}$; *curve 2* and symbol □—$p = 0.3$, $\bar{D}_{\perp} = 1.5 \times 10^{-5}$; *curve 3* and symbol △—$p = 0.5$, $\bar{D}_{\perp} = 2.5 \times 10^{-5}$. The asymptotical PDF $P(x, n)$ (10.58) is plotted by *solid curves*, symbols ⊙, □, and △ correspond to the numerical simulation. Number of trials $n = 10^3$. Averaged over $N = 5 \times 10^5$ orbits

$$P(y,t) = \frac{1}{\sqrt{4\pi\chi_{\|}t}} e^{-y^2/4\chi_{\|}t}, \qquad \chi_{\|} = l_s v_{\|} \frac{1-p}{2p}, \tag{10.59}$$

$$P(x,t) = \frac{1}{\sqrt{4\pi\chi_{\perp}t}} e^{-x^2/4\chi_{\perp}t}, \quad \chi_{\perp} = \frac{\rho^2 v_{\|}}{2l_s} p, \tag{10.60}$$

where $\chi_{\|}$ and $\chi_{\perp}$ are the parallel and the perpendicular diffusion coefficients.

We present an example of the simulation of the random walk process. Figure 10.10 shows the parallel and perpendicular PDFs calculated numerically by large number of trials for the given diffusion coefficients $\bar{D}_{\|}$, $\bar{D}_{\perp}$. As seen the numerical PDFs are well approximated by the asymptotical PDFs (10.57).

10.4.3 Simulation of a Particle Diffusion

For the numerical simulations of particle transport with collisions we need the two parameters, the probability of collision p and the perpendicular displacement ρ.[4] They should be determined by the parallel and the perpendicular diffusion coefficients $\chi_{\|}$ and $\chi_{\perp}$. The first one $\chi_{\|}$ is given by the mean free path of particles, electrons and ions $\lambda_e \approx \lambda_{mfp}$, i.e., $\chi_{\|} = \lambda_{mfp} v_{\|}/2$. From Eq. (10.59) it follows that the probability p should be taken equal to $l_s/(l_s + \lambda_{mfp})$ and the perpendicular displacement ρ is given by $\rho = \sqrt{2\chi_{\perp}/p v_{\|}}$.

---

[4] In a cylindrical plasma the displacement is taken equal to the gyroradius of particles, which gives the classical diffusion coefficient $\chi_{\perp} = \rho^2/2\tau_i$, where τ_i is the collision time. However, the toroidal geometry of plasmas significantly increases the diffusion caused by collisions due to the so-called Prifsch–Schlüter current and the trapped particles. The corresponding theory is known as the *neoclassical transport* of plasmas (see the reviews by Galeev and Sagdeev (1979); Hinton and Hazeltine (1976), also a book by Wesson (2004)).

To be specific we consider the transport of H^+ ion (protons) in the stochastic magnetic field in the ED of Tore Supra. We calculate the local diffusion coefficient D_r of particles near the given magnetic surface as functions of the two parameters, the mean–free path length λ_{mfp} and the perpendicular diffusion coefficient $\chi_\perp$.

The calculations are made by running random walking particles with collisions along the field lines using the forward and backward mappings (6.29)–(6.31) for the model of perturbation magnetic field (9.5) in Sect. 9.1.3. The parameters of the model corresponds to the ED of the Tore Supra. The Poincaré section of field lines are shown Fig. 10.9.

Figure 10.11a shows the typical dependencies of the second moment of radial displacement $\langle(\Delta r)^2\rangle$ on time t for the two different ion temperatures: $T_i = 50\,\mathrm{eV}$ (curve 1) and 700 eV (curve 2). The corresponding mean–free paths are $\lambda_{mfp} =$ 1.06 m and 208.25 m, respectively. The particles are launched from the magnetic

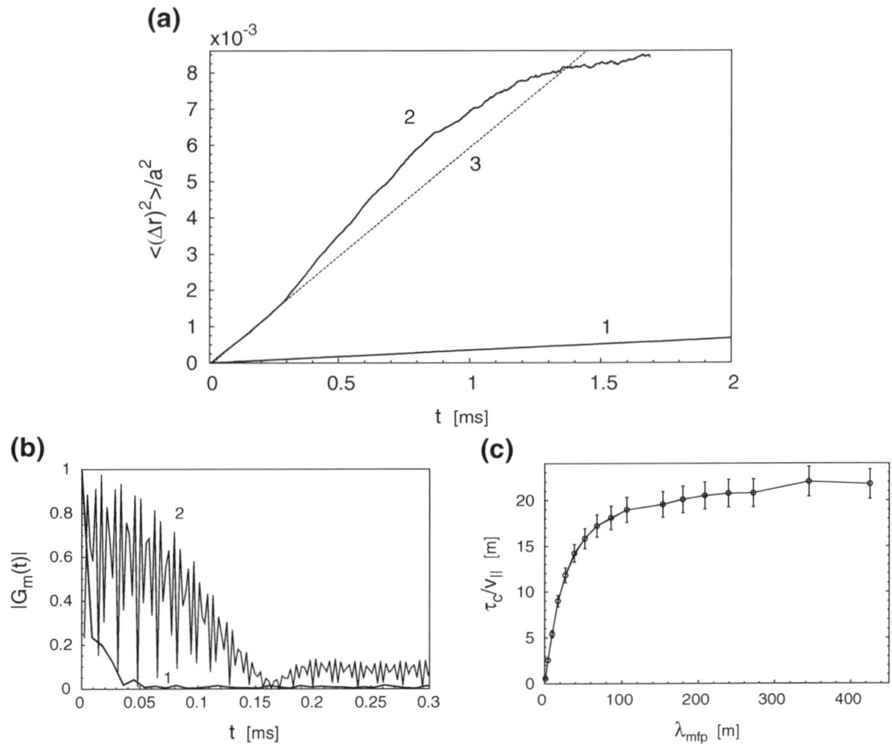

Fig. 10.11 **a** Typical time–dependencies of the second moments of radial displacements $\langle(\Delta r)^2\rangle$: *curve 1* corresponds to the particles of thermal energy $T_i = 50\,\mathrm{eV}$, *curve 2* corresponds—the one of $T_i = 700\,\mathrm{eV}$, *curve 3* corresponds to the *straight line* obtained by fitting the *curve 2* at the initial linear growth interval. The perpendicular diffusion $\chi_\perp = 0.1\ \mathrm{m^2/s}$ and the perturbation amplitude $\epsilon = 10^{-3}$. Averaging is done over $N = 10^4$ orbits. **b** The same as in (**a**) but for the correlation function $G_m(t)$; **c** The dependence of the correlation length $l_{cr} = \tau_c/v_\parallel$ on the mean–free path λ_{mfp}

surface $\psi_t = 0.72\psi_a$. The perpendicular diffusion coefficient $\chi_\perp$ is taken equal to $0.1\ m^2/s$ for the both cases.

The corresponding correlation functions of phase $G_m(t)$ in the presence of collisions are shown in Fig. 10.11b. The correlation time τ_c can be defined as the e-folding time of the correlation function $G_m(t)$. It can be estimated by fitting the function $G_m(t) = e^{-t/\tau_c}$ with the numerical calculated function $G_m(t)$. As seen from the correlation time τ_c of phases depends on the collisionality: with the increasing the collisionality the correlation time τ_c decreases. The dependence of τ_c on the mean–free path λ_{mfp} is plotted in Fig. 10.11c. It shows that the correlation length l_{cr} defined as $l_{cr} = \tau_c/v_\parallel$ grows with λ_{mfp} up to the lengths of order of 100 m, then it saturates reaching the value of order of the connection length $l_c = \pi q R_0$.

The local diffusion coefficient D_r is found by fitting the dependence of the second moment $\langle(\Delta r)^2\rangle$ on the time t with the linear function $2D_r t + a$ at the initial linear growth range of t. Such a fitting is shown by curve 3 in Fig. 10.11. We have performed a large number of numerical calculations of the diffusion coefficient D_r for the different values of $\chi_\perp$, λ_e, and the perturbation amplitude ϵ. The typical dependencies of D_r on the mean free path λ_{mfp} are shown in Fig. 10.12. There the local diffusion coefficients D_r at the given magnetic surfaces ψ_t are plotted as functions of the mean free path λ_{mfp}:

symbols $\odot$, $\square$, and $\triangle$ correspond to the diffusion coefficients at the magnetic surfaces $\psi_t = 0.72\psi_a$, $\psi_t = 0.79\psi_a$, and $\psi_t = 0.85\psi_a$, respectively. The corresponding values of the perpendicular diffusion coefficient $\chi_\perp$ are fixed and taken as $0.1 m^2/s$, $0.1\ m^2/s$, $0\ m^2/s$, respectively.

The numerical simulations reveal the following facts. For the value of the mean free paths λ_e shorter than the connection length $l_c = \pi R_0 q(\psi)$ the diffusion caused

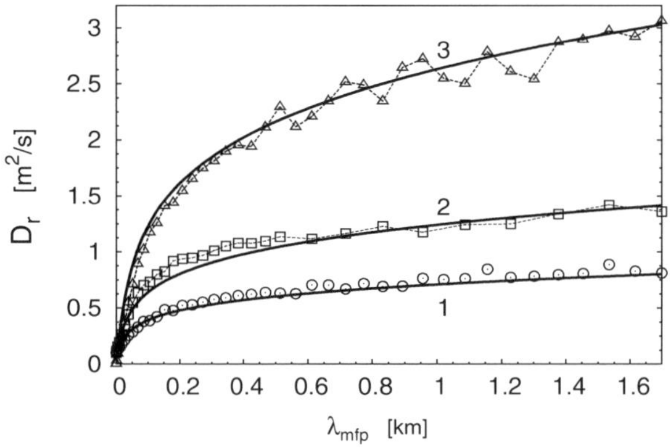

Fig. 10.12 Dependencies of the radial diffusion coefficients D_r on the mean free path length λ_{mfp}. *Curve 1* and symbol $\odot$ correspond to $\chi_\perp = 0.1\ m^2/s$ at the magnetic surface $\psi_t = 0.72\psi_a$; *curve 2* and $\square$ correspond to $\chi_\perp = 0.1\ m^2/s$ at $\psi_t = 0.79\psi_a$; *curve 3* and $\triangle$ correspond to $\chi_\perp = 0\ m^2/s$ at $\psi_t = 0.85\psi_a$. *Solid curves 1, 2,* and *3* describe to the empirical formula (10.62) with the corresponding diffusion coefficients of particles, $\chi_\perp$, and field lines, D_{FL}, and the characteristic lengths $L_c = l_c = \pi R_0 q(\psi_t)$, respectively. The perturbation amplitude $\epsilon = 10^{-3}$

by the chaotic field lines is damped and the diffusion coefficient D_r tends to the perpendicular diffusion $\chi_\perp$ at $\lambda_{mfp} \ll l_c$. With the increasing λ_{mfp} the coefficient D_r monotonically grows and at $\lambda_{mfp} \gg l_c$ it approaches the value $v_T D_{FL}$ of the collisionless case. This growth of D_r is practically independent on $\chi_\perp$ and for finite values of $\lambda_{mfp} \gtrsim l_c$ the diffusion coefficient D_r does not practically depend on $\chi_\perp$.

10.4.4 Empirical Formula for the Diffusion Coefficient D_r in a Stochastic Field

Using the above facts from the numerical simulations and the general arguments one can propose an empirical formula for the local diffusion coefficient D_r of particles in a stochastic magnetic field accounting the collision of particles. We present the increment of a particle's radial coordinate, $\Delta r(t)$ in time as a sum of the two random processes,

$$\Delta r(t) = \Delta r_\perp(t) + \Delta r_\parallel(t),$$

corresponding to the radial increments due to the random kicks $\pm\rho$ at collisions, $\Delta r_\perp(t)$, and the motion along the chaotic field lines, $\Delta r_\parallel(t)$. The latter, however, is modulated by the random changes of the motion direction with the mean free path λ_{mfp}. At small $\rho \ll a$ the functions $\Delta r_\perp(t)$ and $\Delta r_\parallel(t)$ can be considered as independent random processes with $\langle \Delta r_\perp(t) \Delta r_\parallel(t) \rangle = 0$. Then we have

$$\begin{aligned} \langle (\Delta r(t))^2 \rangle &= \langle (\Delta r_\perp(t))^2 \rangle + \langle \left(\Delta r_\parallel(t)\right)^2 \rangle \\ &= 2\chi_\perp t + 2D_\parallel t = 2D_r t, \end{aligned} \tag{10.61}$$

where $D_\parallel$ is the diffusion coefficient of the modulated parallel motion along the chaotic field lines.

The dependence of the coefficient $D_\parallel$ on the mean free path λ_{mfp} can be established using the following arguments. At the values of λ_{mfp} much shorter than the characteristic length L_c which is of order of the connection length $l_c = \pi R_0 q(\psi)$, $\lambda_{mfp} \ll l_c$, the particle moves randomly forth and back along chaotic field lines so often that it is not affected strongly by the stochastic field. Therefore, one expects that $D_\parallel \to 0$ at the limit $\lambda_{mfp} \to 0$. On the other hand at the long mean free paths $\lambda_{mfp} \gg L_c$ the coefficient $D_\parallel$ should approach to the one $v_\parallel D_{FL}$ in the collisionless regime, i.e., $D_\parallel \to v_\parallel D_{FL}$ at the limit $\lambda_{mfp} \to \infty$.

The simplest formula satisfying these conditions is given by

$$D_r = \chi_\perp + \frac{v_\parallel D_{FL}}{1 + L_c/\lambda_{mfp}}. \tag{10.62}$$

This empiric formula describes the local diffusion coefficient of particles D_r in a collisional plasma perturbed by the stochastic magnetic field. It is determined not

Table 10.2 Values of the connection length l_c, the Kolmogorov length L_K, the field line diffusion coefficient D_{FL}, and the characteristic length L_c obtained by fitting of the numerically calculated D_r with Eq. (10.62) at the three different magnetic surfaces

ψ/ψ_a	l_c,	L_K	D_{FL}	L_c
0.72	19.500	78.403	1.625×10^{-6}	19.8
0.79	21.306	73.237	3.044×10^{-6}	15.0
0.85	22.891	68.321	7.028×10^{-6}	33.2

A unit of all lengths are given in m. The perturbation parameter $\epsilon = 10^{-3}$

only by the local diffusion of field lines but also the local mean free path λ_{mfp}. The latter in turn depends on the local temperature T_i and the density n of the plasma, as $\lambda_{mfp} \propto T_i^2/n$.

The comparison of this formula with the numerically calculated coefficients is shown in Fig. 10.12 for three different values of the characteristic length L_c taken equal to the connection lengths l_c. Solid curves 1–3 describe D_r obtained by (10.62) using the field line diffusion coefficients D_{FL} from Fig. 10.4b. The values of L_c found by fitting the numerical data with the formula (10.62) are given in Table 10.2. The corresponding values of the connection length l_c, the Kolmogorov length L_K, the field line diffusion coefficient D_{FL} at the three different magnetic surfaces are listed there. The fitted values of L_c are close to the corresponding values of l_c, but they are smaller than the Kolmogorov length L_K by a factor $2.5 \div 4$. Overall the empirical formula (10.62) for the diffusion coefficient D_r satisfactorily well describes the results of the numerical simulations.

10.4.5 Comparison with the Diffusion Regimes

In the limiting case the empirical formula (10.62) can restore some of the regimes mentioned in Sect. 10.4.1 (see also Table 10.1). One can see that the empirical formula (10.62) formally coincides with the diffusion coefficient in the Laval regime in the long-mean–free–path regime (10.55) if L_c is replaced by the product $l_{c\delta} = L_K \ln|v_\| l_{cr}^{-1}\sqrt{\chi_\|\chi_\perp}|$. In the limit $\lambda_{mfp} \ll L_c$ the formula (10.62) can be approximated by $D_r \approx v_\| D_{FL}\lambda_{mfp}/L_c$ which coincides with the collisional Rechester–Rosenbluth regime (10.52) with L_c is replaced by $L_{K\delta}$ (10.53). However, L_c practically does not depend on $\chi_\perp$ while $l_{c\delta}$ and $L_{K\delta}$ do. The latter have singularities at the zero value of $\chi_\perp$.

One should note that the Kadomtsev–Pogutse regime cannot be recovered from numerical simulations at the typical tokamak plasma parameters.

10.4.6 Example of a Radial Profile of Diffusion Coefficients

Below we calculate the radial profiles of the diffusion D_r at the plasma edge for the inhomogeneous temperature and density profiles. To take into account the ambipolarity effect[5] we replace the parallel velocity $v_\parallel$ in Eq. (10.62) by the speed of sound $c_s = \sqrt{k(T_e + \gamma_i T_i)/m_k}$, where T_e is the electron temperature and γ_i is the adiabatic index.[6] For simplicity we put $T_e = T_i$ and $\gamma_i = 3$. We modeled the radial profiles of the plasma temperature $T_i(r)$ and the plasma density $n(r)$ by the following functions,

$$T_i(r) = T_0 \begin{cases} \exp\left(-r^2/a_T^2\right), & \text{for } r < a_T, \\ e^{-1}(a-r)/\left(a-a_T\right), & \text{for } a_T < r < a_T, \end{cases} \tag{10.63}$$

$$n(r) = \frac{n_0}{\tanh\left[\left(a_N - r\right)/\Delta_N\right]}, \tag{10.64}$$

where a_T, a_N, Δ_N are the parameters. The profiles are shown in Fig. 10.13. The radial profiles of the diffusion coefficients D_r corresponding to these temperature and density profiles are plotted in Fig. 10.14a, b.

As seen from Fig. 10.14a the diffusion coefficients D_r calculated by the formula (10.62) (curve 2 and symbols □) are quite close to the numerically calculated ones (curve 3 and symbols (⊙). They show that the collisions reduce the particle diffusion in a stochastic magnetic field. Particularly, the diffusion is decreased with the lowering the plasma temperature, as well as with the increasing the plasma density. The influence of the plasma density on the diffusion at the given temperature is demonstrated in Fig. 10.14b.

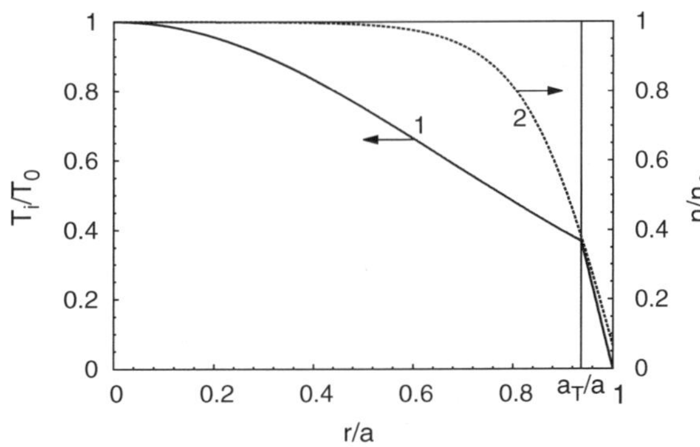

Fig. 10.13 Radial profiles of the plasma temperature $T_i(r)$ (*the left-hand axis*) and the plasma density $n(r)$ (*the right-hand axis*)

[5] In plasmas the negatively charged electrons move faster than the heavier positively charged ions due to their mass differences. It creates the electric field along a magnetic field which slows down the electrons and speeds up the ions. It results in that the electrons and ions move outwardly with the same speed equal to the sound speed c_s. Such a diffusion is known as the ambipolar diffusion.

[6] The plasma can be considered as perfect gas with the equation of state $pV^{\gamma_i} = \text{const}$, where p is the plasma pressure, V is its volume. The parameter γ_i is called the adiabatic index. For the perfect gas $\gamma_i = 1 + 2/n$, where n is the number of degrees of freedom.

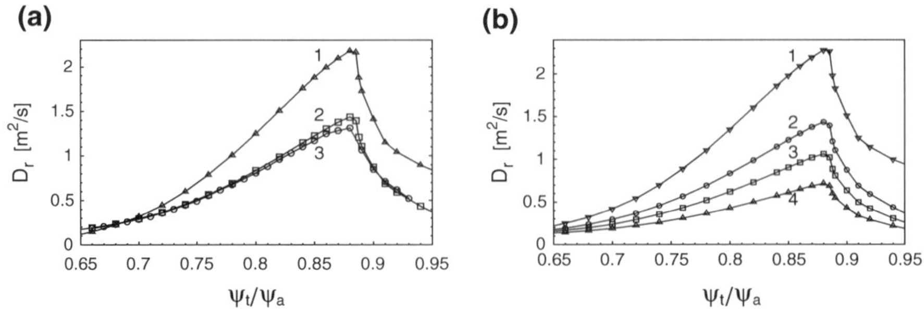

Fig. 10.14 Radial profiles of the radial diffusion coefficient D_r for the different temperatures and densities of the plasma: **a** $T_0 = 1$ keV, $n_0 = 2 \times 10^{19}$ m^{-3}. *Curve 1 with symbol* △ corresponds to the collisionless case, *curve 2* (□) corresponds to the empirical formula (10.62), *curve 3* ⊙ corresponds to the numerical simulation, *curve 4* describes the temperature profile $T_i\psi)$ (*the right-hand axis*). **b** $T_0 = 0.5$ keV, *curve 1 with a symbol* ▽ corresponds to the collisionless case; *curve 2* (⊙) corresponds to $n_0 = 2 \times 10^{19}$ m^{-3}, *curve 3* (□)—$n_0 = 4 \times 10^{19}$ m^{-3}, *curve 4* (△)—$n_0 = 8 \times 10^{19}$ m^{-3}. The perturbation amplitude $\epsilon = 10^{-3}$ and $\chi_\perp = 0.01$ m^2/s

The decrease of the particle diffusion in a stochastic magnetic field with the increasing of the plasma density is in an agreement with the experimental observations in the Tore Supra and the TEXTOR tokamaks Grosman (1999); Lehnen et al. (2005) and in the DIII-D tokamak Jakubowski et al. (2009); Schmitz et al. (2012). In the latter experiments it has been found that the pump–out effect of the electron density at the plasma edge during the operation with the applied magnetic perturbations depends on the plasma collisionality. Namely, the outward particle flux is decreased with increasing the plasma density.

10.5 Bibliographic Comments

Kinetic description. The kinetic equation has been derived by A. Fokker (1914) and M. Planck (1917) for the Brownian motion, and in the more general form by A.N. Kolmogoroff (1931) for the Markovian processes. The derivation of the kinetic equation for the chaotic Hamiltonian systems are given by Filonenko et al. (1967); Zaslavsky and Chirikov (1971); Sagdeev et al. (1988) assuming the smallness of the perturbation. The corresponding equation in the problem of wave–particle interactions is obtained by Elskens and Escande (2002, 2003) in the cases of the non-self-consistent and self-consistent dynamics.

The quasilinear theory of plasmas deals with the simplest description of interacting particle–wave system in a turbulent plasma. It describes a slow evolution of a particle distribution function and amplitudes of unstable (turbulent) waves developed in plasmas [Romanov and Filippov (1961); Vedenov et al. (1962); Drummond and Pines (1962)] (see, for example, books by Wesson (2004); Elskens and Escande (2003); Diamond et al. (2010)).

Transport in a stochastic magnetic field. In early 1960s it has been already realized that any magnetic perturbations in fusion devices which break its toroidal symmetry may destroy magnetic surfaces and lead to the deterioration of plasma confinement [Kerst (1962); Rosenbluth et al. (1966); Filonenko et al. (1967); Ohkawa (1972); Stix (1973); Freis et al. (1973); Hamzeh (1974); Finn (1975); Matsuda and Yoshikawa (1975); Rechester and Stix (1976)]. Therefore, the problem of energy and particle transport in plasmas with destroyed magnetic surfaces became a subject of numerous studies. In early works by Rechester and Rosenbluth (1978); Stix (1978); Krommes (1978); Kadomtsev and Pogutse (1979); Laval (1993) the effect of particle collisions on the heat and particle transport in a stochastic magnetic field have been studied and a several possible regimes have been found. The detailed analysis of transport processes in a stochastic magnetic field were given by Krommes et al. (1983); Isichenko (1991a, b); Laval (1993); Kaganovich and Rozhansky (1998) (see also reviews by Bickerton (1997); Spatschek (2008)).

The collisionless and collisional regimes of transport in a stochastic magnetic field have been also investigated numerically (see, for example, Rax and White (1992); White and Whu (1993); Kanno et al. (2010); Rack et al. (2012) and references therein). The transport of a heat in tokamaks during externally applied resonant magnetic perturbations has been studied by Samain et al. (1993); Tokar (1997); Yu (2006); Tokar et al. (2007, 2008); Hölzl et al. (2008); Yu and Günter (2009). The influence of magnetic stochasticity on heat transport in the reversed field pinch has been analyzed by D'Angelo and Paccagnella (1999).

The transport of energetic particles, for example, alpha–particle and runaway electrons in a stochastic magnetic field has been also a subject of many studies since early 1980s. The deviation of drift surfaces from the magnetic ones plays an important role in these processes [Mynick and Krommes (1979); Mynick and Strachan (1981); Boozer and White (1982)]. The effect of drift motion on the transport of energetic ions in tokamaks caused by internal and externally applied magnetic perturbations has been numerically and analytically studied in numerous works [see, for example, Strumberger et al. (2008) and references therein]. A similar problem in the TEXTOR tokamak with the DED has been considered by Abdullaev et al. (2006a); Fischer et al. (2000). The references on the runaway electron transport in tokamaks in a stochastic magnetic field are given in Sect. 11.3.

Chapter 11
Transport of Particles in a Turbulent Field

In this chapter we study a particle transport in Hamiltonian systems perturbed by a weak turbulent wave field. This problem is a very simplified version of an important but a complex problem of energy and particle transport in magnetically confined turbulent plasmas (Wesson 2004; see also Sect. 11.3 for some bibliographic comments). One expects that this study allows one to understand some features of the complex problem of transport in fusion plasmas. Specifically, we consider the dynamics of a one-degree of freedom Hamiltonian system perturbed by a weak turbulent wave field and closely related problem of a runaway electron transport via micro-turbulence in tokamaks.

The problem itself may also have a theoretical interest as a dynamics of Hamiltonian systems subjected to small amplitude time-dependent perturbations with a wide spectrum of frequencies. We will study the role of rational tori on the chaotic transport in such systems. It will be shown that a profile of diffusion coefficients across tori may have a fractal nature. Particularly, the chaotic transport is reduced near the low-order rational tori thereby forming transport barriers. We will study the condition of forming these transport barriers and their dependence on the mode spectra and correlation time of turbulent field.

The numerical study of particle motion in a turbulent field encounters practical difficulties related with both the accuracy and computational times. The standard methods to integrate the equations of motion containing a large number of fast-oscillating terms require very long computational times (Nishimura and Azumi 1997). Therefore, different computational methods have been proposed to overcome these difficulties. Particularly, the drift wave maps derived by Horton et al. (1998), Kwon et al. (2000) significantly accelerate the calculations. However, the computational capabilities of these mapping models are limited to include only a moderate number of turbulence wave field modes, at most consisting of a few tens of modes. Therefore, they also encounter difficulties in the studies of the particle's motion in typical electromagnetic turbulence fields of tokamak plasmas which may contain up to thousands of small-scale wave modes.

S. Abdullaev, *Magnetic Stochasticity in Magnetically Confined Fusion Plasmas*,
Springer Series on Atomic, Optical, and Plasma Physics 78,
DOI: 10.1007/978-3-319-01890-4_11,

To overcome the above-mentioned difficulties in this problem we apply the full-turn transfer mapping introduced in Sect. 6.4, which drastically reduces the computational times by a several orders. Below we follow the publications by Abdullaev (2011a) and Abdullaev et al. (2012).

11.1 Transport in Hamiltonian Systems Perturbed by a Weak Turbulent Wave Field

11.1.1 Model of Hamiltonian System

Consider a one-degree-of-freedom Hamiltonian system the evolution of which in the presence of time-dependent perturbations is governed by the two-dimensional Hamiltonian system,

$$\frac{d\vartheta}{dt} = \frac{\partial H}{\partial I}, \qquad \frac{dI}{dt} = -\frac{\partial H}{\partial \vartheta}, \tag{11.1}$$

where

$$H(I, \vartheta, t) = H_0(I) + H_1(I, \vartheta, t). \tag{11.2}$$

The Hamiltonian $H_0(I)$ describes the unperturbed motion $I =$ const, $\vartheta = \omega(I)t + \vartheta_0$, $I =$ const, where $\omega(I) = dH_0(I)/dI$ is the frequency of motion. The period (or a transit time) of motion is $T = 2\pi/\omega(I)$.

The perturbation Hamiltonian $H_1(I, \vartheta, t)$ describing the effect of a weak turbulent field can be presented by a Fourier series in ϑ and t,

$$H_1(I, \vartheta, t) = \epsilon \sum_{m,n} H_{mn}(I) \cos(m\vartheta) \cos(\Omega_n t + \chi_{mn}(t)), \tag{11.3}$$

where the Fourier coefficients $H_{mn}(J)$ of the (m, n)-wave mode with the random phases $\chi_{mn}(t)$, Ω_n is the frequency of the nth mode, and ϵ stands for the dimensionless amplitude of the turbulent field. Furthermore, we will set $\Omega_n = n$ which corresponds 2π-time-periodic perturbations. In magnetic fusion devices ϵ is less than or of order of 10^{-4}–10^{-3}.

The turbulent field contains a wide spectrum of modes (m, n) ranging in the interval $m_{\min} \leq m \leq m_{\max}$, $n_{\min} \leq n \leq n_{\max}$ with $m_{\max} \gg 1$, $n_{\max} \gg 1$. We assume that the (m, n) modes with the different m but the same number n are not independent, but the modes belonging to the different mode n are statistically independent. It allows one to assume that the random phases $\chi_{mn}(t)$ depend only on the mode number n, i.e., $\chi_n(t) \equiv \chi_{mn}(t)$. This assumption allows one to present the perturbation Hamiltonian in the form

$$H_1 = \epsilon \sum_{n=n_{\min}}^{n_{\max}} \mathcal{H}_n(\vartheta, I) \cos(nt + \chi_n(t)),$$
$$\mathcal{H}_n(\vartheta, I) = \sum_{m=m_{\min}}^{m_{\max}} H_{mn}(I) \cos(m\vartheta). \tag{11.4}$$

Furthermore we will assume that the turbulent fluctuations as a stationary random process in time with the correlation time τ_c defined as

$$\langle e^{i(\chi_n(t+\tau)-\chi_{n'}(t))}\rangle = \delta_{n,n'} e^{-\tau/\tau_c}, \tag{11.5}$$

where $\langle \cdots \rangle$ stands for the averaging over ensemble of realization of phase χ. The turbulent transport of particles depends on the Kubo number,

$$K = \frac{\tau_c}{T}, \tag{11.6}$$

defined as a ratio of the correlation time τ_c to the transit time T.[1]

For the numerical study of the transport we should specify models for the unperturbed Hamiltonian system and the turbulent field. Below we consider the models corresponding to the particle dynamics in magnetically confined plasmas in the presence of weak magnetic perturbations.

The unperturbed Hamiltonian is characterized by the frequency of motion $\omega(I)$. Furthermore, we will choose the action variable I normalized to its maximal possible value, so that it changes in the interval $0 < I < 1$. We consider the two cases of the dependence of the inverse frequency $\omega(I)$ or the safety factor $q(I) = 1/\omega(I)$ on I:

(i) the monotonic linear dependence $q(I) = q_0 + (q_a - q_0)J$;
(ii) the nonmonotonic dependence $q(I) = q_m + \beta(I - I_m)^2$ with the minimal value at off-axis value I_m.

Here, q_0, q_a, q_m, and β are constant parameters.

The amplitude of turbulent field mode is chosen as a localized in I function in the form

$$H_{mn}(I) = \frac{c_n}{m} \exp\left[-\frac{(md(I) - nq(I))^2}{4w_n^2}\right], \tag{11.7}$$

where w_n describes the width of the mode, c_n is the amplitude of toroidal modes which we will choose in the power-law form $c_n = n^{-\alpha}$ with a constant α.

---

[1] The Kubo number K measures the particle's capacity of exploring the space structure (see, e.g., Bakunin 2008). The definition of K (11.6) is, probably, more adequate in our problem of the test particle transport in a turbulent wave field, particularly, a runaway electron transport in a fluctuating magnetic field which will be studied in Sect. 11.2. Particularly, such a definition (11.6) K has been used by Vlad and Spineanu (2005) and Vlad et al. (2005) in the similar problems of the test particle transport in the turbulent field. There is also another definition of K used by Pommois et al. (2001) for the transport of the magnetic field lines in a turbulent magnetic field.

The function $d(I)$ describes a deviation of the location $I_{\max}$ of the maximal value c_n/m of the amplitude $H_{mn}(I)$ from the corresponding resonant tori I_{mn}, $q(I_{mn}) = m/n$. If $d(I) \equiv 1$ the amplitude reaches it maximum at the resonant tori. We choose the function $d(I)$ as follow

$$d(I) = d_0\left(1 + \kappa(I - I_c)^2\right), \tag{11.8}$$

where d_0, $0 < I_c < 1$, $\kappa > 0$ are constant parameters.

The model (11.7) for the amplitude $H_{mn}(I)$ allows one to obtain the closed analytical formula for the perturbation functions $\mathcal{H}_n(\vartheta, J)$ in the perturbation Hamiltonian (11.4). Using the Poisson summation rule,

$$\sum_{s=-\infty}^{\infty} \delta(x - s) = \sum_{p=-\infty}^{\infty} \cos(2\pi x p), \tag{11.9}$$

the sum over m in Eq. (11.4) can be reduced to

$$\mathcal{H}_n(\vartheta, I) = c_n \sum_{p=-\infty}^{\infty} \int_{m_{\min}}^{m_{\max}} m^{-1} \exp\left[-\frac{(m\gamma - nq(I))^2}{4w_n^2}\right] \times \cos(m\vartheta)\cos(2\pi m p)dm. \tag{11.10}$$

Replacing m by $m = nq/d(I)$ in the pre-exponent of the integral in Eq. (11.10), and taking $m_{\min} = 0$ and $m_{\max} \to \infty$, the function $\mathcal{H}_n(\vartheta, I)$ can be approximated by

$$\mathcal{H}_n(\vartheta, I) \approx 2\sqrt{a\pi}c_n \sum_{p=-\infty}^{\infty} f(a, b, \vartheta - 2\pi p), \tag{11.11}$$

where

$$f(a, b, \vartheta) = e^{-a\vartheta^2}\cos(b\vartheta),$$
$$a = w_n^2/d^2(I), \; b = nq(I)/d(I). \tag{11.12}$$

This analytical formula would accelerate a numerical calculation of the symplectic mappings given by Eqs. (6.9)–(6.11) or (6.29) and (6.47). In practical calculations it is sufficient to take only a few terms $p = 0, \pm 1$ in the sum over p (11.11). However, one should note that such a presentation of $\mathcal{H}_n(\vartheta, I)$ is valid only for the large number modes $n \gtrsim 10$.

11.1.1.1 Full-Turn Transfer Mapping

To avoid difficulties related with computational times we employ the full-turn transfer mapping (6.55) schematically shown in Fig. 6.3a,

$$
\begin{aligned}
h_{k+1} &= h_k - \epsilon \frac{\partial P\left(h_{k+1}, t_k + \pi/\omega(h_k)\right)}{\partial t_k}, \\
t_{k+1} &= t_k + \pi\left[q(h_k) + q(h_{k+1})\right] + \epsilon \frac{\partial P\left(h_{k+1}, t_k + \pi q(h_k)\right)}{\partial h_{k+1}},
\end{aligned}
\tag{11.13}
$$

where (h_k, t_k) are the values of the Hamiltonian $h = H_0(I)$ and time t at the crossing point P_k of the orbit, respectively. The Poincaré integral $P(h, t)$ (6.57) is presented in the form

$$
P(h,t) = \sum_{n=n_{\min}}^{n_{\max}} R_n(h) \exp\left(int + i\chi_n(t)\right). \tag{11.14}
$$

Using the expression for the perturbation Hamiltonian (11.3) the functions $R_n(h)$ can be reduced to:

$$
R_n(h) = \pi q(h) \sum_{s=\pm 1} \sum_{m=m_{\min}}^{m_{\max}} H_{mn}(I) \frac{e^{i x_{mn}^{(s)}} - 1}{i x_{mn}^{(s)}}, \tag{11.15}
$$

where $x_{mn}^{(s)} = (m - snq)2\pi$. Extending the sum in (11.15) from $m_{\min} = 1$ to $m_{\max} = \infty$ the regular and oscillatory parts of $R_n(h)$ (6.82) can be approximated by

$$
\begin{aligned}
R_n^{(reg)} &= (\pi q)\, H_n(n/\omega), \\
R_n^{(osc)} &= \left(\cos^2(\pi n q) - 1 + i \frac{\sin(2\pi n q)}{2}\right) R_n^{(reg)},
\end{aligned}
\tag{11.16}
$$

where $H_n(m) \equiv H_{mn}(I)$ is a function defined by extending the value of integer number m to the continuous ones. We recall that at the resonant tori I_{mn}, $q(I_{mn}) = m/n$, the oscillatory part $R_n^{(osc)} = 0$. The function $R_n^{(reg)}(h)$ for the model (11.7) is reduced to

$$
R_n^{(reg)}(h) = \frac{c_n}{nq(I)} \exp\left[-\frac{n^2 q^2(I)(d(I)-1)^2}{4w_n^2}\right]. \tag{11.17}
$$

From (11.17) follows that for $d(I) \equiv 1$ the function $R_n^{(reg)}(h)$, and therefore the Poincaré integral $P(h,t)$ does not depend on the width w_n of the perturbation amplitude.

The finite correlation time τ_c of the phases χ_n are taken into account by randomly changing them after the M mapping steps equal to the integer part of the Kubo number K, (i.e., $M = [K]$). However, the mapping cannot consider the Kubo number less than one.

11.1.1.2 Comparison of Mapping with Small-Step Integration

For the model described above we compared the results of calculations of the second moment of displacements of the action I,

$$\sigma_0^2(t) = \langle (I(t) - I(0))^2 \rangle = \frac{1}{N_t} \sum_{i=1}^{N_t} (I_i(t) - I(0))^2 , \tag{11.18}$$

obtained by the fast mapping (11.13) and the symplectic mapping (6.29) and (6.47). Here $\langle (\cdots) \rangle$ stands for averaging over the initial phases $\chi_n(0)$, $n_{\min} \le n \le n_{\max}$, randomly chosen in the interval $[0, 2\pi]$, $I_i(t)$ is the orbit corresponding to the ith trial, and N_t is a total number of trials.

The calculations are carried out for the radial profile (i) of $q(I)$, and the values of parameters are chosen $q_0 = 0.8$, $q_a = 5$, in Eq. (11.7) the function $d(I) \equiv 1$ and $w_n = \infty$, the exponent $\alpha = 5/6$, $20 \le n \le 50$, and the turbulence level $\epsilon = 10^{-4}$. For the given parameters the relative accuracy 10^{-4} of $\sigma_0^2(t)$ calculated by the symplectic mapping (6.29) and (6.47) is reached with the time-step $\Delta t = 2\pi/128$.

Figure 11.1 shows the comparison $\sigma_0^2(t)$ calculated by the fast mapping (11.13) (circles and squares) and the symplectic mapping (6.29) and (6.47) (solid curves 1 and 2) with the time-step $\Delta t = 2\pi/128$ at the two positions of tori I_0: solid curve 1 and circles correspond $I_0 \equiv I(0) = 0.52$, solid curve 2 and squares correspond to $I_0 = 0.48$. As seen from Fig. 11.1 the fast mapping quite accurately reproduces the results of the small time-step symplectic integration for the two different type of motion corresponding to the diffusive and regular regimes.

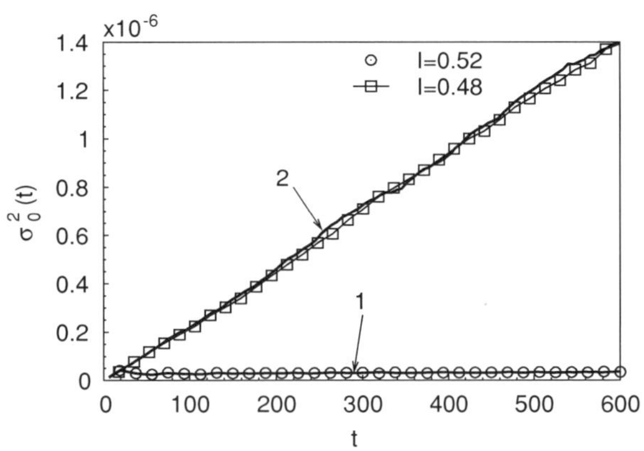

Fig. 11.1 Dependencies of $\sigma_0^2(t)$ on t for the model (11.7) obtained using the symplectic mapping (6.29), (6.47) with the time step $\Delta t = 2\pi/128$ (*solid curves 1* and *2*) and the fast mapping (11.13) (*circles* and *squares*). *Solid curve 1* and *circles* correspond to the tori $I_0 = 0.52$, and *solid curve 2* and *squares* correspond to the tori $I_0 = 0.48$. The parameters of the model are $q_0 = 0.8$, $q_a = 5$, $20 \le n \le 50$, $d(I) \equiv 1$, and $\epsilon = 10^{-4}$

However, the small time-step symplectic integration takes almost three or four orders longer computation time than the fast mapping. It becomes practically impossible to use the symplectic integration in the case with a large number of modes extending up to several thousands. Furthermore, we will use the fast mapping to study our problem.

11.1.2 Quasilinear Approximation for Chaotic Transport

At the large Kubo numbers $K \gg 1$ when the correlation time τ_c is much longer than the transit time T (i.e., $\tau_c \gg T$), the chaotic motion mainly contributes to the radial transport. The latter starts from the certain value of the turbulence level ϵ which depends on the spectrum of the turbulent field. The critical level of the turbulence ϵ_c can be qualitatively described by the overlapping (Chirikov) criterion, $\sigma_{Chir} \approx 1$, where the overlapping parameter σ_{Chir} defined by Eq. (8.30)

$$\sigma_{Chir} = \frac{\Delta I_{mn} + \Delta I_{m'n'}}{2|I_{mn} - I_{m'n'}|}, \tag{11.19}$$

$$\Delta I_{mn} = 4\left|\frac{\epsilon H_{mn}(I_{mn})}{\omega'(I_{mn})}\right|^{1/2}.$$

The transport along I can be characterized by the diffusion coefficient $D(I) = \sigma_0^2(I)/2t$. For the large Kubo numbers $K \gg 1$ and $\sigma_{Chir} \gg 1$, the diffusion coefficient can be described by the following quasilinear formulas (10.18),

$$D(I) = \frac{\pi\epsilon^2}{2\omega}\sum_{m,n} m^2\,|H_{mn}(I)|^2\,\delta\,(m/q - n)$$
$$= \frac{\epsilon^2 q}{2\pi}\sum_n n^2\left|R_n^{(reg)}(I)\right|^2, \tag{11.20}$$

where the second line similar to Eq. (10.41) is obtained using the relations in Eq. (11.16).

The numerical calculations of diffusion coefficients $D(I)$ are carried out, similar to ones in Sect. 10.2.3, by fitting the calculated second moment of radial displacements $\sigma_0^2(t)$ as a function of t with the linear function $2D(I)t$ at the times $t > t_K$, where t_K is the decorrelation time of phases $\phi_n = nt$. The phase decorrelation time t_K is equal to the e-folding time of the phase correlation function $C_n(t) = \langle e^{i[\phi_n(t) - \phi_n(0)]}\rangle \sim \exp(-t/t_K)$. The procedures of the determination of the diffusion coefficients and t_K are shown in Fig. 11.2.

Figure 11.3a, b shows the profiles of the Chirikov parameter σ_{Chir} and diffusion coefficients $D(I)$ along the action I at the turbulence level $\epsilon = 10^{-3}$, respectively. The diffusion coefficients are calculated using the mapping (11.13) with the two

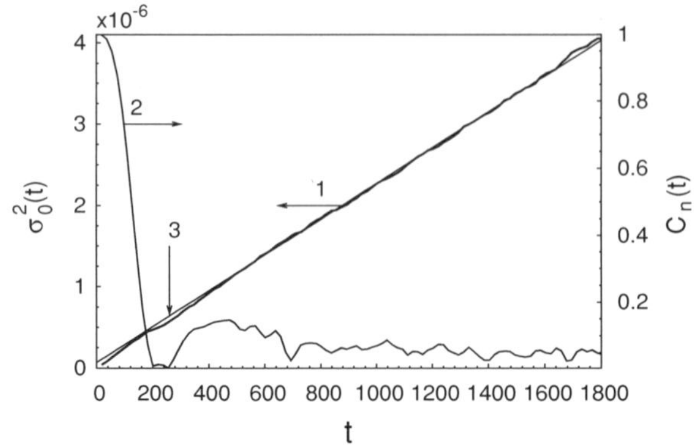

Fig. 11.2 Determination of the diffusion coefficient $D(I)$: *curve 1* (*thick solid line*) corresponds to the calculated $\sigma_0^2(t)$ (*left-hand axis*); *curve 2* corresponds the correlation function of phases $C_n(t)$ for $n = 20$ (*right-hand axis*); *curve 3* (*thin solid line*) corresponds to the linear function $2D(I)t$ obtained by fitting of $\sigma_0^2(t)$ in the interval $600 \leq t \leq 1800$. The other parameters are the same as in Fig. 11.1 and $I_0 = 0.5$

choices of the Poincaré integrals: curve 2 (red) corresponds to the case with $R_n = R_n^{(reg)}$, i.e., when the oscillatory part, $R_n^{(osc)}$, is neglected, curve 3 (blue) corresponds to the case when $R_n^{(osc)}$ is included, i.e., $R_n = R_n^{(reg)} + R_n^{(osc)}$. The quasilinear values are plotted by curve 1 (solid line). The right-hand axis shows the profile of $q(I)$.

As seen from Fig. 11.3b the radial profiles of $D(I)$ along I are described by the fractal-like, irregular curves which closely follow the quasilinear formula (11.20) in the regions of I where the Chirikov parameter takes large values $\sigma_{Chir} > 1$. However, in the narrow regions located near the low-order rational tori I_{mn}, $q(I_{mn}) = m/n$ with the values 1, 3/2, 2, 5/2 the diffusion coefficients become very small or almost negligible, thereby forming the transport barriers. The Chirikov parameter also drops near these rational tori (see Fig. 11.3). We will analyze the nature of these transport barriers in the next sections.

One should note that the approximations of the Poincaré integrals R_n by $R_n = R_n^{(reg)}$ and $R_n = R_n^{(reg)} + R_n^{(osc)}$, given by curves 2 and 3, respectively, give qualitatively close results for the diffusion coefficients. It is related to the fact that for the large Kubo numbers $K \gg 1$ the quasilinear diffusion coefficients (11.20) are determined only by the regular part of the Poincaré integral $R_n^{(reg)}$. Therefore, in these cases one can neglect the oscillatory part $R_n^{(osc)}$ of R_n.

11.1.3 Transition from the Fluctuating to the Chaotic Regime

For the small Kubo number K the transport is mainly caused by the fluctuating in the time turbulent field. For the Kubo number $K = 1$ the diffusion coefficients can be

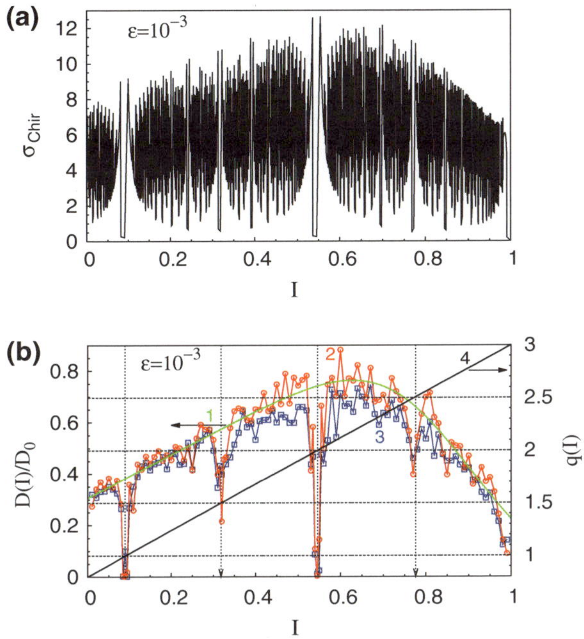

Fig. 11.3 **a** Profile of the Chirikov parameter σ_{Chir} along the action variable I. **b** Radial profiles of diffusion coefficients, $D(I)$ (*left-hand axis*) for the model (11.7), (11.8) ($D_0 = 10^{-7}$). The values of the model parameters are $q_0 = 0.8, q_a = 3, d_0 = 1, \kappa = 0.2, I_c = 0.2, \alpha = 5/6, w_n = 6, \epsilon = 10^{-3}$, the modes $20 \leq n \leq 50$, and the Kubo number $K = 100$. The number of trials $N_t = 2 \times 10^3$. *Curve 1* (*green*) describes the quasilinear approximation (11.20); *curve 2* (*red circles*) and *curve 3* (*blue open squares*) correspond to the fast mapping (11.13) with $R_n = R_n^{(reg)}$ and $R_n = R_n^{(reg)} + R_n^{(osc)}$, respectively; *curve 4* describes the profile of $q(I)$ (*right-hand axis*). The *vertical thin lines* show the locations of main rational resonances, I_{mn}, $[q(I_{mn}) = m/n = 1, 3/2, 2, 5/2]$

obtained from the mapping (11.13). Using the expansion (11.14) and the first equation in Eq. (11.13) the change of the variable h in one mapping turn $\Delta t = 2\pi/\omega(I)$ can be presented as

$$\Delta h_k = h_{k+1} - h_k = \epsilon \sum_n n|R_n| \cos \Phi_n,$$
$$\Phi_n = nt_k + \pi q(h_k) + \chi_n, \tag{11.21}$$

where χ_n are random phases uniformly distributed in the interval $[0, 2\pi]$. The averaged values of Δh_k over phases χ_n are zero $\langle \Delta h_k \rangle$. The second moment of Δh_k is equal to

$$\sigma_h^2 = \langle (\Delta h_k)^2 \rangle = \frac{\epsilon^2}{2} \sum_n n^2 |R_n|^2. \tag{11.22}$$

Using the relations between the variables h and I, that is, $\Delta h = (dh/dI)\Delta I = q^{-1}\Delta I$, and the corresponding diffusion coefficients $D(I)$ and $D(h) = \sigma_h^2/2\Delta t$, we arrive at

$$D(I) = \frac{\epsilon^2 q}{8\pi} \sum_n n^2 |R_n|^2. \tag{11.23}$$

This formula for $D(I)$ similar to Eq. (11.20) in which the integral $R_n^{(reg)}$ is replaced by R_n. The numerical calculations $D(I)$ for the Kubo number $K = 1$ is indeed agreed with the formula (11.23).

One expects that by increasing the Kubo number K the chaotic transport caused by the interaction of resonances starts to contribute to the transport if the turbulence level ϵ exceeds the critical level ϵ_c qualitatively determined by the criteria $\sigma_{Chir}(\epsilon_c) \approx 1$. In order to explore this we consider the dependence of the diffusion coefficient on the Kubo number for the three different values of the turbulence level ϵ: (1) $\epsilon = 10^{-5}$; (2) $\epsilon = 10^{-4}$; (3) $\epsilon = 10^{-3}$ at which the Chirikov parameter σ_{Chir} is (1) less than unity $\sigma_{Chir} \lesssim 1$; (2) slightly greater than 1 $\sigma_{Chir} \gtrsim 1$; and (3) much larger than 1 $\sigma_{Chir} \gg 1$. Figure 11.4 shows the dependencies of the diffusion coefficients $D(I)$ on the Kubo number at the fixed tori $I_0 = 0.2$ for these values of ϵ: Curve 1 corresponds to $\epsilon = 10^{-5}$, curve 2 to $\epsilon = 10^{-4}$, curve 3 to $\epsilon = 10^{-3}$, and curve 4 (a straight dotted line) corresponds to the quasilinear estimation of $D(I)$ (11.20). Since the diffusion coefficient D is proportional to ϵ^2 for the convenience of comparison it is normalized to ϵ^2.

For the small level of turbulence which does not produce the chaotic transport, as in the case $\epsilon = 10^{-5}$ with $\sigma_{Chir} \approx 0.55$ the diffusion coefficient $D(I)$ grows with K as $D(I) \sim K \sim \tau_c$ in the interval $1 \leq K \lesssim 5$, and it reaches the maximum at $K \approx 10$–30 then it decays for $K > 40$. For the moderate and large turbulence

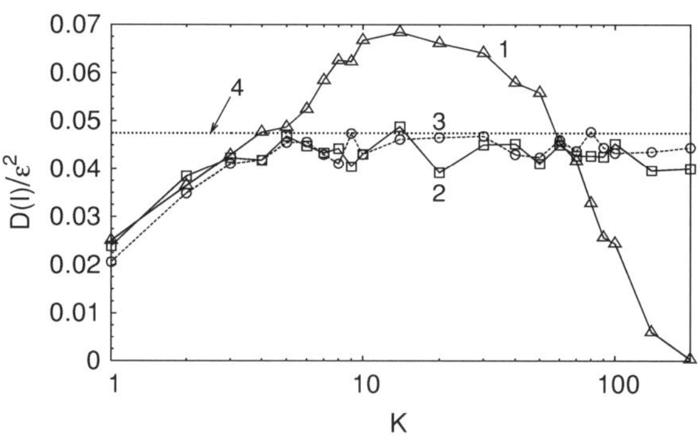

Fig. 11.4 Dependence of the diffusion coefficients $D(I)$ normalized to ϵ^2 on the Kubo number K at the given value $I = 0.2$ for three different values of the turbulence level ϵ: *curve 1* (*up-triangle*) corresponds to $\epsilon = 10^{-5}$; *curve 2* (*square*) corresponds to $\epsilon = 10^{-4}$; *curve 3* (*circle*) corresponds to $\epsilon = 10^{-3}$; and *curve 4* (a *straight dotted line*) corresponds to the quasilinear estimation of $D(I)$ for the large Kubo number $K \gg 1$

levels, $\epsilon = 10^{-4}$, ($\sigma_{Chir} \approx 1.33$) and $\epsilon = 10^{-3}$, ($\sigma_{Chir} \approx 5.5$), which may lead to the chaotic transport, the diffusion coefficient reaches its "stationary" level given by the quasilinear value (11.20) at the Kubo numbers K exceeding 5 to 6.

11.1.4 Density of Rational Tori

As we have shown above at the large Kubo numbers $K \gg 1$ the transport of particles across tori is mainly caused due to their interactions of nonlinear resonances formed at the resonant tori $I = I_{mn}$. The strength of the interaction depends not only on the level of turbulence ϵ, but also on the density of rational tori. In other words, the chaotic transport rate, a diffusion coefficient, depends on how the resonant tori $I = I_{mn}$ are dense along the axis I. Below we calculate the density of rational tori and study its property.

Let I_{mn} be resonant values of tori, $q(I_{mn}) = m/n$, corresponding to the mode numbers (m, n). Assume that the interval of the n modes are $M \equiv n_{\min} \leq n \leq N \equiv n_{\max}$. Let ΔW be a number of rational tori I_{mn} in the interval $I, I + \Delta I$. Then the density $\rho(I, M, N)$ of rational tori is defined as $\rho(I, M, N) = \Delta W/\Delta I$. One can also introduce the density of rational numbers $q = m/n$, with n in the interval $M \leq n \leq N$, and m that are coprime to n i.e., having no common positive factors other than 1. It is defined as $\rho(q, M, N) = \Delta W/\Delta q$. The relation between $\rho(I, M, N)$ and $\rho(q, M, N)$ is given by

$$\rho(I, M, N) = \rho(q, M, N)\frac{dq}{dI}. \tag{11.24}$$

As seen from Eq. (11.24) that the density of rational tori $\rho(I, M, N)$ along the action variable I depends on the shear dq/dI of the inverse frequency of motion $\omega(I)$ as well as on the density of rational numbers $q = m/n$ with $M \leq n \leq N$. The former is prescribed by the unperturbed system $H_0(I)$ while the latter is determined by the minimal $n_{\min}$ and maximal $n_{\max}$ mode numbers of a turbulence field.

In order to study the property of the density function $\rho(q, M, N)$ it is sufficient to calculate it for the rational numbers in the interval $0 \leq q \leq 1$. We have numerically calculated the density function $\rho(q, M = 1, N)$ for the different values N. The numerical calculations shows that there are gaps in the density distribution function $\rho(q, M, N)$ near the low-order rational numbers m/n: 1/1, 1/2, 1/3, and so on.[2]

As an example the gap near $q = 1$ is shown in Fig. 11.5a: the square (blue) corresponds to $N = 400$ and the circle (red) corresponds to $N = 800$. The rescaling parameter $\lambda = N/400$. The widths of gaps $\Delta w_{mn}^{(N)}$ near rational numbers $q = m/n$

[2] The existence of gaps in the density of rational tori has been pointed out already in a number works, for example, by Austin et al. (2006), Brakel et al. (2002), Eriksson et al. (2002), Garbet et al. (2003), Gohil et al. (2003), Hogeweij et al. (2004), Kasilov et al. (2002), Joffrin et al. (2003), Lopes Cardozo (2004), and Razumova et al. (2003, 2008). However, to my knowledge no quantitative estimations have been given.

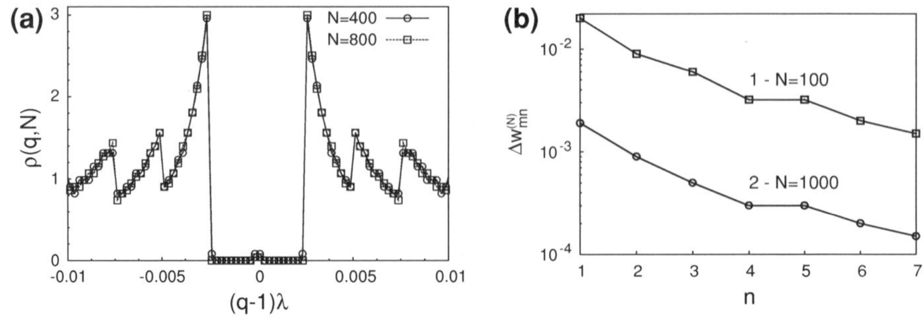

Fig. 11.5 **a** Density distribution function $\rho(q, N)$ of rational numbers $q = m/n$ $(1 \leq n \leq N)$ for the two different numbers N near the $q = 1$: The symbol *square* (*blue*) corresponds to $N = 400$ and the symbol *circle* (*red*) corresponds to $N = 800$. The rescaling parameter $\lambda = N/400$; **b** Gap widths $w_{mn}^{(N)}$ in $\rho(q, N)$ near the rational numbers $1/n$, $n = 1, 2, \ldots, 7$. *Curve 1* corresponds to $N = 100$; *curve 2* to $N = 1000$

decrease with increasing n as shown in Fig. 11.5b for the two different numbers N: curve 1 corresponds to $N = 100$ and curve 2 corresponds to $N = 1000$.

Moreover, it was found that the function $\rho(q, M, N)$ has a self-similar behavior near the low-order rational numbers $q_{mn} = 1, 1/2, 1/3$ and so on. The self-similarity of $\rho(q, N) \equiv \rho(q, M = 1, N)$ is manifested by the rescaling law,

$$\rho\left(q - q_{m,n}, N\right) \approx \rho\left(\frac{q - q_{m,n}}{\lambda}, \lambda N\right), \tag{11.25}$$

where λ is a constant. The rescaling law (11.25) for the density distribution function $\rho(q, N)$ near the low-order rational number $q = 1$ is shown in Fig. 11.5a for the two values of N: $N = 400$ (symbol $\square$, blue) and $N = 800$ (symbol $\odot$, red). The rescaling parameter in this case is $\lambda = N/400$. One expects that these results can also be obtained using the continued fraction expansions of rational numbers or the Euler's totient function (Abramowitz and Stegun 1965).

As we will show in the next section for the large Kubo numbers $K \gg 1$ the gaps in the density of rational tori strongly affect on the chaotic transport near the low-order rational tori, thus forming the transport barriers.

11.1.5 Formation of Transport Barriers

Here we discuss the formation of transport barriers near the low-order rational tori in more derails. We consider the two kind of profiles of $q(I)$ prescribed in Sect. 11.1.1: (i) the monotonic dependence of $q(I)$ on I, and (ii) the non-monotonic dependence of $q(I)$ on I with a minimum at $I = I_m$ located away from the center $I = 0$ and the edge $I = 1$.

11.1.5.1 On the Mechanism of Transport Barrier Formation

The mechanism of transport barrier formation in magnetically confined fusion plasmas has been discussed many publications. According to the one of widely accepted views the radial transport caused by a small-scale turbulent field is due to overlapping of wave functions of neighboring modes (11.7) localized near the corresponding rational tori (magnetic surfaces) I_{mn}. It is schematically shown in Fig. 11.6. The degree of overlapping modes is determined by the widths of modes w_{mn} and the distance between rational surfaces, $\delta I = I_{mn} - I_{m'n'}$. Therefore the radial transport rate depends not only the density of rational surfaces, but also on the width of modes w_{mn}. As we have seen in the previous section the latter has gaps near the low-order rational surfaces which leads to the reduced transport there (the transport barrier).[3]

However, in our mechanism the radial transport is caused not because of overlapping of wave functions but due to overlapping of neighboring islands formed by the resonance interactions of particles with corresponding turbulent modes (m, n), i.e., it depends on the Chirikov parameter (11.19). The transport barriers are formed due to the gaps in density of rational tori, where these interactions are weakened. Unlike the above discussed mechanism the diffusion for the large Kubo number K does not depend on the width of modes w_{mn}.

11.1.5.2 Monotonic Profile

The profile of D along the variable I is shown in Fig. 11.7 in the whole region $0 < I < 1$ for the large Kubo number $K = 100$ (a) and in the region $0.5 < I < 0.55$ located near the rational tori $q = 3$ for the several values of $K = 1, 2, 5$, and 100 (b). The values the other parameters are chosen: $q_0 = 0.8$, $q_a = 5$, $d(I) \equiv 1$, $20 \leq n \leq 50$, and $\alpha = 5/6$, which corresponds to the power spectrum $|H_{mn}|^2 \sim n^{-5/3}$, and

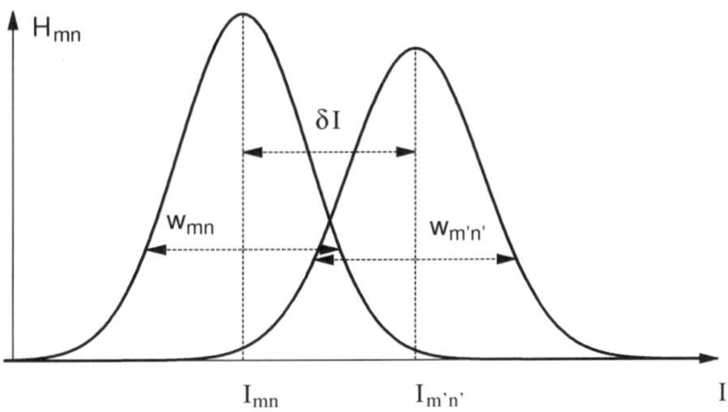

Fig. 11.6 Overlapping of wave functions $H_{mn}(I)$ (11.7) of neighboring modes (m, n) and (m', n')

[3] This mechanism, particularly has been discussed by Austin et al. (2006), Brakel et al. (2002), Eriksson et al. (2002), Garbet et al. (2003), Gohil et al. (2003), Hogeweij et al. (2004), Kasilov et al. (2002), Joffrin et al. (2003), Lopes Cardozo (2004), and Razumova et al. (2003, 2008).

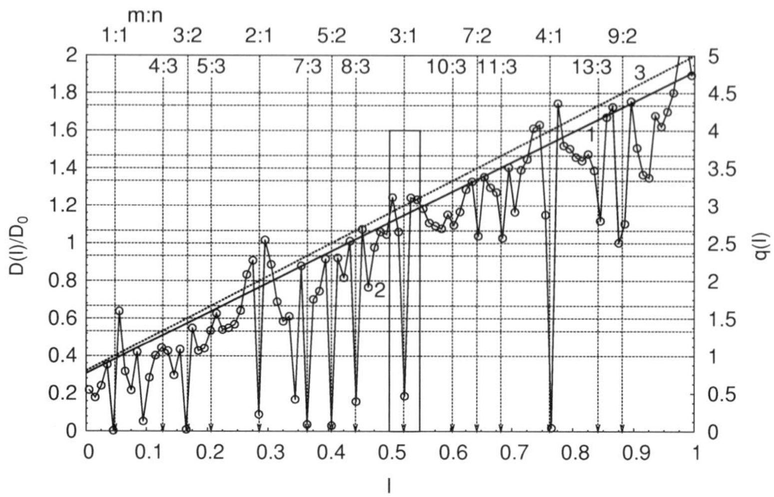

Fig. 11.7 Radial profile of diffusion coefficient for the Kubo number $K = 100$. *Curve 1* corresponds to the quasilinear values; *curve 2* displays the numerical results; *curve 3* corresponds to $q(I)$ (*right-hand axis*). The positions of the low-order rational tori I_{mn}, $(q(I_{mn}) = m/n)$, are shown by *vertical straight lines*. The averaging is done over 2×10^3 orbits. The normalization constant $D_0 = 10^{-9}$

the turbulence level $\epsilon = 10^{-4}$. At the chosen turbulence level ϵ the radial transport is mainly caused by the chaotic diffusion.

As seen from Fig. 11.7 the diffusion coefficient D (curve 2) has a fractal-like dependence which on average follows the quasilinear diffusion coefficient $D(I)$ 11.20 (curve 1). One can see that the diffusion coefficients drop almost to zero at the radial positions $I = I_{mn}$ corresponding to the low-order rational tori $\omega(I_{mn}) = n/m$. Near these resonant tori the second moments $\sigma_0^2(t)$ does not grow with t or grow very slow similar to the one shown in Fig. 11.1 (curve 1).

It is clear that for the large Kubo numbers $K \gg 1$ the absence of chaotic transport near the low-order rational tori is related to the gaps in the density of rational tori there (see Sect. 11.1.4). However, by decreasing the Kubo number the turbulence fluctuations start to cause the transport across low-order rational tori.

The latter is illustrated in Fig. 11.8 by the expanded view of the profiles of $D(I)$ in the region $0.5 \leq I \leq 0.55$ located near the rational tori $I = I_{mn}$, $(m, n) = (3, 1)$, for the different Kubo numbers K. It shows how the transport barrier is changed with the Kubo number K. The maximal depth of the barrier is achieved for the large Kubo number K as shown by curve 2 (dots $\odot$) for $K = 100$. A decrease of the Kubo number K leads to the shrink of the barrier depth without a significant change of its width (see curves 3 (dots $\bullet$), 4 (dots $\square$), and 5 (dots $\blacksquare$) corresponding to the Kubo numbers $K = 5$, 2, and 1, respectively). One can expect that the transport barrier disappears for the values of the Kubo number $K < 1$. The transport barriers do not disappear with increasing the maximal mode number $n_{\max}$ and decreasing $n_{\min}$ up to 1. This changes only affect the depth and the width of transport barriers.

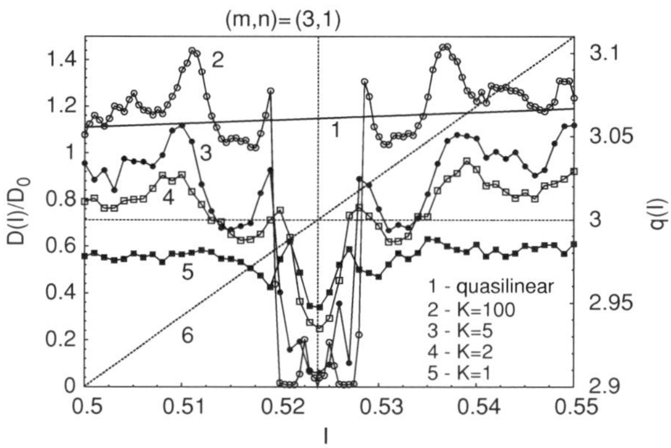

Fig. 11.8 Expanded view of rectangular area in Fig. 11.7 near the resonant tori $\omega = 1 : 3$. *Curve 1* (*solid*) corresponds to the quasilinear prediction; *curve 2* (*open circles*) corresponds to the Kubo numbers $K = 100$; *curves 3* (*solid circles*); *4* (*open squares*); and *5* (*solid squares*) correspond to the Kubo numbers $K = 100$, 5, 2, and 1, respectively. *Dashed curve 6* corresponds to the transit time $T(I)$ (*right-hand axis*)

The transport barrier also does not disappears by increasing the exponent α in the power-law spectrum $c_n = n^{-\alpha}$. However, because of a fast decay of the amplitudes of higher modes n with increasing α the critical turbulence level ϵ for the chaotic transport also grows.

11.1.5.3 Nonmonotonic Profile

From Eq. (11.24) it follows that the width of transport barriers also depends on the shear $dq(I)/dI$. The most pronounced transport barrier appears in the case of the nonmonotonic profile of $q(I)$ near the shearless tori $I = I_m$ where $q(I)$ reaches its minimal value, i.e., $dq/dI = 0$. It corresponds to the reversed magnetic shear configuration. According to (11.24) the density distribution $\rho(I, N)$ near this tori is also low. The formation of the turbulent transport barrier in this case is illustrated in Fig. 11.9 when the minimal value I_m coincides with the $(m = 3, n = 2)$ resonant tori $q(I_{mn}) = 3/2$ for the large, $K = 100$ (curve 2) and small $K = 1$ (curve 3) values of the Kubo number. The parameters of the model are $q_m = 1.5$, $I_m = 0.3516$, $\beta = 4.045$, the interval of modes $10 \le n \le 100$, the perturbation level is $\epsilon = 10^{-4}$, and the values of the other parameters are the same as in Fig. 11.7.

The typical dependencies of the second moment of displacements $\sigma_0^2(t)$ on t at the three positions located near this transport barrier is shown in Fig. 11.10: Curve 1 corresponds to $I = 0.3$ located just outside the barrier, curve 2 corresponds to $I = 0.32$ located inside the barrier and close to the barrier border, and curve 3 corresponds to the resonance value $I_{mn} = 0.3516$, $(m = 3, n = 2)$; curve 4 describes the correlation function of phase $C_n(t)$ for the minimal mode number $n = 10$.

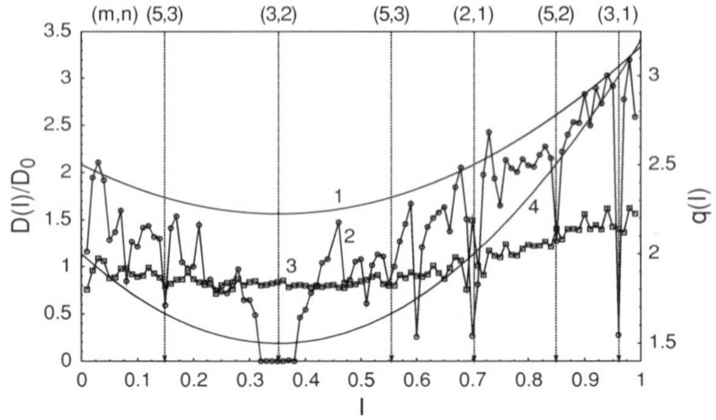

Fig. 11.9 The same as in Fig. 11.7 but for the non-monotonic profile of $q(I) = q_m + \beta(I - I_m)^2$. **a** The turbulence level $\epsilon = 10^{-4}$, the interval of modes $10 \leq n \leq 100$. *Curve 1* corresponds to the quasilinear formula (11.20); *curves 2* (⊙) and *3* (□) correspond to the numerical calculations with the Kubo numbers $K = 100$ and $K = 1$, respectively; *curve 4* describes $q(I)$ (*right-hand axis*). The normalization constant $D_0 = 10^{-9}$

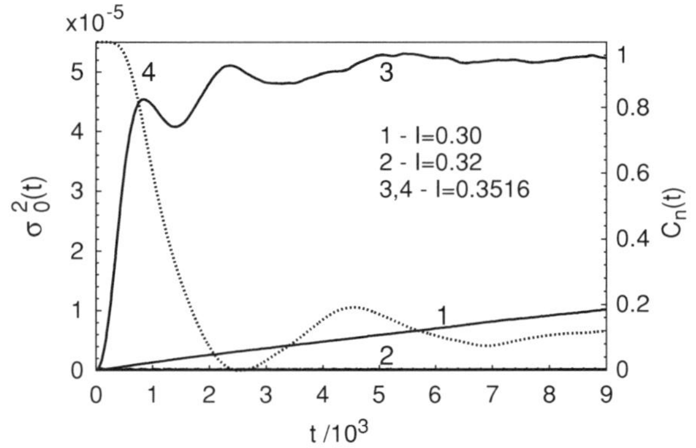

Fig. 11.10 Dependencies of $\sigma_0^2(t)$ on t at the three locations of I: *curve 1* corresponds to $I_0 = 0.3$; *curve 2* to $I_0 = 0.32$; and *curve 3* to $I_0 = 0.3516$ (*left-hand axis*); *curve 4* (*dotted*) corresponds to the phase correlation function $C_n(t)$ at $n = 10$ (*right-hand axis*). The toroidal modes $10 \leq n \leq 100$, the turbulence level $\epsilon = 10^{-4}$

As seen from this figure the motion near the low-order resonant tori is not diffusive, but rather regular. In the transport barrier regions the second moment of displacement, $\sigma_0^2(I)$, do not linearly depend on time t. It grows at the initial stage $t < t_K$, and tend to the constant value at the large $t > t_K$ (curve 4). The correlation time t_K in the transport barrier region is much larger than those in the diffusive regions.

The radial transport barrier in a small-scale turbulent field appears for any turbulence level (but still small) exceeding a certain critical value at which the overlapping of resonances occurs. Only the depth and width of the transport barrier depend on the mode content of the turbulent field. Particularly, the increase of the maximal mode number $n_{\max}$ as well as the increase of the turbulence level ϵ lead

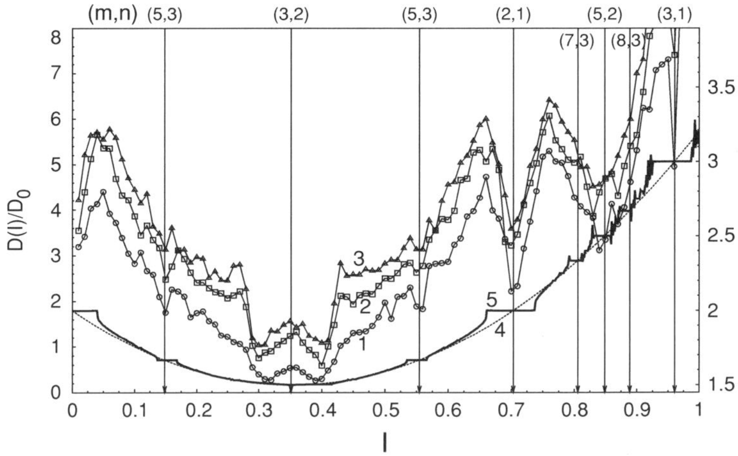

Fig. 11.11 The same as in Fig. 11.9 but for the turbulence level $\epsilon = 10^{-3}$: *curve 1* (⊙) corresponds to the interval of modes $1 \leq n \leq 25$; *curve 2* (□) to $1 \leq n \leq 50$; *curve 3* (△) to $1 \leq n \leq 100$; *curve 4* (*dashed*) and *curve 5* (*solid*) describe $q(I)$ and the inverse of the effective winding number, $q_{eff}(I)$, (11.26) (*right-hand axis*). The normalization constant $D_0 = 10^{-7}$ and the Kubo number $K = 100$

to the growth of diffusion rate at all radial positions due to an increase of the density of rational tori, but it does not eliminate the transport barrier. This is illustrated in Fig. 11.11. Particularly, Fig. 11.11 shows the profiles of diffusion coefficients for the three different values of $n_{\max}$: $n_{\max} = 25$ (curve 1), $n_{\max} = 50$ (curve 2), and $n_{\max} = 100$ (curve 3) at the fixed lowest mode number $n_{\min} = 1$ and the Kubo number $K = 100$. The turbulence level is taken $\epsilon = 10^{-3}$ which is one order higher than in Fig. 11.9.

11.1.5.4 Effect of Low-Order Modes

From the comparison of Figs. 11.11 and 11.9 it is seen that the transport barriers near the shearless tori $(m, n) = (3, 2)$ and the resonant tori $(m, n) = (2, 1)$ and $(5, 2)$ became wider by increasing the turbulence level ϵ and the inclusion of low-order modes $n_{\min} < 10$. At the same time due to a large turbulence level the diffusion inside the barrier does not disappear unlike the case with small ϵ as in Fig. 11.9. Widening the transport barriers near the resonant tori $(m, n) = (2, 1)$ and $(5, 2)$ are due to the presence of low order modes n. These modes alone modify the safety factor $q(I)$ near the low-order rational tori because of the trapping of particles into the corresponding resonances. The motion there can be characterized by the averaged safety factor $q_{eff}(I)$ defined as

$$q_{eff} = \frac{1}{N} \sum_{k=0}^{N} \frac{t_{k+1} - t_k}{2\pi}, \qquad N \gg 1, \tag{11.26}$$

where t_k ($k = 0, 1, 2, \ldots, N$) is a sequence of times in the mapping (11.13), and the averaging is done over a large number of mapping iterations N, $N \gg 1$. The profile of $q_{eff}(I)$ is flattened in the region $I_{mn} - \Delta I_{mn}/2 < I < I_{mn} + \Delta I_{mn}/2$ near the resonant tori I_{mn}, where ΔI_{mn} is the resonance width (11.19). This is shown by curve 5 in Fig. 11.11 calculated for the Hamiltonian system (11.1)–(11.3) with only three lowest modes $n = 1, 2, 3$. According to (11.24), such a profile of q_{eff} leads to widening of the gap in the density of rational tori, and thus reduces the chaotic transport like near the shearless tori.

11.1.6 Summary

The study of the chaotic transport in a Hamiltonian system perturbed by a weak turbulent field with a wide spectrum of modes shows the following:

(i) For the large Kubo numbers $K \gg 1$ the dependence of the diffusion coefficient $D(I)$ on I has an irregular, fractal-like behavior.[4]
(ii) Near the low-order rational tori the diffusion is significantly reduced or vanished thus forming barriers for the radial transport. The width and depth of transport barriers depend on the mode spectra H_{mn}, mode ranges in n, and the Kubo number K.
(iii) For the small Kubo numbers $K \lesssim 1$ the irregular behavior of $D(I)$ is replaced by the more smooth profiles, and particularly, the transport barriers may be disappeared or weakened.
(iv) The formation of transport barriers is due to the gaps of density of rational tori near the low-order rational tori. The latter is determined by the shear of the unperturbed Hamiltonian system and the density of rational numbers, m/n, with n in $n_{\min} \leq n \leq n_{\max}$, where $n_{\min}$ and $n_{\max}$ are the minimal and maximal mode numbers of the turbulent field.
(v) A self-similar behavior of the density of rational tori near the low-order rational numbers has been found.

11.2 Runaway Electron Transport in a Turbulent Field

The runaway electrons are usually generated during the start-up or during disruptions of plasma discharges. They may reach energies from a few MeV up to several tens MeV and may cause a substantial damage of the plasma facing wall components of

[4] The fractal behavior of the diffusion coefficients has been also observed in dynamical systems described by one-dimensional, piecewise linear, chaotic mappings (Klages and Dorfman 1995; Knight and Klages 2011). In these systems the diffusion coefficients have a fractal structure as a function of the mapping parameter.

fusion devices. Therefore, the mitigation of runaway electrons is one of the crucial issues in a fusion research.

Theoretical studies of the runaway electron transport have been based on the assumption that a magnetic turbulence is a main cause of the radial transport of runaway electrons. Due to their high velocity, they are hardly sensitive to electrostatic turbulence like the thermal particles. It is observed that the runaway electrons have longer confinement times than the thermal electrons. It is believed that the main reason of this is related with the large gyro-radii and the deviations of orbits of high energetic electrons from the magnetic surfaces. They cause *orbit-averaging* and *gyro-averaging* of magnetic turbulent field thus decreasing the diffusion of runaway electrons (see, for example, Hauff and Jenko 2009; Myra and Catto 1992; Myra et al. 1993; de Rover et al. 1996a, b, 1999).

In this section we apply the methods and results of the previous section to study the runaway electron transport in a turbulent field of tokamaks. We show that beside of those *orbit-averaging* and *gyro-averaging* effects the formation of transport barriers near the low-order rational drift surfaces may also lead to the improvement confinement of runaway electrons. We mainly follow the publications by Abdullaev et al. (2010, 2012).

11.2.1 Runaway Electron Orbits

Since the runaway electrons are passing electrons we will use the relativistic Hamiltonian equations for a guiding center motion in a toroidal system given in Sect. 4.3.4. From the experimental measurements by Finken et al. (1990) and Jaspers (1995) it follows that the parameter $\lambda_I = I_x/\varepsilon_k$ defined by Eq. (5.3) may reach a value of 0.1. Furthermore we will assume that $\lambda_I \leq 0.1$. Specifically, we consider the runaway electrons in the plasma with the circular cross section with the TEXTOR plasma parameters (see Sect. 5.11). Some examples of the gyrating electron orbits and their guiding-center orbits have been already shown in Fig. 5.3a.

The deviation of the drift surfaces from the magnetic ones is illustrated in Fig. 11.12. It shows the poloidal sections of the drift surfaces corresponding to different electron energies ranging from $E = 1\,\text{MeV}$ up to $E = 20\,\text{MeV}$ (curves 1–5, respectively). They have common points at the section $\vartheta = 0$. The dashed curve 6 corresponds to the magnetic flux surface touching the drift surfaces at $\vartheta = 0$. Due to the toroidal effects the drift surfaces of the runaway electrons do not coincide with the magnetic surface and the geometrical centers of the drift surfaces are shifted outward with respect to the one of the magnetic surface.

Figure 11.12b shows the dependencies of the effective safety factors $q_p(J, E)$ of the runaway electrons on the normalized action variable J/J_a, where J_a is the value of J at the plasma edge $r = a$. Curves 1, 2, and 3 correspond to the electron energies $E = 1$, 10, and 20 MeV, respectively, and curve 4 corresponds to the safety factor $q_M(\psi)$ of the equilibrium magnetic field.

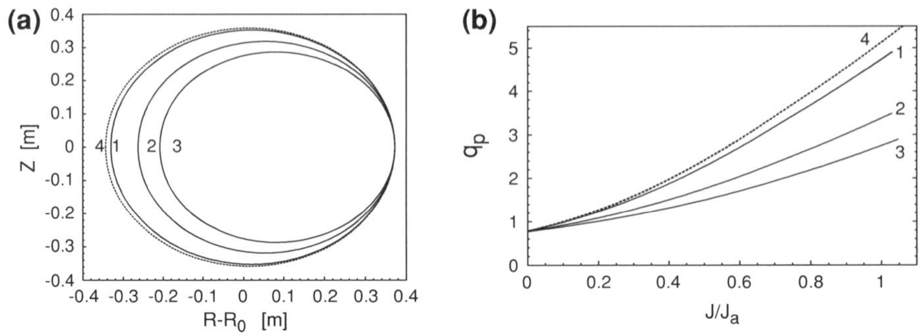

Fig. 11.12 Drift orbits (**a**) and effective safety factors, $q_p(J, E)$ (**b**) of runaway electrons of different energies: *curve 1* corresponds to $E = 1\,\mathrm{MeV}$; *curve 2* to $E = 10\,\mathrm{MeV}$; *curve 3* to $E = 20\,\mathrm{MeV}$. *Curve 4* corresponds to the magnetic surface. The parameter $\lambda_I = 0.1$. The plasma parameters are the same as in Fig. 5.3a, but the plasma current $I_p = 0.3\,\mathrm{MA}$

11.2.2 Turbulent Magnetic Field

The nature of the magnetic fluctuations in tokamak plasmas, as well as electric field fluctuations are not well known. Therefore, we consider the small-scale magnetic fluctuations by presenting them as a sum of a large number of (m, n)-modes radially localized near the corresponding resonant magnetic surfaces without specifying the nature of the magnetic turbulence. Similar presentations of the turbulent magnetic field have been used by Myra and Catto (1992), Myra et al. (1993), and de Rover et al. (1996a, b).

11.2.2.1 Presentation of a Turbulent Magnetic Field

Let $(\psi, \vartheta_M, \varphi)$ be magnetic field coordinates in which the field lines are straight: $\psi = \psi_0, \vartheta_M = \varphi/q(\psi) + \vartheta_{M0}$. We present the toroidal component of the vector potential of the turbulent magnetic field as a sum of (m, n)-modes with random phases χ_{mn},

$$A_\varphi^{(1)}(R, Z, \varphi, t) = \epsilon B_0 \sum_{mn} r m^{-1} b_{mn}(\psi) \cos(m\vartheta_M) \cos(n\varphi + \chi_{mn}), \qquad (11.27)$$

where $\epsilon = \sigma_A/|A_\varphi^{(0)}(R, Z)|$ stands for the dimensionless amplitude of the turbulent magnetic field, i.e., it is determined by the ratio of the amplitude of the magnetic fluctuations, $\delta B = \nabla \times A_\varphi^{(1)}$ to the toroidal field strength, B_0, i.e. $\epsilon \sim \delta B/B_0$. The functions $b_{mn}(\psi)$ describe the radial ($\psi-$) dependence of the (m, n)-mode. They are localized near the resonant magnetic surfaces ψ_{mn}, $q(\psi_{mn}) = m/n$.

The magnetic fluctuations contain a wide range of poloidal and toroidal modes (m, n). In a toroidal system one can assume that for the given toroidal mode n

the different poloidal modes m are not independent but they are coupled due to a toroidicity. On the other hand one can suppose that the modes belonging to the different toroidal modes n are statistically independent. It allows one to assume that the random phases $\chi_{mn}(t)$ depend only on the toroidal mode number n, i.e., $\chi_n(t) \equiv \chi_{mn}(t)$. Furthermore we will assume that the turbulent fluctuations are a stationary random process in time with the correlation time τ_c.

One of the important parameters of the magnetic turbulence is its fluctuation level δB. For the latter we consider the magnetic turbulence caused by micro-tearing modes. According to the theory by Drake et al. (1980) the nonlinear saturation level of δB driven by the electron temperature gradient is given by $\delta B/B_0 \simeq \rho_e/L_{T_e}$, where ρ_e is the thermal electron gyroradius, $L_{T_e} = T_e/(dT_e/dr)$ is the characteristic length of the variation of the electron temperature T_e. The recent gyrokinetic simulations (Doerk et al. 2011) show that the Drake formula quite well describes the level of magnetic fluctuations, and it can be used for rough estimations of δB. For the TEXTOR parameters, $T_e \approx 1\,\text{keV}$, $L_{T_e} = 0.1\,\text{m}$ and $B_0 = 2.2\,\text{T}$ we have $\delta B/B_0 \sim 10^{-4}$.

The transport of particles in a fluctuating magnetic field depends on the Kubo number (11.6), $K = \tau_c/T(J, E) = v_\parallel \tau_c/l_c$, where $T(J, E)$ is the transit time, $v_\parallel$ is the parallel velocity of a particle, and $l_c = 2\pi R_0 q_p$ is the connection length.

The typical value of the correlation time of a turbulent field is of order of $\tau_c \sim 10^{-5}$ s. Since $v_\parallel$ is close to the speed of light c Kubo number $K = K(J, E)$ is of the order of 10^2 for the runaway electrons of energies $E \gtrsim 1\,\text{MeV}$. Such electrons "feel" the magnetic turbulent field as a frozen one and we have seen in Sect. 11.1.3 the radial transport is mainly caused by the interactions of resonances.

11.2.2.2 A Specific Model of Modes

For our calculations we should specify the model for the turbulent magnetic field. Suppose that the amplitude, $b_{mn}(\psi)$, of the (m, n)-mode is described by the following radially dependent function,

$$b_{mn}(\psi) = A(\psi) c_n \exp\left[-(m - nq(\psi))^2/4w_n^2\right], \tag{11.28}$$

where $A(\psi)$ describes the spatial radial dependence of the amplitude of the magnetic fluctuations, c_n is its toroidal spectrum. The poloidal mode spectrum of a given toroidal mode n is localized near the resonant magnetic surface ψ_{mn}, $q(\psi_{mn}) = m/n$. The quantities w_n are the widths of the poloidal mode spectra of a given toroidal mode n. For our model we choose the following power-law distribution of the toroidal modes, $c_n = n^{-\alpha}$, with the constant parameter α. For simplicity we assume that the magnetic fluctuation level $A(\psi)$ is radially homogeneous, i.e. $A(\psi) = \text{constant}$.

According to Eq. (11.27) the magnitude of the radial component of the turbulent magnetic field $B_r = r^{-1}\partial A_\varphi/\partial\theta$ can be reduced to

$$\delta B = \sqrt{\langle B_r^2 \rangle} \approx \frac{\epsilon B_0}{2} \left(\sum_{n=n_{\min}}^{n_{\max}} \sum_m |b_{mn}(\psi)|^2 \right)^{1/2}, \tag{11.29}$$

where $\langle(\cdots)\rangle$ stands for an averaging over random phases χ_n and the poloidal angle ϑ_M. Using Eq. (11.28) and assuming that $w_n \equiv w$ and $n_{\max} \gg n_{\min}$, one can obtain the following estimation $\delta B \simeq \epsilon B_0 C$, where $C = (\sqrt{2\pi w})^{1/2}/2n_{\min}^{\alpha-1/2}$. This estimation is close to the one calculated numerically according to Eq. (11.29). The latter is plotted in Fig. 11.13 for three different values of the exponent α with a fixed mode width w.

11.2.3 Perturbation Hamiltonian

In the presence of magnetic perturbations (11.27) the Hamiltonian (4.54) can be presented in the following form

$$K = K_0(z, t, p_z, p_t) + \epsilon K_1(z, t, p_z, p_t, \varphi), \tag{11.30}$$

where

$$K_0(z, t, p_z, p_t) = \psi_\varphi(x_c, z) - \sigma x_c \left[\varepsilon_0(\gamma^2 - 1) - 2\omega_x I_x\right]^{1/2}, \tag{11.31}$$

is the unperturbed Hamiltonian describing the motion in the equilibrium plasma, and

$$\epsilon K_1(z, t, p_z, p_t, \varphi) = \psi_\varphi^{(1)}(x, z, \varphi, t) = -\frac{RA_\varphi^{(1)}(R, Z, \varphi, t)}{B_0 R_0^2} \tag{11.32}$$

is the perturbation Hamiltonian.

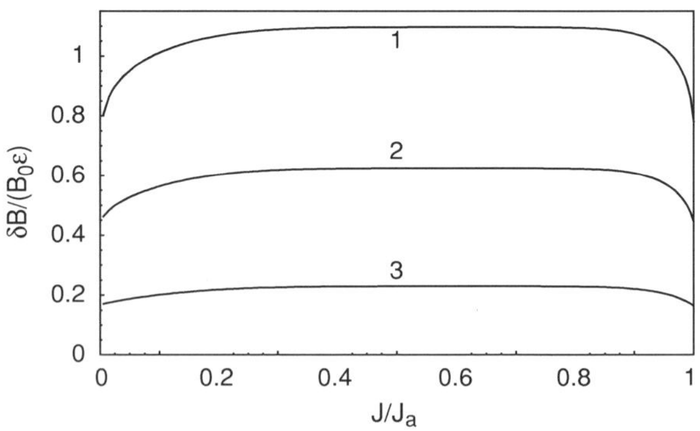

Fig. 11.13 Radial profiles of the magnitude of the turbulent magnetic field δB (11.29) normalized to ϵB_0 for the three different values of the exponent α: *curve 1* corresponds $\alpha = 5/6$; *curve 2* to $\alpha = 1$; and *curve 3* to $\alpha = 4/3$. The width of the mode $w = 6$, and $n_{\min} = 10$

Since the runaway electrons have large gyroradii ρ_e comparable to spatial scales of turbulent modes $2\pi r/m$, $m \gg 1$, we should take the coordinates (x, z, φ) of gyrating electron orbits in the perturbation Hamiltonian K_1, rather than their guiding-center orbits (x_c, z_c, φ_c). Using the relation between these coordinates given by Eq. (4.41) and introducing the action-angle variables (ϑ, J) for the guiding-center motion (5.24) the Hamiltonian (11.30) can be presented in the following form

$$K = K_0(I_x, J, E) + \epsilon K_1(\vartheta_x, \vartheta, I_x, \varphi, J, E),$$
$$\epsilon K_1(\vartheta_x, \vartheta, I_x, \varphi, J, E) = -\frac{R}{B_0 R_0^2} A_\varphi^{(1)}(R, Z, \varphi, t), \tag{11.33}$$

where $K_0(I_x, J, E)$ is the unperturbed Hamiltonian as a function action variables I_x, J and the energy E. The coordinates (R, Z) in the perturbation Hamiltonian K_1 are the functions of action-angle variables $(\vartheta_x, \vartheta, I_x, J)$, (see Eqs. 4.41 and 5.23)

$$R = R_0\left(x_c(J, \vartheta) + \sqrt{\frac{2I_x}{\omega_x}} \sin \vartheta_x\right),$$
$$Z = R_0\left(z_c(J, \vartheta) + \sqrt{\frac{2I_x}{\omega_x}} \cos \vartheta_x\right). \tag{11.34}$$

We recall that ϑ_x stands for the radial gyrophase (4.46).

Since the system in a periodic function of angular variables $(\vartheta, \vartheta_x, \varphi)$ the perturbation Hamiltonian K_1 can be also presented as a Fourier series,

$$\epsilon K_1 = \epsilon \sum_{m,n,s} K_{mns}(I_x, J, E) e^{i(m\vartheta - n\varphi + s\vartheta_x)}, \tag{11.35}$$

where $K_{mns}(J, I_R, E) = |K_{mns}(J, I_R, E)| \exp(i\chi_{mns})$, χ_{mns} are corresponding random phases. Furthermore, we assume that the phases χ_{mns} depend only on the toroidal mode number n, i.e., $\chi_{mns} \equiv \chi_n$. Using Eq. (11.27) the Fourier coefficients $K_{mns}(J, I_R, E)$ can be reduced to

$$K_{mns}(J, I_R, E) = (2\pi)^{-2} e^{i\chi_n} \sum_{m'} \iint_0^{2\pi} f_{m'n} e^{i(m'\vartheta_M - s\vartheta_x - m\vartheta)} d\vartheta \vartheta_x, \tag{11.36}$$

where

$$f_{mn} \equiv f_{mn}(\psi, \vartheta) = \frac{R(\psi, \vartheta)}{R_0} b_{mn}(\psi). \tag{11.37}$$

In Eq. (11.36) the magnetic coordinates (ϑ_M, ψ) are considered as functions of the guiding-center action-angle variables (J, ϑ) and the radial gyromotion variables (I_x, ϑ_x) through the spatial coordinates (R, Z) given by Eq. (11.34): $\vartheta_M = \vartheta_M(R, Z)$, $\psi = \psi(R, Z)$.

Assuming that the gyroradius ρ_g is small in compare to the radial coordinate r we expand the magnetic coordinates (ϑ_M, ψ) in a series of powers ρ_g/r:

$$\vartheta_M(R, Z) \approx \vartheta_M(x_c, z) + \delta \sin(\vartheta_x - \vartheta_{x0}) + O\left(\frac{\rho^2}{R_0^2}\right), \tag{11.38}$$

where (x_c, z) are the guiding center coordinates, the quantities δ and ϑ_{x0} are given by

$$\delta = \rho_g \left[\left(\frac{\partial \vartheta_M}{\partial R}\right)^2 + \left(\frac{\partial \vartheta_M}{\partial Z}\right)^2\right]^{1/2},$$
$$\vartheta_{x0} = \arcsin\left[\rho_g \delta^{-1} \partial \vartheta_M / \partial R\right]. \tag{11.39}$$

One should note that $\delta \sim \rho_g/r$. Similarly one can present the flux variable ψ. One can show that at the values $\vartheta_M = 0, \pi$, the phase $\vartheta_{x0} = 0$, and $\delta = (\rho_g/r)(\partial \vartheta_M/\partial \theta)$.

Substituting the expansion (Eq. 11.38) into the exponents under the integral in Eq. (11.36) and using that

$$\int_0^{2\pi} e^{i(m'\delta \sin(\vartheta_x - \vartheta_{x0}) - s\vartheta_x)} d\vartheta_x = 2\pi e^{-is\vartheta_{R0}} J_s(m'\delta), \tag{11.40}$$

where $J_s(z)$ is the Bessel function, Eq. (11.36) can be reduced to

$$K_{mns}(J, I_x, E) = (2\pi)^{-1} e^{i\chi_n} e^{-is\vartheta_{x0}} \sum_{m'} J_s(m'\delta) \int_0^{2\pi} f_{m'n} e^{i(m'\vartheta_M - m\vartheta)} d\vartheta. \tag{11.41}$$

Furthermore, we will keep only the terms $s = 0$, $K_{mn}(J, I_R, E) \equiv K_{mns=0}(J, I_R, E)$ which gives the main contribution to the diffusion. Equation (11.36) contains the integrals from the fast-oscillating functions which are similar to the ones encountered in Sect. 3.2. Since the perturbation field is localized on the low-field side using the formulas (3.29) and (3.30) the asymptotical formula for $K_{mn}(J, I_R, E)$ at large mode numbers m, $(m \gg 1)$, can be presented by

$$K_{mn}(J, I_x, E) \approx e^{i\chi_n} C J_0(m'\delta) f_{m'n}(0), \tag{11.42}$$

where

$$f_{mn}(0) \equiv f_{mn}(\psi_J, \vartheta_M = 0),$$
$$m' = \frac{m + \xi_c (m\gamma_3/2\gamma_1)^{1/3}}{\gamma_1} \approx \frac{m}{\gamma_1},$$
$$C = \frac{m'}{m'\gamma_1 + \xi_c (m\gamma_3/2\gamma_1)^{1/3}} \approx \frac{1}{\gamma_1}. \tag{11.43}$$

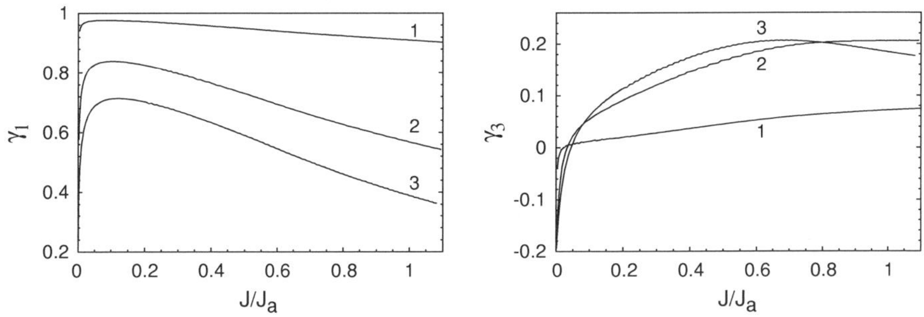

Fig. 11.14 Radial profiles of γ_1 (**a**) and γ_3 (**b**) for the different particle energies: *curve 1* corresponds to $E = 1$ MeV; *curve 2* corresponds to $E = 10$ MeV; *curve 3* to $E = 20$ MeV. The parameter $\lambda_I = 0.1$

Here $\xi_c \approx -1.018793$ is a local maximum of the Airy function $\mathrm{Ai}(\xi)$, $\psi_J \equiv \psi(J, \vartheta_M = 0)$. In Eq. (11.42) γ_1 and γ_3 are the first and third derivatives of magnetic field ϑ with respect to the guiding center ϑ taken on the LFS ($\vartheta_M = 0$) of the torus

$$\gamma_1 = \left.\frac{d\vartheta_M}{d\vartheta}\right|_{\vartheta_M=0}, \qquad \gamma_3 = \left.\frac{d^3\vartheta_M}{d\vartheta_z^3}\right|_{\vartheta_M=0}. \tag{11.44}$$

The typical radial dependencies of the parameters γ_1 and γ_3 for the different values of runaway electron energies are plotted in Fig. 11.14a, b.

The asymptotical formula (11.42) indicates that the (m, n) mode of the perturbation spectrum acting on a runaway particle with a given action J is asymptotically determined by the $(m' \approx m/\gamma_1, n)$ mode of magnetic turbulence spectrum on the magnetic surface ψ which touches the runaway drift orbit on the LFS (see Fig. 11.12).

This effect leads to the dependence of the turbulent transport on the particle energy. Particularly, for the specific form (11.28) of turbulent modes $b_{mn}(\psi)$ the perturbation spectrum (11.42) is reduced to

$$K_{mn}(J, I_x, E) \approx e^{i\chi_n} A(\psi_J)\frac{c_n}{\gamma_1} J_0\left(\frac{m\delta}{\gamma_1}\right) \exp\left[-\frac{1}{4w_n^2}\left(\frac{m}{\gamma_1} - nq(\psi_J)\right)^2\right]. \tag{11.45}$$

In Eq. (11.45) the term with the Bessel function $J_o(m\delta)$ takes into account the gyro-averaging effect of the turbulent field modes m, that effectively cuts off the contributions of the higher poloidal modes $m > m_c$ to the turbulent transport, where $m_c = \gamma_1 r/\rho_g$ is the critical poloidal mode number. The corresponding toroidal mode number n_c is equal to $n_c = m_c/q_p = \gamma_1 r/q_p\rho_g$.

The typical radial dependencies of n_c are shown in Fig. 11.15 for different energies of the runaway electrons.

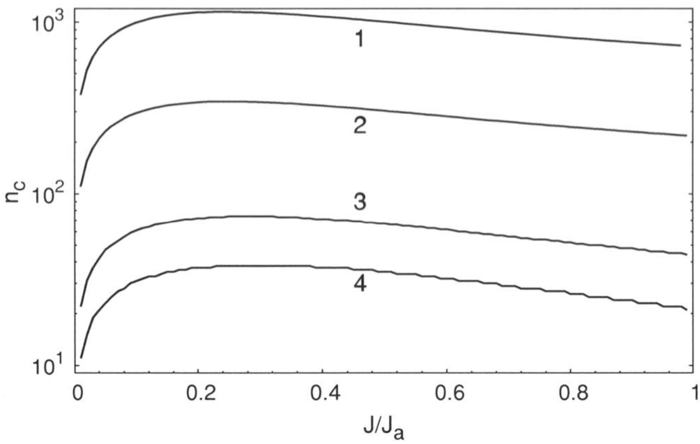

Fig. 11.15 Radial profiles of n_c for runaway electrons: *curve 1* corresponds to $E = 0.1$ MeV; *curve 2* corresponds to $E = 1$ MeV; *curve 3* to $E = 10$ MeV; *curve 4* to $E = 20$ MeV. The plasma parameters are the same as in Fig. 11.12

11.2.4 Full–Turn Transfer Mapping

For numerical simulations we again use the full-turn mapping (6.55) schematically shown in Fig. 6.3. Suppose that $(p_k, h_k, \varphi_k, t_k)$ are the values of the Hamiltonian $p_\varphi = -K_0(I_x, J, E)$, the energy E, the toroidal angle φ and the time t at the crossing point P_k of the orbit, respectively. We suppose that the fluctuating turbulent field does not change the particle's energy E. Then the putting $E = h_k = h_{k+1}$ the full-turn transfer mapping (6.55) and (6.56) can be reduced to

$$
\begin{aligned}
p_{k+1} &= p_k - \epsilon \frac{\partial P_k}{\partial \varphi_k}, \\
\varphi_{k+1} &= \varphi_k + \pi \left[q_p(p_k, E) + q_p(p_{k+1}, E) \right] + \epsilon \frac{\partial P_k}{\partial h_{k+1}}, \\
t_{k+1} &= t_k + \frac{\pi}{\omega(p_k, E_k)} + \frac{\pi}{\omega(p_{k+1}, E_{k+1})},
\end{aligned} \tag{11.46}
$$

where

$$
P_k = P\left(p_{k+1}, E_{k+1}, \varphi_k + \pi q_p(p_k, E_k), t_k + \frac{\pi}{\omega(p_k, E_k)} \right). \tag{11.47}
$$

The Poincaré integral $P(p, E, \varphi, t)$ is given by

$$
P(p, E, t, \varphi) = \int_0^{2\pi/\omega(p,E)} K_1\big(Z(\varphi'), \varphi + \varphi'\big) d\varphi', \tag{11.48}
$$

where $Z(\varphi) = \big(\vartheta_x(\varphi), \vartheta(\varphi), I_x, J, E, t(\varphi')$ is the unperturbed orbit of electrons as a function of φ. Using the expansion (11.35) the integral can be reduced to

$$P(J, E, \varphi, t) = \sum_n R_n(J, I_R, E)\cos(n\varphi + \chi_n), \qquad (11.49)$$

where

$$R_n(J, I_R, E) = \int_0^{2\pi/\omega_z} \sum_{m,s} f_{mn}(\psi, \vartheta) e^{i(m\vartheta + s\vartheta_x - n\varphi)} d\varphi. \qquad (11.50)$$

In Eq. (11.50) the arguments $\psi, \vartheta_x, \vartheta$ are supposed to be functions of the toroidal angle φ on the drift surface with $p = K_0(J, I_R, E)$ and E. One can show that at the resonant drift surfaces $J^{(mn)}$, $[m\omega_z(J^{(mn)}, E) = n]$ the integral $R_n(p, E)$ can be related to the spectrum K_{mn}:

$$R_n(J, E) = \pi q_p(J, E) K_{mn}(J, I_R, E). \qquad (11.51)$$

For the arbitrary values of h (or J) the $R_n(h, E)$ is obtained by extending the relation (11.51) from the discrete values of m to the continuous one $m = n/\omega_z(J, E)$. Then accordingly to Eq. (11.42) we have the following asymptotical formula for $R_n(J, E)$:

$$R_n(J, E) \simeq C F_n\left(\frac{n q_p(J, E)}{\gamma_1}\right). \qquad (11.52)$$

The finite correlation time τ_c of the phases χ_n are taken into account similar as in Sect. 11.1.1.

The optimal number of toroidal modes $n_{\max}$ in Eq. (11.14) which give the main contribution to the Poincaré integral $P(h, E, \varphi)$ is chosen by the following procedure. We calculated the dispersion of the random function $P(h, E, \varphi)$, i.e., $\sigma_P = \sqrt{\langle P^2(h, E, \varphi)\rangle}$ as a function of $n_{\max}$. Initially the dispersion σ_P grows monotonically with $n_{\max}$ and then reaches its saturation level for $n_{\max}$ exceeding a certain critical number n_P. The latter depends on the spectrum of the turbulence $c_n = n^{-\alpha}$ and on the energy of the electrons. In the calculations $n_{\max}$ is taken equal to the critical number n_P.

The mapping (11.46) requires the knowledge of the effective safety factor $q_p(J, E)$ and the Poincaré integral $P(p, E, \varphi, t)$ or its Fourier coefficients $K_{mn}(J, E)$ (11.42) which depends on the parameters $\gamma_1(J, E)$, $\gamma_3(J, E)$, $\delta(J, E)$. Analytical determination of these functions for realistic magnetic configurations is somehow difficult. In order to apply mapping procedure in these cases one can calculate these functions by the numerical integration of the guiding-center equations (4.54) and (4.55) for the equilibrium plasma in a grid of J: $J_i = i\Delta J$, $(i = 1, \ldots, N_J)$ at the given energy E. The functions $q_p(J, E)$, $\gamma_1(J)$, $\gamma_3(J, E)$, $\delta(J, E)$ for the arbitrary values of J can be interpolated by the cubic splines using the pre-calculated values of these functions at J_i), $(i = 1, \ldots, N_J)$.

11.2.5 Quasilinear Diffusion Coefficient

The diffusion coefficient $D_r(J)$ in the quasilinear approximation can be easily found using the corresponding formula (10.41) by replacing the safety factor $q(\psi)$ by the effective safety factor $q_p(J, E)$, the flux coordinate ψ by the action variable J

$$D_r(J) = \frac{\langle (r - r(J))^2 \rangle}{2t} = \frac{v_\parallel}{R_0} \left(\frac{dr}{dJ} \right)^2 D_J, \qquad (11.53)$$

where $v_\parallel$ is the parallel velocity of the runaway electron. The quantity D_J is the quasilinear diffusion coefficient in the action variable J defined as

$$D_J = \frac{\langle (J(\varphi) - J(0))^2 \rangle}{2\varphi} = \frac{\epsilon^2 q_p}{2\pi} \sum_{n=n_{\min}}^{n_{\max}} n^2 \left| R_n^{(reg)}(J, E) \right|^2 .$$

Figure 11.16a, b shows the radial profiles of the quasilinear diffusion coefficients D_J and D_r in Eq. (11.53) for the different electron energies. The intervals of the modes are chosen for $n_{\min} = 10 \le n \le n_{\max} = n_c(J) \sim 10^2$, where $n_c(J)$ is plotted in Fig. 11.15 (curve 2). The exponent α is taken equal to 5/6, which corresponds to the power spectrum $|H_{mn}|^2 \sim n^{-5/3}$ and the turbulence level $\epsilon = 10^{-4}$, which corresponds to the magnetic turbulence level δB shown by curve 1 in Fig. 11.13.

As seen from Fig. 11.16a the diffusion coefficient D_J characterizing the chaotic transport of the orbits takes the highest values for electrons with low energies. At the electron energies between $E = 1$ keV and $E = 10^2$ keV (curve 1) D_J practically coincides with the diffusion coefficients of the field lines. By increasing the electron energies the diffusion coefficients D_J monotonically decrease as shown by the curves 2–4 corresponding to electron energies from 1 to 20 MeV. This is due to a weakening the effect of the magnetic turbulence on the runaway electrons by increasing their energy due to the growth of the deviations of the orbits from the magnetic surfaces and the gyro-averaging.

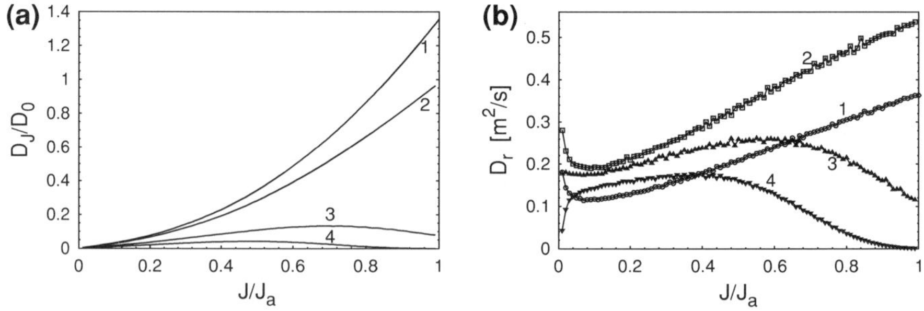

Fig. 11.16 Quasilinear diffusion coefficients: **a** D_J $[D_0 = 10^{-10}]$ (11.53) and **b** $D_r(J) = \langle (r - r(J))^2 \rangle / 2t$ for the different runaway electron energies: *curve 1* corresponds to $E = 10^2$ keV; *curve 2* to $E = 1$ MeV; *curve 3* to $E = 10$ MeV; *curve 4* to $E = 20$ MeV. The turbulent field level $\epsilon = 10^{-4}$, the mode width $w = 6$, and the exponent $\alpha = 5/6$

The diffusion coefficient D_r changes non-monotonically with energy. Since $D_r \sim v_\parallel D_J$ the diffusion coefficient D_r grows in the energy interval from 1 keV to 1 MeV because of the increase of the parallel velocity $v_\parallel$ with the energy E and reaches the maximal value at $E \sim 1$ MeV. If the energy is increased further D_r starts to decrease due to the decay of D_J while $v_\parallel$ approaches the speed of light c.

According to the asymptotical formulas (11.42), (11.51) and (11.52) the physical meaning of the quasilinear diffusion of runaway particles given by Eq. (11.53) is the following. Consider a particle diffusion on the drift surface of radius ρ (or action J). In a quasilinear approximation it is determined by the resonant terms K_{mn}, $m = nq_z(\rho)$, of perturbation Hamiltonian, i.e, by R_n. The latter is asymptotically determined by the mode amplitude $a_{m'n}$, $(m' = m/\gamma_1 = nq_p(\rho)/\gamma_1)$ of magnetic fluctuations localized near the magnetic surface $\psi(\rho)$ which touches the drift surface ρ at the lower field side, i.e., at $\vartheta = 0$. With increasing the energy of runaway electrons the mode number $m' = nq_p(\rho)/\gamma_1$ strongly deviates from the resonant mode $m = nq(\psi)$ of magnetic fluctuations. Since the (m, n)-modes b_{mn} of magnetic fluctuations are localized near the resonant mode $m = nq(\psi)$ the effect of magnetic fluctuations on the runaway electrons are weakened in comparison to low-energetic electrons for which $q_p(\rho) \approx q(\psi)$ and $\gamma_1 \approx 1$. In other words, the decreasing of runaway electron diffusion due to magnetic fluctuations depends on how the ratio $q_p(\rho)/\gamma_1$ deviates from the safety factor $q(\psi)$ and on the width of poloidal modes w_n (see also below Eqs. 11.55–11.57).

11.2.6 Shielding Factor

The decrease of the radial diffusion of runaway electrons with electron energy due to a finite gyroradius and a deviation of runaway electron orbits from magnetic surfaces can be quantified by the so-called shielding factor Υ (see Mynick and Krommes 1980; Myra et al. 1993). It is defined as a ratio of D_r to the collisionless diffusion coefficient $D_r(J) = v_\parallel D_{FL}(J)$ (10.33) determined by the diffusion coefficient of field lines $D_{FL}(J)$ (10.34),

$$\Upsilon = \frac{D_r(J)}{v_\parallel D_{FL}(J)}. \tag{11.54}$$

The diffusion coefficient $D_{FL}(J)$ of field lines in the magnetic turbulence can be easily found from Eq. (11.53) by putting $q_p = q(\psi)$, $\gamma_1 = 1$, $\gamma_3 = 0$, and $J_g = 0$. Then we have

$$\Upsilon = \left(\frac{d\psi}{dJ}\right)^2 \frac{q_p(J)}{q(\psi_J)} \sum_n n^2 \Upsilon_n, \tag{11.55}$$

$$\Upsilon_n = |R_n(J, I_R, E)|^2 \left[\sum_n n^2 |R_n(\psi_J, 0, 0)|^2\right]^{-1}. \tag{11.56}$$

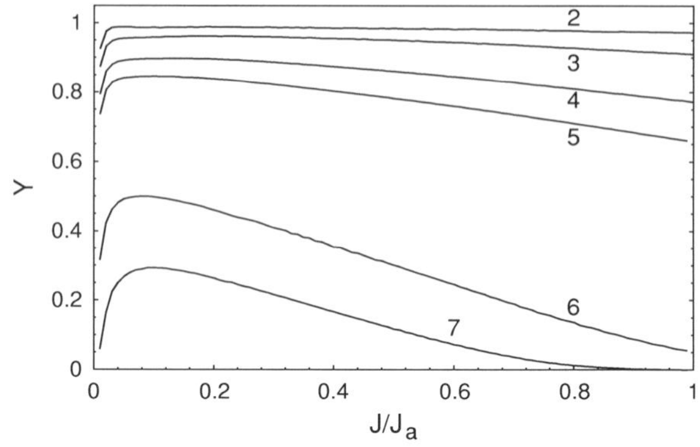

Fig. 11.17 Shielding factor Υ profiles for the different particle energies. Parameters and numbering of curves corresponds to the ones in Fig. 11.16: *curve 2* corresponds to $E = 10\,\text{keV}$; *curve 3* to $E = 10^2\,\text{keV}$; *curve 4* to $E = 5 \times 10^2\,\text{keV}$; *curve 5* to $E = 1\,\text{MeV}$; *curve 6* to $E = 10\,\text{MeV}$; *curve 7* to $E = 20\,\text{keV}$

The first two factors in Eq. (11.55) are related with the modification of particle orbits with respect to magnetic surfaces. The shielding effect of a magnetic turbulence due to a finite gyroradius and the drift of orbits is described by the quantities Υ_n. Particularly, using Eqs. (11.51) and (11.52) we find the following asymptotics of Υ_n for the spectrum of magnetic turbulence (11.28),

$$\Upsilon_n \simeq \tilde{C}_n \left| J_0 \left(\frac{n\delta}{q_p} \right) \right|^2 \exp \left[-\frac{n^2}{2w_n^2} \left(\frac{q_p}{\gamma_1} - q(\psi_J) \right)^2 \right], \qquad (11.57)$$

$$\tilde{C}_n = c_n^2 \Big/ \sum_n n^2 |c_n|^2.$$

As seen from Eq. (11.57) the decay of Υ_n with a particle energy E takes place mainly because of the deviation of the ratio q_p/γ_1 from the safety factor $q(\psi)$ of magnetic field with the increasing energy, which strongly depends on the width of the poloidal mode spectrum w_n at the resonant magnetic surface ψ_{mn}, $(q(\psi_{mn}) = m/n)$. The effect of a finite gyro-radius is described by the Bessel function $J_0\left(n\delta/q_p\right)$.

The radial profiles of the shielding factor Υ for the electrons with different energies are plotted in Fig. 11.17. The parameters and numbering of curves are the same as in Fig. 11.16.

11.2.7 Numerical Diffusion Coefficients

A turbulent magnetic field induces the radial transport of the electrons across the drift surfaces. Such a transport is caused by two effects: the fluctuating field with the finite correlation time τ_c and the overlapping of neighboring islands created by the resonance interaction of orbits with (m, n)-modes of the turbulent field. The latter case known as chaotic transport appears at amplitudes of the turbulent field exceeding a critical level ϵ_c which depends on the mode content and the turbulent field spectrum b_{mn}.

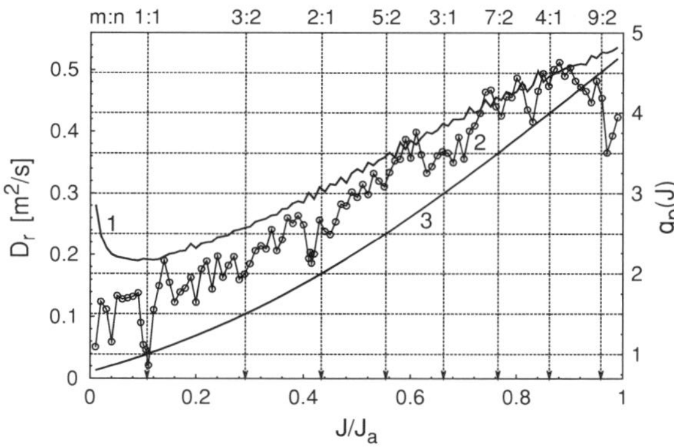

Fig. 11.18 Radial profile of diffusion coefficient D_J for the runaway electron energy $E = 1$ MeV. *Curve 1* corresponds to the quasilinear prediction, *curve 2* corresponds to the mapping results with the Poincaré functions $R_{nc} = \pi q_p K_{nq_p,n}$, $R_{ns} = 0$; *curve 3* describes the effective safety factor profiles, $q_p(J)$ (*right-hand axis*). The positions of the low order resonant drift surfaces, J_{mn} are shown by *vertical straight lines*. The number of orbits $N = 10^3$, the toroidal modes $10 \leq n \leq n_c$, $J_a = 3.32658 \times 10^{-2}$

The radial transport of particles is quantified by the diffusion coefficient defined as $D_r = \sigma_0^2(t)/2t$, where $\sigma_0^2(t)$ is the second moment of the radial displacements given by

$$\sigma_0^2(t) = \langle (r - r(0))^2 \rangle = \frac{1}{N_t} \sum_{i=1}^{N_t} (r_i(t) - r_i(0))^2 \,. \tag{11.58}$$

where the radial coordinate r of the orbit at the outer equatorial plane $\vartheta = 0$ (see Fig. 6.3). The time-dependence of $r(t)$ is calculated through the variable p_k in the mapping (11.46) using the relation $r = r(J(p_\varphi, E), E)$. In Eq. (11.58) $\langle(\cdots)\rangle$ stands for an averaging over the initial phases $\chi_n(0)$, $n_{\min} \leq n \leq n_{\max}$, randomly chosen in the interval $[0, 2\pi]$, $r_i(t)$ is the radial coordinate of an orbit corresponding to the ith trial and N_t is the total number of trials. All initial coordinates $r_i(0)$ are taken on the given drift surface $J =$ constant.

Now we consider the radial profiles of the numerically calculated diffusion coefficients D_r. The dependence of D_r on the action variable J is presented in Fig. 11.18 for the runaway electrons of the energy $E = 1$ MeV at the turbulent field parameters as in Fig. 11.16: curve 1 corresponds to the quasilinear prediction (11.53), curve 2 describes the numerically calculated D_r, curve 3 (right-hand axis) corresponds the radial profile of the effective safety factor $q_p(J)$. The vertical and horizontal lines describe the positions of the low-order rational drift surfaces J_{mn}, $q_p(J_{mn}) = m/n$. The Kubo number is of order of $K \sim 100$. At the chosen turbulence level ϵ and the large Kubo number K of the order of $K \sim 10^2$ the radial transport is mainly caused by the chaotic diffusion due to the overlapping of neighboring resonances.

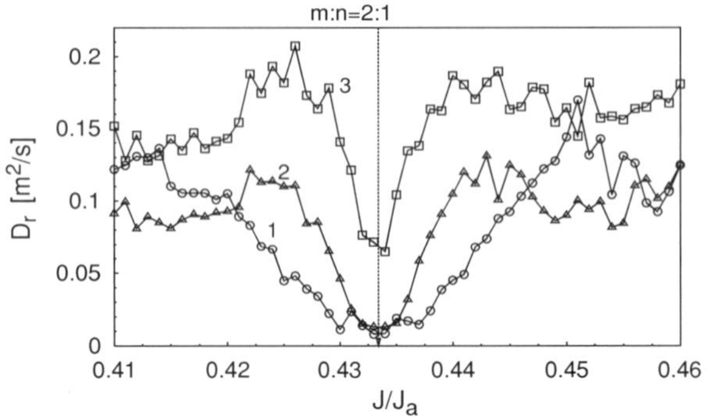

Fig. 11.19 Radial profile of the diffusion coefficient D_r for the runaway electron of energy $E = 1$ MeV near the rational drift surface $q_p = 2$ for the several values of the exponent α in the power-law c_n: *curve 1* ($10 \times D_r$) corresponds to $\alpha = 4/3$ at the perturbation level $\epsilon = 10^{-4}$; *curve 2* corresponds to $\alpha = 4/3$ at the perturbation level $\epsilon = \sqrt{10} \times 10^{-4}$; *curve 3* corresponds to $\alpha = 1$ at the perturbation level $\epsilon = 10^{1/6} \times 10^{-4}$. The position of the rational drift surface $q_p = 2$, J_{mn} is shown by *vertical straight lines*

As seen from Fig. 11.18 the radial profile of D_r (curve 2) has a fractal-like dependence which in average follows the quasilinear prediction (11.53). One can see that near the low-order rational drift surfaces (vertical lines), m/n=1, 3/2, 2/1, 5/2, 3/1, the diffusion coefficient drops to lower values significantly deviating from the quasilinear values. One can say that effective barriers to the radial transport are formed. At the given turbulence level ϵ the width and depth of these barriers depend on the exponent α in the power-law distribution of the toroidal modes $c_n = n^{-\alpha}$. For example, at $\alpha = 5/6$ the barrier near the rational surface $m : n = 2 : 1$ is rather small (see Fig. 11.18). For a larger value of α when the contribution of higher toroidal modes n is decreased the transport barrier becomes wider and deeper. It is shown in Fig. 11.19 where the radial profiles of D_r are plotted for two different values of α the rational drift surface $m : n = 2 : 1$. Curve 1 corresponds to $\alpha = 4/3$ for the turbulence level $\epsilon = 10^{-4}$ (D_r is multiplied by the factor 10), curve 2 corresponds to the same $\alpha = 4/3$ but $\epsilon = \sqrt{10} \times 10^{-4}$ and curve 3 corresponds to the same $\alpha = 1$ for $\epsilon = 10^{1/6} \times 10^{-4}$.

The reduced radial transport near the low-order rational drift surfaces is related to the gaps in their density there. As shown in Sect. 11.1.4 the gap width Δw is inversely proportional to the maximal toroidal mode number $n_{\max} \sim n_c$ for which the corresponding terms R_n in the function (11.14) are not negligible. One would expect that with increasing the energy of the electrons the transport barriers become wider and deeper since the number of the toroidal modes $n > n_c$ contributing to the transport effectively decays. This is shown in Fig. 11.20a, b where the radial profiles of the diffusion coefficients for the runaway energies $E = 10$ MeV and $E = 20$ MeV are plotted.

As seen from Fig. 11.20a, b there are also regions outside the low-order rational surfaces where the diffusion coefficients are much smaller than the quasilinear values.

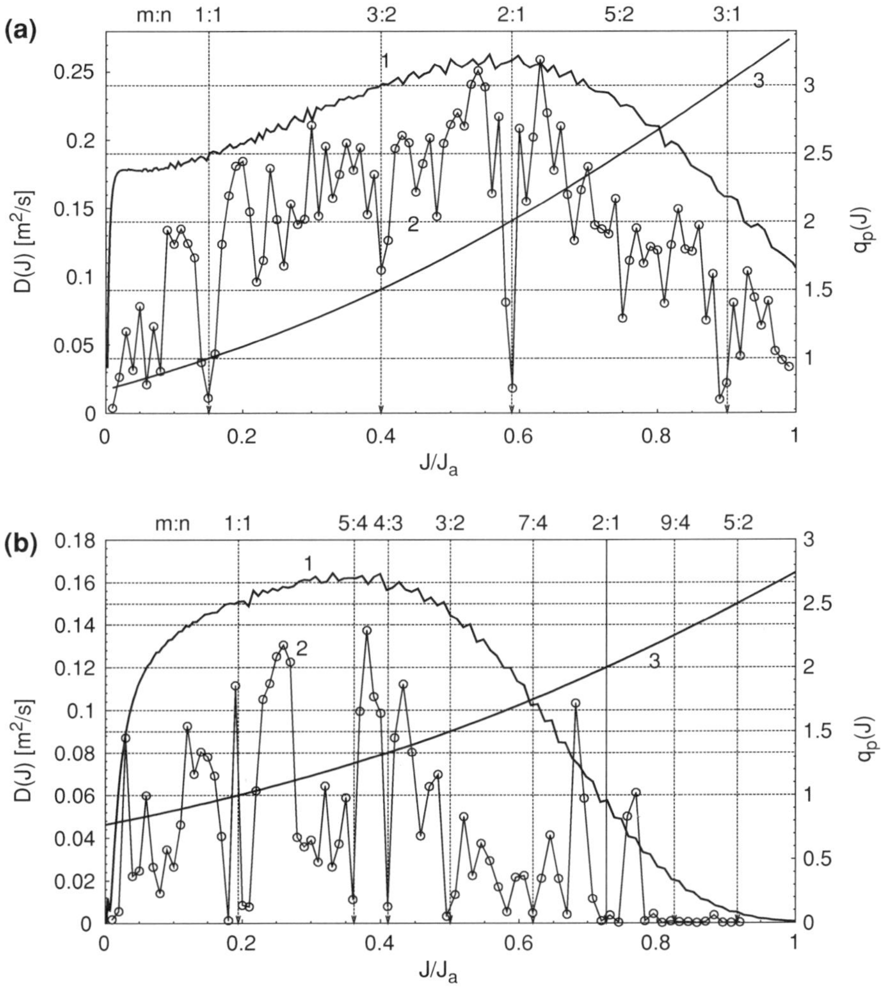

Fig. 11.20 The same as in Fig. 11.18 but for the electron energies $E = 10\,\text{MeV}$ (**a**) and $E = 20\,\text{MeV}$ (**b**). The other parameters are the same as in Fig. 11.18

In these regions the radial transport is caused mainly by the fluctuations of the turbulent field in time. There the chaotic transport is reduced or absent because of the low density of the rational surfaces (not gaps) which is not sufficient for an overlapping of the resonances. However, when one approaches the low-order rational surfaces the diffusion coefficients grow to their local maximal values and then drop drastically. The radial profile of D_r near these areas is similar to the behavior of the density $\rho(q, N)$ near low-order rational numbers $q = m/n$ as it is shown in Fig. 11.5.

11.2.8 Effect of Isolated MHD Modes

MHD modes are frequently excited in real tokamak discharges. Typically the MHD modes with low poloidal and toroidal mode numbers (m, n) are isolated and do not disrupt the plasma. They are localized near the rational magnetic surfaces J_{mn}, $q(J_{mn}) = m/n$. These MHD modes may significantly influence the chaotic transport caused by the micro-turbulence field near the corresponding rational drift surfaces, as it was shown in the Hamiltonian system (see Sect. 11.1.5).

In the presence of the (m, n) MHD mode alone the structures of the field lines and particle orbits near the corresponding rational surfaces are modified. Particularly, it creates an island which traps the particles with the appropriate initial conditions. The average safety factor,

$$q_{av}(J) = \frac{1}{N} \sum_{k=0}^{N} \frac{\varphi_{k+1} - \varphi_k}{2\pi}, \qquad N \gg 1, \tag{11.59}$$

introduced similar to (11.26), is flattened in the region $J_{mn} - W_{mn}/2 < J < J_{mn} + W_{mn}/2$ near the resonant surface J_{mn}, where W_{mn} is the island width. According to Eqs. (11.24) and (11.25) for the density of resonance drift surfaces such a profile of q_{av} leads to a widening of the gap in the density of the rational surfaces, and thus reduces the chaotic transport due to interactions with the modes of the turbulent field.

Figure 11.21 shows the example of a reduced turbulent transport near the $(m : n) = (2 : 1)$ rational surface in the presence of the corresponding MHD mode: curve 1 describes the quasilinear diffusion coefficients D_r, curves 2 and 3 correspond to the numerical D_r without and with MHD mode, respectively. Curves

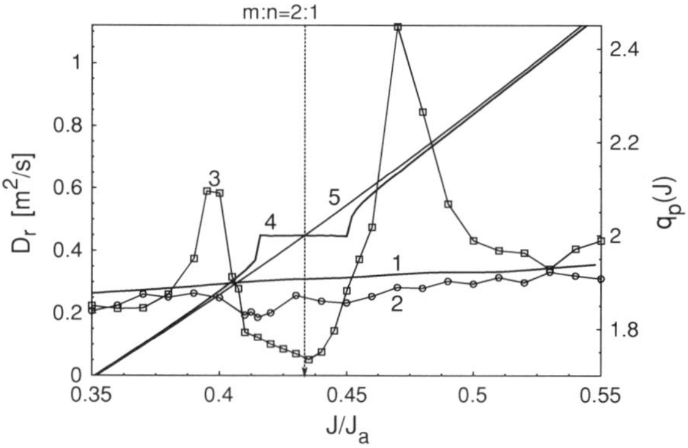

Fig. 11.21 The same as in Fig. 11.18 but in the presence of the isolated $(m, n) = (2 : 1)$ MHD mode: *curve 1* corresponds to the quasilinear prediction, *curves 2* and *3* correspond numerical results without and with MHD mode, respectively. *Curves 4* and *5* (*right-hand axis*) describe the averaged, $q_{av}(J)$, and effective, $q_p(J)$, safety factors, respectively. The amplitude of the MHD mode is taken equal to $\epsilon = 10^{-4}$. The other parameters are the same as in Fig. 11.18

4 and 5 (right-hand axis) describe the averaged, $q_{av}(J)$, and the effective, $q_p(J)$, safety factors, respectively. It is assumed that the MHD mode has a localized radial structure given by Eq. (11.28) and the amplitude of the MHD mode is taken equal to $\epsilon = 10^{-4}$, and the electron energy $E = 1$ MeV.

As seen from Fig. 11.21 there is a strong reduction of the turbulent diffusion near the rational surface where q_{av} is flattened (see curve 4). The width of the transport barrier is of the order of the width of the corresponding island W_{mn}. However, the diffusion has large spikes just outside the island. It is reduced to the quasilinear level (curve 1) in the regions far from the island.

The reduction of runaway electron transport inside the magnetic island is in agreement with the experimental observation in a reversed-field pinch plasma (Clayton et al. 2010). Probably, this effect also explains the experimental observation of electron beams (called 'snakes') inside magnetic islands (Entrop et al. 1999).

11.2.9 Summary

We summarize the assumptions and main results of this section devoted to study the radial transport of runaway electrons in a tokamak plasma in the presence of a background turbulent magnetic field.

(i) It is assumed that a turbulent magnetic field has a wide spectrum of modes with a finite correlation time. The runaway electrons have large Kubo numbers.

(ii) The numerical calculations of diffusion coefficients are carried out using the full-turn transfer mapping constructed from the Hamiltonian equations for the relativistic guiding-center motion of the electrons. The latter takes both effects into account, the deviation of the electron drift orbits from the magnetic surfaces and the gyro-averaging of the turbulent magnetic field.

(iii) The radial transport runaway electrons in a turbulent magnetic field is mainly caused by the chaotic transport due to resonant interactions of particle orbits with the resonant components of the turbulent field. The diffusion coefficient of radial transport in the quasilinear approximation are obtained.

(iv) The numerical calculations show that the radial profiles of the diffusion coefficients of the chaotic transport have fractal-like structures with a reduced diffusivity near the low-order rational drift surfaces thus forming transport barriers for the radial transport. The fractal-like dependence of the runaway electron diffusion stems from the large values of the corresponding Kubo number when high speed electrons "feel" the structure of a turbulent magnetic field. The width and depth of the transport barriers depend on the toroidal mode spectrum $c_n = n^{-\alpha}$ as well as on the energy of the runaway electrons. The barriers become more pronounced by increasing the electron energy. The latter is caused not only by the cutting out of higher modes by gyro-averaging but also due to the increase of the Kubo number.

(v) The described mechanism of the turbulent transport of energetic electron can be one of the main reasons for the improved confinement of runaway electrons in tokamaks additional to the effects of the gyro-averaging and the orbit-averaging mechanisms.

(vi) One can expect that the formation transport barriers near low-order rational drift surface may lead to the formation of the nested beams of runaway electrons. Perhaps, this effect has been observed in a recent experiment by Zhou et al. (2013a) in the EAST tokamak. They experimentally observed the formation of ring-like beams of runaway electrons of energies up to 30–40 MeV. A more detailed analysis of this phenomenon would require the study of the generation and acceleration processes of electron beams in the presence of magnetic micro-turbulence perturbations.

11.3 Bibliographic Comments

Problems of energy and particle transport in fusion plasmas. The main goal of magnetic fusion reactors is to confine the high temperature plasma for a sufficient long time in order to achieve thermonuclear conditions. The confinement time is determined by the loss of the plasma energy. The latter is mainly due to the transport of heat and particles across magnetic surfaces caused by Coulomb collisions and different types of instabilities developing in the plasma. The radiation may be also a source of energy loss (see, e.g., Doyle et al. 2007; Wesson 2004 and references therein).

In the absence of plasma instabilities the heat and particle transport across magnetic surfaces is caused by Coulomb collisions. The corresponding transport model known as *the neoclassical transport model* takes into account the complex toroidal magnetic geometries (see Balescu 1988; Hinton and Hazeltine 1976). However, it's predictions for the transport coefficients and confinement times are much lower than the ones observed in experiments. It is believed that the latter called as the *anomalous transport* are caused by the plasma instabilities developing in the plasma and generating fluctuating magnetic and electric fields or a plasma turbulence.

Understanding the mechanisms of the anomalous transport is one of the important issues in the present-day studies of transport processes in magnetically confined fusion plasmas. Particularly, it is important to understand the improved confinement regimes of plasmas which have been experimentally found by heating plasma (Burrel 1994; Connor and Wilson 2000; Connor et al. 2004; Wagner et al. 1982). Such regimes of the plasma, like the H-mode and the internal transport barrier (ITB), have a crucial importance to achieve and sustain fusion reactions in a magnetically confined plasmas. This problem has been reviewed by Burrel (1994), Connor and Wilson (2000), Connor et al. (2004), Doyle et al. (2007), Garbet (2006), Garbet et al. (2004), Gohil (2006), and Wolf (2003). The role of rational magnetic surface in the formation of transport barriers has been discussed by Austin et al. (2006), Brakel et al. (2002), Eriksson et al. (2002), Garbet et al. (2003), Gohil et al. (2003), Hogeweij et al. (2004), Joffrin et al. (2003), Kasilov et al. (2002), Lopes Cardozo (2004), and Razumova et al. (2003, 2008) (see also references therein).

At present the theoretical studies of the anomalous transport are based on the gyrokinetic simulation of plasma turbulence and associated transport processes in fusion devices (see, e.g., the review by Garbet et al. 2010). The numerical codes developed to implement the gyrokinetic simulation are a massively parallel, particle-in-cell codes in 3D toroidal geometry.

The test models are widely used to study the particle transport caused by turbulent fields in laboratory and space plasmas. One can mention some of these works by Ferreira (2005), Fichtner (2005), Horton et al. (1998), Kwon et al. (2000); Michalek (2001), Miskane et al. (2001), Nishimura and Azumi (1997), de Rover et al. (1999), Vlad and Spineanu (2008), Weyssow et al. (1991) and references therein.

Runaway electron transport. The problem of the runaway electron transport in magnetically confined plasmas has been a subject of many theoretical and experimental studies over three decades (see Fussmann 1979; Gill 1993; Jaspers et al. 1996; Knoepfel and Spong 1979; and theses by Entrop 1999; Jaspers 1995). The problem of improved confinement of runaway electrons in a turbulent plasma has been discussed in many theoretical (see Abdullaev et al. 2010; Eriksson et al. 2004; Hauff and Jenko 2009; Helander et al. 2002; Myra and Catto 1992; Myra et al. 1993; Plyusnin et al. 2006; de Rover et al. 1996a, b, 1999; and references therein). In experimental studies by Entrop et al. (1998, 1999, 2000), Esposito et al. (1996), Kudyakov et al. (2012), Kwon et al. (1988), Zeng et al. (2013), and Zhou et al. (2013a, b) the influence of magnetic turbulence on runaway electron transport has been investigated. Particularly, Entrop (1999), Entrop et al. (1998, 2000), and Kudyakov et al. (2012) have proposed to study a magnetic turbulence in tokamaks by measuring a runaway electron diffusion.

There are a number of theoretical and experimental works on the influence of externally imposed magnetic perturbations on the runaway electrons in tokamaks (Finken et al. 2006, 2007; Forster et al. 2012; Helander et al. 2000; Kawano et al. 1997; Kawashima et al. 1994; Lehnen et al. 2008a, 2009; Martín-Solís and Sánchez 2006, 2008; Martín-Solís et al. 1999, 2000; Tamai et al. 2002; Tokuda and Yoshino 1999; Wingen et al. 2006; Yoshino and Tokuda 2000) and in a stochastic reversed-field pinch plasma Clayton et al. (2010). In the papers by Papp et al. (2011a, 2012) the possible mitigation mechanism of runaway electrons by the RMPs during disruptions in the ITER has been analyzed.

Appendix A
Magnetic Field in Ergodic Divertors

In this chapter I would like to present a detailed analytical calculations of the magnetic field generated by a set of helical coils. The method is demonstrated for the models of the DED of the TEXTOR and the ED of the Tore Supra. The main idea of the method is concluded in the following. Typically a set of coils are located on the surface of a torus. The density of the current flowing in the coils can be described by the delta functions with singularities on the coil locations. Using the Poisson summation rule which gives the relation of the delta functions with the trigonometric functions one can present the current density on the surface of torus a sum of infinity number of helical currents, $j_{mn} \propto \cos(m\theta - n\varphi)$. For the large aspect ratio tokamaks the magnetic field generated by these helical currents can be found by approximating the system by the cylinder (see, e.g., Morozov and Solov'ev (1966b)) and making corrections due to a toroidicity. This approach allows one to qualitative and quantitatively analyze the poloidal and toroidal spectra of the magnetic field generated by a set of coils as well as its radial dependence.

A.1 Magnetic Perturbation in the TEXTOR-DED

The magnetic field perturbations in the TEXTOR-DED by the set of helical coils schematically shown in Fig. 9.4a, b. The geometrical locations of the helical coils in the (φ, θ)-plane are plotted in Fig. A.1a for the ideal configuration and in Fig. A.1b for the real configuration.

A.1.1 Density of Perturbation Currents

It is convenient to introduce the density of DED perturbation currents in order to find the magnetic perturbations. The current density $\mathbf{j}(r, \theta, \varphi)$ is introduced as

$$\mathbf{j} = j_r \mathbf{e}_r + j_\theta \mathbf{e}_\theta + j_\varphi \mathbf{e}_\varphi, \tag{A.1}$$

S. Abdullaev, *Magnetic Stochasticity in Magnetically Confined Fusion Plasmas*, Springer Series on Atomic, Optical, and Plasma Physics 78,
DOI: 10.1007/978-3-319-01890-4,

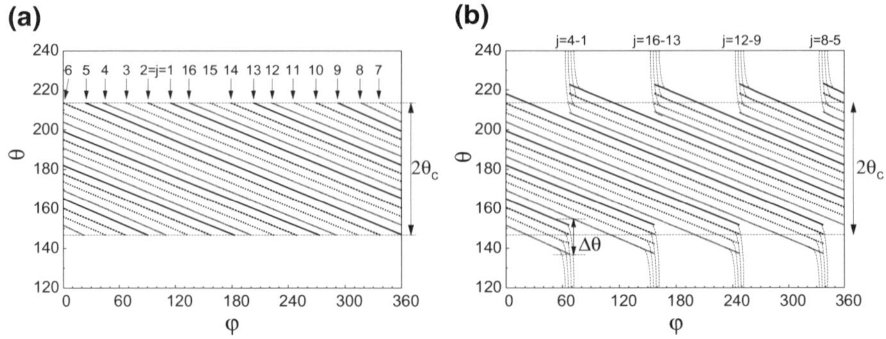

Fig. A.1 Models of the DED coil configuration: **a** the ideal configuration, **b** the real configuration. The coils are numbered by $j = 1, 2, \ldots, 16$

where $\mathbf{e}_r$, $\mathbf{e}_\theta$, $\mathbf{e}_\varphi$ are unit vectors along the coordinates r, θ, φ, respectively. The corresponding components, j_r, j_θ, and j_φ, are defined as

$$(j_r, j_\theta, j_\varphi) = (0, j(r, \theta, \varphi) \sin \alpha_0, j(r, \theta, \varphi) \cos \alpha_0),$$

where

$$j(r, \theta, \varphi) = \delta(r - r_c) \sum_{j=1}^{16} I_j r_c^{-1} \delta(\theta - \theta_j(\varphi)), \tag{A.2}$$

and $\alpha_0 = \theta_c r_c / \pi R_c$ is an angle between current direction and toroidal axis φ, $\theta_j(\varphi)$ is a poloidal position of the jth coil at the toroidal section φ. The current I_j on the jth coil can be expanded into a series

$$I_j = \sum_n \iota_n I_j^{(n)},$$
$$I_j^{(n)} = I_d \sin\left(n \frac{2\pi j}{16} + \chi_n\right), \tag{A.3}$$

where $I_j^{(n)}$ is the basic current distribution which generates the magnetic perturbation with toroidal modes $\bar{n} = n + 4p, (p = 0, 1, 2, \ldots)$. The coefficients ι_n and the phases χ_n can be determined by the currect distribution on coils I_j. They will be listed in Table A.1. The details of calculations of ι_n and χ_n can be found in Finken et al. (2005a).

For the given mode n Eq. (A.2) may be written as

$$j^{(n)}(r, \theta, \varphi) = \delta(r - r_c) g(\theta, \varphi) \frac{I_d}{r_c} \sum_{j=-\infty}^{\infty} \sin\left(n \frac{2\pi j}{16} + \chi_n\right) \delta\Big(\theta - \theta_j(\varphi)\Big), \tag{A.4}$$

where $g(\theta, \varphi)$ is a step function equal to 1 in the area covered by coils and zero elsewhere (see Eqs. (A.14) and (A.19)), the poloidal position $\theta_j(\varphi)$ of the jth coil at the section φ is given by

$$\theta_j = \theta(\varphi) - j\delta\theta, \qquad \theta(\varphi) = \theta_0 - \frac{\theta_c}{\pi}\varphi. \tag{A.5}$$

The angle $\theta_{01} = \theta_0 + \delta\theta$ is a poloidal position of the first coil at the toroidal section $\varphi = 0$, $\delta\theta$ is the angular distance between neighboring coils.

A.1.2 Continuous Current Density

Using the coil positions (A.5), we present the current distribution given by Eq. (A.4) as

$$j^{(n)}(r,\theta,\varphi) = \delta(r - r_c)\frac{I_d}{r_c}g(\theta,\varphi)\sum_{j=-\infty}^{\infty}\sin\left(n\frac{2\pi j}{16} + \chi_n\right)\delta\left(\theta - \theta(\varphi) + j\delta\theta\right). \tag{A.6}$$

Using the representation of delta function

$$\delta(x) = \frac{1}{2\pi}\int_{-\infty}^{\infty} e^{ixp}dp,$$

one obtains

$$j^{(n)}(r,\theta,\varphi) = \frac{I_d\delta(r - r_c)}{2\pi r_c}g(\theta,\varphi)W,$$
$$W = \mathrm{Im}\left\{e^{i\chi_n}\int_{-\infty}^{\infty} dp e^{ip(\theta-\theta(\varphi))}\sum_{j=-\infty}^{\infty}\exp\left[ij\left(\frac{2\pi n}{16} + p\delta\theta\right)\right]\right\}. \tag{A.7}$$

Then using the Poisson rule

$$\sum_{j=-\infty}^{\infty} e^{i2\pi jx} = 2\pi\sum_{s=-\infty}^{\infty}\delta(s - x),$$

we have

$$W = \operatorname{Im}\left\{e^{i\chi_n}\int_{-\infty}^{\infty} dpe^{i(\theta-\theta(\varphi))p}\sum_{s=-\infty}^{\infty}\delta\left(s-\frac{n}{16}-\frac{p\delta\theta}{2\pi}\right)\right\}$$
$$= \frac{2\pi}{\delta\theta}\sum_{s=-\infty}^{\infty}\sin\left(\frac{2\pi(s-n/16)}{\delta\theta}(\theta-\theta_\varphi)+\chi_n\right). \quad \text{(A.8)}$$

Using the dependence of $\theta(\varphi)$ on φ given in Eq. (A.5) we reduce (A.7) to

$$j^{(n)}(r,\theta,\varphi) = \delta(r-r_c)g(\theta,\varphi)J_0$$
$$\times\sum_{s=-\infty}^{\infty}\cos\left(\frac{m_0(16s-n)}{4}\theta+\frac{n_0(16s-n)}{4}\varphi+\chi_s^{(n)}\right), \quad \text{(A.9)}$$

where

$$\chi_s^{(n)} = -\frac{m_0(16s-n)}{4}\theta_0+\chi_n-\frac{\pi}{2}. \quad \text{(A.10)}$$

In (A.9) the following notations are introduced:

$$J_0 = \frac{I_d}{\delta\theta r_c} = \frac{2m_0 I_d}{\pi r_c}, \quad m_0 = \frac{\pi}{2\delta\theta},$$
$$n_0 = \frac{m_0\theta_c}{\pi} = \frac{\theta_c}{2\delta\theta}, \quad \theta_0 = \theta_{01}+\delta\theta. \quad \text{(A.11)}$$

Because of periodicity of $j(r,\theta,\varphi)$ along φ with a period 2π follows that n_0 must be an integer number equal to $n_0 = 4l$, where $l = 1, 2, \ldots$. Putting $n_0 = 4$, the term $s = 0$ in Eq. (A.9) which gives the main contribution the perturbed magnetic field in the plasma can be presented as

$$j_0^{(n)}(r,\theta,\varphi) = \delta(r-r_c)g(\theta,\varphi)J_0\cos\left(nm_0\theta/4+n\varphi-\chi_0^{(n)}\right). \quad \text{(A.12)}$$

A.1.3 Fourier Expansion of the Current Density

For calculations of the magnetic field created by helical coils it is convenient to present the current density $j_0^{(n)}(r,\theta,\varphi)$ in Fourier series in θ, φ:

$$j^{(n)}(r,\theta,\varphi) = \sum_{m,n'} j_{mn'}(r)\cos\left[m\theta+n'\varphi+\chi_{mn'}\right],$$

where

$$j_{mn'}(r)e^{i\chi_{mn'}} = e^{-i\chi_0^{(n)}}\frac{J_0\delta(r-r_c)}{(2\pi)^2} \times \int_0^{2\pi}\int_0^{2\pi} d\theta d\varphi g(\theta,\varphi)\exp\left[-i\left(m-\frac{nm_0}{4}\right)\theta - i(n'-n)\varphi\right]. \tag{A.13}$$

The content of Fourier spectrum, $j_{mn}(r)$, depends on the function $g(\theta, \varphi)$ which determined by the coil configurations. The ideal and real configurations of coils are shown in Fig. A.1a, b, respectively. Below we consider these cases separately.

A.1.3.1 Ideal Coil Configuration

For the ideal coil configuration (see Fig. A.1a) the function $g(\theta, \varphi)$ is given by:

$$g(\theta,\varphi) = \begin{cases} 1 & \text{for } \pi-\theta_c < \theta < \pi+\theta_c, \\ 0 & \text{otherwise.} \end{cases} \tag{A.14}$$

The current density (A.9) can be presented as a Fourier series:

$$j^{(n)}(r,\theta,\varphi) = J_0\delta(r-r_c)\sum_{m=-\infty}^{\infty}\sum_{s=-\infty}^{\infty} g_m^{(s)}\cos\left[m\theta + (16s-n)\varphi + \chi_{ns}\right], \tag{A.15}$$

where

$$g_m^{(s)} = (-1)^m\frac{\sin([m-m_0(n/4-4s)]\theta_c)}{[m-m_0(n/4-4s)]\pi},$$
$$\chi_{ns} = \frac{m_0(16s-n)}{4}(\pi-\theta_0) + \chi_n - \frac{\pi}{2}. \tag{A.16}$$

The main contribution to the magnetic field comes from the term $s = 0$ which can be rewritten as (by changing the summation over m to $-m$)

$$j_0^{(n)}(r,\theta,\varphi) = J_0\delta(r-r_c)\sum_{m=-\infty}^{\infty} g_m\cos(m\theta + n\varphi - \chi_{n0}), \tag{A.17}$$

where

$$g_m \equiv g_m^{(0)} = (-1)^m\frac{\sin[(m+m_0n/4)\theta_c]}{(m+m_0n/4)\pi},$$
$$\chi_{n0} = \chi_n - \frac{m_0n}{4}(\pi-\theta_0) - \frac{\pi}{2}. \tag{A.18}$$

A.1.3.2 Non-ideal Coil Configuration

For the non-ideal configuration of coils (see Fig. A.1b) the step function $g(\theta, \varphi)$ is given by

$$g(\theta, \varphi) = \begin{cases} 1, & \text{for } \pi - \theta_c(\varphi) < \theta < \pi + \theta_c(\varphi), \\ 0, & \text{elsewhere,} \end{cases} \tag{A.19}$$

where $\theta_c(\varphi)$ is the piece-wise function

$$\theta_c(\varphi) = \theta_{c0} - \frac{2\Delta\theta}{\pi}(\varphi - \varphi_l) \quad \text{for } \varphi_l < \varphi < \varphi_{l+1},$$
$$\varphi_l = \varphi_c + (l-1)\frac{\pi}{2}, \quad 0 < \varphi_c < \frac{\pi}{2}, \qquad l = 0, 1, 2, 3, 4. \tag{A.20}$$

For the sake of simplicity we consider the term $s = 0$ in Eq. (A.9), i.e., Eq. (A.12). Furthermore, we need to estimate the integral

$$f_{m,n'} = \frac{1}{(2\pi)^2} \int_0^{2\pi}\int_0^{2\pi} d\theta d\varphi g(\theta, \varphi) e^{-im\theta - in'\varphi}.$$

It is not difficult to show that

$$f_{mn'} = -e^{-i\pi m}\delta_{n',4s} e^{-in'\varphi_c} \frac{2\sin(m\alpha\pi/4)}{\pi^2 m} \frac{2[in'\cos(m\theta_c) + m\alpha\sin(m\theta_c)]}{n'^2 - (m\alpha)^2},$$

where $\alpha = 2\Delta\theta/\pi$, $\theta_c = \theta_{c0} - \Delta\theta/2$, $\delta_{n,k}$ is the Kronecker symbol, i.e., $\delta_{n,k} = 0$ for $n \neq k$ and $\delta_{n,n} = 1$. Substituting this expression into (A.13) we obtain the following expression for the Fourier components of the current density $j_{mn'}(r)$,

$$j_{mn'}(r)e^{i\chi_{mn'}} = \delta(r - r_c)e^{i\chi_0^{(n)}} J_0 \ f_{m-nm_0/4, n'-n}. \tag{A.21}$$

From Eq. (A.21) follows that due to non-ideal configuration the current distribution (A.9) creates the toroidal modes $\hat{n} = n + 4s$, $(s = 0, \pm 1, \pm 2, \ldots)$.

The main contribution to the toroidal spectrum n' comes from the terms $n' = n$. In this case $s = 0$, and one obtains

$$f_{m,0} = e^{-i\pi m} \frac{\sin[\pi m\alpha/4]}{\pi m\alpha/4} \frac{\sin(m\theta_c)}{m\pi},$$

and from Eq. (A.21) we obtain

$$j_0^{(n)}(r, \theta, \varphi) = \delta(r - r_c) \sum_{m=-\infty}^{\infty} \sum_{s=-\infty}^{\infty} J_{m,s} \cos(m\theta + (4s + n)\varphi - \chi_{ms}). \tag{A.22}$$

The Fourier coefficients, $J_{m,0}$, corresponding to the term $s = 0$ which gives the main contribution to the perturbed field is given by

$$J_{m,0} = J_0 g_m C_m, \qquad \chi_{n0} = \chi_n - \frac{m_0 n}{4}(\pi - \theta_0) - \frac{\pi}{2}, \tag{A.23}$$

where g_m is given by Eq. (A.18), and

$$C_m = \frac{\sin[(m + nm_0/4)\Delta\theta/2]}{(m + nm_0/4)\Delta\theta/2} \tag{A.24}$$

is a correction factor due to non-ideal configuration. For the ideal configuration $\Delta\theta = 0$ and therefore $C_m = 1$.

One should note that the current distribution (A.3) with $n = 4$ creates also the toroidal mode $n' = 0$ (see Eq. (A.21)). This mode may disturb the plasma equilibrium. For this reason in the $m : n = 12 : 4$ operational mode of the TEXTOR-DED one applies the compensation coils which annuls the effect of the $n' = 0$ mode.

A.1.4 Magnetic Field Perturbations

In this section we present the formulae for the magnetic field created by the surface current (A.1). Each term in the Fourier expansion (A.22) of this perturbation current describes a helical current on the toroidal surface of radius $r = r_c$. Consider a single helical current vector $\mathbf{j}_{mn}$ corresponding to the (m, n) mode

$$\mathbf{j}_{mn}(r, \theta, \varphi) = \delta(r - r_c) j_{mn} \mathbf{e}_{mn} \cos(m\theta + n\varphi + \phi_{mn}), \tag{A.25}$$
$$\mathbf{e}_{mn} = \Big(0,\ \mathbf{e}_\theta \sin\alpha_{mn},\ \mathbf{e}_\varphi \cos\alpha_{mn}\Big),$$

where $\mathbf{e}_\theta$ and $\mathbf{e}_\varphi$ are unit vectors along the poloidal and toroidal directions, respectively, and $\alpha_{mn} = nr_c/mR_c$, $(m \neq 0)$, is a helicity, i.e., the angle between a helical current direction and toroidal axis.

The total current density (A.1) can be presented as a sum of helical currents (A.25), i.e.,

$$\mathbf{j}_h(r, \theta, \varphi) = \sum_{mn} \mathbf{j}_{mn}(r, \theta, \varphi). \tag{A.26}$$

with the same toroidal components of the vector $\mathbf{j}_{mn}$ but different poloidal components, i.e.,

$$j_{mn} \cos\alpha_{mn} = J_{m,(n-k)/4} \cos\alpha_0,$$
$$j_{mn} \sin\alpha_{mn} \neq J_{m,(n-k)/4} \sin\alpha_0,$$

where k stands for a toroidal mode number n in the basic current distribution (A.3). For the coefficients j_{mn} and the phases, ϕ_{mn}, of the helical current we have

$$j_{mn} = \frac{J_{m,(n-k)/4} \cos \alpha_0}{\cos \alpha_{mn}}, \qquad \phi_{mn} = \chi_{m,(n-k)/4}. \tag{A.27}$$

The difference between $\mathbf{j}_h(r, \theta, \varphi)$ (A.26) and $\mathbf{j}(r, \theta, \varphi)$ (A.1) can be neglected, since the sum of differences of poloidal modes is negligible small, i.e.,

$$\sum_{m=-\infty}^{\infty} \left(j_{mn} \sin \alpha_{mn} - J_{m,\frac{n-k}{4}} \sin \alpha_0 \right) = \sum_{m=-\infty}^{\infty} J_{m,\frac{n-k}{4}} \frac{\sin(\alpha_0 - \alpha_{mn})}{\cos \alpha_{mn}} \approx 0.$$

A.1.5 Cylindrical Approximation

Here we consider the magnetic field created by a single component of the helical current $\mathbf{j}_{mn}(r, \theta, \varphi)$ (A.25) in a cylindrical geometry. The magnetic field $\mathbf{B}$ of this helical current can be expressed by the scalar potential $\Phi(r, \theta, \varphi)$ ($\mathbf{B} = \nabla\Phi(r, \theta, \varphi)$) (see e.g., Morozov and Solov'ev (1966b))

$$\Phi = \begin{cases} a^i I_m\left(\frac{nr}{R_c}\right) \sin(m\theta + n\varphi + \phi_{mn}), & \text{for } r < r_c, \\ a^e K_m\left(\frac{nr}{R_c}\right) \sin(m\theta + n\varphi + \phi_{mn}), & \text{for } r > r_c, \end{cases}$$

where $I_m(z)$ and $K_m(z)$ are modified Bessel functions. Coefficients a^i, a^e are found by the boundary conditions at the $r = r_c$:

$$\begin{aligned} B_r\Big|_{r=r_c-0} - B_r\Big|_{r=r_c+0} &= 0, \\ B_\theta\Big|_{r=r_c-0} - B_\theta\Big|_{r=r_c+0} &= \mu_o j_{mn} \cos(m\theta + n\varphi + \phi_{mn}) \cos \alpha_{mn}. \end{aligned}$$

Using the relations in Eq. (A.27) we have

$$a^i = -\mu_o J_{m,(n-k)/4} r_c \cos \alpha_0 \frac{nr_c}{mR_0} K'_m \left(\frac{nr_c}{R_c}\right).$$

Further we consider only the leading terms $s = 0$ (A.23) for helical currents. For them we have the following formula for the scalar potential $\Phi(r, \theta, \varphi)$ of the magnetic field created by a set of helical currents (A.26) inside the toroidal surface $r < r_c$:

$$\Phi(r, \theta, \varphi) = \sum_m \Phi_{mn}(r) \sin(m\theta + n\varphi + \chi_{n0}),$$

$$\Phi_{mn}(r) = -B_c C_m g_m f_{mn}(r) \frac{r_c}{m}, \tag{A.28}$$

where a quantity

$$B_c = \frac{\mu_\circ I_d m_0 \cos\alpha_0}{\pi r_c} \tag{A.29}$$

is the characteristic value of the DED magnetic field perturbation. For the TEXTOR-DED parameters, $r_c = 53.25\,\text{cm}$, $R_c = 130\,\text{cm}$, $m_0 \approx 20$, $I_d = 15\,\text{kA}$ and $n = 4$ the value of B_c is 0.22535 T (or 2253.5 G).

The radial dependence of magnetic perturbations is described by the function $f_{mn}(r)$:

$$f_{mn}(r) = -\frac{2nr_c}{R_c} K'_m\left(\frac{nr_c}{R_c}\right) I_m\left(\frac{nr}{R_c}\right).$$

Using the asymptotics of the Bessel function $K_m(z), I_m(z)$, one can show the function $f_{mn}(r)$ and its radial derivative have the following asymptotics at $r < r_c$,

$$f_{mn}(r) \approx \left(\frac{r}{r_c}\right)^m, \quad f'_{mn}(r) \approx \frac{m}{r_c}\left(\frac{r}{r_c}\right)^{m-1}.$$

The radial magnetic field $B_r(r, \theta, \varphi)$ is given by

$$B_r(r, \theta, \varphi) = \frac{\partial \Phi}{\partial r} = \sum_m B_{mn}(r) \sin(m\theta + n\varphi + \chi_{mn}),$$

where

$$B_{mn}(r) = -B_c C_m g_m \frac{r_c}{m} \frac{df_{mn}(r)}{dr} \approx -B_c C_m g_m \left(\frac{r}{r_c}\right)^{m-1}.$$

The φ-component of the vector potential A_φ related the magnetic field as

$$B_r(r, \theta, \varphi) = \frac{1}{r}\frac{\partial A_\varphi}{\partial \theta}, \quad B_\theta(r, \theta, \varphi) = -\frac{\partial A_\varphi}{\partial r},$$

is determined by

$$A_\varphi(r, \theta, \varphi) = \sum_m A_{mn}(r) \cos(m\theta + n\varphi + \chi_{mn}),$$

where

$$A_{mn}(r) = -m^{-1} r B_{mn}(r) \approx B_c C_m g_m \frac{r_c}{m}\left(\frac{r}{r_c}\right)^m.$$

A.1.6 Toroidal Corrections

According to Morozov and Solov'ev (1966b) the effect of toroidicity on the magnetic field can be taken into account, multiplying the scalar potential $\Phi(r, \theta, \varphi)$ by a factor $\sqrt{R_0/R}$, if the small corrections $(n_0 r_c/2R_0)^{m+3}$ are neglected for each poloidal component m. In this approximations we have

$$\Phi(r, \theta, \varphi) = \sqrt{\frac{R_0}{R_0 + r\cos\theta}} \sum_m \Phi_{mn}(r) \sin(m\theta + n\varphi + \chi_{n0}), \tag{A.30}$$

where the amplitudes $\Phi_{mn}(r)$ are given by Eq. (A.28). Then, one can show that the vector potential $A_\varphi(r, \theta, \varphi)$ is determined by

$$\begin{aligned} A_\varphi(r, \theta, \varphi) &= \epsilon B_0 R_0 a(r, \theta, \varphi), \\ a(r, \theta, \varphi) &= \sum_m a_{mn}(r, \theta) \cos(m\theta + n\varphi + \chi_{n0}), \end{aligned} \tag{A.31}$$

where ϵ is a *dimensionless perturbation parameter* defined by

$$\epsilon = \frac{B_c}{B_0}, \tag{A.32}$$

B_0 is the toroidal magnetic field at the center of torus R_0, and

$$\begin{aligned} a_{mn}(r, \theta) &= -\frac{1}{B_c R_0} \frac{r}{m} \frac{\partial}{\partial r} \left(\sqrt{\frac{R_0}{R}} \Phi_{mn}(r) \right) \\ &\approx C_m g_m \frac{r_c}{m R_0} \left(\frac{r}{r_c} \right)^m \sqrt{\frac{R_0}{R_0 + r\cos\theta}} \left(1 - \frac{r\cos\theta}{2m(R_0 + r\cos\theta)} \right). \end{aligned} \tag{A.33}$$

For the radial component of the magnetic field, $B_r(r, \theta, \varphi)$, we have

$$\begin{aligned} B_r(r, \theta, \varphi) &= \frac{1}{r} \frac{\partial A_\varphi(r, \theta, \varphi)}{\partial \theta} = \sum_m B_{mn}(r, \theta) \sin(m\theta + n\varphi + \chi_{n0}), \\ B_{mn}(r, \theta) &\approx B_c C_m g_m \left(\frac{r}{r_c} \right)^{m-1} \sqrt{\frac{R_0}{R_0 + r\cos\theta}} \left(1 - \frac{r\cos\theta}{2m(R_0 + r\cos\theta)} \right). \end{aligned} \tag{A.34}$$

The radial dependence of the perturbation field is determined by the poloidal mode spectra, g_m. According (A.24) the latter is localized near the central mode $m_c = (2s - 1)m_0 = nm_0/n_0$, $(s = 1, 2, \ldots)$ and has a width $\Delta m = \pi/\Delta\theta_i$. Therefore,

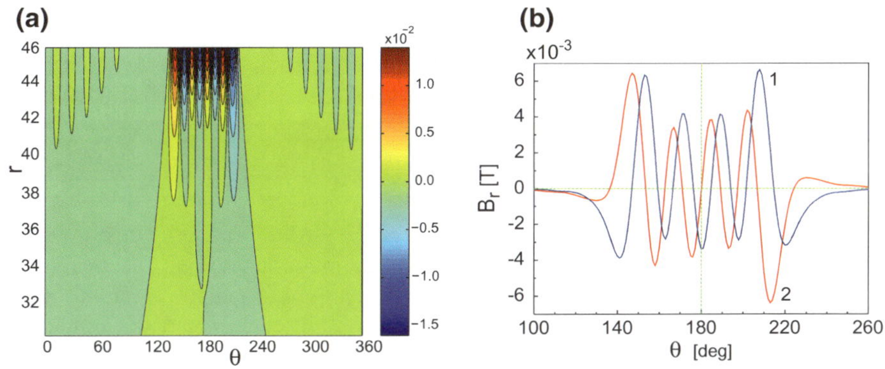

Fig. A.2 **a** Contour plot of the radial magnetic field perturbation $B_r(r, \theta, \varphi)$ in (θ, r) plane at the cross section $\varphi = 204.4°$; **b** Poloidal dependence of the magnetic field perturbation $B_r(r, \theta, \varphi)$ at the given radial coordinate $r = 43$ cm and at the two cross sections: *curve 1*—$\varphi = 0°$, *curve 2*—$\varphi = 204.4°$. The toroidal mode $n = 4$ and the perturbation current is $I_d = 15$ kA

one expects that the perturbation field, $B_r^{(n)}(r, \theta, \varphi)$ of the nth toroidal mode has the following radial dependence,

$$B_r^{(n)}(r, \theta, \varphi) \propto \left(\frac{r}{r_{ci}}\right)^{\gamma_n}, \quad \gamma_n = \frac{nm_0}{n_0} - 1. \tag{A.35}$$

The examples of spatial dependence of the perturbation field are shown in Figs. A.2a, b; (a) the contour plots of $B_r(r, \theta, \varphi)$ in the (θ, r) plane at the fixed toroidal section $\varphi =$ constant; (b) the poloidal variation $B_r(r, \theta, \varphi)$ at the fixed radius r and toroidal angle φ. The perturbation DED current is taken $I_d = 15$ kA.

A.1.7 Summary of Formulas

Below we give a summary of formulas for the perturbation magnetic field of the DED in the DC operation. The toroidal component of the vector potential is given by

$$A_\varphi(r, \theta, \varphi) = -B_c \iota_n \sqrt{\frac{R_0}{R}} \sum_m \frac{r g_{mn}}{|m|} \left(\frac{r}{r_c}\right)^{|m|-1} \cos\left(m\theta + n\varphi + \chi_n\right), \tag{A.36}$$

where the coefficients

$$g_{mn} = (-1)^m C_{mn} \frac{\sin[(m + nm_0/4)\theta_c]}{(m + nm_0/4)\pi},$$
$$C_{mn} = \frac{\sin[(m + nm_0/4)\Delta\theta/2]}{(m + nm_0/4)\Delta\theta/2}, \tag{A.37}$$

Table A.1 Coefficients ϕ_h and ι_n for the different n

n	χ_n	ι_n
1	$3\pi/16$	$\sin(\pi/4)/[4\sin(\pi/16)]$
2	$3\pi/8$	$1/[2\sin(\pi/8)]$
4	$5\pi/4$	$\sqrt{2}$

describes the poloidal mode spectrum at the given toroidal mode n. In Eq. (A.36) the quantity $B_c = \mu_0 m_0 I_d/\pi r_c$ is the characteristic magnitude of the DED magnetic field, I_d is the DED current, the constant m_0 determines the central poloidal mode number $nm_0/4$, r_c is the minor radius of the DED coils, $\chi_n \equiv \chi_{n0}$. The geometrical poloidal angle θ is related to the cylindrical coordinates R, Z as $\theta = \arctan(Z/[R - R_0])$, and the parameter θ_c is the half of the poloidal section covered by the DED coils, $\Delta\theta$ is a geometrical parameter of the coil configuration.

The toroidal mode number n takes the value $n = 4$ for the 12/4 DED mode configuration, $n = 2$ for the 6/2 mode, and $n = 1$ for the 3/1 mode, respectively. The phases χ_n and the factor ι_n in Eq. (A.36) are determined by the coil configuration. For the particular configuration they given by

$$\chi_n = \frac{m_0 n}{4}(\pi - \theta_0) - \chi_n + \frac{\pi}{2}, \tag{A.38}$$

where θ_0 is a poloidal angle of the first coil at the section $\varphi = 0$. The coefficients χ_n and ι_n for the different values of n are given in Table A.1. The parameters r_c, θ_c, θ_0, $\Delta\theta$, and m_0 are determined by the geometry of coil configuration, and take fixed values, $r_c = 0.5325$ m, $\theta_c = 35.49°$, $\theta_0 = 169.35°$, $\Delta\theta = 17.745°$, and $m_0 \approx 20$.

A.2 Magnetic Field of Tore Supra ED

As it was noted in Sect. 9.1.1 and shown in Fig. 9.2 that the ED coil configuration of the Tore Supra consists of six identical modules located on outer board the torus. We model the each module by the coil windings shown in Fig. A.3 where arrows indicate the current direction. The current flows from the feeder located at the beginning of the first section $j = 1$ of the inner side of the winding shown in Fig. A.3a and returns through the outer side of the winding shown in Fig. A.3b. The minor radii of the inner and outer sides are $r_{c1} = 84$ cm $r_{c2} = 86$ cm, respectively Below we calculate the magnetic field, created by the current flowing in this coil system by considering the inner and outer parts of the winding separately, and by summing those parts afterwards.

Suppose that the modules are centered near the toroidal angles $\varphi_k = (k-1)\Delta\varphi$, $\Delta\varphi = 2\pi/6$. The poloidal angle as a function of the toroidal angle of a point located in the jth section in the kth module ($k = 1, 2, \ldots, 6$) is given by

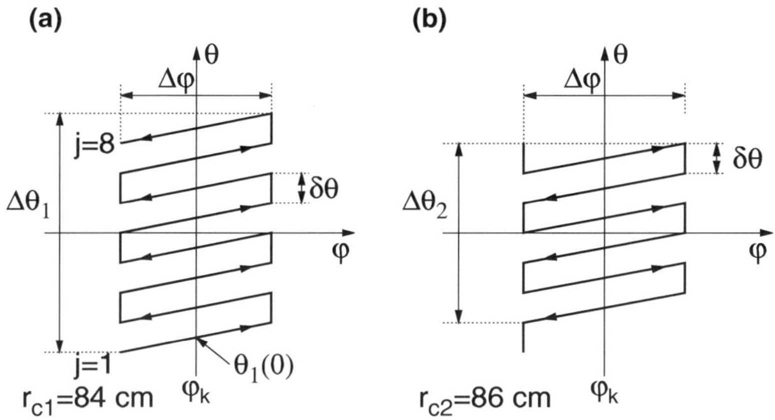

Fig. A.3 Model scheme of one module of the ED coils: **a** the inner winding at $r_{c1} = 84$ cm; **b** the outer winding at $r_{c2} = 86$ cm

$$\theta_j^{(k)}(\varphi) = \theta_1^{(k)}(\varphi) + (j-1)\delta\theta, \quad j = 1, 2, \ldots, N, \tag{A.39}$$

where $N = 8$ for the inner winding and $N = 6$ for the outer winding, and $\delta\theta$ is the poloidal spacing shown in Fig. A.3. The coordinates of the first coils on each module are described by

$$\theta_1^{(k)}(\varphi) = \theta_1(0) + \alpha(\varphi - \varphi_k), \quad \text{for } \varphi_k - \Delta\varphi/2 < \varphi < \varphi_k + \Delta\varphi/2, \tag{A.40}$$

where α is the slope of a coil creating a helical magnetic perturbation.

The poloidal extension, $\Delta\theta$, of a module shown in Fig. A.3 can be expressed as a function of the poloidal position of the first coils, $\theta_1(0)$, at the toroidal section $\varphi = \varphi_k$:

$$\Delta\theta = 2|\theta_1(0)| + \alpha\Delta\varphi. \tag{A.41}$$

One should note that this model of coils is not fully equivalent to the Tore Supra coils. In the latter case the distance between sections of coils in each module are not equidistant along the poloidal angle θ. It slightly decreases with the distance from the equatorial plane $\theta = 0$ (Ghendrih 1995).

A.2.1 Current Density

We describe the current, I_j, which flows in a coil section by

$$I_j^{(i)} = I_d \cos(\pi j) = (-1)^{j+1}, \qquad j = 1, 2, \ldots, N,$$

where I_d is the current flowing in the coil, $i = 1$ for the inner part of the winding and $i = 2$ for its outer part.

Below we shall consider only the long helical section coils since they create the magnetic field perturbations that are resonant with the magnetic field lines of the plasma. The vertical short sections of coils do not contribute to the resonant field, therefore they will not be taken into account.

One can introduce the current density vector $\mathbf{j}(r, \theta, \varphi)$ of the coil system as

$$\mathbf{j}_i(r, \theta, \varphi) = \mathbf{e}^{(i)} \frac{\delta(r - r_{ci})}{r_{ci}} \sum_{k=1}^{6} g_\varphi^{(k)}(\varphi) \sum_{j=1}^{N} I_j^{(i)} \delta\left(\theta - \theta_j^{(k)}(\varphi)\right), \tag{A.42}$$

where $\mathbf{e}^{(i)} = (\mathbf{e}_r, \mathbf{e}_\theta, \mathbf{e}_\varphi) = (0, \sin\alpha_{0i}, \cos\alpha_{0i})$ is a unit vector along the helical section of the coils, $\alpha_{0i} = \alpha r_{ci}/R_c$, $R_c = R_0 + r_{ci}$, $(i = 1, 2)$. Here $g_\varphi^{(k)}(\varphi)$ is a step function of the toroidal angle φ which takes a non-zero value in the areas covered by coils, i.e.,

$$g_\varphi^{(k)}(\varphi) = \begin{cases} 1, & \text{for } \varphi_k - \Delta\varphi/2 < \varphi < \varphi_k + \Delta\varphi/2, \\ 0, & \text{elsewhere,} \end{cases}$$

Introducing the step function $g_i(\theta)$ solely depending on the poloidal angle

$$g_i(\theta) = \begin{cases} 1, & \text{for} - \Delta\theta_i/2 < \theta < \Delta\theta_i/2, \\ 0, & \text{elsewhere,} \end{cases}$$

the current density (A.42) can after some transformations be reduced to

$$\mathbf{j}_i(r, \theta, \varphi) = \mathbf{e}^{(i)} \delta(r - r_{ci}) J_0^{(i)} g_i(\theta) \sum_{k=1}^{6} g_\varphi^{(k)}(\varphi) \times \sum_{s=-\infty}^{\infty} \cos\left\{m_0(2s-1)\left[(\theta - \theta_0) - \alpha(\varphi - \varphi_k)\right]\right\}. \tag{A.43}$$

where

$$m_0 = \frac{\pi}{\delta\theta}, \quad J_0^{(i)} = \frac{m_0 I_d}{\pi r_{ci}}, \quad \theta_0 = \theta_1(0) - \delta\theta.$$

One can show that the current density (A.43) can be expanded into a Fourier series,

$$\mathbf{j}_i(r, \theta, \varphi) = 2\mathbf{e}^{(i)} \sum_{m=-\infty}^{\infty} \sum_{n=-\infty}^{\infty} \sum_{s=1}^{\infty} j_{mn}^{(si)}(r) \cos\left(m\theta - n\varphi + \chi_{mn}^{(s)}\right), \tag{A.44}$$

with the Fourier coefficients

$$j_{mn}^{(si)}(r) = (-1)^q \delta(r - r_{ci}) \, \tilde{J}_0^{(i)} C_n^{(s)} g_{mi}^{(s)},$$
$$\chi_s = m_0(2s-1)\theta_0, \tag{A.45}$$

where

$$\tilde{J}_0^{(i)} = \frac{6J_0^{(i)} \Delta\varphi \Delta\theta_i}{(2\pi)^2} = \frac{m_0 I_d}{\pi r_{ci}} \frac{6\Delta\varphi\Delta\theta_i}{(2\pi)^2}, \tag{A.46}$$
$$g_{mi}^{(s)} = \frac{\sin([m - m_0(2s-1)]\Delta\theta_i/2)}{[m - m_0(2s-1)]\Delta\theta_i/2}, \tag{A.47}$$
$$C_n^{(s)} = \frac{\sin([n - m_0(2s-1)\alpha]\Delta\varphi/2)}{[n - m_0(2s-1)\alpha]\Delta\varphi/2}. \tag{A.48}$$

The toroidal mode number n takes values $n = 6q, q = 0, \pm 1, \pm 2, \ldots$. As one can see from Eqs. (A.45) and (A.48), the biggest effect occurs when the ED coils are designed in such a way that the product $m_0\alpha$ is close to the toroidal mode $n_0 = 6$, i.e., $|m_0\alpha - n_0| \ll \Delta\varphi/2$. Then in the sum (A.44), the main contribution comes from the terms with the toroidal numbers $n = (2s-1)n_0$, $(s = 1, 2, \ldots)$. Leaving in Eq. (A.44) only these terms we have

$$\mathbf{j}_i(r, \theta, \varphi) = 2\mathbf{e}^{(i)} \sum_{m=-\infty}^{\infty} \sum_{s=1}^{\infty} j_{mn}^{(si)}(r) \times \cos(m\theta - (2s-1)n_0\varphi + \chi_s), \tag{A.49}$$

where $n_0 = 6$, $i = 1, 2$.

A.2.2 Magnetic Field

The magnetic field created by the helical currents can be found using the procedure similar in Sect. A.1.6.

$$\Phi(r, \theta, \varphi) = \sqrt{\frac{R_0}{R_0 + r\cos\theta}} \left[\Phi_1(r, \theta, \varphi) + \Phi_2(r, \theta, \varphi)\right], \tag{A.50}$$

where

$$\Phi_i(r, \theta, \varphi) = \sum_{m=-\infty}^{\infty} \sum_n \Phi_{mn}^{(i)}(r) \sin(m\theta - n\varphi + \chi_s). \tag{A.51}$$

The toroidal modes takes $n = (2s-1)n_0$, $(s = 1, 2, \ldots)$ and

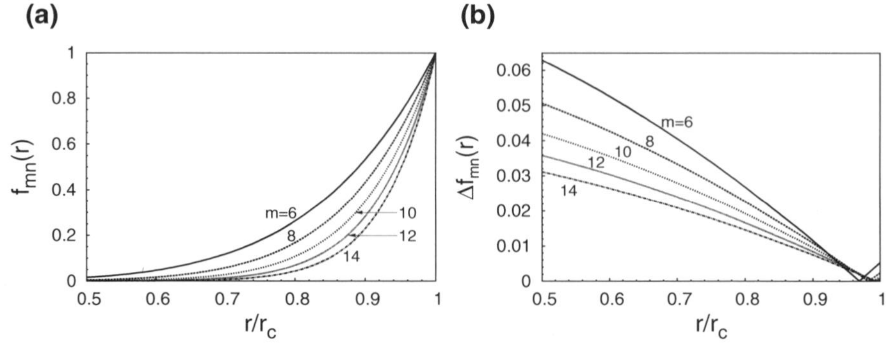

Fig. A.4 **a** Radial dependence of the function $f_{mn}(r)$ for different poloidal modes m. **b** Relative deviation of the $f_{mn}(r)$ from the power law $(r/r_c)^m$: $\Delta f_{mn}(r) = |f_{mn}(r) - (r/r_c)^m|/f_{mn}(r)$. Parameters are $r_c = 85$ cm, $R_0 = 238$ cm, $n = 6$

$$
\begin{aligned}
\Phi_{mn}^{(i)}(r) &= -B_c^{(i)} C_n^{(s)} g_m^{(si)} f_{mn}^{(i)}(r) \frac{r_{ci}}{m}, \\
f_{mn}^{(i)}(r) &= -\frac{2nr_{ci}}{R_{ci}} K_m' \left(\frac{nr_{ci}}{R_{ci}}\right) I_m \left(\frac{nr}{R_{ci}}\right), \\
B_c^{(i)} &= \frac{2\mu_o m_0 I_d \cos(\alpha_{0i})}{\pi r_{ci}} \frac{6\Delta\varphi\Delta\theta_i}{(2\pi)^2},
\end{aligned}
\tag{A.52}
$$

where $I_m(z)$ and $K_m(z)$ are the modified Bessel functions ($K_m'(z) \equiv dK_m(z)/dz$).

Here $B_c^{(i)}$ is the characteristic amplitude of the strength of the perturbation magnetic field.[1] For the typical parameters of the ED of Tore Supra ($r_c = 0.85$ m, $I_d = 22.5$ kA) we have $B_c \approx 425$ G.

The radial dependences of the perturbation field are described by functions $f_{mn}^{(i)}(r)$ which are shown in Fig. A.4 for a several mode numbers m. For large mode number m ($m \geq 4$) the radial dependence is well described by the following asymptotical formula $f_{mn}^{(i)}(r) \approx (r/r_{ci})^m$. Then the radial component of the magnetic field B_r can be represented as

$$
\begin{aligned}
B_r(r, \theta, \varphi) &= \frac{\partial \Phi}{\partial r} = \sum_n B_r^{(n)}(r, \theta, \varphi), \\
B_r^{(n)}(r, \theta, \varphi) &= \sum_{m=-\infty}^{\infty} B_{mn}(r, \theta) \sin\left(m\theta - n\varphi + \chi_s\right),
\end{aligned}
\tag{A.53}
$$

where

[1] We should note that the definition of $B_c^{(i)}$ as well as coefficients $g_m^{(si)}$ are slightly different from the corresponding ones given by Eqs. (A.29) and (A.18).

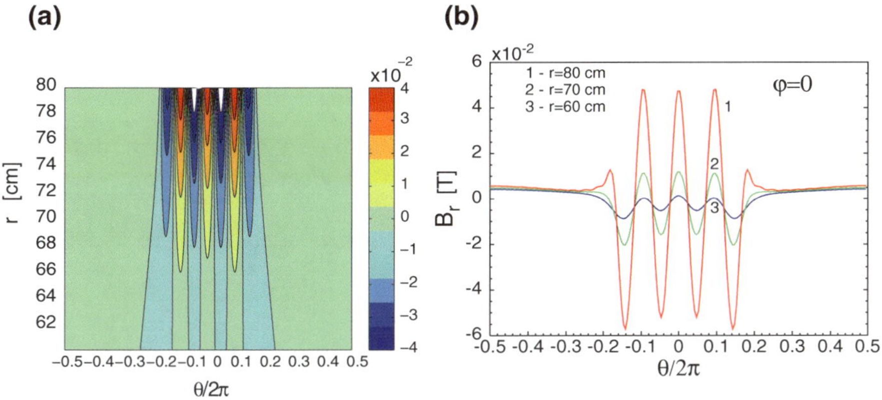

Fig. A.5 Radial component of the perturbation field B_r: **a** contour plot in the (θ, r) plane ($\varphi = 0$); **b** poloidal dependence at different radial positions r: *1*—$r = 80$ cm, *2*—$r = 70$ cm, *3*—$r = 60$ cm. $I_d = 22.5$ kA

$$B_{mn}(r,\theta) = \left[B_c^{(1)} g_{m1}^{(s)} \left(\frac{r}{r_{c1}}\right)^{m-1} + B_c^{(2)} g_{m2}^{(s)} \left(\frac{r}{r_{c2}}\right)^{m-1} \right] \times \sqrt{\frac{R_0}{R_0 + r\cos\theta}} \left(1 - \frac{r\cos\theta}{2m\,(R_0 + r\cos\theta)}\right). \quad \text{(A.54)}$$

where $B_{mns}^{(i)} = -B_c^{(i)} C_n^{(s)} g_m^{(si)}$. The toroidal component of the vector potential of the perturbation field is given by

$$A_\varphi(r,\theta,\varphi) = \epsilon B_0 R_0 \sum_{m=-\infty}^{\infty} \sum_n a_{mn}(r,\theta) \cos(m\theta - n\varphi + \chi_{mn}),$$
$$a_{mn}(r,\theta) = m^{-1} r B_{mn}(r,\theta), \quad \text{(A.55)}$$

where ϵ stands for the dimensionless perturbation parameter, defined as $\epsilon = B_c^{(1)}/B_0$, B_0 is the strength of the toroidal field. The dimensionless Fourier coefficients, $a_{mn}(r,\theta)$, are then given by

$$a_{mn}(r,\theta) = a_{mn}^{(1)}(r,\theta) + \delta a_{mn}^{(2)}(r,\theta),$$

where

$$a_{mn}^{(i)}(r,\theta) = g_m^{(i)} \frac{r_{ci}}{m R_0} \frac{r}{m} \frac{d}{dr} \left(\sqrt{\frac{R_0}{R_0 + r\cos\theta}} \left(\frac{r}{r_{ci}}\right)^m \right). \quad \text{(A.56)}$$

Here $(i = 1, 2)$, $\delta = B_c^{(2)}/B_c^{(1)} = r_{c1}\Delta\theta_2/(r_{c2}\Delta\theta_1)$. The phase $\chi_{mn} = \chi_s$ and toroidal mode number $n = (2s - 1)n_0$.

The angular dependencies of the perturbation field $B_r(r, \theta, \varphi)$ at the fixed values of radial coordinate r and the toroidal angle $\varphi = 0$ are plotted in Fig. A.5. The radial dependence of the perturbation field is determined by (A.35). The power law of the radial decay of perturbation field has the lowest exponent, $\gamma_{n=6} = m_0 - 1$, for the toroidal mode $n = 6$. For the value $\delta\theta = 18°$, one has $m_0 = \pi/\delta\theta = 10$ the exponent $\gamma_{n=6} = 9$. For the next toroidal mode $n = 18$ we have $\gamma_{n=18} = 3m_0 - 1 = 29$.

Appendix B
Magnetic Field of a Set of Saddle Coils

In this Appendix we give some details of calculations of the magnetic field created by a set of saddle coils in a tokamak geometry described in Sect. 3.4.1. The method is similar to the one in the classical magnetostatic to the calculation of the magnetic field created by the circular current loop (see, e.g., Jackson (1998)). But in our case the problem is reduced to the new integrals which can be considered as the generalized elliptic integrals.

As was shown in Sect. 3.4.1 the magnetic field created by the set of saddle coils can be composed a sum of magnetic fields from the horizontal segments lying on the surfaces $Z = \text{const}$ and the ones lying the vertical surfaces (R, Z) (see also Fig. 3.8). We consider the calculations of these magnetic fields separately.

B.1 The magnetic Field of the Current Loop

Let (R, Z, φ) be the cylindrical coordinate system. Consider a current–carrying filament coil given by curve G in the 3D space. Let $d\mathbf{l}$ be an element of this curve, and $\mathbf{J}(R, Z, \varphi)$ be a current flowing in the coil in the form of the circular loop as shown in Fig. 3.8. For the circular loop of radius R_j lying in the horizontal plane $Z_j = \text{constant}$ the current density $\mathbf{j}(R, Z, \varphi)$ has only a component in the φ direction,

$$\begin{aligned} j_\varphi(R, Z, \varphi) &= I_c(\varphi)\delta(R - R_j)\delta(Z - Z_j), \\ j_x(R, Z, \varphi) &= -j_\varphi(R, Z, \varphi)\sin\varphi, \\ j_y(R, Z, \varphi) &= j_\varphi(R, Z, \varphi)\cos\varphi, \end{aligned} \tag{B.1}$$

where $I_c(\varphi)$ is a current depending on the toroidal angle φ. Then, the vector potential $\mathbf{A}$ at the point $P(R, Z, \varphi)$ is determined by the Biot–Savart law,

S. Abdullaev, *Magnetic Stochasticity in Magnetically Confined Fusion Plasmas*, Springer Series on Atomic, Optical, and Plasma Physics 78,
DOI: 10.1007/978-3-319-01890-4, © Springer International Publishing Switzerland 2014

$$\mathbf{A}(R, Z, \varphi) = \frac{\mu_o}{4\pi} \int_G \frac{\mathbf{j}(R', Z', \varphi')dV'}{|\mathbf{r} - \mathbf{r}_j|},$$

$$A_x(R, Z, \varphi) = \frac{\mu_o}{4\pi} \int_G \frac{j_x(R', Z', \varphi')dV'}{|\mathbf{r} - \mathbf{r}_j|} = -\frac{\mu_o R_j}{4\pi} \int\limits_0^{2\pi} \frac{I_\varphi(\varphi_j) \sin \varphi_j d\varphi_j}{|\mathbf{r} - \mathbf{r}_j|},$$

$$A_y(R, Z, \varphi) = \frac{\mu_o}{4\pi} \int_G \frac{j_y(R', Z', \varphi_j)dV'}{|\mathbf{r} - \mathbf{r}_j|} = \frac{\mu_o R_j}{4\pi} \int\limits_0^{2\pi} \frac{I_\varphi(\varphi_j) \cos \varphi_j d\varphi_j}{|\mathbf{r} - \mathbf{r}_j|}, \quad \text{(B.2)}$$

where $dV = RdRdZd\varphi$ is a volume element, $\mathbf{r}_j = (R_j, Z_j, \varphi_j)$ are point coordinates at the curve G,

$$|\mathbf{r} - \mathbf{r}_j| = \sqrt{R^2 + R_j^2 + (Z - Z_j)^2 - 2RR_j \cos(\varphi - \varphi_j)}.$$

The toroidal component of the vector potential A_φ is

$$\begin{aligned} A_\varphi(R, Z, \varphi) &= -A_x(R, Z, \varphi) \sin \varphi + A_y(R, Z, \varphi) \cos \varphi \\ &= \frac{\mu_o R_j}{4\pi} \int_0^{2\pi} \frac{I_\varphi(\varphi_j) \cos(\varphi_j - \varphi)d\varphi_j}{|\mathbf{r} - \mathbf{r}_j|}. \end{aligned} \quad \text{(B.3)}$$

Introducing the notations

$$D_j = \sqrt{(R + R_j)^2 + (Z - Z_j)^2}, \quad k^2 = 4RR_j/D_j^2,$$

and replacing the integration variable from φ_j to ϕ,

$$\phi = (\varphi - \varphi_j + \pi)/2,$$

$$0 \leq \varphi_j \leq 2\pi, \qquad (\varphi + \pi)/2 \geq \phi \geq (\varphi - \pi)/2,$$

we obtain

$$A_\varphi(R, Z, \varphi) = -2\frac{\mu_o R_j}{4\pi D} \int\limits_{-\pi/2}^{\pi/2} \frac{I_c(\varphi - 2\phi + \pi) \cos(2\phi)d\phi}{\sqrt{1 - k^2 \sin^2 \phi}}. \quad \text{(B.4)}$$

Suppose, that the current $I_c(\varphi)$ can be presented as a Fourier series

$$I_c(\varphi) = I_c^{(0)} + \sum_{n=1}^{\infty} I_c^{(n)} \cos(n\varphi + \chi_n) = I_c^{(0)} + \frac{1}{2} \sum_{n=-\infty}^{\infty} I_c^{(n)} e^{in\varphi + i\chi_n}, \quad \text{(B.5)}$$

where

$$I_c^{(n)} e^{i\chi_n} = \frac{1}{\pi} \int_0^{2\pi} I_c(\varphi) e^{-in\varphi} d\varphi.$$

Consider a coil system consisting of N pairs of current loops as shown in Figs. 3.7 and 3.8. The distribution of current $I_c(\varphi)$ corresponding to this system is

$$I_c(\varphi) = I_0 \begin{cases} 0 & \text{for } 2k\pi/N \le \varphi \le 2k\pi/N + \varphi_d, \\ 1 & \text{for } 2k\pi/N + \varphi_d \le \varphi \le (2k+1)\pi/N - \varphi_d, \\ 0 & \text{for } (2k+1)\pi/N - \varphi_d \le \varphi \le (2k+1)\pi/N + \varphi_d, \\ -1 & \text{for } (2k+1)\pi/N + \varphi_d \le \varphi \le 2(k+1)\pi/N - \varphi_d, \\ 0 & \text{for } 2(k+1)\pi/N - \varphi_d \le \varphi \le 2(k+1)\pi/N, \end{cases} \tag{B.6}$$

where $k = 0, 1, \ldots, n-1$, $2k\pi/N \le \varphi \le 2(k+1)\pi/N$. After some lengthy calculations one can obtain the following formula for the Fourier components $I_c^{(n)}$ for this current distribution,

$$\pi I_c^{(n)} e^{i\chi_n}/I_0 = \frac{4i}{n} e^{-in\pi} \sin\left(\frac{n\pi}{2N}\right) \sin\left(\frac{n\pi}{2N} - n\varphi_d\right) \frac{\sin(n\pi)}{\sin(n\pi/N)}.$$

This expression does not vanish only for non-integer numbers $n = (2s+1)N$, $(s = 0, 1, 2, \ldots)$:

$$I_c^{(n)} e^{i\chi_n} = I_c \frac{4}{(2s+1)\pi} e^{i\pi/2} \cos\left[(2s+1)N\varphi_d\right]. \tag{B.7}$$

Finally, we have

$$I_c(\varphi) = \frac{4I_0}{\pi} \sum_{s=0}^{\infty} \frac{\cos\left[(2s+1)N\varphi_d\right]}{2s+1} \sin\left[(2s+1)N\varphi\right]. \tag{B.8}$$

Similar formulas can be obtained when the set of coils consists of odd numbers of saddle coils as the case shown in Fig. 3.6.

B.1.1 Vector Potential

Using (B.5) the vector potential (B.4) can be reduced to

$$A_\varphi(R, Z, \varphi) = -\frac{\mu_o R_j}{\pi D_j} \left\{ I_c^{(0)} a_0(R, Z) + \sum_{n=1}^{\infty} I_c^{(n)} a_n^{(s)}(R, Z) \sin(n\varphi) \right\}, \tag{B.9}$$

where

$$a_0(R, Z) = \int_0^{\pi/2} \frac{\cos(2\phi)d\phi}{\sqrt{1 - k^2 \sin^2 \phi}} = \frac{1}{k^2}\left[(k^2 - 2)K(k) + 2E(k)\right], \tag{B.10}$$

$$a_n^{(s)}(R, Z) = (-1)^n \int_0^{\pi/2} \frac{\cos(2n\phi)\cos(2\phi)d\phi}{\sqrt{1 - k^2 \sin^2 \phi}} = \frac{1}{k^2}\left[(k^2 - 2)K_n(k) + 2E_n(k)\right]. \tag{B.11}$$

Here $K(k)$ and $E(k)$ are the complete elliptic integrals with argument k, and the integral $K_n(k)$ and $E_n(k)$ are defined by

$$K_n(k) = \int_0^{\pi/2} \frac{\cos(2n\phi)d\phi}{\sqrt{1 - k^2 \sin^2 \phi}},$$
$$E_n(k) = \int_0^{\pi/2} \cos(2n\phi)\sqrt{1 - k^2 \sin^2 \phi}\,d\phi. \tag{B.12}$$

They can be called the generalized elliptic integrals. Note, that $K(k) = K_{n=0}(k)$ and $E(k) = E_{n=0}(k)$.

Therefore the vector potential of the magnetic field created by the circular loop and the corresponding normalized perturbation poloidal flux can be presented as

$$A_\varphi(R, Z, \varphi) = \sum_{n=0}^{\infty} A_\varphi^{(n)}(R, Z)\sin(n\varphi),$$
$$\psi^{(pert)}(R, Z, \varphi) = -\frac{RA_\varphi(R, Z)}{B_0 R_0^2} = \epsilon \sum_{n=0}^{\infty} \psi_n(R, Z)\sin(n\varphi), \tag{B.13}$$

where

$$A_\varphi^{(n)}(R, Z) = \frac{\mu_o I_0 i_n D_j}{4\pi R} L_n(k),$$
$$\psi_n(R, Z) = -\frac{i_n D_j}{R_0} L_n(k), \tag{B.14}$$

Here the following notations are introduced

$$L_n(k) = \left(1 - \frac{k^2}{2}\right) K_n(k) - E_n(k),$$

$$i_n = \frac{I_c^{(n)}}{I_0} = (-1)^n \frac{4}{\pi} \frac{\cos\left[(2s+1)N\varphi_d\right]}{2s+1}, \qquad n = (2s+1)N. \tag{B.15}$$

The non-dimensional perturbation parameter ϵ is defined as

$$\epsilon = \frac{\mu_o I_0}{4\pi B_0 R_0}. \tag{B.16}$$

The (R, Z) components of the magnetic field are given by

$$(B_R, B_Z) = \sum_{n=0}^{\infty} \left(B_R^{(n)}(R, Z), B_Z^{(n)}(R, Z)\right) \sin(n\varphi), \tag{B.17}$$

where

$$B_Z^{(n)}(R, Z) = \frac{1}{R} \frac{\partial\left[R A_\varphi^{(n)}(R, Z)\right]}{\partial R} = B_0 \epsilon i_n \left\{ \frac{\partial D_j}{\partial R} L_n(k) + D_j \frac{\partial k_j^2}{\partial R} \frac{dL_n(k)}{dk_j^2} \right\}, \tag{B.18}$$

$$B_R^{(n)}(R, Z) = -\frac{\partial A_\varphi^{(n)}(R, Z)}{\partial Z} = -B_0 \epsilon i_n \left\{ \frac{\partial D_j}{\partial Z} L_n(k) + D_j \frac{\partial k_j^2}{\partial Z} \frac{dL_n(k)}{dk_j^2} \right\}. \tag{B.19}$$

B.1.2 Approximation of the Integrals $L_n(k)$

One can establish the approximation of the integrals $L_n(k)$, $E_n(k)$ (B.15) and its derivative $dL_n(k)/dk^2$ by the series of an expansion in $m = k^2$:

$$L_n(k) = \sum_{k=1}^{M} a_k m^k + \sum_{k=0}^{M} b_k m_1^k \ln \frac{1}{m_1} + R_M, \tag{B.20}$$

$$\frac{dL_n(k)}{dm} = \sum_{k=1}^{M} c_k m^k - \frac{d}{dm_1} \sum_{k=0}^{M} d_k m_1^k \ln \frac{1}{m_1} + R_M, \tag{B.21}$$

where $m_1 = 1 - m$. The coefficients a_k, b_k, c_k, d_k, $(k = 1, 2, \ldots, M)$ found by the fitting with the numerically calculated ones are presented in Tables B.1 and B.2 for the case $M = 3$.

The dependencies of the integrals $L_n(m)$ and and its derivative $dL_n(m)/dm$ on the module m are plotted in Fig. B.1a, b.

Table B.1 The coefficients a_k, b_k in Eq. (B.20) for the case $M = 3$

	$n = 1$	$n = 2$	$n = 3$	$n = 4$	$n = 5$	$n = 6$
a_1	−15.1318	12.614	6.44849	1.05883	8.68808	−20.2141
a_2	24.7153	−22.0231	−7.95135	−5.24451	−9.80602	28.0533
a_3	−9.75115	8.97702	2.30145	3.36486	2.04908	−8.8537
b_0	−0.227805	0.227776	−0.247866	0.227436	−0.226976	0.226234
b_1	1.58137	−0.867444	−2.0157	1.67606	−3.26677	4.89141
b_2	9.47854	−8.3374	−3.3891	−1.38598	−4.72399	12.0068
b_3	3.9163	−3.64604	−0.796313	−1.58287	−0.466855	3.09053
$\|R_M\| <$	3×10^{-3}	3×10^{-3}	2×10^{-3}	2×10^{-3}	10^{-3}	2×10^{-3}

Table B.2 The coefficients c_k, d_k in Eq. (B.21) for the case $M = 3$

	$n = 1$	$n = 2$	$n = 3$	$n = 4$	$n = 5$	$n = 6$
c_1	0.750263	1.16925	−6.19823	14.8482	−27.3604	43.1621
c_2	0.919855	1.28526	−6.18702	14.7706	−27.1634	42.7855
c_3	−1.93061	−0.149124	5.64138	−16.2511	32.8243	−54.8624
d_0	0.248363	−0.248326	0.248192	−0.247847	0.24714	−0.24593
d_1	−0.16046	0.886782	−2.04892	3.56763	−5.3332	7.22269
d_2	0.661822	0.594552	−3.54806	8.851	−16.6575	26.5599
d_3	0.392412	−0.063162	−0.849296	2.67554	−5.61181	9.6156
$\|R_M\| <$	2.5×10^{-6}	4×10^{-6}	$\times 10^{-5}$	2×10^{-5}	3×10^{-5}	4×10^{-5}

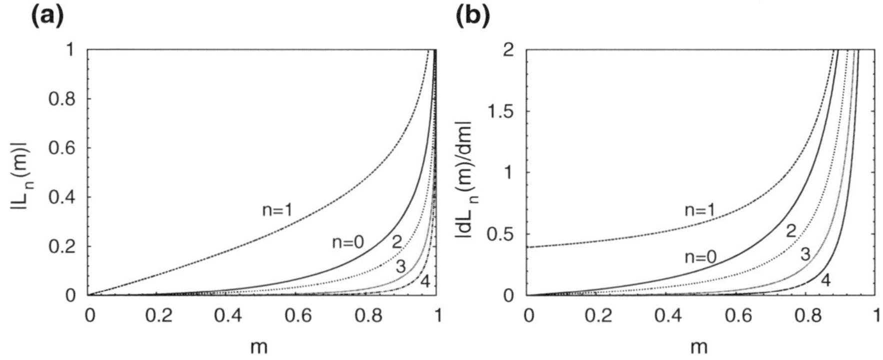

Fig. B.1 The integral $L_n(m)$ and and its derivative $dL_n(m)/dm$ as the functions of $m = k^2$ for the several numbers n

B.2 Magnetic Field of the Vertical Segments of a Set of Saddle Coils

Above we have considered the calculation of the magnetic field created by the horizontal segments of the set of saddle coils. Now consider the magnetic field from the segments of the set on the vertical sections (R, Z).

B.2.1 *Magnetic Field from a Single Vertical Wire*

First we consider a single straight wire located along the vertical lines located at the section (R_c = const, φ_0 = const) between $Z_1 < Z < Z_2$. The current density is given by

$$j_z = I_0 R_0^{-1} \delta(\varphi - \varphi_0)\delta(R - R_0) \begin{cases} 1, & \text{for } Z_1 < Z < Z_2, \\ 0, & \text{for } Z < Z_1, Z > Z_2. \end{cases} \tag{B.22}$$

The magnetic field generated by this wire can be described by the vector potential $(0, 0, A_Z)$ with the non-zero z-component:

$$A_Z(R, Z, \varphi) = \frac{\mu_o}{4\pi} \int_G \frac{j_z(R', Z', \varphi')dV'}{|\mathbf{r} - \mathbf{r}_0|}. \tag{B.23}$$

Using (B.22) it can be calculated directly,

$$\begin{aligned} A_Z(R, Z, \varphi) &= \frac{\mu_o I_0}{4\pi} \int_{Z_1}^{Z_2} \frac{dZ_0}{\sqrt{a^2 + (Z - Z_0)^2}} \\ &= \frac{\mu_o I_0}{4\pi} \left[\ln \frac{Z_2 - Z + \sqrt{a^2 + (Z_2 - Z)^2}}{Z_1 - Z + \sqrt{a^2 + (Z - Z_1)^2}} \right], \\ a^2 &= R^2 + R_0^2 - 2RR_0 \cos(\varphi - \varphi_0), \end{aligned} \tag{B.24}$$

where we used the integral

$$\int_0^x \frac{dx}{\sqrt{a^2 + x^2}} = \ln(x + \sqrt{a^2 + x^2}) - \ln a. \tag{B.25}$$

B.2.2 Set of Straight Wires

Now we consider a set of $2N$ straight wires located at fixed toroidal angles $\varphi_k^{(\pm)} = (\pi/N)k \pm \varphi_d$, $(k = 0, 1, \ldots, 2N-1)$ between circular loops of radii R_1 at Z_1 and radii R_2 at Z_2. Suppose the directions of even/odd currents are positive/negative with the respect to Z axis, and have the same magnitude, i.e., $I_k^{(\pm)} = (-1)^k I_0$. (see Fig. 3.8). Then the current density can be presented as

$$\mathbf{j} = (j_R, j_Z) = (\sin\alpha, \cos\alpha) j(R, Z, \varphi), \tag{B.26}$$

where α is an inclination angle of the wire with respect to the vertical axis Z, i.e.,

$$\tan\alpha = \frac{R_2 - R_1}{Z_2 - Z_1},$$

and the current density

$$j(R, Z, \varphi) = I_0 \delta(R - R_0(Z)) R^{-1} \times \sum_{k=0}^{2N-1} (-1)^k \left[\delta\left(\varphi - \frac{k\pi}{N} - \varphi_d\right) + \delta\left(\varphi - \frac{k\pi}{N} + \varphi_d\right)\right]. \tag{B.27}$$

Using the Poisson summation formula

$$\delta(\varphi) = \frac{1}{2\pi} \sum_{n=-\infty}^{\infty} \cos n\varphi = \frac{1}{2\pi} \sum_{n=-\infty}^{\infty} e^{in\varphi}, \tag{B.28}$$

Eq. (B.26) can be transformed to

$$j(R, Z, \varphi) = I_0 \frac{\delta(R - R_0)}{\pi R} \sum_{n=-\infty}^{\infty} e^{in\varphi} \cos(n\varphi_d) e^{-i\pi n(1-1/2N)} e^{i\pi/2} \frac{\sin(\pi n)}{\cos(\pi n/2N)}. \tag{B.29}$$

All terms in (B.29) vanish except $n_s = (2s+1)N$, $(s = 0, \pm 1, \pm 2, \ldots)$. The latter give

$$j(R, Z, \varphi) = I_0 \delta\left(R - R_0(Z)\right) R^{-1} \sum_{s=0}^{\infty} J_s \cos\left(n_s \varphi\right), \tag{B.30}$$

where

$$J_s = \frac{4N}{\pi} (-1)^{N+1+s} \cos\left[n_s \varphi_d\right]. \tag{B.31}$$

B.2.3 Vector Potential

Using the current density (B.26) and (B.30) the magnetic field created by the vertical segments of the saddle coils can be presented by the R- and Z-components of the vector potential,

$$\begin{aligned}\mathbf{A}(R, Z, \varphi) = (A_R, A_Z) &= \frac{\mu_o}{4\pi} \int \frac{(j_R, j_Z)(R', Z', \varphi') \sin dV'}{|\mathbf{r} - \mathbf{r}_0|} \\ &= \frac{\mu_o I_0}{4\pi} \sum_{s=0}^{\infty} J_s \int \frac{(e_R, e_Z) \cos(n_s \varphi') dZ' d\varphi'}{|\mathbf{r} - \mathbf{r}_0|} \\ &= \frac{\mu_o I_0}{4\pi} \sum_{s=0}^{\infty} J_s \left[\mathbf{a}^{(s)}(R, Z) \cos(n_s \varphi) + \mathbf{b}^{(s)}(R, Z) \sin(n_s \varphi)\right], \qquad \text{(B.32)}\end{aligned}$$

where

$$\begin{aligned}\mathbf{a}^{(s)}(R, Z) &= \int_{Z_1}^{Z_2} \int_0^{2\pi} \frac{(e_R, e_Z) \cos(n_s \phi) dZ' d\phi}{D(\phi, Z')}, \\ \mathbf{b}^{(s)}(R, Z) &= -\int_{Z_1}^{Z_2} \int_0^{2\pi} \frac{(e_R, e_Z) \sin(n_s \phi) dZ' d\phi}{D(\phi, Z')}, \qquad \text{(B.33)}\end{aligned}$$

and

$$\begin{aligned}&e_R = \sin \alpha(Z'), \quad e_Z = \cos \alpha(Z'), \\ &D(\phi, Z') = \sqrt{R^2 + R_0^2(Z') - 2R R_0(Z') \cos \phi + (Z - Z')^2}. \qquad \text{(B.34)}\end{aligned}$$

If the segment of a wire connecting Z_1 and Z_2 is a straight the integral over Z' can be calculated analytically. If this segment is not straight it can be presented as a composed by M number of small straight segments. Each straight segment i connects the points (R_i, Z_i) and (R_{i+1}, Z_{i+1}), $(i = 1, 2, \ldots, M-1)$. Then the vectors $\mathbf{a}_s(R, Z)$, $\mathbf{b}_s(R, Z)$ is given by the sum

$$\begin{aligned}\mathbf{a}^{(s)}(R, Z) &= \sum_{i=1}^{M} (\sin \alpha_i, \cos \alpha_i) \tilde{a}_s^{(i)}(R, Z), \\ \mathbf{b}^{(s)}(R, Z) &= \sum_{i=1}^{M} (\sin \alpha_i, \cos \alpha_i) \tilde{b}_s^{(i)}(R, Z), \qquad \text{(B.35)}\end{aligned}$$

where $\sin \alpha_i = \sin \alpha(Z_i)$, $\cos \alpha_i = \cos \alpha(Z_i)$,

$$\tilde{a}_s^{(i)}(R,Z) = \int\limits_{Z_i}^{Z_{i+1}} \int\limits_0^{2\pi} \frac{\cos(n_s\phi)dZ'd\phi}{D(\phi, Z')},$$

$$\tilde{b}_s^{(i)}(R,Z) = -\int\limits_{Z_i}^{Z_{i+1}} \int\limits_0^{2\pi} \frac{\sin(n_s\phi)dZ'd\phi}{D(\phi, Z')}. \tag{B.36}$$

The function $R_0(Z)$ describes the straight segments between points (R_i, Z_i) and (R_{i+1}, Z_{i+1}), $(i = 1, 2, \ldots, M-1)$, i.e.,

$$R_0(Z) = R_i + \frac{R_{i+1} - R_i}{Z_{i+1} - Z_i}(Z - Z_i) = R_i + \beta_i(Z - Z_i),$$

$$\beta_i = \frac{R_{i+1} - R_i}{Z_{i+1} - Z_i} = \tan\alpha_i, \qquad Z_i \le Z \le Z_{i+1}. \tag{B.37}$$

The denominator $D(\phi, Z')$ in (B.36) can be presented in the form,

$$D(\phi, Z') = \sqrt{\tilde{A}_i\left[a_i^2 + (\tilde{Z}_i - Z')^2\right]},$$

where $\tilde{A}_i$, a_i and $\tilde{Z}_i$ are given by

$$\tilde{A}_i = 1 + \beta_i^2, \qquad \tilde{Z}_i = \frac{1}{1+\beta_i^2}\Big[Z + R\beta_i\cos\phi - (R_i - \beta_i Z_i)\beta_i\Big],$$

$$a_i^2 = C_i - \tilde{Z}_i^2,$$

$$C_i = \frac{1}{1+\beta_i^2}\Big[R^2 + (R_i - \beta_i Z_1)^2 - 2R(R_i - \beta_i Z_i)\cos\phi + Z^2\Big].$$

It allows to integrate (B.36) over Z' thus reducing them to the one-dimensional integral

$$\tilde{a}_s^{(i)}(R,Z) = \frac{1}{\sqrt{\tilde{A}_i}} \int\limits_0^{2\pi} L_g^{(i)}(\phi)\cos(n_s\phi)d\phi,$$

$$\tilde{b}_s^{(i)}(R,Z) = -\frac{1}{\sqrt{\tilde{A}_i}} \int\limits_0^{2\pi} L_g^{(i)}(\phi)\sin(n_s\phi)d\phi, \tag{B.38}$$

where

$$L_g^{(i)}(\phi) = \ln\left[\frac{Z_{i+1} - \tilde{Z}_i + \sqrt{\tilde{a}_i^2 + (Z_{i+1} - \tilde{Z}_i)^2}}{Z_i - \tilde{Z}_i + \sqrt{a^2 + (\tilde{Z}_i - Z_i)^2}}\right].$$

The numerical integration shows that the integral $\tilde{b}_s^{(i)}(R, Z)$ is negligibly small. Furthermore, we neglect this integral.

B.2.4 The Perturbation Poloidal Flux

According to (3.4) the perturbation poloidal flux $\psi^{(pert)}$ created by the vertical segments of saddle coils is determined by the function $g(R, Z, \varphi)$ given by (3.6) and (3.8). Using (B.32) and (3.8) the perturbation flux $\psi^{(pert)}$ on the given magnetic surface ψ =const can be reduced to

$$\psi^{(pert)}(R, Z, \varphi) = \frac{\partial}{\partial \varphi} \int G(R, Z, \varphi)\, d\vartheta, \tag{B.39}$$

where

$$G(R, Z, \varphi) = \epsilon \sum_{n_s}^{\infty} G_{n_s}(R, Z) \cos(n_s \varphi), \tag{B.40}$$

$$G_{n_s}(R, Z, \varphi) = J_s R_0^{-1} \left[a_R^{(s)}(R, Z) \frac{dR}{d\vartheta} + a_Z^{(s)}(R, Z) \frac{dZ}{d\vartheta} \right].$$

The perturbation parameter ϵ is defined by Eq. (B.16). Using the relation $d\vartheta = d\varphi / q(\psi)$ and the equations of magnetic field lines (1.19), the function $G_{n_s}(R, Z)$ is reduced to

$$G_{n_s}(R, Z) = \frac{J_s q(\psi) R}{R_0 B_\varphi} \left(a_R^{(s)}(R, Z) B_R + a_Z^{(s)}(R, Z) B_Z \right), \tag{B.41}$$

where B_R and B_Z are the poloidal components of the equilibrium magnetic field.

On the given magnetic surface ψ = const the function $G_n(R, Z)$ is the 2π–periodic function of the angle variable ϑ, and therefore it can be expanded into the Fourier series,

$$G_n(R, Z) = \sum_m G_{mn}(\psi) \cos(m\vartheta + \phi_{mn}). \tag{B.42}$$

Finally, from (B.40) and (B.42) we obtain the following presentation of the perturbation poloidal flux,

$$\psi^{(pert)}(R, Z, \varphi) = -\epsilon \sum_{m,n} \frac{n}{m} G_{mn}(\psi) \sin(m\vartheta + \phi_{mn}) \sin(n\varphi), \tag{B.43}$$

which can be also rewritten in the form given by Eqs. (3.50) and (3.51).

B.3 Numerical Calculations of Fourier Components

The Fourier expansion of the poloidal fluxes $\psi_n(\psi, \vartheta) \equiv \psi_n(R(\psi, \vartheta), Z(\psi, \vartheta))$ given by Eqs. (3.47) and (3.50) can be presented in the following form

$$\begin{aligned}\psi_n(\psi, \vartheta) &= \sum_{m=0}^{\infty} H_{mn}(\psi)\cos(m\vartheta + \chi_{mn}) \\ &= \sum_{m=0}^{\infty} \left[H_{mn}^{(c)}(\psi)\cos m\vartheta + H_{mn}^{(s)}(\psi)\sin m\vartheta\right],\end{aligned} \tag{B.44}$$

where Fourier components $H_{mn}(\psi)$, $H_{mn}^{(c)}(\psi)$, $H_{mn}^{(s)}(\psi)$ are given by the integrals

$$\begin{aligned}H_{mn}(\psi) &= \left[\left(H_{mn}^{(c)}(\psi)\right)^2 + \left(H_{mn}^{(s)}(\psi)\right)^2\right]^{1/2}, \\ H_{mn}^{(c)}(\psi) &= H_{mn}(\psi)\cos\chi_{mn} = \frac{1}{\pi}\int_{-\pi}^{\pi} \psi_n(\psi, \vartheta)\cos m\vartheta d\vartheta, \\ H_{mn}^{(s)}(\psi) &= -H_{mn}(\psi)\sin\chi_{mn} = \frac{1}{\pi}\int_{-\pi}^{\pi} \psi_n(\psi, \vartheta)\sin m\vartheta d\vartheta, \\ H_{m=0,n}^{(c)}(\psi) &= \frac{1}{2\pi}\int_{-\pi}^{\pi} \psi_n(\psi, \vartheta) d\vartheta.\end{aligned} \tag{B.45}$$

The phases χ_{mn} are found by the following rules

$$\begin{aligned}\chi_{mn}^{(0)} &= \arctan\left(\left|\frac{H_{mn}^{(s)}(\psi)}{H_{mn}^{(c)}(\psi)}\right|\right), \\ \chi_{mn} &= \begin{cases}\chi_{mn}^{(0)}, & \text{if } H_{mn}^{(c)}(\psi) > 0 \text{ and } H_{mn}^{(s)}(\psi) > 0, \\ \pi - \chi_{mn}^{(0)}, & \text{if } H_{mn}^{(c)}(\psi) < 0 \text{ and } H_{mn}^{(s)}(\psi) > 0, \\ \pi + \chi_{mn}^{(0)}, & \text{if } H_{mn}^{(c)}(\psi) < 0 \text{ and } H_{mn}^{(s)}(\psi) < 0, \\ 2\pi - \chi_{mn}^{(0)}, & \text{if } H_{mn}^{(c)}(\psi) > 0 \text{ and } H_{mn}^{(s)}(\psi) < 0.\end{cases}\end{aligned} \tag{B.46}$$

Appendix C
Calculations of the Poincaré Integrals

Here we present the detailed calculations of the Poincaré integrals (6.59) for the perturbation magnetic fluxes $\psi_1(\psi, \vartheta, \varphi)$ of type

$$\psi_1(\psi, \vartheta, \varphi) = \psi_n(\psi, \vartheta) \sin(n\varphi + \chi_n),$$
$$\psi_n(\psi, \vartheta) = \sum_{m=1}^{\infty} H_{mn}(\psi) e^{im\vartheta} = \sum_{m=1}^{\infty} |H_{mn}(\psi)| \, e^{im(\vartheta - \vartheta_0)}. \tag{C.1}$$

We assume that the integration in (6.59) is taken over the unperturbed orbit ($\psi =$ const, $\vartheta(\varphi) = \varphi/q(\psi)$) one poloidal turn starting from and ending at the section Σ_s (see Fig. 2.4). Recall that $\vartheta = \pm\pi$ at the section Σ_s. Using (C.1) the integral (6.59) is reduced to

$$P(\varphi, \psi) = \int_{-\pi q(\psi)}^{\pi q(\psi)} \psi_n\left(\psi, \vartheta(\varphi')\right) \sin\left(n[\varphi + \varphi']\right) d\varphi'$$
$$= K_n(\psi) \sin(n\varphi + \chi_n) + L_n(\psi) \cos(n\varphi + \chi_n), \tag{C.2}$$

where

$$R_n(\psi) = K_n(\psi) + i L_n(\psi) = \int_{-\pi q(\psi)}^{\pi q(\psi)} \psi_n(\psi, \vartheta(\varphi)) \, e^{in\varphi} d\varphi. \tag{C.3}$$

Using a Fourier series of $\psi_n(\psi, \vartheta)$ in ϑ we have

S. Abdullaev, *Magnetic Stochasticity in Magnetically Confined Fusion Plasmas*, Springer Series on Atomic, Optical, and Plasma Physics 78,
DOI: 10.1007/978-3-319-01890-4, © Springer International Publishing Switzerland 2014

$$K_n(\psi) = \frac{1}{2} \sum_{m=1}^{\infty} H_{mn}(\psi) \int_{-\pi q}^{\pi q} \left[e^{i(m/q+n)\varphi} + e^{i(m/q-n)\varphi} \right] d\varphi$$

$$= q(\psi) \sum_{m=1}^{\infty} H_{mn}(\psi) \left[\frac{\sin(\pi[m+nq])}{m+nq} + \frac{\sin(\pi[m-nq])}{m-nq} \right]. \quad \text{(C.4)}$$

Recalling that $H_{mn}(\psi) = |H_{mn}| \exp(-im\vartheta_0)$ one has at the resonant surface $\psi_{m_0,n}$, $(q(\psi_{m_0,n}) = m_0/n)$,

$$K_n(\psi_{m_0,n}) = \pi q(\psi_{m_0,n}) |H_{m_0 n}(\psi_{m_0,n})| \cos(nq\vartheta_0). \quad \text{(C.5)}$$

One can similarly obtain

$$L_n(\psi) = \frac{1}{2i} \sum_{m=1}^{\infty} H_{mn}(\psi) \int_{-\pi q}^{\pi q} \left[e^{i(m/q+n)\varphi} - e^{i(m/q-n)\varphi} \right] d\varphi$$

$$= -iq(\psi) \sum_{m=1}^{\infty} H_{mn}(\psi) \left[\frac{\sin(\pi[m+nq])}{m+nq} - \frac{\sin(\pi[m-nq])}{m-nq} \right], \quad \text{(C.6)}$$

and

$$L_n(\psi_{m_0,n}) = \pi q(\psi_{m_0,n}) |H_{m_0 n}(\psi_{m_0,n})| \sin(nq\vartheta_0). \quad \text{(C.7)}$$

At the arbitrary ψ the function $R_n(\psi) = K_n(\psi) + iL_n(\psi)$ can be presented as

$$R_n(\psi) = R_n^{(reg)}(\psi) + R_n^{(osc)}(\psi), \quad \text{(C.8)}$$

where $R_n^{(reg)}(\psi)$ is the regular part defined as

$$R_n^{(reg)}(\psi) = \pi q(\psi) H_n^*(\psi; m = nq), \quad \text{(C.9)}$$

and $R_n^{(osc)}(\psi)$ is the oscillatory part of the integral $R_n(\psi)$. According to the definition of the $R_n^{(reg)}(\psi)$ the oscillatory part $R_n^{(osc)}(\psi)$ has zeros at the resonant surfaces ψ_{mn}, $q(\psi_{mn}) = m/n$, i.e., $R_n^{(osc)}(\psi_{mn}) = 0$. In Eq. (C.9) the function $H_n(\psi; m)$ is defined by $H_{mn}(\psi)$ by extending the discrete mode number m to the continuous one.

The relation (C.9) can be also obtained from (C.3). Indeed, by replacing the variable φ to ϑ we have

$$R_n(\psi) = q(\psi) \int_{-\pi}^{\pi} \psi_n(\psi, \vartheta(\varphi)) e^{inq\vartheta} d\vartheta. \quad \text{(C.10)}$$

For the integer values of the product nq, $nq = m$, the latter up to the constant factor π coincides with the Fourier coefficients H_{mn}, i.e., $R_n(\psi) = \pi q(\psi) H^*_{mn}(\psi)$.

C.1 Determination of Oscillatory Parts of $R_n(\psi)$

Below we present the method of calculations of $R_n^{(osc)}(\psi)$ based on the contour integration. It can be applied when there exists an analytical formula of the flux $\psi_n(\psi, \vartheta)$. Specifically we use the poloidal fluxes of the perturbation magnetic field for Model I (9.28) and Model II (9.29).

C.1.1 Model I

For Model I with the flux function $\psi_n(\psi, \vartheta)$ (C.1) the integral (C.3) can be taken by a contour integral. Changing the integration variable φ to $\vartheta = \varphi/q$ the integral (C.3) is reduced to

$$R_n(\psi) = e^{-b} \int_{-\pi}^{\pi} \frac{\cos(\vartheta - \vartheta_0) - e^{-b}}{1 + e^{-2b} - 2e^{-b}\cos(\vartheta - \vartheta_0)} e^{inq\vartheta} d\vartheta. \tag{C.11}$$

Introducing a complex integration variable $z = e^{i\vartheta}$, $d\vartheta = dz/iz$, we have

$$\begin{aligned}
&\sin(\vartheta - \vartheta_0) = \frac{e^{i\vartheta_0}}{2zi}\left(z^2 e^{-2i\vartheta_0} - 1\right),\\
&\cos(\vartheta - \vartheta_0) = \frac{e^{i\vartheta_0}}{2z}\left(1 + z^2 e^{-2i\vartheta_0}\right),\\
&1 + e^{-2b} - 2e^{-b}\cos(\vartheta - \vartheta_0) = \frac{z_1}{z}\left[(z - z_1)(z_2 - z)\right],\\
&z_1 = e^{-b+i\vartheta_0}, \qquad z_2 = e^{b+i\vartheta_0}.
\end{aligned}$$

The integral (C.11) can be then rewritten as

$$R_n(\psi) = \frac{e^{2i\vartheta_0}}{2i} \int_{C_\varphi} f(z) dz, \tag{C.12}$$

where C_φ is the segment $-\pi < \theta < \pi$ on the unit circle $|z| = 1$, and

$$f(z) = \frac{z^{nq-1}\left[1 + z^2 e^{-2i\vartheta_0} - 2z_1^* z\right]}{(z - z_1)(z_2 - z)}.$$

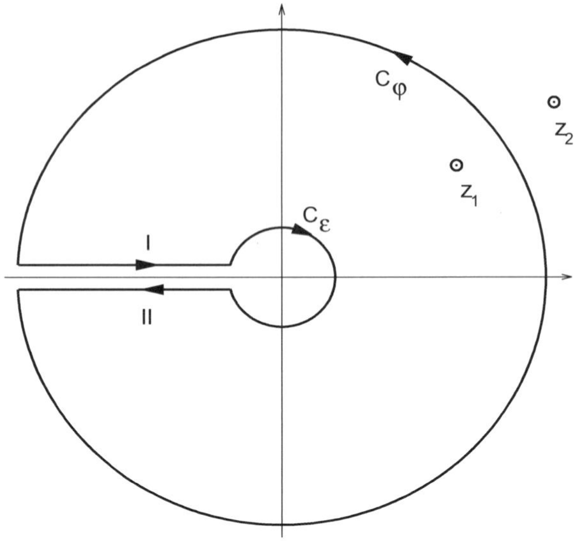

Fig. C.1 Integration contour of the integral (C.13)

The function $f(z)$ has two poles at $z = z_1 = e^{-b+i\vartheta_0}$ and $z = z_2 = e^{b+i\vartheta_0}$ located inside $|z_1| < 1$ and outside $|z_2| > 1$ the circle $|z| = 1$. Moreover, for the noninteger value of the product nq the function $f(z)$ is multivalued because of term z^{nq}, i.e., $f\left(e^{i\theta}\right) \neq f\left(e^{i(\theta+2\pi)}\right)$. To apply the residue formula to integrate (C.12) we choose the closed contour C along which the function $f(z)$ is single-valued. The contour C shown in Fig. C.1 contains the segment C_φ on the unit circle $|z| = 1$ and cuts I and II along the radii $z = r\exp(\pm i\pi)$, $\epsilon < r < 1$, respectively, and the circle C_ϵ of radius $|z| = \epsilon$.

The residue formula for the integral over the function $f(z)$ along the contour C gives,

$$\oint_C f(z)dz = 2\pi i \text{Res}\ [f(z); z = z_1] = 2\pi i e^{-nqb} e^{inq\vartheta_0} e^{-2i\vartheta_0}. \tag{C.13}$$

The left hand side of (C.13) consists of four integrals,

$$\oint_C f(z)dz = \left(\int_{C_\varphi} + \int_I + \int_{II} + \int_{C_\epsilon}\right) f(z)dz. \tag{C.14}$$

It is not difficult to see that at $nq > 0$ the integral along the contour C_ϵ vanish at the limit $\epsilon \to 0$. The integrals along the cuts I and II, where $z_I = r\exp(i\pi)$ and $z_{II} = r\exp(-i\pi)$, respectively, are reduced to

$$\left(\int_I + \int_{II}\right) f(z)dz = 2i\sin(\pi nq) F_I(\psi, \vartheta_0), \tag{C.15}$$

where

$$F_I(\psi, \vartheta_0) = \int_0^1 \phi(r) r^{nq-1} dr,$$
$$\phi^{(I)}(r) = \frac{1 + r^2 e^{-2i\vartheta_0} + 2re^{-i\vartheta_0}e^{-b}}{(r + e^{-b+i\vartheta_0})(r + e^{b+i\vartheta_0})}. \tag{C.16}$$

From Eqs. (C.12), (C.13) and (C.15) it follows that

$$R_n(\psi) = R_n^{(reg)}(\psi) + R_n^{(osc)}(\psi),$$
$$R_n^{(reg)}(\psi) = \pi e^{inq\vartheta_0} e^{-nqb} = \pi e^{-nqb}\left[\cos(nq\vartheta_0) + i\sin(nq\vartheta_0)\right],$$
$$R_n^{(osc)}(\psi) = -e^{2i\vartheta_0}\sin(\pi nq) F_I(\psi, \vartheta_0). \tag{C.17}$$

C.1.2 Asymptotical Expansion

For large $nq \gg 1$ the function $F_I(\psi, \vartheta_0)$ (C.16) can be estimated by the asymptotic expansion in powers of $1/nq$. It can be done using an integration by part,

$$\int_0^1 \phi^{(I)}(r) r^{nq-1} dr = \left[\frac{\phi_0^{(I)}}{nq} - \frac{1}{nq}\int_0^1 \frac{d\phi^{(I)}(r)}{dr} r^{nq} dr\right]$$
$$= \frac{1}{nq}\left\{\phi_0^{(I)} - \frac{1}{nq+1}\phi_1^{(I)} + \cdots\right\}, \tag{C.18}$$

where

$$\phi_0^{(I)} = \phi^{(I)}(1) = \frac{\cos\vartheta_0 + e^{-b}}{\cos\vartheta_0 + \cosh b},$$
$$\phi_1 = \left.\frac{d\phi(r)}{dr}\right|_{r=1} = -i\frac{\sinh b \sin\vartheta_0}{(\cos\vartheta_0 + \cosh b)^2}. \tag{C.19}$$

At $\vartheta_0 = 0$ and $\vartheta_0 = \pi$, we have

$$\phi_0^{(I)}(0) = \frac{1 + e^{-b}}{1 + \cosh b}, \qquad \phi_0^{(I)}(\pi) = -\frac{1 - e^{-b}}{\cosh b - 1}.$$

Near the separatrix $b \to 0$, it has the following asymptotics, $\phi_0(\pi) \approx -(2/b)(1 + O(b))$.

Finally, the leading terms of asymptotical expansion of $R_n^{(osc)}(\psi)$, in $1/nq$ can be presented as

$$K_n^{(osc)}(\psi) = -\frac{\sin(\pi nq)}{nq}\frac{\cos\vartheta_0 + e^{-b}}{\cos\vartheta_0 + \cosh b},$$
$$L_n^{(osc)}(\psi) = \frac{\sin(\pi nq)}{nq(nq+1)}\frac{\sin\vartheta_0 \sinh b}{(\cos\vartheta_0 + \cosh b)^2}. \tag{C.20}$$

C.1.3 Model II

For Model II (9.29), integrating by part one arrives to

$$R_n(\psi) = -nq \int_{-\pi}^{\pi} \arctan\left[\frac{\sin(\vartheta - \vartheta_0)}{\cos(\vartheta - \vartheta_0) - e^b}\right] e^{inq\vartheta} d\vartheta$$
$$= -2\sin(\pi nq) F(\vartheta_0)$$
$$+ ie^{-b} \int_{-\pi}^{\pi} \frac{\cos(\vartheta - \vartheta_0) - e^{-b}}{1 + e^{-2b} - 2e^{-b}\cos(\vartheta - \vartheta_0)} e^{inq\vartheta} d\vartheta, \tag{C.21}$$

where

$$F(\vartheta) = \arctan\left[\frac{\sin\vartheta}{\cos\vartheta + e^b}\right]. \tag{C.22}$$

The integral in the last term of (C.21) coincides with the integral (C.11). Then using (C.17), we obtain

$$R_n(\psi) = R_n^{(reg)}(\psi) + R_n^{(osc)}(\psi),$$
$$R_n^{(reg)}(\psi) = \pi i e^{inq\vartheta_0} e^{-nqb},$$
$$R_n^{(osc)}(\psi) = -\sin(\pi nq)\Big[2F(\vartheta_0) + ie^{2i\vartheta_0} F_I(\psi, \vartheta_0)\Big]. \tag{C.23}$$

Using (C.20) they can be also presented by

$$K_n^{(reg)}(\psi) = -\pi \sin(nq\vartheta_0) e^{-nqb},$$
$$L_n^{(reg)}(\psi) = \pi \cos(nq\vartheta_0) e^{-nqb},$$
$$K_n^{(osc)}(\psi) = -\sin(\pi nq)\left[2F(\vartheta_0) + \frac{1}{nq(nq+1)}\frac{\sin\vartheta_0 \sinh b}{(\cos\vartheta_0 + \cosh b)^2}\right],$$
$$L_n^{(osc)}(\psi) = -\sin(\pi nq)\left[2F(\vartheta_0) + \frac{1}{nq}\frac{\cos\vartheta_0 + e^{-b}}{\cos\vartheta_0 + \cosh b}\right]. \tag{C.24}$$

Appendix D
Advanced Version of the Symplectic Mapping for Hamiltonian Systems

Below we construct the alternative form of the mapping $(\vartheta_k, I_k) \to (\vartheta_{k+1}, I_{k+1})$ for the Hamiltonian system given by Eqs. (6.5) and (6.6). Similar to the method given in Sect. 6.2.2 it is based on the canonical change of variables and the classical perturbation theory in a finite time interval (see Abdullaev (2002, 2006)). In the interval $t_k \leq t \leq t_{k+1}$ we perform a such a canonical transformation of variables $(\vartheta, I) \to (\Theta, J)$ that the new Hamiltonian $\mathcal{H}$ in the new canonical variables (Θ, J) acquires fast oscillating perturbation terms, i.e.,

$$\begin{aligned}
&\mathcal{H} = \mathcal{H}_0(J) + \epsilon \mathcal{H}_1(\Theta, J, t), \\
&\mathcal{H}_1(\Theta, J, t) = -2H_1\,(\Theta, J, t) \sum_{s=1}^{N} \cos\,(sMnt)\,,
\end{aligned} \tag{D.1}$$

where $M \geq 1$ is an integer number, $N \gg 1$ is the number of harmonics. The canonical change of variables $(\vartheta, I) \to (\Theta, J)$ is implemented via the generating function $F(J, \vartheta, t) = J\vartheta + \epsilon S(J, \vartheta, t)$, where the generating function $S = S(J, \vartheta, t)$ satisfies the Hamilton-Jacobi equation

$$H_0\left(\vartheta, J + \epsilon\frac{\partial S}{\partial \vartheta}, t\right) + \epsilon\frac{\partial S}{\partial t} = \mathcal{H}(\vartheta, J, t, \epsilon). \tag{D.2}$$

The generating function $S(J, \vartheta, t)$ is sought as a series in powers of ϵ similar to (6.15). Expanding the Hamilton–Jacobi equation (D.2) in powers of of ϵ one obtains $H_0(J) = \mathcal{H}_0(J)$, and the equations for S_i, $(i = 1, 2, \ldots)$:

$$\frac{\partial S_1}{\partial t} + \frac{\partial H_0}{\partial J}\frac{\partial S_1}{\partial \vartheta} = \mathcal{H}_1(\vartheta, J, t) - H_1(\vartheta, J, t), \tag{D.3}$$

$$\frac{\partial S_j}{\partial t} + \frac{\partial H_0}{\partial J}\frac{\partial S_j}{\partial \vartheta} = -F_j(\vartheta, J, t), \qquad j \geq 2, \tag{D.4}$$

where $F_j(\vartheta, I, t)$ are the polynomial functions of derivatives $\partial S_1/\partial\vartheta, \ldots, \partial S_{j-1}/\partial\vartheta$.

S. Abdullaev, *Magnetic Stochasticity in Magnetically Confined Fusion Plasmas*, Springer Series on Atomic, Optical, and Plasma Physics 78,
DOI: 10.1007/978-3-319-01890-4,

In the first order of the perturbation parameter ϵ the generating function is given by the integral

$$S_1(\vartheta, J, t, t_0) = \int_{t_0}^{t} \Big[\mathcal{H}_1\left(\vartheta(t'), J, t'\right) - H_1\left(\vartheta(t'), J, t'\right)\Big]dt', \tag{D.5}$$

taken along the unperturbed field line $\vartheta(t') = \vartheta + \omega(J)(t - t')$, $J = \text{const}$. The variable t and the free parameter t_0 lie in the intervals: $t_k \leq t \leq t_{k+1}, t_k < t_0 < t_{k+1}$. Using Eq. (D.1) we have

$$\begin{aligned} S_1(J, \vartheta, t, t_0) &= -\int_{t_0}^{t} H_1(\vartheta(t'), J, t') \sum_{s=-N}^{N} \cos\left(sMnt'\right) dt' \\ &= -\frac{2\pi}{Mn} \int_{t_0}^{t} H_1(\vartheta(t'), I, t') \sum_{k=-\infty}^{\infty} \delta\left(t' - t_k\right) dt', \quad \text{at } N \to \infty, \end{aligned} \tag{D.6}$$

where $t_k = k(2\pi/Mn)$, $(k = 0, \pm 1, \pm 2, \ldots)$. From Eq. (D.6) it follows that the generating function S_1 vanishes in the time interval $t_k < t < t_{k+1}$. But it takes non-zero values at $t = t_k$ and $t = t_{k+1}$ given by Eq. (6.47).

Note that the higher order generating functions S_j, $(j \geq 2)$, vanish in the whole time interval $t_k \leq t \leq t_{k+1}$ (see Abdullaev (2006)).

Suppose that $(\Theta(t), J(t))$ is a solution of the new Hamiltonian (D.1) in the interval $t_k < t < t_{k+1}$. Then the mapping can be presented by the following successive canonical transformations

$$\begin{aligned} &J_k = J_k - \epsilon \frac{\partial S_k}{\partial \vartheta_k}, \qquad \Theta_k = \vartheta_k + \epsilon \frac{\partial S_k}{\partial J_k}, \\ &(\Theta_{k+1}, J_{k+1}) = \hat{M}(\Theta_k, J_k), \\ &J_{k+1} = J_{k+1} + \epsilon \frac{\partial S_{k+1}}{\partial \vartheta_{k+1}}, \qquad \vartheta_{k+1} = \Theta_{k+1} - \epsilon \frac{\partial S_{k+1}}{\partial J_{k+1}}, \end{aligned} \tag{D.7}$$

where $(\Theta_k, J_k) = (\Theta(t_k), J(t_k))$. The mapping $(\Theta_{k+1}, J_{k+1}) = \hat{M}(\Theta_k, J_k)$ can be constructed using the regular procedure described in Refs. Abdullaev (2002, 2006). For the Hamiltonian system (D.1) it reads as

$$\begin{aligned}
\bar{J}_k &= J_k - \epsilon \frac{\partial G_k}{\partial \Theta_k}, & \bar{\Theta}_k &= \Theta_k + \epsilon \frac{\partial G_k}{\partial \bar{J}_k}, \\
\bar{\Theta}_{k+1} &= \bar{\Theta}_k + \frac{t_{k+1} - t_k}{q(\bar{J}_k)}, & \bar{J}_{k+1} &= \bar{J}_k, \\
J_{k+1} &= \bar{J}_{k+1} + \epsilon \frac{\partial G_{k+1}}{\partial \bar{\Theta}_{k+1}}, & \Theta_{k+1} &= \bar{\Theta}_{k+1} - \epsilon \frac{\partial G_{k+1}}{\partial \bar{J}_{k+1}},
\end{aligned} \tag{D.8}$$

determined by the generating function $G(\bar{J}, \Theta, t, t_0)$: $G_k = G(\bar{J}_k, \Theta_k, t_k, t_0)$, $G_{k+1} = G(\bar{J}_{k+1}, \Theta_{k+1}, t_{k+1}, t_0)$. In the first order of ϵ it is given by

$$\begin{aligned}
G_1(J, \Theta, t, t_0) &= -\int_{t_0}^{t} \mathcal{H}_1(\Theta(t'), J, t') dt' \\
&= 2 \int_{t_0}^{t} H_1(\Theta(t'), J, t') \sum_{s=1}^{N} \cos\left(sMnt'\right) dt'.
\end{aligned} \tag{D.9}$$

As seen from Eq. (D.9) the generating function G_1 is determined as an integral from the fast oscillating functions. One should expect that G_1 decreases with increasing the number M. Taking the asymptotical expansion of the integral (D.9) one can obtain the following estimation for the generating function $G_1 = G_1(J, \Theta, t_{k+1}, t_k)$:

$$G_1 \approx \frac{4\pi}{Mn} \sum_{s=1}^{N} \left[\sum_{p=1}^{P} \frac{1}{(sMn)^{2p}} f^{(2p)}(t_k) + O([sm\Omega]^{-2P}) \right], \tag{D.10}$$

where $f(t) \equiv H_1(\Theta(t), J, t)$, $f^{(p)}(t) \equiv d^p f(t)/dt^p$, $(p = 1, 2, \ldots)$. Supposing that $n^{-2p} f^{(2p)}(t_k) \sim 1$, one can obtain the following estimation for G_1 at large values $N \gg 1$, $P \gg 1$:

$$|G_1| \lessapprox \frac{2\pi^3}{M^3 n} = \frac{n^2}{12} (\Delta t)^3, \tag{D.11}$$

where $\Delta t = t_{k+1} - t_k = 2\pi/Mn$ is a step of the mapping. At the moderately large values of $M \geq 8 \div 10$ one can neglect the generating function G_1, and the mapping (D.7) is reduced to the form (6.44) with the generating functions (6.47). For instance, for $M = 8$ the divergence of the flux coordinate I of a regular orbit from the one calculated for $M = 128$ is 3.6×10^{-7} per one toroidal turn (for $\epsilon = 2 \times 10^{-3}$). This is sufficiently accurate to plot Poincare sections. We have chosen $M = 16$ for the calculation of diffusion coefficients.

A more detailed study of the accuracy of the described mapping procedure requires a special investigation.

Appendix E
Eigenvalues of the Jacobi Matrix

E.1 Jacobi Matrix

In this appendix we present the calculations of the Jacobi matrix (7.44), its eigenvalues (7.52) and the Lyapunov exponents of the mapping (6.29)–(6.31) [or (7.32)]. We present the latter in the form

$$\hat{M} = \hat{T}_{-}\hat{T}_{0}\hat{T}_{+}, \tag{E.1}$$

of three successive mappings, $\hat{T}_{-}\hat{T}_{0}\hat{T}_{+}$, each of them are given by Eqs. (6.29), (6.30) and (6.31), respectively. Then the Jacobian matrix (7.44) can be written as a product of three Jacobian matrices, corresponding to three successive mappings,

$$\mathbf{J}_k = \hat{M}_{k+1}\hat{M}_0\hat{M}_k, \tag{E.2}$$

where

$$\hat{M}_k = \begin{pmatrix} \dfrac{\partial J_k}{\partial I_k} & \dfrac{\partial J_k}{\partial \vartheta_k} \\ \dfrac{\partial \Theta_k}{\partial I_k} & \dfrac{\partial \Theta_k}{\partial \vartheta_k} \end{pmatrix}, \tag{E.3}$$

$$\hat{M}_0 = \begin{pmatrix} \dfrac{\partial J_{k+1}}{\partial J_k} & \dfrac{\partial J_{k+1}}{\partial \Theta_k} \\ \dfrac{\partial \bar{\Theta}_k}{\partial J_k} & \dfrac{\partial \bar{\Theta}_k}{\partial \Theta_k} \end{pmatrix} = \begin{pmatrix} 1 & 0 \\ w'(J_k)\,(t_{k+1} - t_k) & 1 \end{pmatrix}, \tag{E.4}$$

$$\hat{M}_{k+1} = \begin{pmatrix} \dfrac{\partial I_{k+1}}{\partial J_{k+1}} & \dfrac{\partial I_{k+1}}{\partial \bar{\Theta}_k} \\ \dfrac{\partial \vartheta_{k+1}}{\partial J_{k+1}} & \dfrac{\partial \vartheta_{k+1}}{\partial \bar{\Theta}_k} \end{pmatrix}. \tag{E.5}$$

S. Abdullaev, *Magnetic Stochasticity in Magnetically Confined Fusion Plasmas*, Springer Series on Atomic, Optical, and Plasma Physics 78,
DOI: 10.1007/978-3-319-01890-4,

The derivatives in the matrices (E.3) and (E.5) are easily calculated from the mappings given by Eqs. (6.29) and (6.31):

$$
\begin{aligned}
\frac{\partial J_k}{\partial I_k} &= \frac{1}{1+\epsilon G_{J\vartheta}(\varphi_k)}, \qquad \frac{\partial J_k}{\partial \vartheta_k} = -\frac{\epsilon G_{\vartheta\vartheta}(\varphi_k)}{1+\epsilon G_{J\vartheta}(\varphi_k)},\\
\frac{\partial \Theta_k}{\partial I_k} &= \frac{\epsilon G_{JJ}(\varphi_k)}{1+\epsilon G_{J\vartheta}(\varphi_k)},\\
\frac{\partial \Theta_k}{\partial \vartheta_k} &= 1+\epsilon G_{J\vartheta}(\varphi_k) - \frac{\epsilon^2 G_{JJ}(\varphi_k) G_{\vartheta\vartheta}(\varphi_k)}{1+\epsilon G_{J\vartheta}(\varphi_k)},
\end{aligned}
\tag{E.6}
$$

and

$$
\begin{aligned}
\frac{\partial I_{k+1}}{\partial J_k} &= 1+\epsilon G_{J\vartheta}(\varphi_{k+1}) - \frac{\epsilon^2 G_{JJ}(\varphi_{k+1}) G_{\vartheta\vartheta}(\varphi_{k+1})}{1+\epsilon G_{J\vartheta}(\varphi_{k+1})},\\
\frac{\partial I_{k+1}}{\partial \bar{\Theta}_k} &= \frac{\epsilon G_{\vartheta\vartheta}(\varphi_{k+1})}{1+\epsilon G_{J\vartheta}(\varphi_{k+1})}, \qquad \frac{\partial \vartheta_{k+1}}{\partial J_k} = -\frac{\epsilon G_{JJ}(\varphi_{k+1})}{1+\epsilon G_{J\vartheta}(\varphi_{k+1})},\\
\frac{\partial \vartheta_{k+1}}{\partial \bar{\Theta}_k} &= \frac{1}{1+\epsilon G_{J\vartheta}(\varphi_{k+1})},
\end{aligned}
\tag{E.7}
$$

where

$$
\begin{aligned}
G_{JJ}(\varphi) &\equiv \frac{\partial^2 G}{\partial J^2}, \qquad G_{J\vartheta}(\varphi) \equiv \frac{\partial^2 G}{\partial J \partial \vartheta},\\
G_{\vartheta\vartheta}(\varphi) &\equiv \frac{\partial^2 G}{\partial \vartheta^2}, \qquad G(\varphi) = G(\vartheta, J, \varphi, \varphi_0).
\end{aligned}
\tag{E.8}
$$

E.2 Jacobi Matrix at the Fixed Points

Below we calculate the eigenvalues of the Jacobi matrix J of the mapping (7.32) at the fixed points considered in Sect. 7.2.2. Putting $\varphi_k = \varphi_0 - 2\pi m_0$, $\varphi_{k+1} = \varphi_0$ we reduce the corresponding Jacobi matrix to

$$
\begin{aligned}
\frac{\partial I_{k+1}}{\partial I_k} &= \frac{1}{1+\epsilon G_{J,\vartheta}},\\
\frac{\partial I_{k+1}}{\partial \vartheta_k} &= -\frac{\epsilon G_{\vartheta\vartheta}}{1+\epsilon G_{J\vartheta}},\\
\frac{\partial \vartheta_{k+1}}{\partial I_k} &= \frac{w'(I_{k+1}) 2\pi m_0}{1+\epsilon G_{J,\vartheta}} + \frac{\epsilon G_{JJ}}{1+\epsilon G_{J\vartheta}(\varphi_k)},\\
\frac{\partial \vartheta_{k+1}}{\partial \vartheta_k} &= \frac{\epsilon G_{\vartheta\vartheta} w'(I_{k+1}) 2\pi m_0}{1+\epsilon G_{J,\vartheta}} + 1 + \epsilon G_{J\vartheta} - \frac{\epsilon^2 G_{JJ} G_{\vartheta\vartheta}}{1+\epsilon G_{J\vartheta}}.
\end{aligned}
\tag{E.9}
$$

Using the generating function G (6.34) in the first order of ϵ we have

$$\begin{aligned} G_1 &= G_1(\vartheta_0, J, \varphi_0 - 2\pi m_0, \varphi_0) \\ &= 2\pi m_0 \sum_{m,n} H_{mn}(J)\big[a(x_{mn}) \sin (m\vartheta_0 - n\varphi_0 + \chi_{mn}) \\ &+ b(x_{mn}) \cos (m\vartheta_0 - n\varphi_0 + \chi_{mn})\big]. \end{aligned} \tag{E.10}$$

According to (7.41) the coefficients $a(x_{mn})$, $b(x_{mn})$ near the resonant surfaces $I_{m_0n_0}$, $q(I_{m_0n_0}) = m_0/n_0$, are of order

$$\begin{aligned} &a(x_{mn}) \propto \epsilon^2, \\ &b(x_{mn}) = \begin{cases} 1 + C_0\epsilon & \text{for } m/n = m_0/n_0, \\ C_0\epsilon & \text{for } m/n \neq m_0/n_0, \end{cases} \\ &C_0\epsilon = -\frac{q'm}{q^2}(I_0 - I_{m_0n_0}). \end{aligned} \tag{E.11}$$

Neglecting small terms of order of ϵ and ϵ^2, The generating function G_1 (E.10) at the fixed point I_0 is reduced to

$$G_1 = 2\pi m_0 \sum_{m/n=m_0/n_0} H_{mn}(I_0) \cos (m\vartheta_0 - n\varphi_0 + \chi_{mn})\,. \tag{E.12}$$

The trace TrJ of the matrix can be reduced to

$$\begin{aligned} \text{Tr } J = \frac{\partial I_{k+1}}{\partial I_k} + \frac{\partial \vartheta_{k+1}}{\partial \vartheta_k} &= \frac{1}{1+\epsilon G_{J,\vartheta}} + \frac{\epsilon G_{\vartheta\vartheta} w'(I_{k+1}) 2\pi m_0}{1+\epsilon G_{J,\vartheta}} \\ &+ 1 + \epsilon G_{J\vartheta} - \frac{\epsilon^2 G_{JJ} G_{\vartheta\vartheta}}{1+\epsilon G_{J\vartheta}}. \end{aligned} \tag{E.13}$$

Neglecting the terms of ϵ^2 we obtain Eq. (7.53).

E.3 Eigenvalues of Jacobi Matrix

Consider the linearized equations (7.51) near the fixed point. We introduce new coordinates (x, y) by rotating the coordinates $(\delta J, \delta\vartheta)$ around the point (0, 0), i.e.,

$$\begin{pmatrix} x \\ y \end{pmatrix} = U \begin{pmatrix} \delta J \\ \delta\vartheta \end{pmatrix}, \qquad U = \begin{pmatrix} \cos\phi & -\sin\phi \\ \sin\phi & \cos\phi \end{pmatrix}, \tag{E.14}$$

which would reduce the Eq. (7.51) into form

$$x_{k+1} = \lambda_1 \, x_k,$$
$$y_{k+1} = \lambda_2 \, y_k, \tag{E.15}$$

where λ_i are constants. The angle ϕ are found from the eigenvalue problem of the matrix Λ,

$$\Lambda U = EU, \qquad E = \begin{pmatrix} \lambda_1 & 0 \\ 0 & \lambda_2 \end{pmatrix}. \tag{E.16}$$

The eigenvalues λ of the 2×2 matrix Λ with $\det \Lambda = 1$ are determined by the solution of the quadratic equation

$$\lambda^2 - 2A\lambda + 1 = 0, \qquad A = \frac{1}{2}\mathrm{Tr}\,\Lambda = \frac{1}{2}\left(\frac{\partial I_{k+1}}{\partial I_k} + \frac{\partial \vartheta_{k+1}}{\partial \vartheta_k}\right). \tag{E.17}$$

From Eq. (E.14) we have two solutions:

$$\tan\phi_1 = -\tan\phi_2 = \frac{\lambda_1 - \Lambda_{11}}{\Lambda_{12}} = \frac{\Lambda_{21}}{\Lambda_{22} - \lambda_2}. \tag{E.18}$$

Appendix F
Features of the Perturbation Hamiltonian of Guiding-Center Motion

In this appendix we analysis the perturbation Hamiltonian of guiding-center motion (5.38) in the action-angle variables $(\vartheta, \vartheta_\varphi, J, I_\varphi)$ for the magnetic perturbations created by the set of saddle coils. Furthermore, we consider only time-independent magnetic perturbations and neglect the perturbations of the electric field.

The magnetic perturbation fluxes $\psi_p^{(pert)}(R, Z, \varphi)$ of the set of saddle coils were given in Sect. 3.4 by Eqs. (3.43) and (3.49). Using the latter the perturbation Hamiltonian (5.34) can be reduced to

$$\epsilon h_1 \equiv \epsilon h_1(z, \varphi, p_z, p_\varphi, p_t) = \epsilon \sum_{j=1}^{3} \sum_{n} H_n^{(j)}(R, Z) \sin(n\varphi), \tag{F.1}$$

where

$$H_n^{(j)}(R, Z) = Z_q \frac{u_\varphi \psi_n^{(j)}(R, Z)}{x_c}. \tag{F.2}$$

The dimensionless perturbation parameter ϵ is defined by Eq. (3.46). The functions $\psi_n^{(j)}(R, Z)$, $(j = 1, 2, 3)$, are given by (3.44) and (3.50). We recall that $R = R_0 x$, $Z = R_0 z$.

Using the relations (5.21) and (5.22) between the coordinates $(R = R_o x_c(p_z), Z = R_0 z, \varphi)$ and the action-angle variables $(\vartheta, \vartheta_\varphi, J, I_\varphi)$ the perturbation Hamiltonian (F.1) can be written as

$$\epsilon h_1 = \epsilon \sum_{j=1}^{3} \sum_{n} V_n^{(j)}(\vartheta; J, I_\varphi) \sin\left(n\vartheta_\varphi + nG(\vartheta; J, I_\varphi)\right), \tag{F.3}$$

where

$$V_n^{(j)}(\vartheta; J, I_\varphi) \equiv H_n^{(j)}\left(R(\vartheta; J, I_\varphi), Z(\vartheta; J, I_\varphi)\right).$$

S. Abdullaev, *Magnetic Stochasticity in Magnetically Confined Fusion Plasmas*, Springer Series on Atomic, Optical, and Plasma Physics 78,
DOI: 10.1007/978-3-319-01890-4,

For brevity we notate the first function as $V(\vartheta)$, i.e., $V^{(j)}(\vartheta) \equiv V_n^{(j)}(\vartheta; J, I_\varphi)$. Using the relation

$$\sin\left(n\vartheta_\varphi + nG(\vartheta; J, I_\varphi)\right) = \sin(n\vartheta_\varphi)A(\vartheta) + \cos(n\vartheta_\varphi)B(\vartheta), \tag{F.4}$$

where $A(\vartheta)$ and $B(\vartheta)$ are the 2π–periodic functions of ϑ given by the Fourier series,

$$A(\vartheta) = \cos\left[nG(\vartheta; J, I_\varphi)\right], \qquad B(\vartheta) = \sin\left[G(\vartheta; J, I_\varphi)\right], \tag{F.5}$$

the perturbation Hamiltonian (F.3) can be presented as

$$\epsilon h_1 = \epsilon \sum_{j=1}^{3} \sum_{n} V_n^{(j)}(\vartheta)\left[A(\vartheta)\sin(n\vartheta_\varphi) + B(\vartheta)\cos(n\vartheta_\varphi)\right]. \tag{F.6}$$

To expand the perturbation Hamiltonian h_1 in a Fourier series in ϑ, one can separately find the Fourier expansions of the functions $V_n^{(j)}(\vartheta)$, $A(\vartheta)$, and $B(\vartheta)$, and then multiply them. Below we study the features of these functions and their Fourier expansions.

F.1 Fourier Asymptotics of $V(\vartheta)$

Suppose that the function $V(\vartheta)$ is given by the following Fourier series

$$V(\vartheta) = \sum_m |V_m| \cos(m\vartheta + \chi_m) = \frac{1}{2}\sum_{m=1}^{\infty}\left(V_m e^{im\vartheta} + V_m^* e^{-im\vartheta}\right),$$
$$V_m = |V_m| e^{i\chi_m}, \tag{F.7}$$

with the Fourier coefficients V_m.

The typical dependencies of the function $V(\vartheta)$ on the angle variable $\vartheta = 2\pi t/T$ for the different types of guiding center orbits of α-particle are plotted in Fig. F.1 for the magnetic perturbation created by the one horizontal segment of the saddle coils in the ITER-like plasma. The corresponding orbits were shown in Fig. 5.8a, b. The energy of the α-particle is $E = 1$ MeV, and the ratio $\lambda_I = 0.8$. The plasma parameters are given in Sects. 2.3.1 and 3.4, and Table 2.2 and Fig. 3.6. The corresponding perturbation parameter ϵ is equal to $3.043 \times 10^{-3} I_c$, where the perturbation current I_c is in MA. As seen from Fig. F.1a the perturbation function $V(\vartheta)$ has a pulse like behavior along the orbit. Such a dependence is due to the similar behavior of the perturbation poloidal flux $\psi_n^{(j)}(R, Z)$ as was shown in Fig. 3.9. For the passing orbits (curves 1 and 2) $V(\vartheta)$ has only one peak, while the trapped orbits (curve 3 and 4) it has the two peaks (see Fig. F.1a, b).

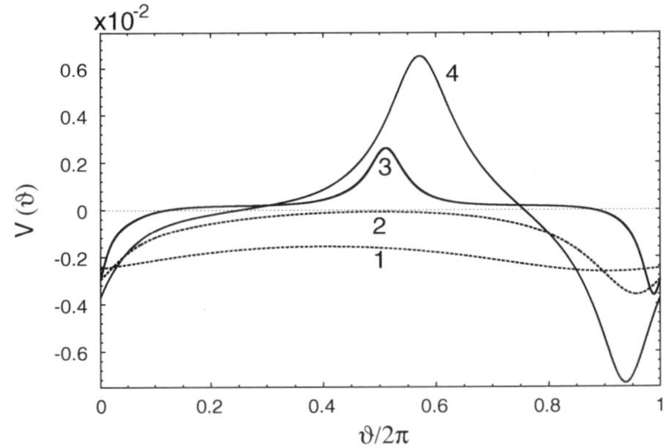

Fig. F.1 Dependencies of the perturbation function $V(\vartheta)$ (F.1) on the angle variable ϑ along the several guiding–center orbits of α-particle. *Curve 1* and *2* correspond co-passing and barely co-passing orbits, respectively; *curve 3* and *4* correspond to the barely trapped and the trapped orbits, respectively. The magnetic perturbation created by the one single horizontal segment of the set of saddle coils located at (R_2, Z_2). The energy $E_k = 1$ MeV, the ratio $\lambda_I = 0.8$. The plasma parameters and the RMP coils positions are given in Fig. 3.6. The toroidal mode of the perturbation magnetic field is $n = 3$. The perturbation current $I_c = 1$ kA

Fig. F.2 Fitting of $V(\vartheta)$ with the Lorentzian pulse function (F.8) near the one its peak. *Curve 1* (*solid line*) corresponds to $V(\vartheta)$ and *curve 2* (*dashed line*) corresponds to $V_L(\vartheta)$ with the appropriate fitting parameters a, σ, and ϑ_0

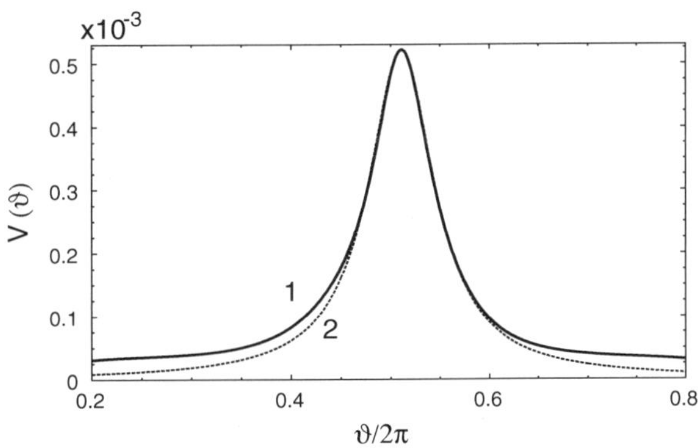

Each of peaks in the dependence $V(\vartheta)$ versus ϑ similar to the one for $\psi_n^{(j)}(R, Z)$ can be approximated by the Lorentzian pulse form (3.52),

$$V_L(\vartheta) = \frac{a}{(\vartheta - \vartheta_0)^2/\sigma^2 + 1}, \tag{F.8}$$

with the appropriate amplitude a, the pulse width σ, and its location ϑ_0. The fitting of the one of peaks by the Lorentzian pulse form is shown in Fig. F.2.

Therefore using the arguments similar to the ones in Sect. 3.4.4 we can obtain the asymptotical formulas for the poloidal spectra V_m in Eq. (F.7) similar to Eq. (3.54).

However, unlike from the latter the spectrum V_m for the trapped motion will contain the contributions from two Lorentzian peaks.

According to the localization principle[2] the asymptotic estimation of the Fourier integral

$$V_m = \frac{1}{2\pi} \int_0^{2\pi} V(\vartheta) e^{-im\vartheta} d\vartheta, \tag{F.9}$$

for large m is given by the sum of the contributions from each critical points of the function $V(\vartheta)$ corresponding its peaks. Near these points $V(\vartheta)$ is well approximated by the Lorentzian function (F.8) whose Fourier transform is given by the exponential form (3.53). Therefore, the asymptotical formula for V_m at large m will be given by

$$V_m \approx \frac{1}{2T} \begin{cases} A^{(o)} \exp\left(-\frac{mC^{(o)}}{T} - im\vartheta_0^{(o)} \right) - A^{(i)} \exp\left(-\frac{mC^{(i)}}{T} - im\vartheta_0^{(i)} \right), \\ \qquad\qquad \text{for trapped orbits,} \\ A^{(o)} \exp\left(-\frac{mC^{(o)}}{T} - im\vartheta_0^{(o)} \right), \qquad \text{for passing orbits.} \end{cases} \tag{F.10}$$

The parameters $A^{(o,i)}$, $C^{(o,i)}$, and $\vartheta_0^{(o,i)}$ depend on the orbit characteristics. The subscripts (o) and (i) stand for the outer and inner branches of the orbit. Particularly, $A^{(o,i)}$ and $C^{(o,i)}$ are similar to the parameters A and C in the asymptotical formula (3.54) and take finite values at the separatrix of the banana orbit as well as at the magnetic separatrix where the period of motion (transit time) T has singularities as was shown in Figs. 5.12b and 5.13b. One should notice the period T in the asymptotical formula (F.9) plays the similar role as the safety factor $q(\psi)$ in Eq. (3.54).

In order to find the parameters $A^{(o,i)}$, $C^{(o,i)}$, and $\vartheta_0^{(o,i)}$ it is not necessary to calculate the numerical Fourier transform of $V(\vartheta)$ (F.9) as it has been used in Sect. 3.4.4. It is sufficient to find the amplitude a, the width σ, and the peak's location ϑ_0 by fitting the function $V(\vartheta)$ near its peaks with the Lorentzian function (F.8). According to Eq. (3.53) the relations between (a, σ) and (A, C) are $A = a\sigma T$, $C = \sigma T$.

Figure F.3a, b shows the typical dependencies of the exponents $C^{(o,i)}$ and the amplitudes $A^{(o,i)}$ on the toroidal momentum $p_\varphi = I_\varphi$. Curve 1 corresponds to the passing orbits at $p_\varphi > p_s$ and to the outer branch of the trapped orbits at $p_\varphi < p_s$, while curve 2 corresponds to the inner branch of the trapped orbits.

The phases $\vartheta_0^{(o,i)}$ are also similar to the ones in Eq. (3.54). They determine the angular positions of the horizontal segments of the saddle coils. For the orbits near the separatrix their distance $\Delta\vartheta$ measured from the outer equatorial plane $\vartheta = 0$ goes to zero inversely proportional to the period of motion T. Therefore, the product $\Delta\vartheta T$ has a finite nonzero value at the separatrix. The typical dependence of the product $\Delta\vartheta T$ on the toroidal momentum of a particle $p_\varphi = I_\varphi$ is shown in Fig. F.4.

---

[2] The localization principle in the asymptotic expansion of integral can be found in many books devoted to asymptotic methods in analysis, for example, by Fedoryuk (1989).

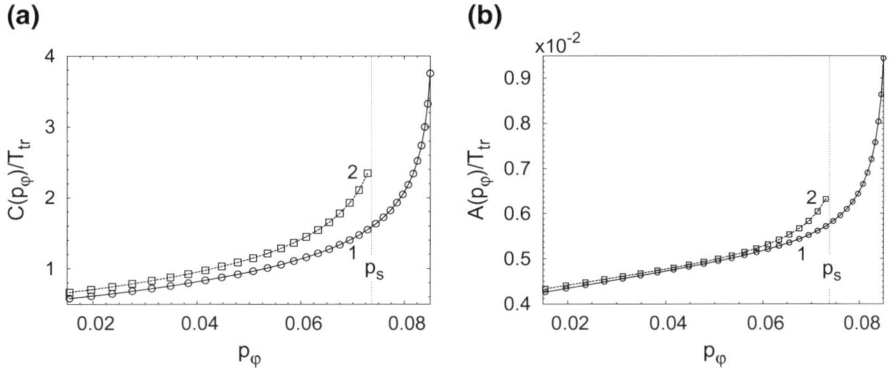

Fig. F.3 Typical dependencies of the exponents $C^{(o,i)}(p_\varphi)$ (**a**) and the amplitudes $A^{(o,i)}(p_\varphi)$ (**b**) on the momentum $p_\varphi = I_\varphi$ in the asymptotical formula (F.9) ($T_{tr} = 2\pi R_0/\sqrt{2m_a E_k}$ is the characteristic transit time). The plasma parameters and the RMP coils positions are given Fig. F.1. *Curve 1* corresponds to the passing orbits at $p_\varphi > p_s$ and to the outer branch of the trapped orbits at $p_\varphi < p_s$, *curve 2* corresponds to the inner branch of the trapped orbits

Fig. F.4 Typical dependencies of the phases $\vartheta_0^{(o,i)}$ on the momentum $p_\varphi = I_\varphi$ in the asymptotical formula (F.9). The plasma parameters and the RMP coils positions are given Fig. F.1. Numbering of curves are the same as in Fig. F.3

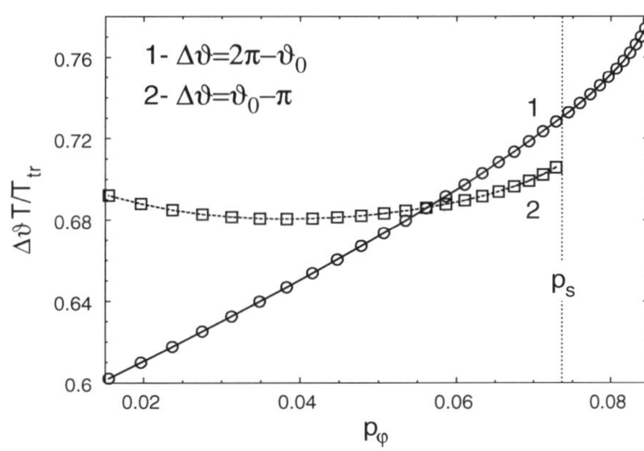

Using Eqs. (F.7) and (F.10) we can obtain the asymptotic formula for the function $V(\vartheta)$ similar to the perturbation poloidal flux $\psi_n^{(j)}$ (3.67),

$$V_{asym}(\vartheta) = \begin{cases} V^{(o)}(\vartheta) - V^{(i)}(\vartheta), & \text{for trapped orbits,} \\ V^{(o)}(\vartheta), & \text{for passing orbits,} \end{cases} \tag{F.11}$$

where

$$\begin{aligned} V^{(l)}(\vartheta) &= \frac{A^{(l)}}{2T\Lambda^{(l)}}\left(1 - e^{-2\alpha^{(l)}}\right), & l &= (o,i), \\ \Lambda^{(l)} &= 1 + e^{-2\alpha^{(l)}} - 2e^{-\alpha^{(l)}}\cos\left(\vartheta - \vartheta^{(l)}\right), & \alpha^{(l)} &= C^{(l)}/T. \end{aligned} \tag{F.12}$$

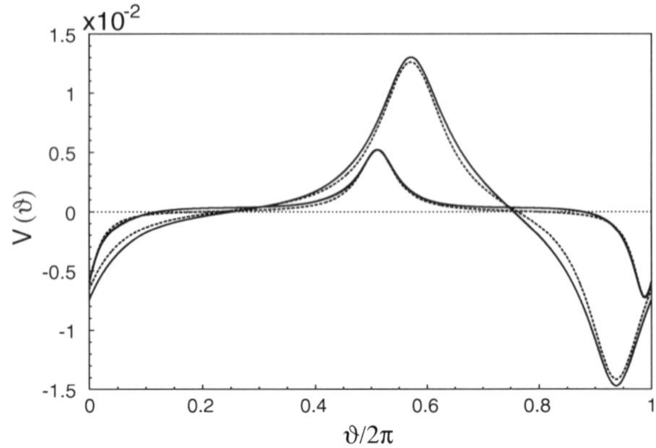

Fig. F.5 Comparison of $V(\vartheta)$ with the asymptotic presentation (F.11) for the two trapped orbits. *Solid curves* correspond to $V(\vartheta)$ and *dashed curves* correspond to $V_{asym}(\vartheta)$ with the fitting parameters A, C, and ϑ_0

The comparison of the asymptotic formula (F.11) for the perturbation function $V(\vartheta)$ with its numerically calculated values is shown in Fig. F.5. It shows that the asymptotic formula well describes the positions, the amplitudes, and the widths of peaks of $V(\vartheta)$.

F.2 Fourier Expansions of the Functions $G(\vartheta)$, $A(\vartheta)$, $B(\vartheta)$

We present the function $G(\vartheta)$ in a Fourier series,

$$G(\vartheta; J, I_\varphi) = \sum_{s=1}^{M} g_s \sin(s\vartheta), \tag{F.13}$$

where $g_s \equiv g_s(J, I_\varphi)$ is the Fourier coefficients depending on the action variables (J, I_φ). Typical dependencies of the first three coefficients g_s, $s = 1, 2, 3$ on the toroidal momentum $p_\varphi = I_\varphi$ are plotted in Fig. F.6. The coefficient g_1 is of a two order larger than the ones g_s, $(s = 2, 3, \ldots)$ for the both trapped, $p_\varphi < p_s$, and passing, $p_\varphi > p_s$, orbits. They reach their maximal values only at $p_\varphi \to p_s$, i.e., for the barely trapped and passing orbits.

The functions $A(\vartheta)$ and $B(\vartheta)$ in (F.5) are the 2π-periodic functions of ϑ given by the Fourier series,

$$A(\vartheta) = \sum_{m=0}^{\infty} a_m \cos(m\vartheta), \qquad B(\vartheta) = \sum_{m=1}^{\infty} b_m \sin(m\vartheta). \tag{F.14}$$

Figure F.7 show the dependencies of the functions $A(\vartheta)$, $B(\vartheta)$, and $G(\vartheta)$ on the angle variable ϑ for the four different types of orbits: (a) passing; (b) barely passing; (c) barely trapped; (d) trapped.

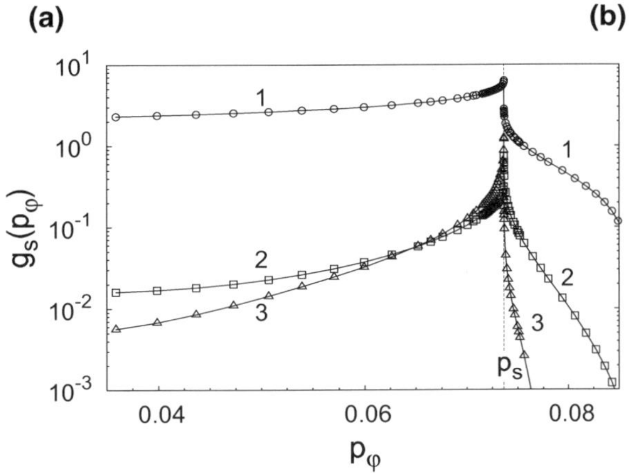

Fig. F.6 Fourier coefficients $g_s(p)$ as functions of the toroidal momentum p_φ: **a** $E = 1.0$ MeV, **b** $E = 3.5$ MeV: *curve 1* $s = 1$, *curve 2* $s = 2$, *curve 3* $s = 3$

To find the Fourier coefficients a_m, b_m, in (F.5) one can use the following formulae (Abramowitz and Stegun (1965), page 361)

$$\cos(z\cos\theta) = J_0(z) + 2\sum_{k=1}^{\infty} J_{2k}(z)\cos(2k\theta),$$

$$\sin(z\cos\theta) = 2\sum_{k=0}^{\infty} J_{2k+1}(z)\sin[(2k+1)\theta]. \tag{F.15}$$

where $J_k(z)$ are the Bessel functions. Then presenting the coefficients $A(\vartheta)$ and $B(\vartheta)$ in the complex form,

$$A(\vartheta) + iB(\vartheta) = \exp\left(in\sum_{s=1}^{M} g_s \sin(s\vartheta)\right) = \prod_{s=1}^{M} \exp\left(ing_s \sin(s\vartheta)\right), \tag{F.16}$$

and expanding the product in (F.16) in a trigonometric series of $\sin(k\vartheta)$, $\cos(k\vartheta)$, one finds the coefficients a_m, b_m. Particularly, the first three nonzero coefficients $m = 0, 1, 2$ are given by

$$a_0 = \prod_{s=1}^{M} J_0(ng_s), \qquad b_1 = 2J_1(ng_1)\prod_{s=2}^{M} J_0(ng_s),$$

$$a_2 = 2J_2(ng_1)\prod_{s=2}^{M} J_0(ng_s), \qquad b_2 = 2J_0(ng_1)J_1(ng_2)\prod_{s=3}^{M} J_0(ng_s),$$

$$b_3 = 2J_3(ng_1)\prod_{s=2}^{M} J_0(ng_s) + J_0(ng_1)J_0(ng_2)J_2(ng_3)\prod_{s=4}^{M} J_0(ng_s).$$

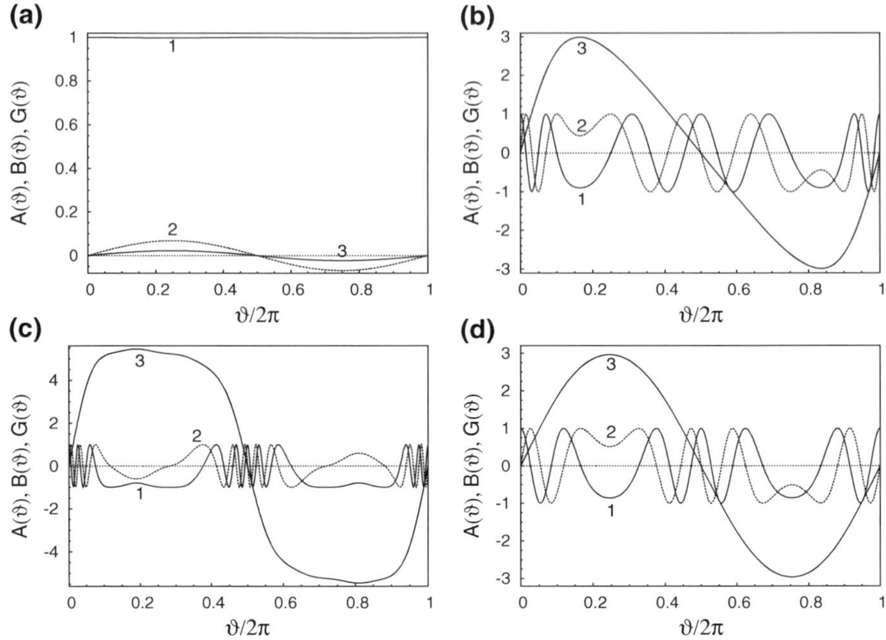

Fig. F.7 Dependence of the coefficeints $A(\vartheta)$, $B(\vartheta)$, $G(\vartheta)$ on the angle variable ϑ for α-particle of energy $E = 1.0$ MeV: **a** passing orbit; **b** barely passing; **c** barely trapped orbit; **d** trapped orbit. *Curve 1* corresponds to $A(\vartheta)$, *curve 2* corresponds to $B(\vartheta)$, *curve 3* corresponds to $G(\vartheta)$. The toroidal mode $n = 3$

For *deeply passing orbits* the largest coefficient $g_1 \ll 1$. Therefore $a_0 \lessapprox 1$, but the coefficients $a_m, b_m, (m > 1)$ are much smaller, and can be neglected. Then we have $A(\vartheta) \approx 1$, $B(\vartheta) \approx 0$ (see Fig. F.7a).

For *deeply trapped orbits* the largest coefficient $g_1 \sim 1$ and one can neglect others $g_s, (s > 1)$. Then only the coefficient g_1 can be retained giving $G(\vartheta) \approx g_1 \sin\vartheta$ (see Fig. F.7d). Therefore we have $A(\vartheta) \approx \cos(ng_1 \sin\vartheta)$, $B(\vartheta) \approx \sin(ng_1 \sin\vartheta)$.

For the barely trapped orbits the large number of coefficients g_s give the contributions to $G(\vartheta)$. The latter becomes a steep function of ϑ at the low field side $\vartheta = 0, \pi$ as shown in Fig. F.7c. The functions $A(\vartheta)$ and $B(\vartheta)$ become also fast-oscillating there.

Appendix G
Derivation of the Many-Dimensional Full-Turn Transfer Mapping

Below we present the detailed derivation of the full-turn transfer mapping (6.53). First we rewrite the mapping (6.9)–(6.11) in the following form

$$I_{k+1} = I_k + \epsilon \left(\frac{\partial S_{k+1}}{\partial \vartheta_{k+1}} - \frac{\partial S_k}{\partial \vartheta_k} \right), \tag{G.1}$$

$$\vartheta_{k+1} = \vartheta_k + w(J_k, \epsilon)(\tau_{k+1} - \tau_k) - \epsilon \left(\frac{\partial S_{k+1}}{\partial J_k} - \frac{\partial S_k}{\partial J_k} \right), \tag{G.2}$$

where $S_k \equiv S(\vartheta_k, J_k, \tau_k, \tau_0, \epsilon)$, $S_{k+1} \equiv S(\vartheta_{k+1}, J_k, \tau_{k+1}, \tau_0, \epsilon)$ are the values of the generating function $S(\vartheta, J, t, \tau_0, \epsilon)$ at the time instants $\tau = \tau_k$ and $\tau = \tau_{k+1}$, respectively. The intermediate action variables J are given by

$$J_k = I_k - \epsilon \frac{\partial S_k}{\partial \vartheta_k} = I_{k+1} - \epsilon \frac{\partial S_{k+1}}{\partial \vartheta_{k+1}}. \tag{G.3}$$

In general the mapping (G.1)–(G.3) is exact and valid for arbitrary magnitude of the perturbation parameter ϵ.

We intend to derive the simplified form of the Poincaré mapping to the section Σ of the phase space where the angle variable $\vartheta \equiv \vartheta_1$ takes constant value $\pm\pi$, mod 2π. Furthermore we introduce the following notations for the rest of angle variables $\Theta = (\vartheta_2, \ldots, \vartheta_N, \vartheta_{N+1} = t)$, and the frequencies $\omega(J) = \omega_1(J)$, $\Omega = (\omega_2, \ldots, \omega_N, \omega_{N+1} = 1)$. We consider the cases of small perturbation parameter $\epsilon \ll 1$.

In the first order of the perturbation parameter ϵ the generating function $S(\vartheta, J, t, \tau_0, \epsilon)$ is determined by the integral (6.21). Without loosing the generality one can put $H_1(J) = 0$. Then according to (6.26) $w(J, \epsilon) = \omega(J)$ and Eq. (6.21) is reduced to

$$S(\vartheta, \Theta, J, \tau, \tau_0) = -\int_{\tau_0}^{\tau} H_1(\vartheta_0(\tau'), \Theta_0(\tau')J) d\tau', \tag{G.4}$$

S. Abdullaev, *Magnetic Stochasticity in Magnetically Confined Fusion Plasmas*, Springer Series on Atomic, Optical, and Plasma Physics 78,
DOI: 10.1007/978-3-319-01890-4, © Springer International Publishing Switzerland 2014

where integration is taken along the unperturbed trajectory $\vartheta_0(\tau') = \vartheta_0(\tau) + \omega(J)(\tau' - \tau)$, $\Theta_0(\tau') = \Theta_0(\tau) + \Omega(J)(\tau' - \tau)$, $[\vartheta \equiv \vartheta_0(\tau), \Theta \equiv \Theta_0(\tau)]$.

Since $\vartheta_{k+1} = \vartheta_k + 2\pi$ for the angle variable ϑ from the mapping (G.2) it follows that

$$\tau_{k+1} - \tau_k = \frac{2\pi}{\omega(J_k)} + O(\epsilon), \tag{G.5}$$

where the terms of order of $O(\epsilon)$ are much smaller than the period of motion along the angle ϑ and they can be neglected. The unperturbed orbits $\vartheta_0(\tau')$, $\Theta_0(\tau')$ in (G.4) can be presented as

$$\vartheta_0(\tau') = \vartheta_{k+1} + \omega(J_k)(\tau' - \tau_{k+1}) = \vartheta_k + \omega(J_k)(\tau' - \tau_k),$$
$$\Theta_0(\tau') = \Theta_{k+1} + \Omega(J_k)(\tau' - \tau_{k+1}) = \Theta_k + \Omega(J_k)(\tau' - \tau_k).$$

In (G.4) we introduce the new integration variable $\tau = \tau' - \tau_0$, where τ_0 is the time instant when the angle variable $\vartheta_0(\tau)$ crosses the section $\vartheta = 0$, i.e.,

$$\tau_0 = \frac{\tau_{k+1} + \tau_k}{2} = \tau_k + \frac{\pi}{\omega(J_k)} = \tau_{k+1} - \frac{\pi}{\omega(J_k)}. \tag{G.6}$$

Then $\vartheta_0(\tau') = \vartheta_0(\tau + \tau_0) = \vartheta_0(\tau) = \omega(J)\tau$, $\vartheta_0(\tau_k) = -\pi$, and $\vartheta_0(\tau_{k+1}) = \pi$. Thus the generating functions S_k in (G.1)–(G.3) are reduced to

$$S_k \equiv S^{(-)}(\Theta_k, J_k) = \int\limits_{-\pi/\omega(J_k)}^{0} H_1(\vartheta_0(\tau), \Theta_0(\tau + \tau_0), J_k)d\tau,$$
$$S_{k+1} \equiv S^{(+)}(\Theta_{k+1}, J_k) = -\int\limits_{0}^{\pi/\omega(J_k)} H_1(\vartheta_0(\tau), \Theta_0(\tau + \tau_0), J_k)d\tau, \tag{G.7}$$

where $\Theta_k \equiv \Theta(\tau_k)$. The argument $\Theta_0(\tau + \tau_0)$ of the perturbation Hamiltonian H_1 under the integrals (G.7) is a function of Θ_k or Θ_{k+1}, i.e.,

$$\Theta_0(\tau + \tau_0) = \Theta_0(\tau) + \Theta_0(\tau_0) - \tilde{\Theta}_0, \tag{G.8}$$

where $\tilde{\Theta}_0$ is the constant initial phase, and $\Theta_0(\tau_0)$ is given by

$$\Theta_0(\tau_0) = \Theta_k + \frac{\pi\Omega(J_k)}{\omega(J_k)} = \Theta_{k+1} - \frac{\pi\Omega(J_k)}{\omega(J_k)}. \tag{G.9}$$

Equation (G.9) gives the relation between Θ_k and Θ_{k+1} in the zero-order of ϵ. Using this relation the derivative $\partial S_{k+1}/\partial\Theta_{k+1}$ can be presented as

$$\frac{\partial S_{k+1}}{\partial \Theta_{k+1}} = \frac{\partial S_{k+1}}{\partial \Theta_k}\frac{\partial \Theta_k}{\partial \Theta_{k+1}} = \frac{\partial S_{k+1}}{\partial \Theta_k} + O(\epsilon).$$

Then the mapping (G.1)–(G.3) for the action variables $I = (I_2, \ldots, I_N, I_{n+1} = H)$ and the angle variables Θ can be reduced to

$$I_{k+1} = I_k - \epsilon \frac{\partial P_k}{\partial \Theta_k}, \tag{G.10}$$

$$\Theta_{k+1} = \Theta_k + \frac{2\pi \Omega(J_k)}{\omega(J_k)} + \epsilon \frac{\partial P_k}{\partial J_k}, \tag{G.11}$$

where the function $P_k = S^{(-)} - S^{(+)}$, according to (G.7), is given by the integral

$$P_k \equiv P\left(\Theta_k + \frac{\pi\Omega(J_k)}{\omega(J_k)}, J_k\right) = S^{(-)}(\Theta_k, J_k) - S^{(+)}(\Theta_{k+1}, J_k)$$
$$= \int\limits_{-\pi/\omega(J)}^{\pi/\omega(J)} H_1\left(\vartheta(\tau), \Theta_0(\tau) + \Theta_k + \frac{\pi\Omega(J_k)}{\omega(J_k)} + \tilde{\Theta}_0, J_k\right) d\tau. \tag{G.12}$$

We will call the function $P(\Theta, J)$ the *Poincaré integral*. It depends on the intermediate action variables J which are related to the action variables I_k and I_{k+1}:

$$J_k = I_k - \epsilon \frac{\partial S_k}{\partial \Theta_k} = I_{k+1} - \epsilon \frac{\partial S_{k+1}}{\partial \Theta_{k+1}}. \tag{G.13}$$

The system of Eqs. (G.10)–(G.13) with the generating functions (G.7) and (G.12) describe the full–turn transfer mapping $(\Theta_k, I_k) \to (\Theta_{k+1}, I_{k+1})$ to the section Σ. This form of the mapping is an invariant with respect to time–reverse $\tau \to -\tau$, i.e., $k \leftrightarrow k+1$. This property of the mapping (G.10)–(G.13) is due to the intermediate action variables J_k: the generating functions S_k, S_{k+1}, and P_k are the functions of these intermediate action variables.

The mapping (G.10)–(G.13) can be simplified by replacing the intermediate variables J by the action variables (I_k, I_{k+1}) using the relations (G.13). However, such a replacement should preserve the two properties of the mapping: (i) the invariance with respect to time–reverse, and (ii) the area–preserving (or the symplectic) (6.8). These requirements can be satisfied under the following transformations. The ratio of frequencies $\alpha(J) = \pi\Omega(J)/\omega(J)$ in (G.11) can be transformed to the following symmetric form

$$\alpha(J_k) \approx \frac{1}{2}\left[\alpha(I_k) + \alpha(I_{k+1})\right] + \frac{1}{2}\frac{\partial \alpha(J)}{\partial J} \cdot (2J_k - I_k - I_{k+1})$$
$$= \frac{1}{2}\left[\alpha(I_k) + \alpha(I_{k+1})\right] - \frac{\epsilon}{2}\frac{\partial \alpha(J)}{\partial J} \cdot \left(\frac{\partial S_k}{\partial \Theta_k} + \frac{\partial S_{k+1}}{\partial \Theta_{k+1}}\right). \tag{G.14}$$

Using (G.9) and (G.13) the Poincaré integral $P(\Theta_0, J_k)$ can be transformed as

$$P(\Theta_0, J_k) = P\left(\Theta_k + \alpha(J_k), I_{k+1} - \epsilon\frac{\partial S_{k+1}}{\partial \Theta_{k+1}}\right)$$
$$\approx P\left(\Theta_k + \alpha(I_k), I_{k+1}\right) - \epsilon\frac{\partial P}{\partial \Theta_k}\cdot\frac{\partial \alpha(J)}{\partial J}\cdot\frac{\partial S_k}{\partial \Theta_k} - \epsilon\frac{\partial P}{\partial J_k}\cdot\frac{\partial S_{k+1}}{\partial \Theta_{k+1}}. \quad \text{(G.15)}$$

Neglecting the terms of order of ϵ in (G.14) and (G.15) the mapping (G.10) and (G.11) is reduced to

$$I_{k+1} = I_k - \epsilon\frac{\partial P_k}{\partial \Theta_k}, \quad \text{(G.16)}$$

$$\Theta_{k+1} = \Theta_k + \frac{\pi\Omega(I_k)}{\omega(I_k)} + \frac{\pi\Omega(I_{k+1})}{\omega(I_{k+1})} + \epsilon\frac{\partial P_k}{\partial I_{k+1}}, \quad \text{(G.17)}$$

where

$$P_k \equiv P\left(\Theta_k + \frac{\pi\Omega(I_k)}{\omega(I_k)}, I_{k+1}\right). \quad \text{(G.18)}$$

It is not difficult to show that the mapping in the form (G.16) and (G.17) is invariant with respect to the transformation $k \leftrightarrow k+1$ and the area–preserving, i.e.,

$$\left|\frac{\partial\left(\Theta_{k+1}, I_{k+1}\right)}{\partial\left(\Theta_k, I_k\right)}\right| = 1.$$

References

Abdullaev, S. S. (1994). Two-dimensional model of kicked oscillator: motion with intermittency. *Chaos*, *4*(3), 569–581.
Abdullaev, S. S. (1999). A new integration method of Hamiltonian systems by symplectic maps. *Journal of Physics A: Mathematical, Nuclear and General*, *32*, 2745–2766.
Abdullaev, S. S. (2000). Structure of motion near saddle points and chaotic transport in Hamiltonian systems. *Physical Review E*, *62*, 3508–3528.
Abdullaev, S. S. (2002). The Hamilton–Jacobi method and Hamiltonian maps. *Journal of Physics A: Mathematical, Nuclear and General*, *35*, 2811–2832.
Abdullaev, S. S. (2004). Canonical maps near separatrix in Hamiltonian systems. *Physical Review E*, *70*, 046202.
Abdullaev, S. S. (2004). On mapping models of field lines in a stochastic magnetic field. *Nuclear Fusion*, *44*, S12–S27.
Abdullaev, S. S. (2005). Asymptotical forms of canonical mappings near separatrix in Hamiltonian systems. *Physical Review E*, *72*, 046202.
Abdullaev, S. S. (2006). *Construction of Mappings for Hamiltonian Systems and Their Applications*. Berlin: Springer-Verlag.
Abdullaev, S. S. (2007). Canonical stochastic web map. *Physical Review E*, *76*, 026216.
Abdullaev, S. S. (2009). Model of a magnetic field in poloidal divertor tokamaks affected by resonant magnetic perturbations. *Physics of Plasmas*, *16*, 030701.
Abdullaev, S. S. (2010). Generic magnetic field model in poloidal divertor tokamaks in the presence of resonant magnetic perturbations. *Nuclear Fusion*, *50*, 034001.
Abdullaev, S. S. (2011). Chaotic transport in Hamiltonian systems perturbed by a weak turbulent wave field. *Physical Review E*, *84*(2), 026204.
Abdullaev, S. S. (2011). Universal asymptotics of poloidal spectra of magnetic perturbations of saddle coils in tokamaks. *Physics of Plasmas*, *18*, 020701.
Abdullaev, S. S. (2012). Universal asymptotical behavior of poloidal spectra of resonant magnetic perturbations created by a set of saddle coils in tokamaks. *Nuclear Fusion*, *52*, 054002.
Abdullaev, S. S. (2013). On collisional diffusion in a stochastic magnetic field. *Physics of Plasmas*, *20*, 082507.
Abdullaev, S. S., Eich, T., & Finken, K. H. (2001). Fractal structure of the magnetic field in the laminar zone of the dynamic ergodic divertor of the torus experiment for technology-oriented research (TEXTOR-94). *Physics of Plasmas*, *8*, 2739–2749.
Abdullaev, S. S., Finken, K., & Forster, M. (2012). New mechanism of runaway electron diffusion due to mircoturbulence in tokamaks. *Physics of Plasmas*, *19*(7), 072502.

S. Abdullaev, *Magnetic Stochasticity in Magnetically Confined Fusion Plasmas*, Springer Series on Atomic, Optical, and Plasma Physics 78,
DOI: 10.1007/978-3-319-01890-4,

Abdullaev, S. S., & Finken, K. H. (1998). Widening the magnetic footprints in a poloidal divertor tokamak: a proposal. *Nuclear Fusion, 38*, 531–544.
Abdullaev, S. S., & Finken, K. H. (2002). Hamiltonian guiding center equations in a toroidal system. *Physics of Plasmas, 9*, 4193–4204.
Abdullaev, S. S., Finken, K. H., Jakubowski, M., et al. (2003). Overview of magnetic structure induced by the TEXTOR-DED and the related transport. *Nuclear Fusion, 43*, 299–313.
Abdullaev, S. S., Wingen, A., & Spatschek, K. H. (2006a). Mapping of drift surfaces in toroidal systems with chaotic magnetic fields. *Physics of Plasmas, 13*(4), 042509.
Abdullaev, S. S., Finken, K. H., Jakubowski, M., & Lehnen, M. (2006b). Mapping of stochastic field lines in poloidal divertor tokamaks. *Nuclear Fusion, 46*(4), S113–S126.
Abdullaev, S. S., Finken, K. H., Jakubowski, M., et al. (2004). Magnetic field structure in the DED of TEXTOR and its experimental observation. In J. Connor, O. Sauter, & E. Sindoni (Eds.), *Theory of fusion plasmas* (pp. 373–379). SIF, Bologna: International Workshop, Societa Italiana di Fisica.
Abdullaev, S. S., Finken, K. H., Kaleck, A., & Spatschek, K. H. (1998). Twist mapping for the dynamics of magnetic field lines in a tokamak ergodic divertor. *Physics of Plasmas, 5*, 196–210.
Abdullaev, S. S., Finken, K. H., & Spatschek, K. H. (1999). Asymptotical and mapping methods in study of ergodic divertor magnetic field in a toroidal system. *Physics of Plasmas, 6*, 153–174.
Abdullaev, S. S., Finken, K., Kudyakov, T., & Lehnen, M. (2010). Asymptotical theory of runaway electron diffusion due to magnetic turbulence in tokamak plasmas. *Contributions to Plasma Physics, 50*(10), 929–941.
Abdullaev, S. S., Jakubowski, M., Lehnen, M., Schmitz, O., & Unterberg, B. (2008). On description of magnetic stochasticity in poloidal divertor tokamaks. *Physics of Plasmas, 15*(4), 042508.
Abdullaev, S. S., Van Rompuy, T., Finken, K., et al. (2008). Modelling of magnetic field structure of the ergodic divertor of Tore Supra and comparison to the dynamic ergodic divertor of TEXTOR. arXiv:0805.1477v1 (physics.plasm-ph).
Abdullaev, S. S., & Zaslavsky, G. M. (1994). New properties of the ergodic layer. *Bulletin of the American Physical Society, 39*, 1659.
Abdullaev, S. S., & Zaslavsky, G. M. (1995). Self-similarity of stochastic magnetic field lines near the X-point. *Physics of Plasmas, 2*, 4533–4540.
Abdullaev, S. S., & Zaslavsky, G. M. (1996). Application of the separatrix map to study perturbed field lines near the separatrix. *Physics of Plasmas, 3*, 516–528.
Abramowitz, M., & Stegun, I. (Eds.). (1965). *Handbook on mathematical functions*. New York: Dover Publications.
Aguirre, J., Giné, J., & Peralta-Salas, D. (2008). Integrability of magnetic fields created by current distributions. *Nonlinearity, 21*(1), 51–69.
Aguirre, J., & Peralta-Salas, D. (2007). Realistic examples of chaotic magnetic fields created by wires. *Europhysics Letters, 80*(6), 60007.
Aguirre, J., Viana, R. L., & Sanjuán, M. A. F. (2009). Fractal structures in nonlinear dynamics. *Reviews of Modern Physics, 81*, 333–386.
Ahn, T., & Kim, S. (1994). Separatrix map analysis of chaotic transport in planar periodic vortical flows. *Physical Review E, 49*, 2900–2911.
Ahn, T., Kim, S., & Kim, G. (1996). Analysis of the separatrix map in hamiltonian systems. *Physica D, 89*, 315–328.
Altmann, E. G., Portela, J. S. E., & Tél, T. (2013). Leaking chaotic systems. *Reviews of Modern Physics, 85*, 869–918.
Anosov, D. V. (1967). Geodetic flows on closed Riemanian manifolds of negative curvature. *Proceedings of the Steklov Institute of Mathematics, 90*, 210.
Arnold, V. I. (1963a). Proof of a theorem of A. N. Kolmogorov's on the invariance of quasi-periodic motions under small perturbations of the Hamiltonian. *Russian Mathematical Surveys, 5*, 9–36.
Arnold, V. I. (1963b). Small denominators and the problems of stability of motion in classical and celestial mechanics. *Russian Mathematical Surveys, 6*, 85–191.

Arnold, V. I. (1964). Instability of dynamical systems with several degrees of freedom. *Soviet Mathematics—Doklady*, *5*, 581–585.

Arnold, V. I. (1989). *Mathematical methods of classical mechanics* (3rd ed.). Berlin: Springer.

Arnold, V. I., Kozlov, V. V., & Neishtadt, A. I. (2006). *Mathematical aspects of classical and celestial mechanics* (3rd ed.). Berlin: Springer.

Austin, M. E., Burrell, K., Waltz, R., et al. (2006). Core barrier formation near integer q surfaces in DIII-D. *Physics of Plasmas*, *13*(8), 082502.

Bakunin, O. (2008). *Turbulence and diffusion: scaling versus equations*. Berlin: Springer.

Balescu, R. (1988). *Transport processes in plasmas: neoclassical transport theory*. Amsterdam: North-Holland Press.

Balescu, R. (1998). Hamiltonian non-twist map for magnetic field lines with locally reversed shear in a toroidal geometry. *Physical Review E*, *58*, 3781–3792.

Balescu, R., Vlad, M., & Spineanu, P. (1998). Tokamap: a Hamiltonian twist map for magnetic field lines in a toroidal geometry. *Physical Review E*, *58*, 951–964.

Bazzani, A., DiSebastiano, A., & Turchetti, G. (1998). Diffusion of magnetic field lines a confined RFP plasma. *Il Nuovo Cimento*, *20*, 1795–1818.

Bazzani, A., Malavasi, M., & Siboni, M. (1989). Poincaré map and anomalous transport in a magnetically confined plasma. *Il Nuovo Cimento*, *103*, 659–668.

Berg, S. J., Warnock, R. L., Ruth, R. D., & Forest, E. (1994). Construction of symplectic maps for nonlinear motion of particles in accelerators. *Physical Review E*, *49*, 722–739.

Bickerton, R. J. (1997). Magnetic turbulence and the transport of energy and particles in tokamaks. *Plasma Physics and Controlled Fusion*, *39*, 339–365.

Birkhoff, G. D. (1927). *Dynamical systems*. New York: AMS College Publications. Reprinted 1966.

Bleher, S., Grebogi, C., & Ott, E. (1990). Bifurcation to chaotic scattering. *Physica D*, *46*, 87–121.

Bleher, S., Ott, E., & Grebogi, C. (1989). Routes to chaotic scattering. *Physical Review Letters*, *63*, 919–922.

Bogoliubov, N. N. (1950). *Perturbation theory in nonlinear mechanics*. Kiev, pp. 9–34 (in Russian).

Boozer, A. H. (1983). Evaluation of the structure of ergodic zone. *Physics of Fluids*, *26*, 1288–1291.

Boozer, A. H. (1984). Time-dependent drift Hamiltonian. *Physics of Fluids*, *27*(10), 2441–2445.

Boozer, A. H. (2004). Physics of magnetically confined plasmas. *Reviews of Modern Physics*, *76*, 1071–1141.

Boozer, A. H., & Rechester, A. B. (1978). Effect of magnetic perturbations on divertor scrape-off width. *Physics of Fluids*, *21*, 682–689.

Boozer, A. H., & White, R. B. (1982). Particle diffusion in tokamaks with partially destroyed magnetic surfaces. *Physical Review Letters*, *49*(11), 786–789.

Brakel, R., & the W7-AS Team (2002). Electron energy transport in the presence of rational surfaces in the Wendelstein 7-AS stellarator. *Nuclear Fusion*, *42*(7), 903–912.

Brizard, A. (2011). Compact formulas for guiding-center orbits in axisymmetric tokamak geometry. *Physics of Plasmas*, *18*, 022508.

Burrel, K. H. (1994). Summary of experimental progress and suggestions for future work. *Plasma Physics and Controlled Fusion*, *36*, A291–A306.

Caldas, I. L., Pereira, J. M., Ullmann, K., & Viana, R. L. (1996). Magnetic field line mappings for a tokamak with ergodic limiters. *Chaos, Solitons and Fractals*, *7*, 991–1010.

Caldas, I. L., Viana, R. L., Araujo, M. S., et al. (2002). Control of chaotic magnetic fields in tokamaks. *The Brazilian Journal of Physics*, *32*, 980–1004.

Carlsson, J. (2001). Breakdown of adiabatic invariance in spherical tokamaks. *Physics of Plasmas*, *8*(11), 4725–4728.

Cary, J. R., & Brizard, A. J. (2009). Hamiltonian theory of guiding-center motion. *Reviews of Modern Physics*, *81*(2), 693–738.

Cary, J. R., & Littlejohn, R. G. (1983). Noncanonical Hamiltonian mechanics and its applications to magnetic field line flow. *Annals of Physics*, *151*, 1–34.

Cerfon, A., & Freidberg, J. P. (2010). 'One size fits all' analytic solutions to the Grad–Shafranov equation. *Physics of Plasmas*, *17*, 032502.

Chapman, I. T., Kirk, A., Ham, C. J., et al. (2013). Towards understanding edge localised mode mitigation by resonant magnetic perturbations in MAST. *Physics of Plasmas*, *20*, 056101.
Chirikov, B. V. (1959). Resonance processes in magnetic traps. *Soviet Journal of Atomic Energy*, *6*(6), 464–470.
Chirikov, B.V. (1969). Studies on the theory of nonlinear resonances and stochasticity, Technical Report 71–40. USSR Academy of Science, Siberian Branch. Report No. 267, CERN translation.
Chirikov, B.V. (1977). *Nonlinear resonance*. Novosibirsk State (Gos.) University, Novosibirsk.
Chirikov, B. V. (1979). A universal instability of many-dimensional oscillator systems. *Physics Reports*, *52*, 263–379.
Chudnovskiy, A. (2004). Model magnetic field configurations with islands. *Plasma Physics Reports*, *30*(11), 907–917.
Clayton, D. J., Chapman, B. E., O'Connell, R., et al. (2010). Observation of energetic electron confinement in a largely stochastic reversed-field plasmas. *Physics of Plasmas*, *17*, 012505.
Connor, J. W., Fukuda, T., Garbet, X., et al. (2004). A review of internal transport barrier physics for steady-state operation of tokamaks. *Nuclear Fusion*, *44*, R1–R49.
Connor, J. W., & Wilson, H. R. (2000). Review of theories of the L-H transition. *Plasma Physics and Controlled Fusion*, *42*, R1–R47.
Constantinescu, D. (2005). Non-linear dynamics and the structure of the magnetic field in tokamak. *Romanian Journal of Physics*, *50*(3–4), 325–336.
Constantinescu, D., & Constantinescu, R. (2005). Transport barriers and diffusion phenomena for the magnetic field lines in tokamak. *Physica Scripta*, *T118*, 244–250.
Constantinescu, D., & Firpo, M.-C. (2012). Modifying locally the safety profile to improve the confinement of magnetic field lines in tokamak plasmas. *Nuclear Fusion*, *52*(5), 054006.
Constantinescu, D., Misguich, J. H., Pavlenko, I., & Petrisor, E. (2005). Internal transport barrier in some Hamiltinian systems modeling the magnetic field lines in tokamak. *Journal of Physics: Conference Series*, *7*, 233–238.
Constantinescu, D., Misguich, J. H., Reuss, J.-D., & Weyssow, B. (2007). The influence of the safety factor on the formation of the internal transport barrier in the Tokamap-model. *Annals of the University of Craiova Physics AUC*, *17*, 190–200.
Cooper, W. A., Cooper, G. A., Graves, J. P., & Isaev, M. Y. (2011). Full-field drift Hamiltonian particle orbits in axisymmetric tokamak geometry. *Physics of Plasmas*, *18*, 052507.
Cooper, W. A., Jucker, M., Cooper, G. A., Graves, J. P., & Isaev, M. Y. (2007). Exact canonical drift Hamiltonian formalism with pressure anisotropy and finite perturbed fields. *Physics of Plasmas*, *14*, 102506.
Cushman, R. H., & Bates, L. M. (1997). *Global aspects of classical integrable systems*. Basel: Birkhäuser.
Da Silva, E. C., Caldas, I. L., & Viana, R. L. (2001a). Field line diffusion and loss in a tokamak with an ergodic magnetic limiter. *Physics of Plasmas*, *8*, 2855–2865.
Da Silva, E. C., Caldas, I. L., & Viana, R. L. (2001b). The structure of chaotic magnetic field lines in a tokamak with the external non-symmetric magnetic perturbations. *IEEE Transactions on Plasma Sciences*, *29*, 617–631.
da Silva, E. C., Caldas, I. L., & Viana, R. L. (2002a). Bifurcations and onset of chaos on the ergodic magnetic limiter mapping. *Chaos, Solitons and Fractals*, *14*, 403–423.
da Silva, E. C., Caldas, I. L., Viana, R. L., & Sanjuán, M. F. (2002b). Escape patterns, magnetic footprints, and homoclinic tangles due to ergodic magnetic limiter. *Physics of Plasmas*, *9*, 4917–4928.
D'Angelo, F., & Paccagnella, R. (1996). The stochastic diffusion process in reversed-field pinch. *Phys. Plasmas*, *3*, 2353–2364.
D'Angelo, F., & Paccagnella, R. (1999). Stochastic diffusivity and heat transport in the presence of a radial dependence of the perturbed magnetic field in the reversed field pinch. *Plasma Physics and Controlled Fusion*, *41*, 941–954.
de Michelis, C., Grosman, A., Garbet, X., et al. (1995). Characteristics of ergodic divertor plasmas in the TORE SUPRA tokamak. *Nuclear Fusion*, *35*, 1133–1153.

de Rover, M., & Lopes Cardozo, N. J. (1996). Hamiltonian description of the topology of drift orbits of relativistic particles in a tokamak. *Physics of Plasmas*, *3*, 4468–4477.
de Rover, M., LopesCardozo, N. J., & Montvai, A. (1996). Motion of relativistic particles in axially symmetric and perturbed magnetic fields in a tokamak. *Physics of Plasmas*, *3*, 4478–4488.
de Rover, M., Schilham, A. M., & Montvai, A. (1999). Test particle transport in perturbed magnetic fields in tokamaks. *Physics of Plasmas*, *6*, 2443–2451.
del Castillo-Negrete, D. (1998). Asymmetric transport and non-gaussian statistics of passive scalars in vortices in shear. *Physics of Fluids*, *10*, 576–594.
del Castillo-Negrete, D., Greene, J. M., & Morrison, P. J. (1996). Area preserving non-twist maps: periodic orbits and transition to chaos. *Physica D*, *91*, 1–23.
del Castillo-Negrete, D., Greene, J. M., & Morrison, P. J. (1997). Renormalization and transition to chaos in area preserving non-twist maps. *Physica D*, *100*, 311–329.
del Castillo-Negrete, D., & Morrison, P. J. (1993). Chaotic transport by rossby waves in shear flow. *Physics of Fluids A*, *5*, 948–965.
D'haeseleer, W., Hitchon, W., Callen, J., & Shohet, J. (1991). *Flux coordinates and magnetic field structure: a guide to a fundamental tool of plasma theory*. Berlin: Springer-Verlag.
Diamond, P., Itoh, S.-I., & Itoh, K. (2010). *Plasma physics. Physical kinetics of turbulent plasmas*. Cambridge: Cambridge University Press.
Dirickx, M., & Weyssow, B. (1998). Hamiltonian theory of guiding-centre motion in an electric field with strong gradients. *Journal of Plasma Physics*, *59*, 211–242.
Doerk, H., Jenko, F., Pueschel, M. J., & Hatch, D. R. (2011). Gyrokinetic microtearing turbulence. *Physical Review Letters*, *106*(15), 155003.
Doyle, E., et al. (2007). Progreess in the ITER Physics Basis. Chapter 2: Plasma confinement and transport. *Nuclear Fusion*, *47*(6), S18–S127.
Dragt, A. J. (1979). A method of transfer maps for linear and nonlinear beam elements. *IEEE Transactions on Nuclear Science*, *NS–26*(3), 3601.
Dragt, A.J. (1996). Summary of the working group on maps. *Particle Accelerators*, *55*, [499–530]/253–281.
Dragt, A.J. (2000). *Lie methods for nonlinear dynamics with application to accelerator physics*. Center for Theoretical Physics, Department of Physics, University of Maryland, College Park.
Dragt, A. J., & Abel, D. T. (1996). Symplectic maps and computation of orbits in particle accelerators. In G. P. J. E. Marsden & W. Shedwick (Eds.), *Integration algorithms for classical mechanics*, vol. 10 (pp. 59–85). Rhode Island: Fields Institute Communications.
Drake, J. F., Gladd, N. T., Liu, C. S., & Chang, C. L. (1980). Microtearing modes and anomalous transport in tokamaks. *Physical Review Letters*, *44*(15), 994–997.
Drummond, W. E., & Pines, D. (1962). Nonlinear stability of plasma oscillations. *Nuclear Fusion Supplements*, *3*, 1049–1057.
Duistermaat, J. J. (1980). On global action-angle coordinates. *Communications on Pure and Applied Mathematics*, *33*, 687–706.
Eckhard, B. (1988). Irregular scattering. *Physica D*, *33*, 89–98.
Edwards, D. A., & Syphers, M. J. (1993). *An Introduction to the Physics of High Energy Accelerators*. New York: Wiley.
Egedal, J. (2000). Drift orbit topology of fast ions in tokamaks. *Nuclear Fusion*, *40*(9), 1597–1610.
Eich, T., Reiser, D., & Finken, K. H. (2000). Two dimensional modeling approach to transport properties of the TEXTOR-DED laminar zone. *Nuclear Fusion*, *40*, 1757–1772.
El Mouden, M., Saifaoui, D., Dezairi, A., et al. (2007). Transport barriers with magnetic shear in a tokamak. *Journal of Plasma Physics*, *73*(4), 439–453.
Elskens, Y., & Escande, D. F. (2002). Proof of quasilinear equations in the chaotic regime of the weak warm beam instability. *Physics Letters A*, *302*, 110–118.
Elskens, Y., & Escande, D. F. (2003). *Microscopic dynamics of plasmas and chaos*. Bristol: Institute of Physics.
Engelhardt, W., & Feneberg, W. (1978). Influence of an ergodic magnetic limiter on the impurity content in a tokamak. *The Journal of Nuclear Materials*, *76—-77*, 518–520.

Entrop, I. (1999). Confinement of relativistic runaway electrons in tokamak plasmas. PhD thesis, Technical University of Eindhoven.

Entrop, I., Jaspers, R., Cardozo, N. J. L., & Finken, K. H. (1999). Runaway snakes in TEXTOR-94. *Plasma Physics and Controlled Fusion, 41*, 377.

Entrop, I., & Lopes Cardozo, N. J. (1998). Diffusion of runaway electrons in TEXTOR-94. *Plasma Physics and Controlled Fusion, 40*, 1513.

Entrop, I., & Lopes Cardozo, N. J. (2000). Scale size of magnetic turbulence in tokamaks probed with 30-MeV electrons. *Physical Review Letters, 84*, 3606–3609.

Eriksson, L.-G., Fourment, C., Fuchs, V., et al. (2002). Discharges in the JET tokamak where the safety factor profile is identified as the critical factor for triggering internal transport barriers. *Physical Review Letters, 88*(14), 145001.

Eriksson, L. G., Helander, P., Andersson, F., Anderson, D., & Lisak, M. (2004). Current dynamics during disruptions in large tokamaks. *Physical Review Letters, 92*, 205004.

Eriksson, L.-G., & Porcelli, F. (2001). Dynamics of energetic ion orbits in magnetically confined plasmas. *Plasma Physics and Controlled Fusion, 43*, R145–R182.

Escande, D. F. (1982). Large-scale stochasticity in Hamiltonian systems. *Physics Scripta, T2/1*, 126–141.

Escande, D. F. (1988). Hamiltonian chaos and adiabaticity. In V. G. Bar'yakhtar, V. M. Chernousenko, N. S. Erokhin, A. G. Sitenko, & A. V. Zakharov (Eds.), *Proceedings of international workshop on plasma theory and nonlinear and turbulent processes in Physics, Kiev, 1987* (pp. 398–430). Singapore: World Scientific.

Esposito, B., Martin-Solis, R., van Belle, P., et al. (1996). Runaway electron measurements in the JET tokamak. *Plasma Physics and Controlled Fusion, 38*, 2035–2049.

Evans, T., Degrassie, J. S., garner, H. R., et al. (1989). Resonant helical divertor experiments in ohmic and auxiliary heated JIPP T-IIU plasmas. *The Journal of Nuclear Materials, 162 & 164*, 636–642.

Evans, T. E., Fenstermacher, M. E., Moyer, R. A., et al. (2008). RMP ELM suppression in DIII-D plasmas with ITER similar shapes and collisionalities. *Nuclear Fusion, 48*, 024002.

Evans, T. E., Moyer, R. A., Burrel, K. H., West, W. P., et al. (2006). Edge stability and transport control with resonant magnetic perturbations in collisionless tokamak plasmas. *Nature Physics, 2*, 419–423.

Evans, T. E., Moyer, R. A., & Monat, P. (2002). Modeling of stochastic magnetic flux loss from the edge of poloidally diverted tokamak. *Physics of Plasmas, 9*(12), 4957–4967.

Evans, T. E., Moyer, R. A., Thomas, P. R., et al. (2004). Suppression of large edge-localized modes in high-confinement DIII-D plasmas with a stochastic magnetic boundary. *Physical Review Letters, 92*, 235003.

Evans, T. E., Moyer, R. A., Watkins, J. G., et al. (2005). Suppression of large edge localized modes with edge resonant magnetic fields in high confinement DIII-D plasmas. *Nuclear Fusion, 45*, 595–607.

Evans, T. E., Roeder, R. K., Carter, J. A., & Rapoport, B. I. (2004). Homoclinic tangles, bifurcations and edge stochasticity in diverted tokamaks. *Contributions to Plasma Physics, 44*, 235–240.

Evans, T. E., Roeder, R. K., Carter, J. A., Rapoport, B. I., Fenstermacher, M. E., & Lasnier, C. J. (2005). Experimental signatures of homoclinic tangles in poloidally diverted tokamaks. *Journal of Physics: Conference Series, 7*, 174–190.

Evans, T. E., Wingen, A., Watkins, J., & Spatschek, K. H. (2010). A conceptual model for the nonlinear dynamics of edge-localized modes in tokamak plasmas. In T. Evans (Ed.), *Nonlinear Dynamics* (pp. 59–78). Croatia: InTech.

Fedoryuk, M.V. (1989). Asymptotic methods in analysis. In *Analysis I. Encyclopaedia of mathematical sciences* (vol. 13, pp. 83–191). Berlin: Springer.

Feneberg, W., & Wolf, G. H. (1981). A helical magnetic limiter for boundary-layer control in large tokamaks. *Nuclear Fusion, 21*, 669–676.

Feng, Y., Kobayashi, M., Morisaki, T., et al. (2008). Fluid features of the stochastic layer transport in LHD. *Nuclear Fusion, 48*, 024012.

Feng, Y., Grigull, P., et al. (2002). Transport in island divertors: physics, 3D modelling and comparison to first experiments on W7-AS. *Plasma Physics and Controlled Fusion*, *44*, 611–625.
Feng, Y., Sardei, F., & Kisslinger, J. (2005). A simple highly accurate field-line mapping technique for three-dimensional Monte Carlo modeling of plasma edge transport. *Physics of Plasmas*, *12*, 052505.
Ferreira, S. (2005). The transport of galactic and jovian cosmic ray electrons in the heliosphere. *Advances in Space Research*, *35*(4), 586–596.
Fichtner, H. (2005). Cosmic rays in the heliosphere: progress in the modelling during the past 10 years. *Advances in Space Research*, *35*(4), 512–517.
Filonenko, N. N., Sagdeev, R. Z., & Zaslavsky, G. M. (1967). Destruction of magnetic surfaces in tokamaks by magnetic field irregularities: part II. *Nuclear Fusion*, *7*, 253–266.
Finken, K. H., Abdullaev, S. S., De Bock, M. F. M., et al. (2005). Background and initial experiments with the dynamic ergodic divertor on TEXTOR. *Fusion Science and Technology*, *47*(2), 87–96.
Finken, K. H., Abdullaev, S. S., Jakubowski, M., Jaspers, R., Lehnen, M., Schlikeiser, R., Spatschek, K. H., Wolf, R. & the TEXTOR Team (2007). Runaway losses in ergodized plasmas. *Nuclear Fusion*, *47*, 91–102.
Finken, K. H., Abdullaev, S. S., Jakubowski, M., Jaspers, R., Lehnen, M., & Zimmermann, O. (2006). Losses of runaway electrons during ergodization. *Nuclear Fusion*, *46*, S139–S144.
Finken, K. H., Abdullaev, S. S., Jakubowski, M., Lehnen, M., Nicolai, A. & Spatschek, K. H. (2005). The structure of magnetic field in the TEXTOR-DED, Forschungszentrum Jülich, Jülich, Germany. http://wwwzb1.fz-juelich.de/contentenrichment/onlinepublikationen/00312_Finken.pdf.
Finken, K. H., Abdullaev, S. S., Kaleck, A., & Wolf, G. H. (1999). Operating space of the Dynamic Ergodic Divertor for TEXTOR-94. *Nuclear Fusion*, *39*, 637–661.
Finken, K. H., Watkins, J. G., Rüsbuldt, D., Corbett, W. J., Dippel, K. H., Goebel, D. M., et al. (1990). Observation of sychrotron radiation from tokamak runaway electrons in TEXTOR. *Nuclear Fusion*, *30*, 859.
Finken (ed.), K. H. (1997). Special issue: the dynamic ergodic divertor. *Fusion Engineering and Design*, *37*, 335–448.
Finn, J. M. (1975). The destruction of magnetic surfaces in tokamaks by current perturbations. *Nuclear Fusion*, *15*, 845–854.
Fischer, O., & Cooper, W. A. (1998). Mapping of a stochastic magnetic field in toroidal systems. *Plasma Physics Reports*, *24*, 727–731.
Fischer, O., Cooper, W. A., & Villard, L. (2000). Magnetic topology and guiding centre drift orbits in a reversed shear tokamak. *Nuclear Fusion*, *40*, 1453–1462.
Fokker, A. (1914). Die mittlere energie rotierender elektrischer dipole im strahlungsfeld. *Annalen der Physik*, *43*, 810–820.
Forster, M., Abdullaev, S., Finken, K., et al. (2012). Runaway electron transport in turbulent and resonantly perturbed magnetic topologies of TEXTOR. *Nuclear Fusion*, *52*, 083016.
Fowler, R., Lee, D., Gaffney, P., & Rome, J. (1978). Floc-field line and orbit code for the study of ripple beam injection into tokamaks, Technical Report TM-6293. ORNL, Oak Ridge.
Freis, R. P., Hartman, C. W., Hamzeh, F. M., & Lichtenberg, A. J. (1973). Magnetic island formation and destruction in a levitron. *Nuclear Fusion*, *13*, 533–548.
Frerichs, H., Clever, M., Feng, Y., Lehnen, M., Reiter, D., & Schmitz, O. (2012). Numerical analysis of particle recycling in the TEXTOR helical divertor. *Nuclear Fusion*, *52*(3), 023001.
Frerichs, H., Reiter, D., Schmitz, O., Harting, D., Evans, T., & Feng, Y. (2012). On gas flow effects in 3D edge transport simulations for DIII-D plasmas with resonant magnetic perturbations. *Nuclear Fusion*, *52*(5), 054008.
Frerichs, H., Reiter, D., Schmitz, O., Schmitz, T., & Feng, Y. (2010). Three-dimensional edge transport simulations for DIII-D plasmas with resonant magnetic perturbations. *Nuclear Fusion*, *50*(3), 034004.
Fussmann, G. (1979). Motion of runaway electrons in momentum space. *Nuclear Fusion*, *19*(3), 327–334.

Galeev, A. A., & Sagdeev, R. Z. (1979). Theory of Neoclassical Diffusion. In M. Leontovich (Ed.), *Reviews of plasma physics* (pp. 257–343). New York: Consultants Bureau.
Garbet, X. (2006). Introduction to turbulent transport in fusion plasmas. *Comptes Rendus Physique* (France), *7*(6), 573–583.
Garbet, X., Baranov, Y., Bateman, G., et al. (2003). Micro-stability and transport modelling of internal transport barriers on JET. *Nuclear Fusion*, *43*(9), 975.
Garbet, X., Idomura, Y., Villard, L., & Watanabe, T. (2010). Gyrokinetic simulations of turbulent transport. *Nuclear Fusion*, *50*(4), 043002.
Garbet, X., Mantica, P., Angioni, C., et al. (2004). Physics of transport in tokamaks. *Plasma Physics and Controlled Fusion*, *46*(12B), B557.
Gardner, C. S. (1959). Adiabatic invariants of periodic classic systems. *Physical Reviews*, *115*(4), 791–794.
Gardner, C. S. (1966). Magnetic moment to second order for axisymmetric static field. *Physics of Fluids*, *9*(10), 1997.
Gates, D., Ahn, J., Allain, J., Andre, R., et al. (2009). Overview of results from the National Spherical Torus Experiment (NSTX). *Nuclear Fusion*, *49*(10), 104016.
Gentle, K. W. (1981). The Texas experimental tokamak (TEXT) facility. *Nuclear Technology—Fusion*, *1*(4), 479–485.
Ghendrih, P. (1995). *Résonance du divertor ergodique, Technical Report Report EUR-CEA-FC-1537*. Cadarache: CEA.
Ghendrih, P., Capes, H., Nguyen, J., & Samain, A. (1992). Control of the edge transport with the Ergodic Divertor. *Contributions to Plasma Physics*, *32*, 179–191.
Ghendrih, P., Grossman, A., & Capes, H. (1996). Theoretical and experimental investigations of stochastic boundaries in tokamaks. *Plasma Physics and Controlled Fusion*, *38*, 1653–1724.
Ghendrih, Ph. & Grosman, A. (2009). Main scientific achievements of the TORE SUPRA ergodic divertor experiments, *Fusion Science and Technology*, *56*, 1432–1444.
Gill, R. D. (1993). Generation and loss of runaway electrons following disruptions in JET. *Nuclear Fusion*, *33*(11), 1613–1625.
Gohil, P. (2006). Edge transport barriers in magnetic fusion plasmas. *Comptes Rendus Physique*, *7*(6), 606–621.
Gohil, P., Kinsey, J., Parail, V., et al. (2003). Increased understanding of the dynamics and transport in ITB plasmas from multi-machine comparisons. *Nuclear Fusion*, *43*(8), 708.
Goldstein, H. (1980). *Classical mechanics* (2nd ed.). Reading: Addison-Wesley.
Goldston, R. J., & Rutherford, P. H. (1995). *Introduction to Plasma Physics*. Bristol: Institute of Physics.
Grad, H., & Rubin, H. (1958). Hydrodynamic equilibria and force-free fields. *Proceedings of 2nd United Nations international conference on the peaceful uses of atomic energy, Geneva*. New York: Columbia University Press.
Greene, J. M. (1979). A method for determining a stochastic transition. *Journal of Mathematical and Physical Sciences*, *20*, 1183–1201.
Greene, J. M., & Johnson, J. L. (1962). Stabilty criterition for arbitrary hydromagnetic equilibria. *Physics of Fluids*, *5*(5), 510–517.
Grosman, A. (1999). Review of experimental achievements with stochastic boundaries. *Plasma Physics and Controlled Fusion*, *41*, A185–A194.
Guazzotto, L., & Freidberg, J. P. (2007). A family of analytic equilibrium solutions for the Grad–Shafranov equation. *Physics of Plasmas*, *14*(11), 112508.
Guckenheimer, J., & Holmes, P. (1983). *Nonlinear oscillations, dynamical systems, and bifurcations of vector fields*. New York: Springer.
Hamada, S. (1959). Notes on megneto-hydrodynamic equilibrium. *Progress of Theoretical Physics*, *22*(1), 145–146.
Hamada, S. (1962). Hydromagnetic equilibria and their proper coordinates. *Nuclear Fusion*, *2*(1–2), 23.

Hamzeh, F. M. (1974). Magnetic surface destruction in toroidal systems. *Nuclear Fusion*, *14*, 523–536.
Hattori, K., Seike, Y., Yoshida, Z., et al. (1984). Shrinkage of tokamak current channel by external ergodization. *Journal of Nuclear Materials*, *121*, 368–373.
Hauff, T., & Jenko, F. (2009). Runaway electron transport via tokamak microturbulence. *Physics of Plasmas*, *16*(10), 102308.
Hayashi, T., Sato, T., Gardner, H. J., & Meiss, J. D. (1994). Evolution of magnetic islands in a Heliac. *Physics of Plasmas*, *2*(3), 752–759.
Helander, P., Eriksson, L.-G., & Andersson, F. (2000). Supperssion of runaway electron avalanches by radial diffusion. *Physics of Plasmas*, *7*(10), 4106–4111.
Helander, P., Eriksson, L.-G., & Andersson, F. (2002). Runaway acceleration during magnetic reconnection in tokamaks. *Plasma Physics and Controlled Fusion*, *44*, B247–B262.
Hinton, F. L., & Hazeltine, R. D. (1976). Theory of plasma transport in toroidal confinement systems. *Reviews of Modern Physics*, *48*, 239–308.
Hogeweij, G., Dumortier, P., Eester, D. V., et al. (2004). Confinement and transport in EC heated RI-mode discharges in TEXTOR. *Nuclear Fusion*, *44*(4), 533–541.
Hölzl, M., & Günter, S. (2008). Heat diffusion across magnetic islands and ergodized plasma regions in realistic tokamak configurations. *Physics of Plasmas*, *15*, 072514.
Horton, W., Park, H. B., Kwon, J. M., Strozzi, D., Morrison, P. J., & Choi, D. I. (1998). Drift wave test particle transport in reversed shear profile. *Physics of Plasmas*, *5*(11), 3910–3917.
Hosoda, M., Miyaguchi, T., Imagawa, K., & Nakamura, K. (2009). Ubiquity of chaotic magnetic-field lines generated by three-dimensionally crossed wires in modern electric circuits. *Physical Review E*, *80*, 067202.
Howard, J. E., & Hohs, S. M. (1984). Stochasticity and reconnection in Hamiltonian systems. *Physical Review A*, *29*, 418–421.
Huba, D. (2000). NRL Plasma, Formulary, NRL/PU/6790-09-523.
Ichikawa, Y. H., Kamimura, T., & Hirose, K. (1987). Stochastic diffusion in the standard map. *Physica D*, *29*, 247–255.
Isichenko, M. B. (1991). Effective plasma heat conductivity in "braided" magnetic field. I. Quasi-linear limit. *Plasma Physics and Controlled Fusion*, *33*(7), 795–807.
Isichenko, M. B. (1991b). Effective plasma heat conductivity in "braided" magnetic field. II. Percolation limit. *Plasma Physics and Controlled Fusion*, *33*(7), 809–826.
Jackson, G. L. (2003). Initial results from the new internal magnetic field coils for resistive wall mode stabilization in the DIII-D tokamak. In R. Koch & S. Lebedev (Eds.), *Europhysics Conference Abstracts* (p. P-4.47). Mulhouse: European Physics Society. Proceedings of 30th EPS conference on controlled fusion and plasma physics, St. Petersburg, 7–11 July 2003.
Jackson, J. D. (1998). *Classical electrodynamics* (3rd ed.). New York: Wiley.
Jakubowski, M., Abdullaev, S. S., & Finken, K. H. (2004). Modelling of the magnetic field structures and first measurements of heat fluxes for TEXTOR-DED operation. *Nuclear Fusion*, *44*, S1–S11.
Jakubowski, M. W., Abdullaev, S. S., Finken, K. H., Lehnen, M., & TEXTOR Team, (2005). Heat deposition patterns on the target plates dynamic ergodic divertor. *Journal of Nuclear Materials*, *337—-339*, 176–180.
Jakubowski, M. W., Evans, T. E., Fenstermacher, M., et al. (2009). Overview of the results on divertor heat loads in RMP controlled H-mode plasmas on DIII-D. *Nuclear Fusion*, *49*(9), 095013.
Jakubowski, M. W., Lehnen, M., Finken, K. H., et al. (2007). Influence of the dynamic ergodic divertor on the heat deposition pattern in TEXTOR at different collisionalities. *Plasma Physics and Controlled Fusion*, *49*, S109–S121.
Jakubowski, M. W., Schmitz, O., & Abdullaev, S. S. (2006). Change of the magnetic-field topology by an ergodic divertor and the effect on the plasma structure and transport. *Physical Review Letters*, *96*, 035004.
Jakubowski, M. W., Wingen, A., Abdullaev, S. S., et al. (2006). Observation of the heteroclinic tangles in the heat flux pattern of the ergodic divertor at TEXTOR. *The Journal of Nuclear Materials*, *363–365*, 371–376.

Jaspers, R. J .E. (1995). Relativistic runaway electrons in tokamak plasmas. PhD thesis, Eidenhoven University of Technology, The Netherlands.
Jaspers, R., Lopes Cardozo, N. J., et al. (1996). Disruption generated runaway electrons in TEXTOR and ITER. *Nuclear Fusion*, *36*, 367.
Joffrin, E., Challis, C., Conway, G., et al. (2003). Internal transport barrier triggering by rational magnetic flux surfaces in tokamaks. *Nuclear Fusion*, *43*(10), 1167–1174.
Kadomtsev, B. B. (1988). *Collective phenomena in plasmas*. Moscow: Nauka (in Russian).
Kadomtsev, B. B., Pogutse, O. P. (1979). Electron heat conductivity of the plasma across a "braided" magnetic field. In Nuclear Fusion Supplement I: Proceedings of 7th IAEA Conference on Plasma Physics Controlled Fusion. Innsbruck, 1978, pp. 649–663.
Kaganovich, I., & Rozhansky, V. (1998). Transverse conductivity ia a braided magnetic field. *Physics of Plasmas*, *5*, 3901–3909.
Kaleck, A., Hassler, M., & Evans, T. (1997). Ergodization of the magnetic field at the plasma edge by the dynamic ergodic divertor. *Fusion Engineering and Design*, *37*, 353–378.
Kanno, R., Nunami, M., Satake, S., et al. (2010). Modelling of ion energy transport in perturbed magnetic field in collisionless toroidal plasma. *Plasma Physics and Controlled Fusion*, *52*, 115004.
Kapitsa, P. L. (1951). Dynamic stabililty of a pendulum for oscillating point of suspension. *Zh. Eksp. Teor. Fiz.*, *21*(5), 588–598. in Russian.
Kasilov, S. V., Kernbichler, W., Nemov, V. V., & Heyn, M. F. (2002). Mapping techniques for stellarators. *Physics of Plasmas*, *9*(8), 3508–3525.
Kasilov, S. V., Moiseenko, V. E., & Heyn, M. F. (1997). Solution of the drift kinetic equation in the regime of weak collisions by stochastic mapping techniques. *Physics of Plasmas*, *4*, 2422–2435.
Kasilov, S. V., Reiter, D., Runov, A. M., Kernbichler, W., & Heyn, M. F. (2002). On the 'magnetic' nature of electron transport barrier in tokamaks. *Plasma Physics and Controlled Fusion*, *44*, 985–1004.
Kawamura, T., Abe, Y., & Tazima, T. (1982). Formation of magnetic islands and ergodic magnetic layers in wall-lapping plasma as a non-divertor concept for a reactor-relevant tokamak. *Journal of Nuclear Materials*, *111—-112*, 268–273.
Kawano, Y., Yoshino, R., Kondoh, T., et al. (1997). Suppression of runaways-electrons generation during disruptive discharge–termination in JT-60U. *Controlled fusion and plasma physics. Proceedings of 24th European conference, Berchtesgaden, 1997* (pp. 501–504). Geneva: European Physical Society.
Kawashima, H., Nagashima, K., Tamai, H., et al. (1994). Study of runaway electron transport in edge stochastic magnetic field in the JFT-2M tokamak. *Journal of Plasma and Fusion Research*, *70*, 868–876.
Kennedy, J., & Yorke, J. A. (1991). Basins of wada. *Physica D*, *51*, 213–225.
Kerst, D. W. (1962). The influence of errors on the plasma-confining magnetic fields. *Journal of Nuclear Energy. Part C*, *4*, 253–262.
Kirk, A., Harrison, J., Liu, Y., Nardon, E., Chapman, I. T., & Denner, P. (2012). Observation of lobes near the *X* point in resonant magnetic perturbation experiments on MAST. *Physical Review Letters*, *108*, 255003.
Kirk, A., Chapman, I. T., Liu, Y., et al. (2013). Understanding edge-localized mode mitigation by resonant magnetic perturbations on MAST. *Nuclear Fusion*, *53*, 043007.
Klages, R., & Dorfman, J. (1995). Simple maps with fractal diffusion coefficients. *Physical Review Letters*, *74*(3), 387–390.
Knight, G., & Klages, R. (2011). Linear and fractal diffusion coefficients in a family of one dimensional chaotic maps. *Nonlinearity*, *24*(1), 227–241.
Knoepfel, H., & Spong, D. A. (1979). Runaway electrons in toroidal discharges. *Nuclear Fusion*, *19*(6), 785–829.
Kobayashi, M., Feng, Y., Masuzaki, S., et al. (2008). Modelling of impurity transport in ergodic layer of LHD. *Contributions to Plasma Physics*, *48*, 255–259.
Kobayashi, M., Feng, Y., Sardei, F., et al. (2004). Implementation of the EMC3-EIRENE code on TEXTOR-DED: accuracy and convergence study. *Contributions to Plasma Physics*, *44*, 25–30.

Kolmogoroff, A. (1931). Über die analytischen Methoden in der Wahrscheinlichkeitsrechnung. *Mathematische Annalen*, *104*, 415–458.

Kolmogorov, A. N. (1954). On conservation of conditionally periodic motions under small perturbations of the hamiltonian. *Doklady Akademii Nauk SSSR*, *98*, 527–530.

Kolmogorov, A. N. (1958). A new metric invariant of transient dynamical systems and automorphisms in lebesque spaces. *Doklady Akademii Nauk SSSR*, *119*, 861–864.

Kolmogorov, A. N. (1959). Entropy per unit time as a metric invariant of automorphism. *Doklady Akademii Nauk SSSR*, *124*, 754–755.

Kozlov, V. V. (1983). Integrability and non-integrability in Hamiltonian mechanics. *Russian Mathematical Surveys*, *38*, 1–76.

Kozlov, V. V. (1996). *Symmetries, topology and resonances in Hamiltonian mechanics*. Berlin: Springer.

Kroetz, T., Roberto, M., Caldas, I., Viana, R., Morrison, P., & Abbamonte, P. (2010). Integrable maps with non-trivial topology: application to divertor configurations. *Nuclear Fusion*, *50*(3), 034003.

Krommes, J. A. (1978). Plasma transport in stochastic magnetic fields. II. Principles and problems of test particle transport. *Progress of Theoretical Physics Supplement*, *64*, 137–149.

Krommes, J. A., Oberman, C., & Kleva, R. G. (1983). Plasma transport in stochastic magnetic fields. 3. Kinetics of test particle diffusion. *Journal of Plasma Physics*, *30*, 11–56.

Kruskal, M., & Kulsrud, R. S. (1958). Equilibrium of a magnetically confined plasma in a toroid. *Physics of Fluids*, *1*, 265–274.

Kudyakov, T., Abdullaev, S., Bozhenkov, S., et al. (2012). Influence of B_t on the magnetic turbulence and on the runaway transport in low-density discharges. *Nuclear Fusion*, *52*(2), 023025.

Kulsrud, R. S. (1957). Adiabatic invariant for the harmonic oscillator. *Physical Review*, *106*, 205–207.

Kuznetsov, L. (2003). Similarity of homoclinic tangles in open planar flows. *Communications in Nonlinear Science and Numerical Simulation*, *8*(3–4), 183–187.

Kuznetsov, L., & Zaslavsky, G. M. (1998). Regular and chaotic advection in the flow field of a three-vortex system. *Physical Review E*, *58*, 7330–7349.

Kuznetsov, L., & Zaslavsky, G. M. (2002). Scaling invariance of the homoclinic tangle. *Physical Review E*, *66*, 046212.

Kwon, J. M., Horton, W., Zhu, P., Morrison, P. J., Park, H.-B., & Choi, D. I. (2000). Global drift wave map test particle simulations. *Physics of Plasmas*, *7*, 1169–1180.

Kwon, O., Diamond, P., Wagner, F., et al. (1988). A study of runaway electron confinement in the ASDEX tokamak. *Nuclear Fusion*, *28*, 1931–1943.

La Haye, R. J. (1991). Calculations of the effects of field errors on diverted magnetic-field lines in the DIII-D tokamak. *Nuclear Fusion*, *31*(8), 1550–1555.

Landau, L. D., & Lifshitz, E. M. (1976). *Mechanics* (3rd ed.). New York: Pergamon Press.

Lao, L., John, H., Stambaugh, R., Kellman, A., & Pfeiffer, W. (1985). Reconstruction of current profile parameters and plasma shapes in tokamaks. *Nuclear Fusion*, *25*, 1611–1622.

Latka, M., & West, B. J. (1995). Structure of the stochastic layer of a perturbed resonant triad. *Physical Review E*, *52*, 3252–3255.

Lau, Y. T., & Finn, J. M. (1991). 3-Dimensional kinematic reconnection of plasmoids. *The Astrophysical Journal*, *366*, 577–591.

Laval, G. (1993). Particle diffusion in stochastic magnetic fields. *Physics of Fluids B*, *5*(3), 711–721.

Lehnen, M., Abdullaev, S., Biel, W., et al. (2005). First results from the dynamic ergodic divertor at TEXTOR. *The Journal of Nuclear Materials*, *337–339*, 171–175.

Lehnen, M., Abdullaev, S. S., Arnoux, G., et al. (2009). Runaway generation during disruptions in JET and TEXTOR. *Journal of Nuclear Materials*, *390—-391*, 740–746.

Lehnen, M., Bozhenkov, S. A., Abdullaev, S. S., The TEXTOR Team, & Jakubowski, M. W. (2008). Suppression of runaway electrons by resonant magnetic perturbations in TEXTOR disruptions. *Physical Review Letters*, *100*, 255003.

Lehnen, M., Abdullaev, S. S., Brezinsek, S., et al. (2008). The DED at TEXTOR: transport and topological properties of a helical divertor. *Plasma and Fusion Research (Japan)*, *3*, S1039.

Liang, Y., Gimblett, C. G., & Browning, P. K. (2010). Multiresonance effect in type-I edge-localized mode control with low *n* fields on JET. *Physical Review Letters*, *1056*, 065001.

Liang, Y., et al. (2007a). Active control of type-I Edge-Localized Modes with $n = 1$ perturbation fields in the JET tokamak. *Physical Review Letters*, *98*, 265004.

Liang, Y., et al. (2007b). Active control of type-I edge-localized modes on JET. *Plasma Physics and Controlled Fusion*, *49*, B581–B589.

Lichtenberg, A. (1992). Arnold diffusion in a torus with time-varying fields. *Physics of Fluids B*, *4*, 3132–3137.

Lichtenberg, A. J., & Lieberman, M. A. (1992). *Regular and Stochastic Motion*. New York: Springer.

Lichtenberg, A. J., & Wood, B. (1989). Diffusion through a stochastic web. *Physical Review A*, *39*, 2153–2159.

Litaudon, X. (1998). Profile control for steady-state operation. *Plasma Physics and Controlled Fusion*, *40*, A251–A268.

Littlejohn, R. G. (1981). Hamiltonian formulation of guiding center motion. *Physics of Fluids*, *24*(9), 1730–1749.

Littlejohn, R. G. (1983). Variational principles of guiding centre motion. *The Journal of Plasma Physics*, *29*, 111–125.

Lopes Cardozo, N. (2004). Anomalous transport due to magnetic turbulence. *Transfusion Science and Technology*, *45*(2T), 321–329.

Luo, A. C. J. (2002). Resonant layers in a parametrically excited pendulum. *International Journal of Bifurcations and Chaos*, *12*, 409–419.

Luo, A. C. J., & Han, R. P. S. (2000). The dynamics of stochastic and resonant layers in a periodically driven pendulum. *Chaos, Solitons and Fractals*, *11*, 2349–2359.

Luo, A. C. J., & Han, R. P. S. (2001). The resonance theory for stochastic layers in nonlinear dynamic systems. *Chaos, Solitons and Fractals*, *12*, 2493–2508.

MacKay, R. S. (1983). A renormalization approach to invariant circles in area-preserving maps. *Physica D*, *9*, 289–300.

MacKay, R. S. (1992). Greene's residue criterion. *Nonlinearity*, *5*, 161–187.

Mandelbrot, B. (1982). *The fractal geometry of nature*. New York: W. H. Freeman.

Martín-Solís, J. R., & Sánchez, R. (2006). Estimation of synchrotron radiation and limiting energy of high-energy runaway electrons in tokamak stochastic magnetic fields. *Physics of Plasmas*, *13*, 012508.

Martín-Solís, J. R., & Sánchez, R. (2008). Pitch angle scattering and synchrotron radiation of relativistic runaway electrons in tokamak stochastic magnetic fields. *Physics of Plasmas*, *15*, 112505.

Martín-Solís, J. R., Sánchez, R., & Esposito, B. (1999). Effect of magnetic and electrostatic fluctuations on the runaway electron dynamics in tokamak plasmas. *Physics of Plasmas*, *6*(10), 3925–3933.

Martín-Solís, J. R., Sánchez, R., & Esposito, B. (2000). Productions on the runaway current and energy during disruptions in tokamak plasmas. *Physics of Plasmas*, *7*(8), 3369–3377.

Martin, T. J., & Taylor, J. B. (1984). Ergodic behavior in a magnetic limiter. *Plasma Physics and Controlled Fusion*, *26*, 321–334.

Matsuda, S., & Yoshikawa, M. (1975). Magnetic island formation due to error field in the JFT-2 tokamak. *Japanese Journal of Applied Physics*, *14*, 87.

McCool, S. C., Wootton, A. J., Kotschenreuther, M., et al. (1990). Particle transport studies with the applied resonant fields on TEXT. *Nuclear Fusion*, *30*, 167–173.

Meiss, J. D., & Hazeltine, R. D. (1990). Canonical coordinates for guiding center particles. *Physics of Fluids B*, *2*(11), 2563–2567.

Melnikov, V. K. (1962). Lines of force of a magnetic field. *Soviet Physics—Doklady*, *7*(6), 502–504.

Melnikov, V. K. (1963a). The magnetic field lines of force of produced by helical currents flowing on the surface of a torus. *Soviet Physics—Doklady*, *8*(4), 362–364.

Melnikov, V. K. (1963b). On the stability of the center for time-periodic perturbations. *Transactions of the Moscow Mathematical Society*, *12*, 1–56.
Mendonca, J. T. (1991). Diffusion of magnetic field lines in a toroidal geometry. *Physics of Fluids B*, *3*, 87–94.
Meyer, R. E. (1973). Adiabatic variation. Part I. Exponential property for the simple oscillator. *Zeitschrift fur Angewandte Mathematik und Physik*, *24*, 293–303.
Michalek, G. (2001). On the cosmic ray cross-field diffusion for solar wind turbulence. *Astronomy and Astrophysics*, *376*, 667–671.
Miller, R. L., Chu, M. S., Greene, J. M., et al. (1998). Noncircular, finite aspect ratio, local equilibrium model. *Physics of Plasmas*, *5*(4), 973–978.
Misguich, J. H. (2001). Dynamics of chaotic magnetic field lines: intermittency and noble internal barrier in the tokamap. *Physics of Plasmas*, *8*, 2132–2138.
Misguich, J. H., Reuss, J. D., Constantinescu, D., Steinbrecher, G., Vlad, M., Spineanu, F., et al. (2003). Noble internal transport barriers and radial subdiffusion of toroidal magnetic lines. *Annales Physics France*, *28*, 1–101.
Miskane, F., Dezairi, D., Saifaoui, D., Imzi, H., Imrane, H., & Benharraf, M. (2001). Contribution of electrostatic and magnetic turbulence to anomalous transport in tokamak. *The European Physical Journal: Applied Physics*, *13*, 205–223.
Miyaguchi, T., Hosoda, M., Imagawa, K., & Nakamura, K. (2011). Topology of magnetic field lines: chaos and bifurcations emerging from two-action systems. *Physical Review E*, *83*(1), 016205.
Miyamoto, K. (2005). *Plasma Physics and Controlled Nuclear Fusion*. Berlin: Springer.
Montvai, A., & Düchs, D. F. (1993). *Proceedings of Physics Computing'92. Prague, 1992*. Singapore: World Scientific.
Morozov, A. I., & Solov'ev, L. S. (1966a). Motion of charged particles in electromagnetic fields. In M. Leontovich (Ed.), *Reviews of plasma physics* (pp. 201–297). New York: Consultants Bureau.
Morozov, A. I., & Solov'ev, L. S. (1966b). The structure of magnetic fields. In M. Leontovich (Ed.), *Reviews of plasma physics* (pp. 1–101). New York: Consultants Bureau.
Morrison, P. J. (2000). Magnetic field lines, Hamiltonian dynamics, and nontwist systems. *Physics of Plasmas*, *7*, 2279–2289.
Moser, J. (1962). On invariant curves of area-preserving mappings of an annulus. *Nachr. Akad. Wiss., Goettingen, Math.-Physics Kl.*, *1*, 1–20.
Moser, J. (1973). Stable and random motion in dynamical systems, number 77. In *Annals of mathematics studies*. Princeton: Princeton University Press.
Mynick, H. E. (1979). Guiding-center Hamiltonian for arbitrary gyration. *Physical Review Letters*, *43*, 1019–1022.
Mynick, H. E. (1980a). Guiding-center Hamiltonian for figure-8 particles in axisymmetric field-reversed configurations. *Physics of Fluids*, *23*, 1897–1902.
Mynick, H. E. (1980b). Guiding-center Hamiltonian for large gyroexcursion particles in mirror configuration. *Physics of Fluids*, *23*, 1888–1896.
Mynick, H. E., & Krommes, J. A. (1979). Particle diffusion by magnetic perturbations of asymmetric geometries. *Physical Review Letters*, *43*, 1506–1509.
Mynick, H. E., & Krommes, J. A. (1980). Particle stochasticity due to magnetic perturbations of asymmetric geometries. *Physics of Fluids*, *23*, 1229–1237.
Mynick, H. E., & Strachan, J. (1981). Transport of runaway and thermal electrons due to magnetic microturbulence. *Physics of Fluids*, *24*, 695–702.
Myra, J. R., & Catto, P. (1992). Effects of drifts on the diffusion of runaway electrons in tokamak stochastic magnetic fields. *Physics of Fluids B*, *4*(1), 176–186.
Myra, J. R., Catto, P. J., Mynick, H. E., & Duvall, R. E. (1993). Quasilinear diffusion in stochastic magnetic fields: Reconciliation of drift-orbit modification calculations. *Physics of Fluids B*, *5*(5), 1160–1163.
Nasi, L., & Firpo, M.-C. (2009). Enhanced confinement with increased extent of the low magnetic shear region in tokamak plasmas. *Plasma Physics and Controlled Fusion*, *51*, 045006.

Nguyen, F., Ghendrih, P., & Samain, A. (1995). Calculation of magnetic field topology of ergodized zone in real tokamak geometry. Application to the tokamak TORE-SUPRA through the MASTOC code, number DFRC/CAD Preprint EUR-CEA-FC-1539, CEA, Cadarache.

Nishimura, Y., & Azumi, M. (1997). Stochastic particle transport in a magnetic island due to electrostatic drift waves. *Physics of Plasmas*, *4*(7), 2365–2375.

Northrop, T. (1963). *Adiabatic Motion of Charged Particles*. New York: Wiley.

Nusse, E., & Yorke, J. A. (1996a). Basins of attraction. *Science*, *271*, 1376–1380.

Nusse, E., & Yorke, J. A. (1996b). Wada basin boundaries and basin cells. *Physica D*, *90*, 242–261.

Nusse, E., & Yorke, J. A. (1998). *Dynamics: numerical explorations*. New York: Springer.

Ohkawa, T. (1972). The enhancement factor of collisional diffusion near the rational surface. *Physics Letters A*, *38*, 21–22.

Okabayashi, M., Bialek, J., Bondeson, A., et al. (2005). Control of the resistive wall mode with internal coils in the DIII-D tokamak. In *Proceedings of 20th IAEA fusion energy conference*, Vilamoura, Portugal, 1–6 November 2004. IAEA-CSP-25/CD, IAEA, Vienna. File No. IAEA-CN-116/EX/3-1Ra.

Ott, E. (2002). *Chaos in dynamical systems* (2nd ed.). Cambridge: Cambridge University Press.

Ott, E., & Tel, T. (1993). Chaotic scattering: an introduction. *Chaos*, *3*(4), 417–426.

Papp, G., Drevlak, M., Fülök, T., Helander, P., & Pokol, G.-I. (2011). Runaway electron losses caused by resonant magnetic perturbations in ITER. *Plasma Physics and Controlled Fusion*, *53*, 095004.

Papp, G., Drevlak, M., Fülök, T., & Pokol, G.-I. (2012). The effect of resonant magnetic perturbations on runaway electron transport in ITER. *Plasma Physics and Controlled Fusion*, *54*, 125008.

Pavlenko, I., Rapoport, B., Weyssow, B., & Carati, D. (2003). Hamiltonian mapping of magnetic reconnection during the crash stage of the sawtooth instability. *Physics of Plasmas*, *10*, 1083–1091.

Perkins, R., & Bellan, P. (2010). Wheels within wheels: Hamiltonian dynamics as hierarchy of action variables. *Physical Review Letters*, *105*, 124301.

Petrisor, E. (2007). On model maps of field lines in reversed shear configurations. *Annals of the University of Craiova Physics AUC*, *17*, 164–171.

Piftankin, G. N., & Treshchev, D. V. (2007). Separatrix maps in hamiltonian systems. *Russian Mathematical Surveys*, *62*(2), 219–322.

Pires, C. J. A., Saettonne, E. A. O., et al. (2005). Magnetic field structure in the TCABR tokamak due to ergodic limiters with a non-uniform current distribution: theoretical and experimental results. *Plasma Physics and Controlled Fusion*, *47*, 1609–1632.

Pires, F., Coda, S., Firno, I., et al. (2009). Snowflake divertor plasmas on TCV. *Plasma Physics and Controlled Fusion*, *51*, 055009.

Planck, M. (1917). Über einen satz der statistichen dynamik und eine erweiterung in der quantumtheorie. Sitzungsber. d. Preuss. Akad. pp. 324–341.

Plyusnin, V. V., Riccardo, V., Jaspers, R., et al. (2006). Study of runaway electron generation during major disruptions in JET. *Nuclear Fusion*, *46*, 277–284.

Poincaré, H. (1892/1893/1899). Les méthodes de la mécanique céleste, vol. 1–3, Paris: Gauthier-Villars (New methods of celestial mechanics, vols. 1–3 (AIP, New York, 1992)).

Poincaré, H. (2009). *Science and method*. New York: Cosimo (originally published in 1914).

Pommois, P., Veltri, P., & Zimbardo, G. (2001). Kubo number and magnetic field line diffusion coefficient for anisotropic magnetic turbulence. *Physical Review E*, *63*(6), 066405.

Pomphrey, N., & Reiman, A. (1992). Effect of nonaxisymmetric perturbations on the structure of a tokamak poloidal divertor. *Physics of Fluids B*, *4*, 938–948.

Portela, J. S. E., Viana, R. L., & Caldas, I. L. (2003). Chaotic magnetic field lines in tokamaks with ergodic limiter. *Physica A*, *317*, 411–431.

Press, W. H., Teukolsky, S. A., Vetterling, W. T., & Flannery, B. P. (1992). *Numerical recipes in C. The art of scientific computing* (2nd ed.). Cambridge: Cambridge University Press.

Prudnikov, A. P., Brichkov, Y. A., & Marichev, O. I. (1986). *Integrals and series. Elementary functions*. New York: Gordon and Breach.

Punjabi, A., Verma, A., & Boozer, A. H. (1992). Stochastic broadening of the separatrix of tokamak divertor. *Physical Review Letters*, *69*, 3322–3325.
Punjabi, A., Ali, H., & Boozer, A. H. (1997). Symmetric simple map for a single-null divertor tokamak. *Physics of Plasmas*, *4*, 337–345.
Punjabi, A., Ali, H., & Boozer, A. H. (2003). Effects of dipole perturbation on the stochastic layer and magnetic footprint in single-null divertor tokamaks. *Physics of Plasmas*, *10*, 3992–4003.
Punjabi, A., Ali, H., Evans, T., & Boozer, A. H. (2007). Effect of low and high mode number tearing modes in divertor tokamaks. *Physics of Plasmas*, *14*, 082503.
Punjabi, A., Ali, H., Evans, T., & Boozer, A. H. (2008). Modeling of stochastic broadening in a poloidally diverted discharge with piecewise analytic symplectic mapping flux functions. *Physics of Plasmas*, *15*, 082507.
Rack, M., Spatschek, K., & Wingen, A. (2012). Diffusion in a collisional standard map. *Chaos*, *22*, 023114.
Ram, A., & Dasgupta, B. (2010). Dynamics of charged particles in spatially chaotic magnetic fields. *Physics of Plasmas*, *17*(12), 122104.
Rax, J. M., & White, R. B. (1992). Effective diffusion and nonlocal heat transport in a stochastic magnetic field. *Physical Review Letters*, *68*, 1523–1526.
Razumova, K. A., Andreev, V., Borshchegovskii, A., et al. (2003). Effect of the $q(r)$ profile on electron transport barrier formation in the T-10 tokamak. *Plasma Physics and Controlled Fusion*, *45*, 1247–1260.
Razumova, K., Andreev, V., Dnestrovskij, A., et al. (2008). The main features of self-consistent pressure profile formation. *Plasma Physics and Controlled Fusion*, *50*(10), 105004.
Rechester, A. B., & Rosenbluth, M. N. (1978). Electron heat transport in a tokamak with destroyed magnetic surfaces. *Physical Review Letters*, *40*, 38–41.
Rechester, A. B., Rosenbluth, M. N., & White, R. B. (1979). Calculation of the Kolmogorov entropy for motion along a stochastic magnetic field. *Physical Review Letters*, *42*, 1247–1250.
Rechester, A. B., Rosenbluth, M. N., & White, R. B. (1981). Fourier-space paths applied to the calculation of diffusion for the Chirikov–Taylor model. *Physical Review A*, *23*, 2664–2672.
Rechester, A. B., & Stix, T. H. (1976). Magnetic braiding due to weak asymmetry. *Physical Review Letters*, *36*, 587–591.
Rechester, A. B., & White, R. B. (1980). Calculation of turbulent diffusion for the Chirikov–Taylor model. *Physical Review Letters*, *44*, 1586–1589.
Regianni, N., & Sakanaka, P. H. (1994). The effect of the magnetic limiter current on the peripheral chaotic region of a tokamak. *Plasma Physics and Controlled Fusion*, *36*, 513–522.
Reiman, A. (1996). Singular surfaces in the open field line region of a divertor tokamak. *Physics of Plasmas*, *3*, 906–913.
Roberto, M., da Silva, E. C., Caldas, I. L., & Viana, R. L. (2004). Magnetic trapping caused by resonant perturbations in tokamaks with reversed magnetic shear. *Physics of Plasmas*, *11*, 214–225.
Roberto, M., da Silva, E. C., Caldas, I. L., & Viana, R. L. (2005). Derivation of an analytical area-preserving map to describe transport barriers in tokamaks. *Journal of Physics: Conference Series*, *7*, 163–173.
Roeder, R. K. W., Rapoport, B. I., & Evans, T. (2003). Explicit calculations of homoclinic tangles in tokamaks. *Physics of Plasmas*, *10*, 3796–3799.
Rom-Kedar, V. (1994). Homoclinic tangles—classification and applications. *Nonlinearity*, *7*, 441–473.
Rom-Kedar, V. (1995). Secondary homoclinic bifurcation theorems. *Chaos*, *5*, 385–401.
Romanov, Y., & Filippov, G. F. (1961). The interaction of fast electron beams with longitudinal plasma waves. *Soviet Physics, JETP*, *40*, 123–132.
Rome, J. A., & Peng, Y.-K. (1979). The topology of tokamak orbits. *Nuclear Fusion*, *19*(9), 1193–1205.

Rompuy, T. V. (2009). Study of ergodic divertor edge regimes on the tokamaks Tore Supra and TEXTOR, and sensitivity of tunnel probe electron temperature measurements to a suprathermal electron component. PhD thesis, University of Gent, Gent, Belgium.

Rosenbluth, M. N., Sagdeev, R. V., Taylor, J. B., & Zaslavsky, G. M. (1966). Destruction of magnetic surfaces by magnetic field irregularities. *Nuclear Fusion*, *6*, 297–300.

Runov, A., Kasilov, S., McTaggart, N., Schneider, R., Bonnin, X., Zagorski, R., et al. (2004). Transport modelling for ergodic configurations. *Nuclear Fusion*, *44*, S74–S82.

Runov, A. M., Reiter, D., Kasilov, S. V., Heyn, M. F., & Kernbicher, W. (2001). Monte Carlo study of heat conductivity in stochastic boundaries: application to the TEXTOR ergodic divertor. *Physics of Plasmas*, *8*, 916–930.

Ryutov, D. D. (2007). Geometrical properties of a "snowflake" divertor. *Physics of Plasmas*, *14*, 064502.

Ryutov, D. D., Cohen, R. H., Rognlien, T. D., & Umansky, M. V. (2008). The magnetic field structure of a snowflake divertor. *Physics of Plasmas*, *15*, 092501.

Sagdeev, R. Z., Usikov, D. A., & Zaslavsky, G. M. (1988). *Nonlinear physics. From the pendulum to turbulence and chaos*. Chur: Harwood Academics.

Samain, A., Blenski, T., Ghendrih, P., & Grossman, A. (1990). Theory of the ergodic divertor. *Contributions to Plasma Physics*, *30*, 157–165.

Samain, A., Capes, H., Ghendrih, P., & Nguyen, F. (1993). Particle-flux across a stochastic magnetic layer. *Physics of Fluids B*, *5*(2), 471–476.

Samain, A., Grosman, A., & Feneberg, W. (1982). Plasma motion and purification in an ergodic divertor. *Journal of Nuclear Materials*, *111 & 112*, 408–412.

Samain, A., Grossman, A., Blenski, T., Fuchs, G., & Steffen, B. (1984). An ergodic divertor for TORE-SUPRA. *Journal of Nuclear Materials*, *128 & 129*, 395–399.

Sanz-Serna, J. M. (1991). Symplectic integrators for Hamiltonian problems: an overview. In *Acta numerica* (pp. 243–286). Cambridge: Cambridge University Press.

Schaffer, M. J., Menard, J. E., Aldan, M. P., et al. (2008). Study of in-vessel nonaxisymmetric ELM suppression coil concepts for ITER. *Nuclear Fusion*, *48*, 024004.

Schelin, A., Caldas, I., Viana, R., & Benkadda, S. (2011). Collisional effects in the tokamap. *Physics Letters A*, *376*(1), 24–30.

Schmitz, O., Evans, T. E., Fenstermacher, M. E., et al. (2012). Resonant features of energy and particle transport during application of resonant magnetic perturbation fields at TEXTOR and DIII-D. *Nuclear Fusion*, *52*(4), 043005.

Schmitz, O., Evans, T. E., Fenstermacher, M. E., et al. (2009). Resonant pedestal pressure reduction induced by a thermal transport enhancement due to stochastic magnetic boundary layers in high temperature plasmas. *Physical Review Letters*, *103*(16), 165005.

Shafranov, V. (1958). On magnetohydrodynamical equilibrium configurations. *Soviet Physics: JETP*, *6*, 545.

Shafranov, V. (1966). Plasma equilibrium in a magnetic field. In M. Leontovich (Ed.), *Reviews of plasma physics* (p. 102). New York: Consultants Bureau.

Shafranov, V. D. (1964). Equilibrium of a plasma column that does not possess axial symmetry. II. *Nuclear Fusion*, *4*(3), 232.

Shen, Y., Miyake, M., Takamura, S., Kuroda, T., & Okuda, T. (1989). Ergodic magnetic limiter experiments in the HYBTOK-II tokamak. *Journal of Nuclear Materials*, *168*, 295–303.

Shepelyansky, D. (2009). Chirikov criterion. Scholarpedia 4(9). http://www.scholarpedia.org/article/Chirikov_criterion

Shevchenko, I. I. (1998). Marginal resonances and intermittent behavior in the motion in the vicinity of a separatrix. *Physics Scripta*, *57*, 185–191.

Shevchenko, I. I. (1999). The separatrix algorithmic map: application to the spin-orbit motion. *Celestial Mechanics and Dynamical Astronomy*, *73*, 259–268.

Shevchenko, I. I. (2000). Geometry of a chaotic layer. *The Journal of Experimental and Theoretical Physics*, *91*, 615–625.

Shevchenko, I. I. (2008). The width of a chaotic layer. *Physical Letters A*, *372*, 808–816.

Shevchenko, I. I. (2012). Width of the chaotic layer: maxima due to marginal resonances. *Physical Review E, 85*, 066202.
Shevchenko, I. I., & Scholl, H. (1997). Intermittent trajectories in the 3/1 jovian resonance. *Celestial Mechanics and Dynamical Astronomy, 68*, 163–175.
Shi, B. (2005). Analytic description of high poloidal beta equilibrium with a natural inboard poloidal field null. *Physics of Plasmas, 12*(12), 122504.
Shoji, T., Tamai, H., Miura, Y., et al. (1992). Effects of ergodization on plasma confinement in JFT-2M. *Journal of Nuclear Materials, 196 & 198*, 296–300.
Simó, C., & Treschev, D. V. (2008). Stability islands in the vicinity of separatrices of near-integrable symplectic maps. *Discrete and Continuous Dynamical Systems: Series B, 10*(2&3), 681–698.
Simó, C., & Vieiro, A. (2011). Some remarks on the abundance of stable periodic orbits inside homoclinic lobe. *Physica D, 240*, 1936–1953.
Sinai, Y. G. (1959). On the notion of entropy of a dynamical system. *Doklady Akademii Nauk SSSR, 124*(4), 768–771.
Sinai, Y. G. (2010). Chaos theory yesterday, today and tomorrow. *The Journal of Statistical Physics, 138*, 2–7.
Skinner, D. A., Osborne, T. H., Prager, S. C., & Park, W. (1987). Stochasticity about a poloidal divertor separatrix. *Physics of Fluids, 30*(4), 1218–1221.
Smale, S. (1998). Finding a horseshoe on the beaches of Rio. *Mathematical Intelligencer, 20*, 39–44.
Solov'ev, L. S. (1968). The theory hydrodynamic stability of toroidal plasma configurations. *Soviet Physics: JETP, 26*(2), 400–407.
Soskin, S. M. (1994). Nonlinear resonance for the oscillator with a nonmonotonic dependence of eigenfrequency on energy. *Physical Review E, 50*, R44–R46.
Soskin, S. M., & Mannella, R. (2009). Maximal width of the separatrix chaotic layer. *Physical Review E, 80*, 066212.
Soskin, S. M., Mannella, R., & McClintock, P. V. E. (2003). Zero-dispersion phenomena in oscillatory systems. *Physics Reports, 373*, 247–408.
Soukhanovskii, V., Ahn, J.-W., Bell, R., et al. (2011). Taming the plasma-material interface with the 'snowflake' divertor in nstx. *Nuclear Fusion, 51*(1), 012001.
Spatschek, K. H. (2008). Aspects of stochastic transport in laboratory and astrophysical plasmas. *Plasma Physics and Controlled Fusion, 50*(12), 124027.
Srinivasan, R., Avinash, K., & Kaw, P. K. (2001). High beta compact toroidal equilibria. *Physics of Plasmas, 8*(10), 4483–4488.
Srinivasan, R., Lao, L. L., & Chu, M. S. (2010). Analytical description of poloidally tokamak equilibrium with linear stream functions. *Plasma Physics and Controlled Fusion, 52*(3), 035007.
Stephenson, A. (1908). On a new type of dynamical stability. *Memoir: Manchester Society, 52*, 1–10.
Stix, T. H. (1973). Magnetic braiding in toroidal plasma. *Physical Review Letters, 30*, 833–835.
Stix, T. H. (1978). Plasma transport across a braided magnetic field. *Nuclear Fusion, 18*(3), 353–358.
Strumberger, E., Günter, S., et al. (2008). Fast particle losses due to NTMs and magnetic field ripple. *New Journal of Physics, 10*, 023017.
Sugiyama, L. E. (2012). Intrinsic stochasticity in fusion plasmas. *Physica Scripta, 86*, 058205.
Suttrop, W., Gruber, O., Günter, S., et al. (2009). In-vessel saddle coils for MHD control in ASDEX upgrade. *Fusion Engineering and Design, 84*, 290–294.
Tabet, R., Imrane, h, Saifaoui, D., Dezairi, D., & Miskane, F. (2000). Stochastic magnetic field lines diffusion in a toroidal configuration (tokamak). *The European Physical Journal: Applied Physics, 12*, 145–153.
Tabet, R., Saifaoui, D., Dezairi, D., & Raouak, A. (1998). Contribution to the study of the non-gaussian dynamics of stochastic magnetic field lines in a toroidal geometry. *European Physics Journal of Applied Physics, 4*, 329–336.
Takamura, S., Ohnishi, N., Yamada, H., & Okuda, T. (1987). Electric and magnetic structure of an edge plasma in a tokamak with a helical magnetic limiter. *Physics of Fluids, 30*, 144–147.

Takamura, S., Shen, Y., Yamada, H., et al. (1989). Electric and magnetic structure of tokamak edge plasma with static and rotating helical magnetic limiter. *The Journal of Nuclear Materials, 162—-164*, 643–647.

Tamai, H., Yoshino, R., Tokuda, S., et al. (2002). Runaway current termination in JT-60U. *Nuclear Fusion, 42*, 290–294.

Tokar, M. Z. (1997). Modelling of plasma and impurity behavior in a tokamak with a stochastic layer. *Fusion Engineering and Design, 37*, 417–426.

Tokar, M. Z., Evans, T. E., Gupta, A., et al. (2007). Mechanisms of edge-localized-mode motigation by external-magnetic-field perturbations. *Physical Review Letters, 98*, 095001.

Tokar, M. Z., Evans, T. E., Singh, R., & Unterberg, B. (2008). Particle transfer in edge transport barrier with stochastic magnetic field. *Physics of Plasmas, 15*, 072515.

Tokuda, S., & Yoshino, R. (1999). Simulation study on collisionless loss of runaway electrons by magnetic perturbations in a tokamak. *Nuclear Fusion, 39*, 1123–1132.

Tomita, Y., Momota, H., & Itatani, R. (1978). Destruction of magnetic surfaces near a separatrix of stellarator attributed to perturbations of magnetic fields. *Journal of the Physical Society of Japan, 44*(2), 637–642.

Tomita, Y., Seki, S., & Momota, H. (1977). Destruction of magnetic surfaces in a divertor region attributed to a discrete structure of magnetic coils. *Journal of the Physical Society of Japan, 42*(2), 687–693.

Tongue, B. H. (1987). On obtaining global nonlinear system characteristics through interpolated cell mapping. *Physica D, 28*, 401–408.

Treschev, D. (2002). Multidimensional symplectic separatrix maps. *The Journal of Nonlinear Science, 12*, 27–58.

Treschev, D. (2004). Evolution of slow variables in a priori unstable hamiltonian systems. *Nonlinearity, 17*, 1803–1841.

Treschev, D. (2006). Hamilton dynamics. Steklov Mathematical Intstitute of Russian Academy of Sciences, Moscow (in Russian).

Treshev, D. (1998). Width of stochastic layers in near-integrable two-dimensional symplectic maps. *Physica D, 116*, 21–43.

Ullmann, K., & Caldas, I. L. (2000). A symplectic mapping for the ergodic magnetic limiter and its dynamical analysis. *Chaos, Solitons and Fractals, 11*, 2129–2140.

Vecheslavov, V. V. (1996). Formation of a chaotic layer of nonlinear resonance by a two-frequency perturbation. *JETP Letters, 63*, 1047–1053.

Vecheslavov, V. V. (1999). Chaotic layer of nonlinear resonance driven by quasiperiodic perturbation. *Physica D, 131*, 55–67.

Vecheslavov, V. V. (2002). The chaotic layer of a nonlinear resonance under low-frequency perturbation. *Technical Physics, 47*, 160–167.

Vecheslavov, V. V., & Chirikov, B. V. (1998). High-precision measurement of separatrix splitting in a nonlinear resonance. *The Journal of Experimental and Theoretical Physics, 87*(4), 823–831.

Vedenov, A. A., Velikhov, E. P., & Sagdeev, R. G. (1962). Quasilinear theory of plasma oscillations. *Nuclear Fusion, 2*, 465–475.

Viana, R. L. (2000). Chaotic magnetic field lines in a tokamak with resonant helical windings. *Chaos, Solitons and Fractals, 11*, 765–778.

Viana, R. L., & Caldas, I. L. (1992). Peripheral stochasticity in tokamak with an ergodic magnetic limiter. *Zeitschrift für Naturforschung A, 47*, 941–944.

Viana, R. L., Silva, E. C. D., Kroetz, T., et al. (2011). Fractal structures in nonlinear plasma physics. *Philosophical Transactions of the Royal Society A, 369*(1935), 371–395.

Viana, R. L., & Vasconcelos, D. B. (1997). Field-line stochasticity in tokamak with an ergodic magnetic limiter. *Dynamics Stability Systems, 12*, 75.

Vlad, M., & Spineanu, F. (2005). Larmor radius effects on impurity transport in turbulent plasmas. *Plasma Physics and Controlled Fusion, 47*, 281–294.

Vlad, M., Spineanu, F. (2008). Test particles, test modes and drift turbulence, vol. 1061 of AIP Conference Series, AIP, pp. 24–33.

Vlad, M., Spineanu, F., Itoh, S.-I., Yagi, M., & Itoh, K. (2005). Turbulent transport of ions with large larmor radii. *Plasma Physics and Controlled Fusion*, *47*, 1015–1029.

von Hertweck, F., & Schlüter, A. (1957). Die "adiabatische invarianz" des magnetischen bahnmomentes geladener teilchen. *Z. Naturforsch. A*, *12*, 844–849.

Wagner, F., Becker, G., Behringer, K., et al. (1982). Regime of improved confinement and high beta in neutral-beam-heated divertor discharges of the asdex tokamak. *Physical Review Letters*, *49*, 1408–1412.

Wang, S. (2006). Canonical hamiltonian theory of the guiding-center motion in an axisymmetric torus, with the different time scales well separated. *Physics of Plasmas*, *13*, 052506.

Warnock, R. L., & Berg, S. J. (1996). Fast symplectic mapping, quasi-invariants, and long-term stability in the LHC. *Particle Accelerators*, *54*, 213–222.

Weeks, E., Urbach, J. S., & Swinney, H. L. (1996). Anomalous diffusion in asymmetric random walks with a quasi-geostrophic flow example. *Physica D*, *97*, 291–310.

Weiss, J. B., & Knobloch, E. (1989). Mass transport and mixing by modulated traveling waves. *Physical Review A*, *40*, 2579–2589.

Weitzner, H. (1981). Motion of a charged particle in a nearly axisymmetric magnetic field. *Physics of Fluids*, *24*(12), 2280–2294.

Weitzner, H. (1995). Single-particle motion in nonaxisymmetric static electric and magnetic fields. *Physics of Plasmas*, *2*(10), 3595–3604.

Weitzner, H., & Pfirsch, D. (1999). Nonperiodicity in space of the magnetic moment series. *Physics of Plasmas*, *6*(1), 420–423.

Wesson, J. (2004). *Tokamaks* (3rd ed.). Oxford: Clarendon Press.

Weyssow, B., Misguich, J. H., & Balescu, R. (1991). Chaotic diffusion across a magnetic island due to a single low-frequency electrostatic wave. *Plasma Physics and Controlled Fusion*, *33*(7), 763–793.

White, R. B. (1990). Canonical Hamiltonian guiding center variables. *Physics of Fluids B*, *2*(4), 845–872.

White, R. B. (2001). *The theory of toroidally confined plasmas* (2nd ed.). London: Imperial College Press.

White, R. B., Boozer, A. H., & Hay, R. (1982). Drift Hamiltonian in magnetic coordinates. *Physics of Fluids*, *25*(3), 575–576.

White, R. B., & Chance, M. S. (1984). Hamiltonian guiding center drift orbit calculation for plasmas of arbitrary cross section. *Physics of Fluids*, *27*(10), 2455–2467.

White, R. B., & Wu, Y. (1993). Collisionless transport in a stochastic magnetic field. *Plasma Physics and Controlled Fusion*, *35*, 595–599.

White, R. B., & Zakharov, L. E. (2003). Hamiltonian guiding center equations in toroidal magnetic configurations. *Physics of Plasmas*, *10*(3), 573–577.

Wiggins, S. (2003). *Introduction to applied nonlinear dynamical systems and chaos* (2nd ed.). New York: Springer.

Wingen, A., Abdullaev, S. S., Finken, K. H., & Spatschek, K. H. (2006). Influence of stochastic fields on relativistic electrons. *Nuclear Fusion*, *46*, 941–952.

Wingen, A., Evans, T. E., & Spatschek, K. H. (2009). Footprint structures due to resonant magnetic perturbations in DIII-D. *Physics of Plasmas*, *16*, 042504.

Wingen, A., Jakubowski, M., Spatschek, K. H., et al. (2007). Traces of stable and unstable manifolds in heat flux patterns. *Physics of Plasmas*, *14*, 042502.

Wingen, A., & Spatschek, K. H. (2008). Ambipolar stochastic particle diffusion and plasma rotation. *Physics of Plasmas*, *15*, 052305.

Wingen, A., Spatschek, K. H., & Abdullaev, S. S. (2005). Stochastic transport of magnetic field lines in the symmetric tokamap. *Contributions to Plasma Physics*, *45*, 500–513.

Wobig, H. (1987). Magnetic surfaces and localized perturbations in the Weldelstein VII-A stellarator. *Z. Naturforsch. 42a*, 1054–1066.

Wobig, H., & Pfirsch, D. (2001). On guiding centre orbits of particles in toroidal systems. *Plasma Physics and Controlled Fusion*, *43*, 695–716.

Wolf, R. (2003). Internal transport barriers in tokamak plasmas. *Plasma Physics and Controlled Fusion*, *45*, R1–R91.

Yamagishi, T. (1995). Evaluation of the stochastic layer near separatrix in toroidal divertor. *Fusion Technology*, *27*, 505–508.

Yan, L. W., & Evans, T. E. (2007). Stochastic boundary modeling by resonant magnetic perturbations on DIII-D. *Journal of Nuclear Materials*, *363—-365*, 723–727.

Yavorskij, V. A., Darrow, D., Goloborod'ko, et al. (2002). Fast ion non-adiabaticity in spherical tokamaks. *Nuclear Fusion*, *42*, 1210–1215.

Yoshida, Z. (1993). A remark on the Hamiltonian form of the magnetic-field-line equations. *Physics of Plasmas*, *1*(1), 208–209.

Yoshino, R., & Tokuda, S. (2000). Runaway electrons in magnetic turbulence and runaway current termination in tokamak discharge. *Nuclear Fusion*, *40*(7), 1293–1309.

Yu, Q. (2006). Numerical modeling of diffusive heat transport across magnetic islands and local stochastic field. *Physics of Plasmas*, *13*(6), 062310.

Yu, Q., & Günter, S. (2009). Effect of resonant magnetic perturbations on particle confinement. *Nuclear Fusion*, *49*(6), 0621001.

Zaslavsky, G. M., & Abdullaev, S. S. (1995). Scaling properties and anomalous transport of particles inside the stochastic layer. *Physical Review E*, *51*, 3901–3910.

Zaslavsky, G. M., & Chirikov, B. V. (1971). Stochastic instability of non-linear oscillations. *Soviet Physics: Uspekhi*, *14*(5), 549–568.

Zaslavsky, G. M., Sagdeev, R. Z., Usikov, D. A., & Chernikov, A. A. (1991). *Weak chaos and quasi-regular patterns*. Cambridge: Cambridge University Press.

Zehrfeld, H., Fussmann, G., & Green, B. J. (1981). Electric field effects on relativistic charged particle motion in tokamaks. *Plasma Physics*, *23*(5), 473.

Zeng, L., Koslowski, H.R., Liang, et al. (2013). Experimental observation of a magnetic-turbulence threshold for runaway generation in the TEXTOR tokamak. Physical Review Letters (to be published).

Zheng, S. B., Wootton, A. J., & Solano, E. R. (1996). Analytical tokamak equilibrium for shaped plasmas. *Physics of Plasmas*, *3*(3), 1176–1178.

Zhou, R. J., Hu, L. Q., Li, E. Z., et al. (2013a). Investigation of ring-like runaway electron beams in the EAST tokamak. *Plasma Physics and Controlled Fusion*, *55*(5), 055006.

Zhou, R. J., Hu, L., Li, E., et al. (2013b). Effect of magnetic fluctuations on the confinement and dynamics of runaway electrons in the HT-7 tokamak. *Physics of Plasmas*, *20*(3), 032511.

Index

S. Abdullaev, *Magnetic Stochasticity in Magnetically Confined Fusion Plasmas*,
Springer Series on Atomic, Optical, and Plasma Physics 78,
DOI: 10.1007/978-3-319-01890-4,

Printed by Printforce, the Netherlands